NUMBER 361

THE ENGLISH EXPERIENCE

ITS RECORD IN EARLY PRINTED BOOKS
PUBLISHED IN FACSIMILE

M. BLUNDEVILE

HIS
EXERCISES CONTAINING
SIXE TREATISES

LONDON 1594

DA CAPO PRESS
THEATRVM ORBIS TERRARVM LTD.
AMSTERDAM 1971 NEW YORK

The publishers acknowledge their gratitude
to the Master and Fellows of Trinity College, Cambridge,
for their permission to reproduce
the Library's copy (Shelfmark:Syn.6.59.17)
and to the Trustees of the Maritiem Museum 'Prins Hendrik'
Rotterdam, for their permission to reproduce
the movable volvelles from the copy
in the W.A.Engelbrecht Collection.
Library of Congress Catalog Card Number:
78-171736

S.T.C. No. 3146
Collation: A-Z^8, Aa-Xx^8, Yy^6 + 3 folding plates

Published in 1971 by
Theatrum Orbis Terrarum Ltd.,
O.Z. Voorburgwal 85, Amsterdam
&
Da Capo Press
- a division of Plenum Publishing Corporation -
227 West 17th Street, New York, 10011
Printed in the Netherlands
ISBN 90 221 0361 7

A

M. BLVNDEVILE
Syn. 6. 59. 17.
His Exercises, containing

fixe Treatiſes, the titles wherof are ſet down in the next printed page: which Treatiſes are verie neceſſarie to be read and learned of all yoong Gentlemen that haue not bene exerciſed in ſuch diſciplines, and yet are deſirous to haue knowledge as well in Coſmographie, Aſtronomie, and Geographie, as alſo in the Arte of Navigation, in which Arte it is impoſſible to profite without the helpe of theſe, or ſuch like inſtructions.

To the furtherance of which Arte of Navigation, the ſaid M. Blundevile ſpeciallie wrote the ſaid Treatiſes and of meere good will doth dedicate the ſame to all the young Gentlemen of this Realme.

LONDON.
Printed by John Windet, dwelling at the ſigne of the croſſe Keies, neere Paules wharfie, and are there to be ſolde. 1594.

The Titles of the Treatiſes con-
tayned in this booke.

Irſt, a verie eaſie Arithmeticke ſo plainlie written as any man of a mean capacitie may eaſilie learn the ſame without the helpe of any teacher.

Item the firſt principles of Coſmographie, and eſpecially a plaine treatiſe of the Spheare, repreſenting the ſhape of the whole world, together with the chiefeſt and moſt neceſſarie vſes of the ſaid Spheare.

Item a plaine and full deſcription of both the Globes, aſwell Terreſtriall as Celeſtiall, and all the chiefeſt and moſt neceſſary vſes of the ſame, in the end whereof are ſet downe the chiefeſt vſes of the Ephemerides of *Iohannes Stadius*, and of certaine neceſſarie Tables therein contained for the better finding out of the true place of the Sunne and Moone, and of all the reſt of the Planets vpon the Celeſtiall Globe.

Item a plaine and full deſcription of *Petrus Plancius* his vniverſall Mappe, lately ſet forth in the yeare of our Lord 1 5 9 2. contayning more places newly found, aſwell in the Eaſt and Weſt Indies, as alſo towards the North Pole, which no other Map made heretofore hath, whereunto
A 3   is

The Contentes

is also added how to find out the true diſtance betwixt a-
nie two places on the land or ſea, their longitudes and latitudes being firſt knowne, and thereby you may correct
the skales or Tronkes that be not trulie ſet downe in anie
Map or Carde.

Item, A briefe and plaine deſcription of M. *Blagraue*
his Aſtrolabe, otherwiſe called the Mathematicall Iewel,
ſhewing the moſt neceſſary vſes thereof, and meeteſt for
ſea men to know.

Item the firſt & chiefeſt principles of Navigation more
plainlie and more orderly taught than they haue bene
heretofore by ſome that haue written thereof, lately collected out of the beſt modern writers, and treaters of that
Arte.

And moreover, I haue thought good to adde vnto
mine Arithmeticke, as an appendix depending thereon,
the vſe of the Tables of the three right lines belonging
to a circle, which lines are called Sines, lines tangent, and
lines ſecant, whereby many profitable and neceſſarie concluſions aſwell of Aſtronomie, as of Geometrie are to be
wrought only by the help of Arithmeticke, which Tables are ſet downe by *Clauius* the Ieſuite, a moſt excellent
Mathematician, in his booke of demonſtrations made vpon the Sphericks of *Theodoſius*, more trulie printed than
thoſe of *Monte Regio*, which booke whileſt I read at mine
owne houſe, together with a loving friend of mine, I took
ſuch delight therein, as I mind (God willing) if God giue
me life, to tranſlate all thoſe propoſitions, which *Clauius*
himſelfe hath ſet downe of his owne, touching the quantitie of Angles, and of their ſides, as well in right line triangles

of this booke.

angles, as in Sphericall triangles: of which matter, a *Monte Regio* wrote diffufedlie and at large, so *Copernicus* wrote of the same brieflie, but therewith somewhat obscurelie, as *Clauius* saith. Moreover, in reading the Geometrie of *Albertus Durerus*, that excellent painter, and finding manie of his conclusions verie obscurelie interpreted by his Latine interpreter (for he himselfe wrote in high Dutch) I requested a friend of mine, whome I knewe to haue spent some time in the studie of the Mathematicals, not onelie plainelie to tranflate the foresaide *Durerus* into Englifh, but also to adde thereunto manie neceffary propofitions of his owne, which my requeft he hath (I thanke him) verie well perfourmed, not onely to my satisfaction, but also to the great commoditie and profite of all thofe that defire to bee perfect in Architecture, in the Arte of Painting, in free Mafons craft, in Ioyners craft, in Carvers craft, or anie such like Arte commodious and serviceable in any common Wealth, and I hope that he will put the fame in print ere it be long, his name I conceale at his owne earneft intreatie, although much againft my will, but I hope that he will make himfelfe known in the publifhing of his Arithmeticke, and the great Arte of *Algebra*, the one being almoft finifhed, and the other to bee vndertaken at his beft leafure, as alfo in the printing of *Durerus*, vnto whom he hath added many neceffary Geometrical conclufions, not heard of heretofore, together with divers other of his workes as wel in Geometrie as as in other of the Mathematicall sciences, if he be not called away from thefe his ftudies by other affaires. In the mean time I pray al young Gentlemen and feamen to take thefe my labours already ended in good part, whereby I seeke neither praife nor glorie, but onely to profite my countrey.

A 4   To

To the Reader.

I Greatlie reioyce to see so manie of our English Gentlemen, both of the Court and Countrie in these dayes so earnestlie giuen to trauell aswell by sea as land, into straunge and vnknowne countries, and speciallie into the East and West Indies, following therein the good example of diuers worthy knightes and Gentlemen that haue ventured their liues to discouer strange countries to the great honour of their countrie, and to their owne immortall fame. And because that to trauell by sea requireth skil in the Art of Nauigation, in which it is vnpassible for any man to be perfect vnles he first haue his Arithmetick, and also some knowledge in the principles of Cosmographie, and specially to haue the vse of the Spheare, of the two Globes, of the Astrolabe, and crosse staffe, and such like instruments belonging to the Arte of Nauigation, I thought good therefore to write the Treatises before mentioned, to serue as an introduction for such yoong Gentlemen as haue not bene exercised in such kind of studies, which Treatises if they shall vouchsafe to read with attentiue minde, and in such order as they are before set downe, I doubt not but that it will cause them hereafter to seeke for further knowledge therein. And in any wise I wish them to begin with my Arithmeticke, the contents whereof are declared in the next chapter following. In the meane time I do earnestly request all yoong Gentlemen to take these my simple pamphlets no lesse thankfullie than they haue done my horse booke, and in so doing I shall haue iust cause to thinke my labour well bestowed.

What

What cause first mooved the Author to write this Arithmeticke, and with what order it is here taught, which order the contents of the chapters therof hereafter following doe plainly shew.

Began this Arithmeticke more than seuen yeares since for a vertuous Gentlewoman, and my verie deare frend *M.* Elizabeth Bacon, *the daughter of Sir* Nicholas Bacon *Knight, a man of most excellent wit, and of most deepe iudgment, and sometime Lord Keeper of the great Seale of* England, *and latelie (as shee hath bene manie yeares past) the most loving and faithfull wife of my worshipfull friend* M. *Iustice* Wyndham, *not long since deceased, who for his integritie of life, and for his wisedome and iustice daylie shewed in gouernement, and also for his good hospitalitie deserued great commendation. And though at her request I had made this Arithmeticke so plaine and easie as was possible (to my seeming) yet her continuall sicknesse would not suffer her to exercise her selfe therein. And because that diuerse hauing seene it, and liking my plaine order of teaching therein, were desirous to haue copies thereof, I thought good therefore to print the same, and to augment it with many necessarie rules meet for those that are desirous to studie any part of Cosmographie, Astronomie, or Geographie, and speciallie the Arte of Nauigation, in which without Arithmeticke, as I haue said before, they shall hardly profit. But nowe to returne to my*

mat-

matter, the contents of this Arithmeticke are these here following. First, hauing defined what Arithmeticke is, and what numeration or numbring is, and of what parts it consisteth, and how to number any great summe written in many figures, I deale with the foure speciall kindes of Arithmeticke, that is, Addition, Subtraction, multiplication, and diuision. Then I shewe the order of working in whole numbers, called in Latine Integra, by the rule of proportion, otherwise called the golden rule, or the rule of three. Then I treat of broken numbers called Fractions, setting downe seuen necessarie rules belonging thereunto, by the helpe whereof you shall be the better able to adde, subtract, to multiplie and divide the same. That done, I shew how to vse the rule of three aswell in dealing with sole fractions as with fractions annexed to integrums, which rule of three is threefold, that is, the common rule, the rule reuerse, and the double rule, the order of all which three kindes I doe plainlie teach by examples, shewing wherein, how, and when, they are to be vsed. Next to that I set downe the rule of fellowship, a necessarie rule for those that haue to traffique in anie trade of merchandize, giuing divers examples thereof. Next to that I treat of Arithmeticall, and Geometricall progression, and also of proportion, and of the three kindes thereof, that is, of proportion Arithmeticall, Geometricall, and musicall. Then I shew how to finde out the square roote of anie number, and also the vse thereof in setting of Battels, and also how to finde out the cubique roote of any number. And last of all I treat of Astronomicall fractions, shewing how to adde, to subtract, to multiply and divide the same. And also to take the square roote thereof, without the knowledge of which fractions, you can neuer calculate any thing trulie out of the Astronomicall Tables.

    Some perhaps do look here that I should speak somewhat of the rules of Algebra, whereby all subtill and intricate questions of Arithmeticke are to be vnfolded, wherewith I leaue indeede deale, partlie, for that I haue not of long time exercised my selfe therein, and partlie becauſe I knowe that one hath begun to write thereof, whoſe booke being once ended, I doubt not but that hee will ſhortlie after print the ſame to the great profite and furtherāce of al thoſe that delight in ſuch good exerciſes. But in the mean time I haue thought good to adde (as I haue ſaide before) vnto this Arithmeticke a plaine deſcription, together with the vſe of the Tables of Sines, of lines tangent and ſecant, which Tables will pleaſure many that would gladlie know how and wherein to vſe them, and ſpecially ſuch ſeamen as haue ſome taſte of Arithmeticke, without the which no good almoſt is to bee done in anie ſcience.

This

This Treatise of Arithmeticke containeth 26. Chapters as followeth.

WHat Arithmetique is, what numeration is, and of what partes it consisteth, and what signification euerie digit hath according to his place, and how to expresse or tell a great number written in manie figures. Chapter. 1.

Of the foure speciall kindes of Arithmetique, and first of Addition, with examples thereof. Chap. 2.

Of Subtraction, with examples thereof, and how to trie the same. Chap. 3.

Of Multiplication and certaine Tables belonging thereunto, together with the vse thereof, and what is to be obserued therein, with examples and triall thereof. Chap. 4.

Of Diuision, and what is to be obserued therein, with examples and triall thereof, and of halfing any number. Chap. 5.

Of the rule of three, called the Golden rule, and what order is to bee obserued in working thereby, and of the three kindes thereof. Chap. 6.

Of Fractions what they be, with a Demonstration thereof, together with seuen necessarie rules belonging to the same, and what euerie rule teacheth. Chap. 7.

Of Addition, Subtraction, Multiplication, and Diuision of Fractions. Chap. 8.

Of the common rule of three belonging to Fractions with examples. Chap. 9.

Of

Of the rule reuerſe, *called in Latine* Regula euerſa, *and the order of working thereby with examples. Chap. 10.*

Of the double rule called in Latine Regula duplex, *and the order of working thereby with examples, and in working thereby how to know when you haue to vſe* Regula everſa, *or the common rule of three. Chap. 11.*

Of the rule of Fellowſhip, and the order of working thereby, with divers examples thereof. Chap. 12.

Of Progreſſion, what it is, and of the two kindes thereof, that is, Arithmeticall and Geometricall. Chap. 13.

Of Addition belonging to progreſſion Arithmeticall, and the order thereof, with examples. Chap. 14.

Of Addition belonging to progreſſion Geometricall, and the order thereof, with examples. Chap. 15.

Of proportion, what it is, and of the three kindes thereof, that is, Arithmeticall, Geometricall and Muſicall. Chap. 16.

Of Arithmeticall proportion, what it is, and how it is diuided. Chap. 17.

Of proportion Geometricall, what it is, and how it is divided. chap. 18.

Of the chiefe and ſpeciall kinds of Geometrical proportion, that is, of equalitie and inequalitie both greater and leſſer. Chap. 19.

Of proportion of the greater inequalitie, and of the 2. kinds thereof, that is, Simplex, *and* Multiplex, *and of their diuers kindes, with certaine Tables belonging thereunto Chap. 21.*

Of proportion of the leſſer inequality. Chap. 21.

Of Muſicall proportion, vvhat it is, and of the two kindes thereof, that is, Simple and compound, of which compound there be alſo 2. kyndes that is proper and vnproper. Chap. 22.

How proportions are to bee ſet downe in writing, and how they are to be added, ſubtracted, multiplyed, and divided euen like to fractions in all reſpectes. Chap. 23.

How

How to finde out the square roote of any number with examples. Chap. 24.

The vse of the square roote in setting of Battels, which according to the Italian vse are to be set foure manner of wayes, the order whereof is heere set downe with examples. Chap. 25.

How to finde the cubique roote of any number, and the order therof with examples. Chap. 26.

Of Astronomicall Fractions, vvhereto they serue, and how to adde, subtract, multiplie, and to diuide the same. Chap. 27.

How to diuide such fractions vvhen the diuisor is greater than the dividend. Chap. 28.

How to take the square roote of Astronomicall Fractions. Chap. 29.

And immediately after these Chapters doe followe the vse of the Tables of Sines, lines tangent, and lines secant before mentioned, together with the Tables themselues.

To

To the Reader touching the order of finding
and correcting the faults escaped
in printing.

First, I pray you vouchsafe with your penne to correct the numbers of the leaues of these bookes, which numbers are not in some leaues trulie set downe, that done, for the rest of the faults resort to the Table following, which consisteth of fiue collums, whereof the first on the left hand sheweth the true number of the leafe, the seconde collum sheweth the page both first and second: the third sheweth the line: in counting whereof you must begin at the head of everie page, and so proceede downward, leaving no printed line vntold, though it contayneth but one word, omitting notwithstanding such lines as doe containe the title of any chapter, so shall you bee sure to finde any fault mentioned in the sayde Table, and the correction thereof hard by it on the right hand : for the fourth collum of this Table containeth the faults, and the fift collum the correction thereof: by helpe of which Table you may correct your owne booke, where need is, in setting downe the correction in the margent right against the fault found.

Errata.

Leafe	Page	Line	The faultes.	The Correction.
1	2	27	for, 4270570. read	4320570.
2	1	15	for, who read	Why.
3	2	5	for, calling. read	Culling.
5	2	9	for, squarewise. read	squirewise.
10	1	32	for, limediate, read	immediate.
24	2	33	for, the former 4. read	the former doubled 4.
29	1	29	for, 2650. read	2550.
32	1	4	for, the productes, read	the particular productes.
32	1	27	for, containe manie, read	containe so manie.
32	2	31	for, ouer 1. read	over 1′′′.
32	2	32	for, appointing to his, read	appointing to the product his
34	1	1	for, the nūber must be 24. r.	the number 24. must be
37	1	18	for, 53′′. read	50′′.
42	1	4	for, 10. read	16.
42	2	23	for, 20. read	29.
53	1	16	for, 890. read	899.
56	2	13	for, 23. degr. 38′. read	23. degr. 28′.
131	1	30	for, procesion, read	precesion.
137	2	13	for, euer the heauen, read	euerie heauen.
139	2	3	for Anges, read	Auges.
169	1	6	for, returning, read	returneth.
173	1	16	for, or, read	and.
181	1	32	for, 2160. miles. read	21600. miles.
193	2	6	for, Iland, read	Island.
199	2	5	for, Aphiscii. read	Amphiscii.
210	1	31	for, before fastned, read	before taken.
215	2	2	for, the hour of the north, r.	the houre of the night.
216	2	2	for, dawning twilight, read	dawning and twilight.
217	1	28	for, the Celestial globe. read	the terrestriall Globe.
227	2	20	for, meere lately, read	more lately.
229	2	7	for, in a great spheare, reade	in a right spheare.
233	1	32	for, the 7. of Scorpio. read	the 8. of Scorpio.
236	2	16	for, 30. deg. 47′. of, read	the 3. deg. 47′. of
249	1	22	for, Equinoctionall, read	Equinoctiall.
253	1	11	for, that is also, read	it is also.
253	1	25	for, the yeare 1542. read	1452.
254	1	25	for, Burghes, read	Bruges.
255	2	15	for Sinacum regio, read	Sinarum regio.
255	2	17	for, Oceanus Sicicus, read	Oceanus Siricus.
256	1	13	for, that is the selfe same, read	it is the selfe same.
269	2	8	for, should by. read	should be.
269	2	14	for, of summe, read	or summe.
274	1	11	for, the houre leagues, read	the houre leagues for Marin.
275	1	7	for, Moro	Morea.
278	1	23	for, 2184. miles, read	1856. Italian miles.
281	1	35	for, flat both together, read	clapt flat together.
284	1	6	for, taken from, read	taken for the pole.
284	1	15	for, distinction cause, read	for distinction sake.
289	1	2	for, at the time, read	at that time.
297	2	31	for, the eleuenth Meridian, r	the 111. Meridian.
300	1		In the figure, for, 1. ♉. read	♉. 1.
310	2	30	for, beginneth, read	beginning.
317	2	30	for, greater error, read	great error.
318	2	22	for, of the Zenith on both sides, read	on both sides of the Zenith.
320	2	5	for, Loadstarre, read	loadstone.
328	2	31	for, vnto the 80. degrees, read	vnto the 8. degree.
329	2	26	for, at the fit widenes, read	at a fit widenes.
333	1	15	for, making any charge, read	making any change.
340	1	15	for, about the sawe, read	about the same.
344	2	last	for, ♃ read	♄.

Of Arithmetike.

Cap. 1.

What is Arithmetike?
 It is the art of counting or numbring by figures.
 What is to number by figures?
 It is to expresse the value of any number in his proper Characters and figures, which is called by a Latine name Numeration.

What belongeth to Numeration?
 Two things, to know the shapes of the figures, and the signification of their places.

How many figures are there?
 These ten. 1.2.3.4.5.6.7.8.9.0. Whereof the tenth made like an. o. as you see here, is called a Cypher, which is no number of it selfe, but serueth onely to fill vp a number.

What is number?
 Number is a collection or somme of many ones added together.

How is number deuided?
 Into three kinds, that is Digit, Article and compound.

Which be they?
 The Digit is any of the first 9. figures before set downe. Article is any number ending in ten, as ten is one Article, twentie is two Articles, thirtie is three Articles, &c. Compound is that which is compounded of Article & Digit, as 13.14.17.24.&c.

Shew what signification euery Digit hath according to his place,

Of Arithmetike.

place, and in what order such place is to be considered in expressing any number?

The order as touching the place, is to beginne at the right hand, and so to procæde towardes the left hand. For any of the 9. Digits whatsoeuer, standing in the first place, which is on the right hand, signifieth the value of himselfe onely, in the second place ten times himselfe, in the third place a Hundreth times himselfe, in the fourth place a Thousand times himselfe, in the fift place ten Thousand times himself, in the sixt place a hundreth Thousand times himselfe, in the seuenth place a Myllion, in the eight place ten Myllions, in the ninth place a hundreth Myllions in the tenth place a thousand Myllions, &c.

Doth the Cypher signifie nothing?

Yes it maketh a place wheresoeuer it standeth, so as it be not the outermost on the left hand: for there it hath no place at all, as here you may sée in this number, 04500. whereof the first Cypher on the right hand, signifieth the first place, the second Cypher the second place, but the last and outermost Cypher on the left hand, signifieth no place at all, because it hath no Digit standing before it towards the left hand, and therefore though in this number there be 5. figures, yet it signifieth no more but foure Thousand and fiue Hundreth.

By what meanes may a great number written in many figures, be readily expressed or tolde?

By diuiding the same into diuerse parts with strækes, or prickes made at the ende of euery third figure, beginning to tell from the right hand towards the left, as in this number, 4/320/570. In which, beginning with the Cypher on the right hand, I tell one, two, and thrée, and there make a Stræke, and so procéding forth still towards the left hand, I make a Stræke at the ende of euery thirde figure, by which Strækes or partitions I make them now seuerall numbers, and euery Stræke must bé named by this word Thousand. Notwithstanding, in expressing this number being thus deuided, or any other suchlike, you must begin at the left hand, and say thus foure Thousand thousand, thrée Hundreth twentie Thousand, fiue Hundreth and seuentie: for by reason of this Diuision, the figure 4. standeth here alone, and in the first place, and in déed signifieth 4. Myllions, and by

that

Of Arithmetike.

that means, you may moꝛe fitly expꝛesse the said number in saying thus, foure Myllions, thꝛée Hundꝛeth and twentie Thousand, fiue Hundꝛeth and seuentie: And the better to discerne the Myllions from ẏ rest in great numbers, it shal not be a misse to set an M. signifiing Myllions, right ouer the head of ẏ Stréek which is dꝛawne betwixt the sixth and seuenth figures, as in this example cõtaining a leauen figures, 34/545ᵐ/678/694. which is to be vttered thus, thirtie foure Thousand, fiue Hundꝛeth foꝛtie fiue Myllions, sixe Hundꝛeth seuentie eight Thousand, sixe Hundꝛeth nintie and foure Crownes, oꝛ Pounds, oꝛ whatsoeuer other denomination oꝛ name, it shall please you to giue them.

Of the foure speciall kinds
or partes of Arithmetike.

Cap. 2.

Hich are those foure kindes? These, Addition, Subtraction, Multiplication and Deuision.

Whi. is not Numeration also counted as a parte?
Becaufe Numeration together with the figures, and places whereof it confifteth, are counted rather as firft Elements, and principles of Arithmetike, then as partes or fpeciall kindes thereof.

Of Addition.

Hat is Addition?
It is that which teacheth to bꝛing many seuerall Sommes into one somme.

How is that done?
First by placing euery seuerall number one right vnder another, vnder which you must dꝛawe a line, that done, you must adde together the numbers of the first rancke, beginning on the right hand

Of Addition.

hand with the lowest figure of the same rancke, and so going vpward to the highest figure of the same rancke, and so from rancke to rancke, til you come to the last, and if the Somme of any rancke doe not excæde the number of any of the foresaide 9. Digits, then set downe that Digit which comprehendeth that number right vnder his proper rancke, beneath the line, but if the Somme of that rancke, excædeth the number of any one Digit by reason that it consisteth of Articles and Digits, then set downe the Digit and kæpe the Article or Articles in your minde, to be added to the first figure of the next rancke on the left hand, but if the Somme be an euen Article or Articles: then set downe a Cypher, kæping the number of Article or Articles in your minde, be it, one, two, or thræ, to be added to the next rancke, all which things you shall better vnderstand by this example here following. As for example, I spent in one yeare 125.£. in another yeare 234.£, and in another peare 240.£. Now to knowe the totall Somme of all this, I place these seueral Sommes one right vnder another, and then I drawe a line vnder them as here you sæ.

Then beginning on the right hand with the lowest figure of the first rancke aboue the line, I say that a Cypher and 4. is but 4. Againe 4. and 5. maketh 9. which I set down vnder þ line, then procæding to the second rancke towards the left hand, I say that 4. and 3. maketh 7. and 7. and 2. maketh 9. which I also set downe, then remoouing to the thirde rancke, I say that 2. and 2. maketh 4. and 4. and 1. maketh 5. which I also set downe as you sæ in the former example, so as the totall Somme vnder the line is. 599.£.

£
125.
234.
240.
599.

Another example hauing Cyphers mixt with Digits.

Here I say that 9. and 8. maketh 17. and 17. and 7. maketh 24. wherefore I set downe the Digit 4. and kæpe 2. Articles in minde, which being added to the lowest figure of the second rancke, which is 4. maketh 6. then 6. and 4. maketh 10. here I set downe a Cypher, kæping one Article in minde, which being added to the figure 5. of the third ranck maketh 6. which I also set downe, then I say that 3. and 4. maketh 7. and 7. and 3. maketh 10. for the which I set down first a Cypher, and then because there is no more to be added, I set downe on the left hand the one Article which I had in minde, so as the whole Somme

3047.
4508.
3049.
10/604.

Of Addition.

Somme commeth to 10/604. as in the former example.

How are pounds, shillings, pence, halfe pence, and farthings and all other numbers of diuerse Denominations to be added?

You must deuide euery seuerall name into diuerse Collums or Spaces by themselues, and then beginning with the first on the right hand, you must adde euery Collum by it selfe, bringing farthings to halfe pence, and halfe pence to pence, pence to shillings, and shillings to poundes, setting the Somme of euery Collum vnder the nether line as you see in this example following.

Here first beginning with the Collum of farthings, I finde therein 3. farthings which is one halfe peny and one farthing. Wherefore I set downe the od farthing as you see, and keepe the halfe peny in minde:

l	s	d	ob.	q
345	13	1	0	1
234	11	0	1	1
45	14	9	1	0
320	6	8	1	1
946	5	8	0	1

then adding the halfe pennie in minde to the lowest halfe pennie of the second Collum, I say that 1. in minde and 1. maketh 2. and 2. and 1. maketh 3. then 3. and 1. maketh 4. which 4. halfe pence because they make iust two pence, I set downe a Cypher keeping the two pence in minde, which two pence being added to 8. maketh 10. then 10. and 9. maketh 19. and 19. and 1. maketh 20: Now because that 20. d. maketh one shilling and 8. d. I set down the 8. d. keeping the shilling in minde, which one shilling being added to the 6. of the next Collum maketh 7. then 7. and 4. maketh 11. and 11. and 1. maketh 12. then 12. & 3. maketh 15. wherefore I set downe 5. keeping the Article in minde, which being added to one of the next Collum maketh 2. and 2. and 1. maketh 3. and 3. and one maketh 4. Articles, which 4. Articles maketh 40. s. which is two pound which I keepe in minde, and therefore I adde the 2. l. to the Collum of pounds, saying that 2. and 5. maketh 7. and 7. and 4. maketh a 11. and 11. and 5. maketh 16. wherefore I set downe 6. keeping the one Article in mind, which being added to 2. of the next Collum, maketh 3. then 3. and 4. maketh 7. and 7. and 3. maketh 10. then 10. and 4. maketh 14. wherefore I set downe 4. keeping one Article in mind, which being added to 3. of the next Collum maketh 4. then 4. and 2. maketh 6. and 6. and 3. maketh 9. which I also set downe, so as the totall

Of Subtraction.

totall Somme amounteth to 946.℔.5.ß.8.d. no halfe pennie, one farthing as you ſée in the former example.

How ſhall I know whether theſe ſeuerall Sommes be truely added or not?

Some doe teach it to be done by caſting out all the nines, which way is more tedious then ſure: for the ſureſt tryall indéede is to be done by Subtracting the ſeuerall Sommes out of the totall Somme, of which Subtraction we come now to ſpeake, for all the foure ſpeciall kindes are tryed one by an other.

Of Subtraction.

Cap. 3.

What doth Subtraction teach?

It teacheth to take a leſſer number out of a greater and to ſée what remayneth.

What is to be obſerued in this kind?

Firſt, you muſt ſet downe your greater number aboue, and then the leſſer number right vnder the ſame. As for example, I haue lent to one 564.℔. and hée hath paide me thereof 57.℔. Here to knowe what remaineth, I firſt ſet downe the number lent, and vnder that the number paid, and then drawe a line as you ſée in this manner.

Here beginning on the right hand, I firſt ſay, take 7. out of 4. that cannot bée: wherfore I take one Article of the next figure or place of the lent number, which Article being added to 4. maketh 14. then I ſay take 7. out of 14. and there remaineth 7. which I ſet downe vnder the 4. then I adde that one Article which I borrowed, to the ſecond figure of the paide number which is 5. ſaying that 5. and 1. in minde maketh 6. then take 6. out of 6. and there remaineth nothing, wherefore I ſet downe a Cypher, vnder the 5. of the paide number, then I procéede to the third figure of the lent number,

Lent. 564.℔.
Paide. 57.℔.

which

Of Subtraction. 4

which is 5. and because I finde nothing written vnder it, nor haue nothing in minde to take out of it, I say, take nothing out of 5. and their remaineth still 5. so as the remainder is 507.£. as you sée in this example following.

How shall I knowe whether this bee right or not?

Lent.	564.£.
Paid.	57.£.
Remaine.	507

By adding the remainder and the number paide together, the Somme whereof (if you haue done well) wilbe all one with the number lent, as in the former example, I first adde 7. and 7. together and that maketh 14. wherefore according to the precepts of Addition before taught, I set downe 4. kéeping the Article in minde, then I say, one in mind and 5. maketh 6. which I also set downe, then I say nothing and 5. is 5. which I set downe in the third place, which in all maketh 564. a number equall to the number lent, as you sée here following.

You may perceiue by this, that if any figure of the paid number be greater then the figure ouer him, out of the which it is to bée Subtracted, you must alwaies borrow one

Lent.	564.£.
Paid.	57:
Remaine.	507.
Proofe.	564.

Article of his next fellow, to be added againe to him in his proper place. But you haue to note, that hauing to deale with numbers of diuerse Denominatiōs, thē in borrowing any number, you must alwaies haue respect to the Denomination or name of the thing, from whence you borrow, as in borrowing from shillings you borrow 12. and not 10. from pounds you borrow not one Article but 2. Articles which doe make 20.s. but when the whole number is altogether of one selfe Denomination, then you must alwaies borrow one Article which is 10. to make vp your nūber yͤ wanteth. As you shall more plainely perceiue by this example contayning numbers of diuerse Denominations, as of pounds, shillinges and pence, halfe pence, and farthings. Suppose therefore that you haue lent to one 467.£.13.s. 4.d.ob.q. and hee hath paide you againe thereof, 89.£.16.s.9.d.ob.q. Here hauing set downe the Somme lent in seuerall Collums, according to their diuerse names, and then the Somme paide, right vnder the same, drawe a line as you sée in this example following.

B 4   Heere

Of Subtraction.

	l	s	d	ob	q
Lent.	467	13	4	1	1
Paide.	89	16	9	1	0
Remaine.	377	16	7	0	1

Here beginning with the first Collum on the right hand, I say take nothing out of one, and one stil remaineth, which I set down, then procæding to the next, I say take one out of one and nothing remaineth, wherefore I set downe a Cypher, then procæding to the next Collum, I say take 9. out of 4. that cannot be, wherefore I borrow a shilling of the next Collum that is 12. d. which being added to 4. maketh 16. pence, then I say take 9. out of 16. and there remaineth 7. which I set downe, then procæding to the next, I adde the one shilling which I borrowed, to the 16. which maketh 17.s. then I say take 17.s. out of 13.s. that cannot be. wherefore I borrow one pound which is 2. Articles of the next rancke, ȳ is 20.s. which being added to the 13.s. maketh 33.s. then I say take 17. out of 33.s. and there remaineth 16.s. which I set downe, then I adde the one pound which I borrowed, to 9. and that maketh 10. then I say take 10. out of 7. that cannot be, wherefore I borrow one Article out of the next 6. which being added to the 7. maketh 17. then I say take 10. out of 17. and there remayneth 7. which I set downe, then the one, which I borrowed, I adde to the 8. of the next rancke, and that maketh 9. Againe I say take 9 out of 6. that cannot be: wherfore I borrow one Article of the next 4. which being added to 6. maketh 16. then I say take 9. out of 16 and there remaineth 7. which I set downe, then I take the one which I borrowed out of the 4. & there remaineth 3. so as the remainder is as you sée in ȳ former example. 377.l.16.s.7.d.0.q.

How shall I trie whether this be true or not?

By adding the remainder and the Somme paid together: as in the former example, and of that Addition, will rise if you haue done truely a Somme like in euery condition to the Somme lent: In making which prœfe or triall you cannot lightly erre, if you remember to reduce pence to shillings, and shillings to poundes, and therefore in the Collum of pence, no particular Somme can be aboue 11.d, nor in the Collum of shillings no particular somme can be aboue 19.s, for if it be 20.s, then it is a pound and must be brought to the Collum of poundes.

Of

Of Multiplication.

Cap. 4.

What is Multiplication?
It is the producing or bringing foorth of a third number, by Multiplying two other numbers the one into the other: And it consisteth of three numbers, that is the multiplycand, the multiplyer, and the product.

What signifie those names?

The Multiplicand is that number which is to be multiplyed, and the multiplyer is that whereby the same is multiplied, & the product is the Somme of such Multiplicatiō: As for example, if I would multiply 4. by 3. as in saying 3. times 4. maketh 12. here the number of 4. is the Multiplycand, and the number 3. is the multiplyer, and the number 12. is the product of that Multiplycation.

What order is to be obserued in multiplying, and how are those numbers to be set?

Before I teach you the true order of multiplying, I thinke it good to set you downe a table of Multiplycation, which vnlesse you learne perfectly by hart, you shal neuer multiply readily nor quicklie.

2	2	4	3	5	15	4	9	36	6	8	48	9	9	81
2	3	6	3	6	18	4	10	40	6	9	54	9	10	90
2	4	8	3	7	21	5	5	25	6	10	60	10	10	100
2	5	10	3	8	24	5	6	30	7	7	49			
2	6	12	3	9	27	5	7	35	7	8	56			
2	7	14	3	10	30	5	8	40	7	9	63			
2	8	16	4	4	16	5	9	45	7	10	70			
2	9	18	4	5	20	5	10	50	8	8	64			
2	10	20	4	6	24	6	6	36	8	9	72			
3	3	9	4	7	28	6	7	42	8	10	80			
3	4	12	4	8	32									

How

Of Multiplication.

How is this Table to be read?

In this manner, 2. times 2. maketh 4. and 2. times 3. maketh 6. and 2. times 4. maketh 8. and so foorth: multiplying still one Digit by another, vntill you come to a 100. for this Table serueth onely for Digits, which may bee made to extend so farre as you will, and vntil you haue learned the foresaid Table without booke, you may helpe your selfe with this other Table of Digits made squire wise as you sée here.

In the front of which Table are set downe the 9. Digits beginning on the left hand, and so procéeding to the right hand, from 1. to 9. Againe on the right side of the said Table are set downe the foresaid 9. Digits, beginning aboue, and so procéeding right downe from 1. to 9. the vse whereof

1	2	3	4	5	6	7	8	9	0
1	2	3	4	5	6	7	8	9	1
	4	6	8	10	12	14	16	18	2
		9	12	15	18	21	24	27	3
			16	20	24	28	32	36	4
				25	30	35	40	45	5
					36	42	48	54	6
						49	56	63	7
							64	72	8
								81	9

is thus, first séeke the Digit to be Multiplyed in the front, and séek the other wherby you haue to multiply the same on the right hand, and the square Angle aunswering to these 2. Digits will shew the product of such Multiplycation. As for example hauing to multiply 7. by 6. I first séeke the greater Digit which is 7. in the front, and the lesser which is 6. on the right hand, the product whereof I finde in the square Angle answering both to 7. and to 6. to be 42. & the like is to be obserued in any 2. Digits that are of like value as 7. times 7. the product whereof is 49. But the first Table being perfectly learned without booke, whensoeuer you haue to multiply one number by another, you must obserue these rules hére following.

First that you set downe the first figure of your multiplyer right vnder the first figure of the number that is to be multiplyed, on the right hand, and then orderly to place the rest of the figures of your multiplyer, be they few or many, towards your left hand, directly vnder the rest of the figures of the number that is to bé multiplyed, for if the figures stand not in order one right ouer another

Of Multiplication.   6

other, it will bréede a confusion in your working. Secondly you must not forget to multiply all the figures of the number that is to be multiplyed, by the first figure of your multiplyer before you deale with the next multiplyer, beginning alwaies on your right hand, and so to procéed from one to another, wherby you shal make as many seuerall products as there be figures in your multiplyer. Thirdly you must remember to set downe the first figure of euery seuerall product, right vnder the figure of that multiplyer, wherby you doe multiply, and hauing ended your Multiplication, draw a line vnder the seuerall products: ÿ done, ad the seuerall products together according to the rules of Addition, & the Somme therof shalbe the total or general product of that Multiplycation: al which rules you shall the better vnderstand by working this example following. Suppose thē that you would know how many houres there are in a peare, knowing first ÿ a peare consisteth of 365. daies, here because that euery naturall day containeth 24. houres comprehending both day & night, you haue to multiply 365. by 24. and therefore you must first set down 365. because it is the greater number, and is the multiplycand, which must alwaies stand aboue, & right vnder ÿ, your multiplyer which is 24. is to be set downe in his due place, according to the Rules before taught thus as you sée here. Then say thus 4. times 5. maketh 20. & hauing set down a Cypher right vnder the 4. kéep the 2. Atticles in mind, thē say 4. times 6. is 24. & 2. in mind is 26. here set down 6.

   365. the multiplicād
    24. the multiplier.

& kéepe 2. in mind, then say 4. times 3. is 12. which with the 2. in minde maketh 14. here first set downe the 4. vnder the 3. and because you can procéede no further: you must therefore set downe hard by the 4. on the lefthand the one Article which you had in minde, then hauing cancelled the first figure of the multiplyer, by making a Dash through it with your pen, as you sée in the example following: Procéede with the other figure of the multiplyer, saying that 2. times 5. maketh 10. wherefore set downe a Cypher right vnder the said 2. kéeping the one Article in minde, then say 2. times 6. is 12. and one in minde maketh 13. wherfore set downe 3. and kéepe one in minde, then say 2. times 3. is 6. which with one in minde maketh 7. the which you must set downe, and because you haue made an end of your Multiplycation, cancell

Of Multiplication.

the 2. and draw another line vnder the 2. seuerall productts, that done, ad together whatsoeuer is contained betwixt the two lines, and you shall finde the generall product to be 8760. houres.

How shall I knowe whether the last Multiplycation be right or not? By diuiding the generall product by the Multiplyer, for in so doing your quotient will bee like vnto the first Somme that was Multiplyed, which you cannot do, vntill such time as you haue learned to deuide, and therefore hauing first shewed certaine compendious waies of Multiplycation, I will then procæde to Diuision.

The multiplycand,	365.
The multiplyer.	2 4
The seuerall products.	1460.
	730.
The generall product or total Som.	8760.

Certaine compendious waies of Multiplycation.

VVHen is any such way to be vsed?

When the Multiplyer beginning on the right hand with one Cypher or with many, endeth on the left hand with the digit 1. as these numbers following, 10/100/1000/ &c.

Why what is then to be done?

If you haue to Multiply by 10. then you haue no more to doe, but to set downe on the right hand of the number that is to bæ Multiplyed, one Cypher, if by a 100. then 2. Cyphers, if by a 1000. then 3. Cyphers, as for example: if you would Multiply 365. by 10. then by setting downe on the right hand one Cypher as hath bæne said, the product will be 3/650. if you set downe 2. Cyphers the product wil be 36/500. if you set down 3. Cyphers, then the product will be 365/000.

What if the number do end on the left hand with any other Digit, as 2. 3. or more, as 200. 300. 400?

Then you must Multiply the number of the Multiplycand first by that Digit, and then adde to the end of the product on the right hand all the Cyphers annexed to the said Digit, as if you woulde Multiply 365. by 200. First say 2. times 5. maketh 10. and 2. 6. and one in minde is 13. and so to procæd in Multiplying euery figure of the Multiplycand by 2. and you shall finde the product

Of Diuision. 7

duct to be 730. whereunto if you adde on the right hand 2. Cyphers the whole product wil be 73/000.

Of Diuision.

Cap. 5.

WHat is Diuision?
Diuision is that whereby any number is diuided into as many parts as you will.

How many numbers are incident to Diuision and how are they called?

These 4. that is to say, the number which is to be diuided which is called the diuidend, the second number whereby you doe diuide which is called the Diuisor, the third number is called the quotient, which sheweth how many times the Diuisor is comprehended in the diuidend, and the fourth number is called the remainder, if any be.

What order is to be obserued in Diuision?

This here following, first set downe your diuidend and directly vnder that. beginning on the left hand, set downe your Diuisor, that is to say the first figure of your Diuisor right vnder the first figure of your diuidend on the left hand, and so consequently one after another proceeding towardes the right hand, which is alwaies to be done, so often as your Diuisor doth not exceed in quantitie, the figure standing right ouer his head, for if it doe, then you must remoue your Diuisor one figure further towardes the right hand. As for example if you would diuide 487. by 53. you must not set the first figure of your Diuisor which is 5. vnder the first figure of your diuidend, which is but 4. but vnder the second figure of your diuidend which is 8. for you cannot take 53. out of 48. and therefore you must set the first figure of your Diuisor vnder the second figure of the diuidend, and so follow on with the rest and then draw a line as you see here in this example following.

What is thē to be done? The diuidend .487. (9. the quotient
The diuisor.   53.

Then you must aske how many times 5. is comprehended in 48. and you shall finde that 5. is comprehended in 48. 9. times

which

Of Diuision.

which 9. being the quotient, must be placed on your right hand, behind a crooked line made like a halfe Moone as you ſée in the example aboue, then Multiply the firſt figure of the Diuiſor by the quotient 9. and the product thereof ſhal be 45. which being taken out of 48. there will remaine 3. which you muſt ſet downe ouer the head of 8. and ſtréeke out the 48. and alſo the firſt figure of your Diuiſor which is 5. that done Multiply the ſecond figure of your diuiſor which is 3. by the foreſaid quotient which is 9. & the product therof wil be 27. which being taken out of 37. ther remaineth 10. which you muſt ſet ouer the head of 37. and cancell the 37. and alſo the 3. beneath, as you ſée in this example following.

Wherein you ſée   1   the remainder.
that the diuidend is   3 0
487. the Diuiſor 53   The diuidend.   4 8 7 (9 the quotient.
the quotient 9. and   The diuiſor.   5 3
the remainder 10 which if you will apply to any vſe you may imagine that there is 487. ℔. to be diuided amongſt 53. ſoldiours, and by working as before, you ſhal find that euery Souldiour muſt haue 9. ℔. and there ſhal be remaining 10. ℔. to be diuided amongſt the foreſaid 53. Souldiours, which 10. ℔. being reduced to ſhillings, may bee diuided amongſt the Souldiours as well as the pounds: As for example if you Multiply 10. ℔. by 20. ſ. it will make 200. ſ. which being diuided by 53. the firſt Diuiſor, the quotient will be 3. ſ. and the remainder 41. ſ. and that being reduced into pence, which is done by Multiplying 41. by 12. the quotient, will be 9. d. and the remainder 15. d. which if you Multiply by 4. you ſhall reduce them to farthings, the product whereof will be 60. which product being diuided by 53. the firſt Diuiſor you ſhall find in the quotient 1. farthing and the remainder to be $\frac{7}{53}$ of a farthing, the value of which Fraction how to find out is taught hereafter when wee come to ſpeake of Fractions. So as you ſée by this meanes that euery Souldiour ſhall haue for his ſhare ℔. 3. ſ. 9. d. q. and ſome what more a thing of no moment.

How many things are to be remembred in Diuiſion?

Theſe fiue Rules here following. Firſt that you put no number at one time in the quotient aboue 9. Secondly to moderate your quotient in ſuch ſorte, as hauing Multiplyed the firſt figure of the Deuiſor by the quotient, there may remaine ſufficient for the

Of Diuision. 8

the next figure of the Diuisor being Multiplyed by the quotient to be deducted out of that number which standeth right ouer his head. Thirdly that you Multiply euery figure contained in the Diuisor by the quotient. Fourthly that if at any time in working it happeneth so as your Diuisor is not comprehended in the number ouer his head, then to put a Cypher in the quotient and to remoue your Diuisor one figure further towardes the right hand, and that done to worke as before. Fiftly to sée that the last remainder, if there be any left, doe not excéede in quantitie the Diuisor, all which things you shall better vnderstand by this one example following: Suppose then that you haue to diuide 819096. by 92. here hauing set downe your diuidend and Diuisor in such order as is before taught, and as you sée here in this example.

First aske howe
many times 9. is in   8
81. and you shal find   9 3
the quotient to be 8. *The diuidend.* 8 1 9 0 9 6 (8 the quotiēt
which being Multi= *The diuisor.*   9 2
plyed into 9. maketh
72. which if you take out of 81. there shall remaine 9. which must be set downe ouer the head of the figure 1. and cancel the said 81. together with the first figure of the Diuisor, which is 9. then say 2. times 8. is 16. which being Subtracted out of 99. there remaineth 83. which you must set ouer the head of the 99. and cancell the 99. that is aboue, and also the 2. beneath, as you sée in the former example. That done remoue your Diuisor towardes the right hand, that is to say, by setting the last figure of your Diuisor which is 2. vnder the next Cypher on the right hand, and place the first figure of your Diuisor which is 9. next to that towardes the left hand: then aske how many times 9. is comprehended in 83. and you shall find 9. times, which 9. must be set downe in the quotient next to the 8. then say 9. times 9. is 81. which being taken out of 83. there remaineth 2. then hauing set downe the 2. ouer the head of 3. cancel the 83. aboue & also the 9. beneath, then Multiply the last figure of the Diuisor which is 2. by the 9. which is in the quotient, and the product thereof is 18. which being taken out of 20. there remaineth 2. which 2. must be set downe ouer the Cypher, and the Cypher cancelled and also the 2. beneath, as

you

Of Diuision.

you see in this example following.
 That doone remooue
your Diuisor one figure The diuidend. $\overset{2}{8}\overset{2}{\cancel{X}}2$
further towards the right The diuisor. $\cancel{9}\,\cancel{2}$ $\cancel{8}\,\cancel{X}\,\cancel{9}\,096\ (89.$
hand by setting the 2. vn-
der the 9. which is the last figure saue one of the diuidend, and
place the first figure of your Diuisor which is 9. next to that on the
left hand, then aske how many times 9. is conteined in 2. and you
shall finde none, wherefore you must set downe a Cypher in the
quotient and cancell the Diuisor as you see in this example
 That done remoue   2
your Diuisor againe by The diuidend. $\cancel{8}\,\cancel{X}\,\cancel{9}\,096\ (890.$
setting the 2. vnder the The diuisor. $\cancel{9}\,\cancel{2}$
last figure of the diuidēd
and place the 9. next to that on the left hand, then aske how many
times 9. is contained in 29. and you shall find 3. times, which 3.
you must set downe in the quotient as you see in the example following, then Multiply the first figure of your Diuisor which is 9.
by 3. and the product thereof will be 27. which being taken out of
29. there will remaine 2. which 2. you must set ouer the head of 9
and cancell the 9. aboue, together with the 9. beneath, then Multiply the 3. which is in the quotient by the last figure of your Diuisor which is 2. and the product thereof will be 6. which being taken out of 26. there remaineth 20. wherefore you must set downe
a Cypher ouer the 6. which is the last figure of the diuidend, and
cancell the said 6. aboue, and also the 2. beneath and then you
haue done rightly, and in such order as all the former examples
and this also next following doe plainely shew.

 For heere
you see that y^e the remainder.
diuidende is $\overset{2}{\cancel{X}}\,20$ |20.
819/096. the   The diuidend. $\cancel{8}\,\cancel{X}\,\cancel{9}\,\cancel{9}\,\cancel{6}\ (8903|92.$
Diuisor 92.   The diuisor. $\cancel{9}\,\cancel{2}$ the quotient.
the quotient
8903. and the remainder 20. which remainder must be set on the
right hand of the quotient ouer the Diuisor, hauing a line drawne
betwixt them as you see in the last example.

                   How

Of Diuision. 9

How shall I know whether I haue diuided truely or not?

By Multiplying the quotient by your Diuisor & by adding to the product thereof the remainder, for so shall you haue a number like to the first diuidend if you haue wrought well, if not, then your Somme will be either more or lesse then the diuidend.

Certaine compendious waies of Diuision.

IS there no shorter way of Diuision?

No, vnlesse the Diuisor hath one Cypher or more on the right hand, for then you may vse a breefer way by cutting off the last figure, or figures of the diuidend: As for example if you had to diuide 3708. by 10. heere cut off the last figure 8. with a dash of your penne in this wise 370/8. and the quotient will be 370. and the remainder shall be 8. Againe if you had to diuide the foresaide number by 100. then cut off the 2. last figures of the diuidend, that is the 8. & the cypher standing next it, and then the quotient wil be but 37. and the remainder 8. as before. but if you haue to diuide any number by 1000. then you must cut off 3. figures of the diuidend, and so forth, remembring alwaies for euery Cypher in the Diuisor, to cut off one figure or Cypher of the diuidend.

Of halfing any number.

HOw is that done?

By diuiding the number by 2. As for example, if you would knowe the halfe of 3708. diuide the same by 2. and you shall finde the quotient to be 1854. which is the iust one halfe of the diuidend.

Of the Rule of three otherwise called *the golden Rule whereof there be* three kinds, that is, the common Rule, the rule Reuerse, and the Double rule.

Cap. 6.

C Wherefore

The rule of three.

WHerefore is it called the Rule of three?
Becauſe that by 3. knowne numbers, it teacheth you to finde out a fourth number vnknowne.
How are ſuch 3. numbers to be placed?
The firſt is to be placed on the left hand, the third on the right hand, and the ſecond in the middeſt betwixt both.
How ſhall I know in what place each one is to be ſet?
By marking to what number the queſtion is annexed, for ẏ muſt alwaies be the third number, and hauing the third number, you ſhall quickely haue the firſt, becauſe the firſt and third muſt alwaies be of one ſelfe Denomination or name, betwixt which two, the ſecond number is to be placed: As for example let this bee the queſtion. If I pay 35.s. for 13. yards of Linnen cloth, how much ſhall I pay for 3. yards, here becauſe the queſtion is annexed to the number 3. yards, that muſt bee the third number and is to be placed on the right hand, then becauſe that the number 13. yards is of the ſelfe ſame Denomination, that muſt bee the firſt number, and is to be placed on the left hand, & the number 35.s. (which is the price) is to be placed in the middeſt betwixt the other two, as you ſee here following.

Now muſt you make your queſtion in this manner, if 13. yards did coſt me 35.s. what ſhall 3. yards coſt me, and ſo the firſt and laſt ſhal be of one Denomination or name, that is to ſay yards.

yards, s. yards:
 13.   35.   3.

What order is to be obſerued in working by this rule?
You muſt Multiply the third and ſecond numbers the one into the other, and diuide the product therof by the firſt, and the quotient will ſhew you what the fourth number ſhoulde bee, which fourth number is alwaies of like Denomination to the ſecond or middle number, as in the former example, firſt Multiply 35. by 3. and the product will be 105. which if you diuide by 13. the quotient will be 8.s. and one third of a ſhilling which is 4.d. and that is the price of 3. yards which is the number that you ſeek to know. And this is the common kind of working by the Rule of 3. whereof it is called the common Rule of three.

What if there be diuerſe Denominations either of Coyne as of pounds, ſhillings, and pence, or parts of yards, as quarters, and halfe quarters, and ſuch like?

Then

Of Fractions.

Then you must reduce them all to the smallest Denomination which belongeth to Fractions, whereof we come now to treate of.

Of Fractions.

Cap. 7.

Hat are Fractions?

They are broken partes of some whole thing called in Latine Integrum.

What is Integrum?

Any thing that is whole, and not broken, or diuided into parts: As one whole yard, a pound, a shilling.

Of how many numbers doth a Fraction consist?

Of two, that is the Numerator, and the denominator, whereof the Numerator is alwaies set aboue, and the denominator beneath, hauing a little line drawne betwixt them thus $\frac{1}{2}$ which signifieth one second or one halfe, againe two thirds and three fourths are written thus, $\frac{2}{3}\frac{3}{4}$ whereof the first signifieth two thirds, and the other three fourths and such like.

What is ment by these two words Numerator, and Denominator, and whereto serue they?

The denominator is as much to say as the namer of the parts, which sheweth into how many parts the Integrum is to be diuided, and the Numerator is as much to say as the Numberer, and sheweth how many of those parts are to be taken: As for example $\frac{2}{3}$ of a shilling, here the nethermost called the denominator, sheweth that the shilling is to be diuided into three parts, as into three groats, and the vpper number called the Numerator, sheweth that you must take 2. out of those 3. parts, so as $\frac{2}{3}$ of a shilling is as much to say as 2. groats or 8. d. Againe Fractions may be diuided into two kinds, that is, into simple Fractions, and into Fractions of Fractions.

Define those two kinds?

Simple Fractions are such as I speake of before which are the first & immediate parts of any Integrum that is diuided into parts,

C 2 but

Of Fractions.

but if those Fractions bee diuided into moze Fractions, then are they called Fractions of Fractions, as when I say thzee fourthes of two thirds, of sir seuenthes and such like.

How are such Fractions to be set downe in writing?

In manner of simple Fractions thus $\frac{1}{4}$ of $\frac{2}{3}$ of $\frac{6}{7}$.

Make some demonstration of this example that I may the better vnderstand it?

Imagine that there is a whole peece of Golde of vii.s. as our Angel in times past hath bene, which vii.s. be the vii. partes of that peece of Gold, foz tryall whereof laie downe befoze you vii. twelue pennie peeces of Siluer, oz in steede thereof vii counters, to signifie those vii. partes, of which vii. parts you must first take away vi. reiecting oz laying aside the od seuenth part, then diuide those 6. parts into 3. parts, and euery such part will be iust 2.s. of which 3. parts you must take away 2. partes that is 4.s. reiecting the other third part which remaineth, that done, diuide those 2. parts which you haue taken away into 4. parts, which is 4.s. and take 3. of them reiecting the fourth part, so shall you finde that $\frac{1}{4}$ of $\frac{2}{3}$ of $\frac{6}{7}$ of the foresaid peece of Golde is iust 3.s. and there remaineth still of the parts reiected 4.s. which being added to 3.s. that was taken away, doe make vp againe the whole Integrum of 7.s. Notwithstanding Fractions of Fractions doe seeldome chaunce in the Diuision of any number, but if they doe, you must reduce them into simple Fractions, befoze you can deale with them any manner of way, and because there are diuerse rules belonging to Fractions, without the knowledge whereof you can nether adde, noz subtract them, noz yet ioyne them with any Integrum, I wil bziefly set downe here seauen necessarie rules foz the same.

Seauen necessarie Rules belonging to Fractions.

What doth the first rule teach?

To bzing Fractions of Fractions into simple Fractions.

How is that done?

Thus, Multiply the first Numeratoz into the second, and if there be any moze Fractions, then Multiply also the said pzoduct of the first two Numeratozs into the third Numeratoz, and that shall

Of Fractions.

shall be the Numerator of the simple Fraction, then Multiply the denominators in like manner, and the Somme thereof shall be the denominator of the simple Fraction, as in the former example $\frac{1}{4}$ of $\frac{2}{3}$ of $\frac{6}{7}$ here I Multiply the first 2. Numerators together, in saying 2. times 3. maketh 6. now because I haue a third Numerator, I multiply the product of the first 2. Numerators, which is 6. into the third Numerator which is also 6. saying, 6. times 6. maketh 36. and that shall be the Numerator of the simple Fraction, which I set downe. Then by Multiplying the 3. denominators in like manner, I finde the denominator of the simple Fraction to be 84. the which I set downe vnder the 36. and draw a line betwixt them so as now I finde that $\frac{1}{4}$ of $\frac{2}{3}$ of $\frac{6}{7}$ do make $\frac{36}{84}$ which indeed is no more but $\frac{3}{7}$ as you shall learne hereafter by the sixt rule.

What doth the second rule teach?

To bring Fractions being more in value then Integrums into integrums.

When are Fractions said to be more then integrums?

When the Numerator is a greater number then the denominator, for if they be both of like value as $\frac{4}{4}$ and such like, then such Fraction is an euen integrum.

How shall I know how many Integrums such Fractions as be more then Integrums doe containe?

By diuiding the Numerator by the denominator, and if any thing remaine, write that aboue the denominator, as in this example $\frac{806}{7}$ here if you diuide 806. by 7. which is the denominator, you shall finde the quotient to containe 115. Integrums, and the remainder to be 1. which is $\frac{1}{7}$.

What doth the third rule teach?

To bring Integrums into parts by Multiplying the number of the Integrums by the denominator of those partes, as if you would bring 64. yards into quarters, Multiply 64. by 4. and there will arise thereof 256. quarters.

What doth the fourth rule teach?

It teacheth to bring Integrums hauing Fractions annexed to them, into one Fraction.

How is that done?

Multiply the number of the Integrums by the denominator of the Fraction, and then adde to the product the Numerator of the

C 3   said

Of Fractions.

saide Fraction, and that somme shall be the Numerator, vnder which write the denominator aforesaid, and so you shall finde that 23. Integrums hauing $\frac{2}{3}$ annexed thereunto shall make $?\frac{1}{3}$.

What doth the fift rule teach?

It teacheth to expresse a Fraction written with many figures, in so fewe as may be.

How shall I doe that?

Thus, finde out some number that may first diuide the Numerator, and then the denominator seuerally by themselues, without leauing in either of them any remainder, and the quotient of the first Diuision will shew the Numerator, and the quotient of the second Diuision will shew the denominator, but if you cannot readily finde out a number that will diuide them both without leauing some remainder, then leaue not to Subtract the lesser number out of the greater, vntill you finde two like numbers, by one of the which two like numbers diuide both the Numerator, and also the denominator seuerally as before, and the quotient will shew that which you seeke: but if such 2. like numbers cannot be found, then you may assure your selfe that the Fraction cannot be written in fewer figures then they alreadie be.

Giue examples of these two waies?

For example of the first way, suppose that you would expresse $\frac{9}{12}$ in lesser numbers, here seeke out some number that may diuide euenly both the Numerator, and also the denominator, without leauing in either of them any remainder, which by diuiding each of them by the number of 3. you may doe: For by asking how many times 3. is contained in 9. the quotient will be 3. and in asking how many times 3 is conteined in 12. the quotient shall be 4. so shall you find $\frac{9}{12}$ to be no more then $\frac{3}{4}$. Now for example of the second way, let $\frac{27}{81}$ be the Fraction which you would set down in fewer figures, here because you cannot readily finde out a number that wil euenly diuide both the Numerator and the denominator, Subtract the lesser out of the greater, that is to say 27. out of 81. and there will still remaine 54. from which Subtract 27. and there will remaine 27. which 2. numbers because they are both like, diuide the Numerator and denominator of the foresaid Fraction, each of them by one of these numbers, that is to say, by 27. and the quotient of the first Diuision will shew you the Numerator which

Of Fractions.

is 1. and the quotient of the second Diuision wil shew you the denominator which is 3. so as you shal find $\frac{1}{3}$ to be as much in value as $\frac{27}{81}$ and note that when there be Cyphers both in the Numerator and in the denominator standing in such sorte as they may bee euenly cut off, the remainder will shew the fewest figures, wherin the Fraction may be written as $\frac{200}{500}$ here by cutting off the two Cyphers as well beneath as aboue the line, with a dash of your penne in this manner $2/\frac{00}{500}$ the remainder shall bee $\frac{2}{5}$ which is as much in value as $\frac{200}{500}$.

What doth the sixt rule teach?

It teacheth to find out the value of the Fraction of any integrum.

How is that done?

Thus, Multiply the Numerator by the knowne partes of the Integrum, and diuide the product thereof by the denominator, so shall you haue the value of the Fraction: As for example, if you would know the value of $\frac{3}{4}$ of an Angel, consider first what partes an Angel hath, and you shall find the parts thereof to be 10.s. or 30. groats, here if you Multiply 3. by 10.s. it will make 30.s. which being diuided by 4. you shall finde in the quotient 7.s. and the remainder to be $\frac{1}{2}$ or one half of a shilling, which is 6.d. Againe if you Multiply 3. by 30. groates it will make 90. groats, which being diuided by 4. you shall finde in the quotient 22. groats, and the remainder to be $\frac{1}{4}$ or one halfe of a groat which is 2.pence, making in all 7.s.6.d. like vnto the first Somme, so as you see that $\frac{3}{4}$ of an Angel is 7.s.6.d. and thus you may deale with the Fractions of any other Integrum that hath knowne parts.

What doth the seauenth rule teach?

It teacheth to bring Fractions of diuerse denominations to one selfe denomination, without the which nether Addition, nor Subtraction of Fractions can be made : As for example if you would adde $\frac{2}{3}$ and $\frac{1}{5}$ together, you must first bring them to one selfe denomination thus, Multiply the denominators the one into the other, and the product thereof shall be a common denominator to both the Fractions, wherefore say 3. times 5. doe make 15. which must be set downe in 2. seuerall places by them selues thus, 15. and 15. then Multiply the Numerator of the first Fraction into the denominator of the second, as 2. into 5. maketh 10. which set downe

C 4 ouer

Of Fractions.

ouer the first common denominator thus $\frac{10}{15}$ then Multiply the Numerator of the second Fraction into the denominator of the first Fraction, as 4. into 3. which maketh 12. and that must bee likewise set downe ouer the second common denominator thus $\frac{12}{15}$ so shall you find that $\frac{2}{3}$ and $\frac{4}{5}$ are all one in value with $\frac{10}{15}$ and $\frac{12}{15}$.

Addition of Fractions.

Cap. 8.

How are Fractiōs to be added together? Fractions being first brought to one denomination are easily added, for then you haue no more to doe but to adde the Numerators together, and to write the common denominator vnder the Somme of such Addition, as in the former example, $\frac{10}{15}$ being added to $\frac{12}{15}$ maketh $\frac{22}{15}$ but if the denominators be not like, then you must make them like by the seuenth rule before taught.

What if there be three Fractions of diuerse denominations to be added together as $\frac{2}{3} \frac{3}{4} \frac{4}{5}$?

Then hauing reduced the 2. first to one self denomination adde the 2. Numerators together and write vnder the Somme thereof the common denominator, that done, deale with that Fraction last found, and with the third as you did before with the 2. Fractions that had diuerse denominations, and so you shall find that $\frac{2}{3}$ and $\frac{3}{4}$ being first brought to one denomination, and then added together doe make $\frac{17}{12}$. whereto if you will adde $\frac{4}{5}$. then you must bring them againe to one selfe denomination, and so you shall finde that $\frac{17}{12}$. and $\frac{4}{5}$ doe make $\frac{85}{60}$. and $\frac{48}{60}$. which last 2. Numerators being added together doe make in all $\frac{133}{60}$. which is 2. Integrums and $\frac{13}{60}$.

Subtraction of Fractions.

How are Fractions to be Subtracted, one out of another, or out of Integrums?

first

Of Fractions. 13

First make the denominators like as you did before in Addition, then take the lesser Numerator out of the greater, and vnder the remainder thereof write the common denominator, so shall you find that $\frac{3}{7}$. being taken out of $\frac{6}{7}$. there remaineth $\frac{3}{7}$. but if you would Subtract Fractions out of Integrums, then you must take one of the Integrums and breake it into partes. As for example, to Subtract $\frac{3}{7}$. out of 9. Integrums take one from 9. and breake it into parts making it $\frac{7}{7}$. for that is one Integrum, for whensoeuer the Numerator is made like and equall to the denominator, it signifieth one Integrum, then take $\frac{3}{7}$. out of $\frac{7}{7}$. and there remaineth $\frac{4}{7}$. and 8. integrums.

What is to be obserued in breaking the integrum?

In breaking the Integrum, you are to be directed alwaies by the denominator of the Fraction which you haue to Subtract, for the Integrum which should be the Numerator, must be equall in value to the said denominator, as before is said.

Multiplycation of Fractions.

HOw are Fractions to be Multiplyed by Fractions, or Integrums by Fractions?

As touching Fractions, Multiply the Numerators one into another, and the product thereof shall bee the Numerator, then Multiply the denominators in like manner, and the product thereof shall be the denominator, so shall you find that $\frac{5}{7}$. being Multiplyed by $\frac{3}{4}$. doe make $\frac{15}{28}$. But if you would Multiply integrums by Fractions, then Multiply the integrums by the Numerator of the Fraction, and vnder the product thereof, set downe the denominator of the same Fraction, drawing a line betwixt them, so shal you find that 20. integrums being Multiplyed by $\frac{5}{12}$. doe make $\frac{100}{12}$. that is to say 8. integrums and $\frac{4}{12}$. which is $\frac{1}{3}$. &c.

Diuision of Fractions.

HOw are Fractions to be diuided by Fractions, or integrums by Fractions, or Fractions by integrums?

The diuision of Fractions is done by Multiplycation thus, first set downe your diuidend on the left hand, and the Diuisor alwaies on

Of Fractions.

on the right hand, and then draw 2. crosse lines like a Saint Andrews crosse betwixt them, which shal direct you in your working: As for example, if you woulde diuide $\frac{2}{3}$. by $\frac{4}{5}$. set them downe thus $\overset{dividend}{\underset{}{\frac{2}{3}}}\times\overset{diuisor}{\underset{}{\frac{4}{5}}}$ and worke as followeth, first Multiply the Numerator of the dividend, by the denominator of the Diuisor, and the product thereof shalbe the Numerator, then Multiply the denominator of the dividend, by the Numerator of the Diuisor, and that shall be the denominator, so shall you find that $\frac{2}{3}$. being divided by $\frac{4}{5}$. there remaineth $\frac{10}{12}$. that is $\frac{5}{6}$. but if you would diuide Integrums by Fractions, or cōtrarywise Fractions by Integrums, then make of the Integrums a Fraction, by setting down 1. in the place of a denominator vnder the Integrums, and worke as before, so shall you find that $\frac{2}{1}$. Integrums being divided by $\frac{1}{4}$. doe make $\frac{8}{1}$. contrarily if you diuide $\frac{3}{4}$. by $\frac{2}{1}$. Integrums there will arise $\frac{3}{28}$.

What if I haue to diuide Fractions annexed to integrums?

Then you must first reduce each Integrum with his Fraction annexed, into one selfe Fraction by the fourth rule of Fractions before taught. As for example, if you would diuide 9. Integrums hauing $\frac{3}{4}$. thereunto annexed, by 3. Integrums and $\frac{1}{7}$. heere the fourth rule of Fractions teacheth you first to Multiply the Integrum 9. by 4. which is the denominator of the Fraction annexed, the product whereof is 36. whereunto by adding the Numerator of the same Fraction which is 3. you shall make the Numerator of the Diuidend to be 39. vnder which you must set the denominator 4. thus $\frac{39}{4}$. and that is your whole dividend, then hauing in like manner brought the Diuisor which is 3. Integrums and $\frac{1}{7}$. into one selfe Fraction, worke as the former rule of Diuision teacheth you, and you shall produce $\frac{117}{40}$. which is 2. Integrums and $\frac{17}{40}$.

The rule of three belonging
to Fractions.

Cap. 9.

WHat order is to bee obserued in this rule hauing to deale with Fractions onely?

The

The rule of three. 14

The selfe same order that hath beene taught before touching Integrums: for in working with Fractions you must haue also 3. seuerall numbers, and you must sée that the first and third numbers be of one selfe denomination, and that number to bee placed alwaies in the third place whereunto the question is annexed, and then to Multiply the second by the third, and to diuide by the first, and so the fourth number which you séeke to know, shall appeare: As for example if $\frac{1}{4}$. of an ell of fine Holland cost me $\frac{2}{3}$. of an English crowne, in value 15. groats, what shall $\frac{5}{6}$. of an ell cost me, here first you must set downe your 3. seuerall numbers in order thus, $\frac{1}{4}$ ell $\frac{2}{3}$ $\frac{5}{6}$ ell. so as the first and third may be of one selfe denomination, then Multiply the second and third Fraction the one into the other, which will make $\frac{10}{18}$. and that being diuided by $\frac{1}{4}$. which is the first Fraction will produce $\frac{40}{54}$. of a crowne, the value of which Fraction if you séeke to knowe by helpe of the sixt rule of Fractions, teaching you to Multiply the Numerator of the Fraction, by the knowne parts of a crown which are 5 s. or 15. groats, you shall finde the value of that Fraction to be 11. groats and $\frac{6}{54}$. of a groat, which is one farthing and somewhat more, supposing the least knowen parts of a groat to be 16. farthings.

The golden Rule reuerse called in
Latine regula euersa that is to say
turned back-ward.

Cap. 10.

What is the order of this rule? Multiply the first by the second, and diuide the product therof by ye third, as if a pennie Loafe must waigh 2. P. Wheate being at 3. s. the bushell, what shall a pennie Loafe waigh when Wheate is at 2. s. the bushell, the question must be framed thus, if 3. s. require 2. P. waight, what shall 2. s. haue, then by working according to this rule you shall finde that the pennie Loafe must waigh. 3. P.

Ano-

The golden rule.

Another example of the same rule.

I would know how many yards of Bayes bearing in breadth ¾. wil suffice to line 7. yards of Silke bearing in breadth 3. quarters & a halfe. Here you must frame your question thus, if ¾. and ½. require 7. yards, what shall ¾. require, but because the first and third number of this question are not of one selfe denomination by reason of the fraction annexed to the first number, you shall doe well to reduce the first and third number both into halfe quarters, and then to worke as though they were all Integrums, which is more easie then to make al the numbers Fractions, wherefore say thus: If 7. halfe quarters doe require 7. yards, what shall 14. halfe quarters require, and in working by the rule Reuerse, you shall finde in the quotient 3. yards of Bayes and a halfe.

The double rule called in Latine regula duplex.

Cap. 11.

Hereto serues this rule and what order is to be obserued therein?

This rule serueth to vnfold two questions wrapt in one, as thus. If I pay 4.d. for the carriage of 20.l. waight 30. miles, what shal I pay for the carriage of 50.l. waight 60. miles, here of this and such like demaunds, you must make 2. sundrie questions, and the fourth Somme of the first question being found, shall be the second or middle number of the second question: wherefore frame your first question thus, if 20.l. cost 4.d. what shall 50.l. cost, and you shall find that it will cost you 10.d. then say if 30. miles cost 10.d. what shall 60. miles cost, and you shall finde that it will cost 20.d. And note that each of these 2. questions is to bee wrought by the common rule of 3. that is to say by Multiplying the second into the third, and by diuiding the product thereof by the first, and the fourth found number

The double rule. 15

ber of the first question must be the second, or middle number of the second question, as in the former example, you sée that 10.ß. which was the fourth found number, is here the middle number of the second question.

Another example.

If 25.℔. doe gaine me 8.℔. in 4. yeares, how much shall a 100.℔. win one in 10. yeares, both these questions are also to bee wrought by the common rule of 3. Wherefore set downe the first question thus, if 25.℔. yéldeth 8.℔. what shall 100.℔. yélde, and you shall finde 32.℔. then say 4. yeares yéldeth 32.℔. what shall 10. yeares yélde, and you shall finde 80.℔. But note that these double questions, may be put in such sort as you must worke ye first or second question, sometimes by the rule reuerse. As in this question here following, if 6.℔. win 8. Crownes in 10. yeares, in how many yeares shall 3.℔. win 12. Crownes, here frame your first question thus, if 6 ℔. require 10. yeares how many yeares shall 3.℔. require: And in working this question by the rule Reuerse, you shall finde 20. yeares, then for the second question say thus, if 8. Crownes require 20. yeares, how many yeares shal 12. Crownes require: Here if you worke by the common rule of 3. you shall finde 30 yeares.

Another example.

If 7. horses doe eate 12. bushels of Oates in 20. daies, how many bushels shall 14. horses eate in 15. daies, here frame your first question thus, if 7. horses doe eate 12. bushels, what will 14 horses eate, and in working by the common rule of 3. you shall finde in the quotient 24. bushels, then frame your question thus, if 20. daies require 24. bushels, what will 15. daies require, here in working by the common rule of 3. you shall finde in the quottent 18. bushels.

Another example.

If ten reapers reape 15. Acres in 7. dayes, in how many daies shall 16. reapers reape 20. Acres: Here frame your first question thus, if 10. reapers require 7. daies, how many daies shall 16. reapers require, which question must be wrought by the rule Reuerse, and so you shall finde 4. daies and ½. of a day which is 9. houres:

The double rule.

houres, then say, if 15. Acres require 4. daies and ½. of a day, how many daies shall 20. Acres require, and in working this second question by the common rule of 3. belonging to Fractions as is before taught, you shall finde that 20. Acres wil require 5. daies and 10/12. or ⅚. of a day which is 20 houres. But it were more easie in this second question to reduce the daies into houres by Multiplying the 4. daies by 24. houres, the product whereof wil be 96. houres, whereunto if you adde the od 9. houres it will make in all 105. houres, which being Multiplyed by the third number of this second question, which is 20. the product shall be 2100. houres, which diuided by the first number of the said question which is 15. you shall finde in the quotient 140. houres, which if you diuide againe by 24. you shall finde in the quotient 5. daies, and the remainder to be 20. houres, which agréeth in all pointes with the first manner of working by Fractions, and is the easier way of the two.

How shall I know hauing to worke by this Double rule, when to vse the rule reuerse?

By considering whether the third number requireth more or lesse of time, or of any other measure or quantitie, as in the former example, of Bayes for lining, the more breadth it had, the lesse did serue for lining. Againe in the example of the gaine by yeares, of 6.£. and 3.£. you did sée that 3.£. require more yeares then 6.£. and therefore that first question was wrought by the rule reuerse: Also in this last example of ẏ reapers, the more reapers, the lesse time they require, & therefore that question was wrought by the rule reuerse.

The rule of Fellowship

Cap. 12.

Hat doth this rule teach?

To knowe the gaine or losse of such as doe make a stocke and doe occupie together in the trade of Marchandize: As for example, foure Marchants did put their mony in lot in this manner.

The rule of Fellowship.   16

ner. The firſt brought 30. Crownes, the ſecond 50. the third 60 and the fourth 100. and with theſe portions they gained 3000. Crownes, the queſtion is how much euery one ſhall haue to his ſhare of that gaine according to the portion which he brought. To know this, you muſt firſt gather all the ſeuerall portions together by Addition into one Somme, and you ſhall finde the ſomme of the portions to be 240. Crownes, then ſay if 240. Crownes do gaine 3000. Crownes, what ſhall 30. gaine, and after this manner worke by the common rule of 3. with all the reſt of the portions, and ſo you ſhall finde that he which brought 30. ought to haue 375. Crownes, and hee that brought 50. ought to haue 625. Crownes, and hee that brought 60. ought to haue 750. Crownes, and he that brought 100. ought to haue 1250. which maketh in all 3000. Crownes, for that is the very Somme of the gaine before ſet downe, the order of working whereof, this figure plainely ſheweth.

$$240. \longrightarrow 3000. \longrightarrow \begin{cases} 30. \longrightarrow 375. \\ 50. \longrightarrow 625. \\ 60. \longrightarrow 750. \\ 100. \longrightarrow 1250. \end{cases}$$

The common Diuiſor which is the Somme of the particular portiós added together. } the generall gaine. } the particuler portions. } euery mans ſingle ſhare.

A like example of loſſe receiued by Shipwracke.

Three Marchants doe venture their goods in one Shippe, the goods of the firſt were worth 300. Crownes, the ſecond 400. the third 500. there were as much goods caſt out as was worth 100. Crownes, the queſtion is how much euery one ſhould loſe according to his portion, here worke as before, and you ſhall finde that euery one ſhall loſe ſo much as this figure following ſheweth.

$$1200. \longrightarrow 100. \longrightarrow \begin{cases} 300. \longrightarrow 25. \\ 400. \longrightarrow 33.\tfrac{1}{3} \text{ which is } 5. \text{groats} \\ 500. \longrightarrow 41.\tfrac{2}{3} \text{ which is } 10. \text{groats} \end{cases}$$

The cómon diuiſor which is the Som of the particular portiós added together. } the generall loſſe. } the particular portions. } euery mans ſeuerall loſſe

Here

The rule of Fellowship.

Here to know whether the 3. seuerall losses doe make vp the generall losse, do thus, first adde the Integrums of the seuerall losses together, the Somme whereof will bee 99. whereunto if you adde the 2. Fractions which doe make one whole Crowne, the Somme will be 100. a number like vnto the generall losse.

Another example.

Three Marchants haue bought 1000.℔. of Pepper for 300. crownes, the first taketh 200.℔. the second 350.℔. and the third 450.℔. what shal euery man pay according to the portion which he hath receiued, then say if 1000.℔. be worth 300. Crownes, what is 200.℔. worth, here working by the common rule of 3. with euery mans seuerall portion receiued, you shall finde that the first must pay 60. Crownes, the second 105. and the third 135. Crownes, as this figure here following sheweth.

```
                    ℔.         ⎧ 200.——60.
       1000.———————300.———————⎨ 350.——105.
                               ⎩ 450.——135.
```

The common ⎱ the gene- ⎰ the seueral por- ⎱ euery mãs par-⎰
Diuisor.   ⎰ rall price. ⎱ tions receiued. ⎰ ticular paimẽt.⎱

Another like example of diuerse distance of time.

Three Marchaunts occupying together did gaine 2345. Crownes, the first put in 40. Crownes for the space of 14. Moneths, the second put in 50. Crownes for the space of 8. Moneths, and the third put in 85. Crownes for the space of 6. Moneths, the question is, how much euery one shall haue after the rate of his mony, and according to the quantitie of time: This is to bee wrought according to the rule of Fellowship thus, Multiply euery mans mony by his time, either of which, that is to say, the mony and the time must be of one selfe denomination, then adde the Sommes of those seuerall portions together, and the total somme of such Addition shall be the first number, and the common gaine shall be the second number, and the third number shall bee euery mans mony Multiplyed by his time, then in working by the common rule of 3. you shall finde that euery man shall haue such share as this figure here following sheweth.

147.

The rule of Fellowship. 17

$$1470. \underline{\qquad} 2345. \underline{\qquad} \begin{cases} 560. \underline{\qquad} 893.\tfrac{1}{3} \\ 400. \underline{\qquad} 638.\tfrac{2}{21} \\ 510. \underline{\qquad} 813.\tfrac{4}{7} \end{cases}$$

The common Diuisor which is the totall Somme of the seuerall portions of mony added together. } the general Somme of the gaine. { the particular portions of the mony multiplyed by his time. { euery mans single share.

Now if you would know whether the single shares in this example doe make vp the generall Somme of the gaine, then adde together the Integrums or whole Sommes of euery mans single share, and you shall finde the Somme to bee but 2344. which wanteth one whole Integrum of the second or middle number of the question, which you shal easily supply by adding the 3. seuerall Fractions together according to the rule of Addition of Fractions before taught, for so shall you finde the said 3. Fractions to make in all one whole Integrum, which being added to 2344. will be aunswerable to the second number of the question, which as you see in the example is 2345. And remember in adding together the said 3. Fractions, to set them in this order $\tfrac{1}{3}/\tfrac{2}{21}/\tfrac{4}{7}$. and then to bring them to one selfe denomination, as the seuenth rule of Fractions teacheth you. But first because you shal find $\bar{y}$ 3. remainders after the first 3. Diuisions of the 3. seuerall portions, made by the first common Diuisor of the question to be written in many figures, you must set them downe in fewer figures according as the fift rule of Fractions teacheth you, so shall you finde the first remainder containing $\tfrac{490}{1470}$. to be no more but $\tfrac{1}{3}$. and the second remainder containing $\tfrac{140}{1470}$. to be $\tfrac{2}{21}$. and the third remainder containing $\tfrac{840}{1470}$. to be $\tfrac{4}{7}$. which 3. Fractions being added together according to the seuenth rule of Fractions, wil make $\tfrac{441}{441}$. which is one whole Integrum and must be added by the name of 1 to the middle or second number of the question, as I haue said before: you may also make vp the foresaid number by seeking to knowe the value of euery Fraction annexed to the Integrums, according as the sixt rule of Fractions teacheth, for so shall you finde the value of the first Fraction to be fiue groats, & the value of the second Fractiō to be one groat 6. farthings, and $\tfrac{18}{21}$. of a farthing

D

The rule of Fellowship.

thing, & you shal find the value of þ third Fraction to be 8.groats 9.farthings, and $\frac{1}{7}$.of a farthing, & if you adde $\frac{1}{2}\frac{8}{7}$ & $\frac{1}{7}$ of a farthing together, you shall find that it will make $\frac{147}{147}$. which is one whole Integrum or one whole farthing. Now if you adde 6.9.and 1.farthings together, it wil make in al 16.farthings, and þ is one groat which being added to þ 14 groats before found out, will make in al 15.groats which is one Crowne, and þ being added to the Somme 2344. will make it 2345. Crownes, which is a number agrææble to the middle Somme of the question. Truely if you exercise your selfe in this and such like questions, it will make you perfect not onely in Addition, Subtraction, Multiplycation, and Diuision of whole numbers, but also of Fractions, and almost in al the other rules belonging as wel to Fractions, as to Integrums: Wherefore I would wish you often to vse your penne therein.

Hauing hitherto treated of the 4. speciall kinds of Arithmetike, that is of Addition, Subtraction, Multiplycation and Diuision, as well belonging to whole numbers, as to Fractions, and also shewed the vse of the Golden rule, otherwise called the rule of 3. and of all the three kindes thereof, that is the common rule, the rule Reuerse, and the Double rule, and giuen examples how and when euery one is to be vsed, together with the rule of Fellowship, necessarie for them that vse any trade of Marchandize, I thinke good now to speake somewhat of Arithmeticall and Geometricall progression, and also of Proportion, and of the three kinds thereof, that is Arithmeticall, Geometricall, and Musicall proportion, and then of the extraction of rootes both square and Cubicall, and last of all, of Astronomicall Fractions, shewing how they are to be added, Subtracted, Multiplyed & diuided.

Of Progression and how manifold it is.

Cap. 13.

WHat is Progression?

It is a certaine order of procéeding with diuerse numbers in such sorte as euery one may excéede each other, either by like difference of quantitie, or else by like likenesse of Proportion, whereof springeth 2. kinds of Progression, the one called Arithmeticall, and the other Geometricall.

What is Arithmeticall Progression?

It is that which procéedeth by like difference of quantitie, as thus, 3.5.7.9.11.13. whereof euery one excéedeth his fellow by the difference of 2.

What is Progression Geometricall?

It is that wherein euery mumber excéedeth his fellow by like proportion, for as 6. containeth 3. twise, so doth 12. containe 6. twise, &c.

Of Addition belonging to *Progression Arithmeticall.*

Cap. 14.

How are numbers being set in order according to Progression Arithmeticall, to be added together and to be reduced into one totall somme?

Thus, first sée how many seuerall numbers there be in all, and note the Somme by it selfe, then adde the first number of the Progression to the last thereof, and note that Somme by it selfe, that done, Multiply the one of these 2. reserued Sommes by the one halfe of the other, and you shall haue the totall Somme. As for example let this be your Progression 6.10.14/18/22/26/30/34/ wherein euery number excéedeth his fellow by the difference of 4. here hauing tolde the numbers, I finde them to be 8. which I set downe by it selfe, then I adde 6. which is the first number to 34. which is the last, and the Somme thereof is 40. which I also set downe by it selfe, then in Multiplying 40. by 4. which is the one halfe of 8. before set downe a part by it selfe, I finde the totall Somme of the Progression to be 160.

Of Progreſsion.

Of Addition belonging to progreſsion Geometricall.

Cap. 15.

How are numbers being ſet in order according to progreſsion Geometricall, to be added and to bee brought into one Somme?

Thus, Multiply the laſt Somme of the progreſſion by the number of the proportiõ whereby ſuch progreſſion procædeth, and from the product of that Multiplycation, Subtract the firſt number of the progreſſion, that done, diuide that which remaineth by a number which is leſſe then the proportion by one, and the quotient of ſuch Diuiſion ſhall ſhew you the totall Somme of the ſaid progreſſion. As for example, let this bee the progreſſion 2.6.18.54.162.486.1458. the proportion of which progreſſion is trypple, wherefore accordyng to the rule, I Multiply the laſt number of the progreſſion by 3. and the product of ſuch Multiplycation amounteth to 4/374. out of which Somme I Subtract the firſt number which is 2. and there remaineth 4372. which I diuide by 2. (for it is leſſe then the proportion 3. by 1.) and ſo I finde in the quotient 2186. which is the totall Somme of the progreſſion Geometricall.

Is there no brieffer way of adding ſuch kinds of progreſsion?

Yes indæde, but not ſo plaine as this way is, and therefore I thinke not good to trouble your memorie therewith.

Of Proportion.

Cap. 16.

What is Proportion?

Proportion is taken generally for the comparing of 2. diuerſe quantities or numbers together, ſhewing what likenes is betwixt them. But before we

Of Arithmeticall proportion.

we deale with Proportion, and with the 3. kinds thereof, that is Proportion Arithmeticall, Geometricall, and Musicall: You haue first to vnderstand, that of numbers some are said to be abstract, and some concreate.

The abstract are such as are not tyed to any denomination, and such are twofold, that is absolute and relatiue.

The absolute, are simply pronounced without hauing any relation to any other number, measure, or quantitie, as 2.3.4. &c. and all numbers whatsoeuer, that are without denomination and are not attributed to any other thing.

The relatiue are those which haue relation one to another, which may be three manner of waies. First in respect of difference which is found by Subtraction, secondly in respect of the quotient found by Diuision, thirdly in respect of both. Of the first way springeth Arithmeticall proportion: Of the second way Geometricall proportion: And of the third Musicall proportion.

Cap. 17.

What is Arithmeticall proportion?

Arithmeticall proportion vnproperly so called, (because it is no proportion indeed) is when many seuerall numbers haue one selfe and like difference, as 8.12.16. which doe onely differ one from another by 4. and this Proportion is twofold, that is continual and disiunct.

Continuall is when many numbers procéede with like difference, as hath bene said before whē we spake of Arithmetical Progression, as 8.12.16.20. &c. whose difference betwixt euery 2. numbers is 4.

The Disiunct, is when many numbers being seuerally propounded, the difference of the first 2. numbers is not like to the differēce that is betwixt the second and the third, and so forth as 5.8.4.7. for 8. differeth from 5. by 3. and 4. from 8. by 4. and 7. from 4. by 3. Now of the second way of comparing which is done by Diuision, springeth as hath beene saide before, Geometricall proportion.

Of Geometricall proportion.

Cap. 18.

What is Geometricall proportion?
Geometricall proportion is that which sheweth what part or parts one number is of another number, as 3. is the halfe of 6. which proportion is found by Diuision, wherein you haue to note that if the Diuisor be greater then the diuidend, then it is to be made a Fraction, as in the former example, if you would diuide 3. by 6. then you must make it a Fraction thus, $\frac{3}{6}$. or $\frac{1}{2}$, and this kinde of proportion which may be truely and properly called a proportion indéed or rather a proportionalitie, is said to be twofold, that is, direct, and reuerse, and the direct is either Coniunct, or Disiunct.

Coniunct differeth not from Geometricall Progression before taught.

Disiunct proportion Geometricall, consisting most commonly of 4. numbers or of 3. at the least, is when there is not like proportion betwixt the second and the third, that is betwixt the first and the second, or betwixt the third and the fourth, as 3. 6. 4. 8. for here 6. containeth 4. once, and one halfe thereof, which is called Proportio Sesquialtera, and 6. containeth 3. twise, which is called Proportio dupla: and so is 8. to 4.

Cap. 19.

Againe, proportion Reuerse differeth not from the rule of 3. called Regula euersa. But you haue to vnderstand that the two chéefe and speciall kinds of Geometricall proportion, are these, that is, proportion of equalitie, and proportion of inequalitie.

Proportion of equalitie, is when 2. numbers compared together, are equall the one to the other, as 3. to 3. 4. to 4.

The proportion of inequalitie, is when 2. vnequal numbers are compared together, as 6. to 5. 4. to 9. and of this there are two kinds,

Of Geometricall proportion.

kinds, that is proportion of the greater inequalitie, and proportion of lesser inequalitie.

Cap. 20.

Roportion of the greater inequalitie, is when the greater number is compared to the lesser number, as 6. to 5.

Proportion of the lesser inequality, is whē the lesser is compared to the greater, as 5. to 6.

Of proportion of the greater inequalitie there be two kindes, Simplex and Multiplex, that is to say, simple and manifold.

Simpler, is when the Antecedent, that is to say the former number containeth the consequent, that is to say the latter number once and somewhat more, which ouerplus must alwaies bee lesse then the consequent it selfe, as 5. containeth 4. once and one parte thereof more, for if you diuide 5. by 4. the quotient will be 1. and $\frac{1}{4}$ ouer. Againe, this proportion is twofold, that is Superparticuler, and Superpartient: Superparticular, is when the Antecedent containeth the consequent once and some one part thereof, as 3. to 2. for 3. containeth 2. once and one halfe thereof, which is called Sesquialtera, or as 4. to 3. for'4. containeth 3. once and one third part thereof, and is called Sesquitertia, the like is to be said of Sesquiquarta, Sesquiquinta, Sesquisexta, and so forth infinitely as this Table sheweth.

Superparticular proportiōs are these & such like.		which is as much as.	
	Sesquialtera as 3. to 2. 6. to 4. 9. to 6.		$1\frac{1}{2}$
	Sesquitertia as 4. to 3. 8. to 6. 12. to 9.		$1\frac{1}{3}$
	Sesquiquarta as 5. to 4. 10. to 8. 15. to 12.		$1\frac{1}{4}$
	Sesquiquinta as 6. to 5. 12. to 10. 18. to 15.		$1\frac{1}{5}$
	Sesquisexta as 7. to 8. 14. to 12. 21. to 18.		$1\frac{1}{6}$
	Sesquiseptima as 8. to 7. 16. to 14. 24. to 21.		$1\frac{1}{7}$
	Sesquioctaua as 9. to 8. 18. to 16. 27. to 24.		$1\frac{1}{8}$
	Sesquinona as 10. to 9. 20. to 18. 30. to 27.		$1\frac{1}{9}$
	Sesquidecima as 11. to 10. 22. to 20. 33. to 30		$1\frac{1}{10}$
	Sesquiūdecima as 12. to 11. 24. to 22. 36. to 33		$1\frac{1}{11}$
	Sesqiuduodec. as 13. to 12. 26. to 24. 39. to 36		$1\frac{1}{12}$

But

Of Geometricall proportion.

But superpartient is when the Antecedent containeth the consequent once and some parts thereof, that is to say more parts then one, as 5. to 3. for 5. containeth 3. once and 2. third parts thereof, which is called Superbipartiens tertias, of which kinds are these set downe in the Table following.

			which is as much as.	
Proportions superpartient are these and such like.	Superbipartiens.	Tertias as 5. to 3. 10. to 6. 15. to 9. Quintas as 7. to 5. 14. to 10. 21. to 15. Septimas as 9. to 7. 18. to 14. 27. to 21. Nonas as 11. to 9. 22. to 18. 33. to 27. Vndecimas as 13. to 11. 26. to 22. 39. to 33. Decimas tertias as 15. to 13. 30. to 26. 45. to 39.	which is as much as.	$1\frac{2}{3}$ $1\frac{2}{5}$ $1\frac{2}{7}$ $1\frac{2}{9}$ $1\frac{2}{11}$ $1\frac{2}{13}$
	Supertripartiens.	Quartas as 7. to 4. 14. to 8. 21. to 12. Quintas as 8. to 5. 16. to 10. 24. to 15. Septimas as 10. to 7. 20. to 14. 30. to 21. Octauas as 11. to 8. 22. to 16. 33. to 24. Decimas as 13. to 10. 26. to 20. 39. to 30. Vndecimas as 14. to 11. 28. to 22. 42. to 33.	which is as much as.	$1\frac{3}{4}$ $1\frac{3}{5}$ $1\frac{3}{7}$ $1\frac{3}{8}$ $1\frac{3}{10}$ $1\frac{3}{11}$
	Superquadripartiens.	Quintas as 9. to 5. 18. to 10. 27. to 15. Septimas as 11. to 7. 22. to 14. 33. to 21. Nonas as 13. to 9. 26. to 18. 49. to 27. Vndecimas as 15. to 11. 30. to 22. 45. to 33. Decimas tertias as 17. to 13. 34. to 26. 51. to 39. Decim. quintas as 19. to 15. 38. to 30. 57. to 45.	which is as much as.	$1\frac{4}{5}$ $1\frac{4}{7}$ $1\frac{4}{9}$ $1\frac{4}{11}$ $1\frac{4}{13}$ $1\frac{4}{15}$
	Superquintupartiens.	Sextas as 11. to 6. 22. to 12. 33. to 18. Septimas as 12. to 7. 24. to 14. 36. to 21. Octauas as 13. to 8. 26. to 16. 39. to 24. Nonas as 14. to 9. 28. to 18. 42. to 27. Vndecimas as 16. to 11. 32. to 22. 48. to 33. Duodecimas as 17. to 12. 34. to 24. 51. to 36.	which is as much as.	$1\frac{5}{6}$ $1\frac{5}{7}$ $1\frac{5}{8}$ $1\frac{5}{9}$ $1\frac{5}{11}$ $1\frac{5}{12}$
	Supersextupartiens.	Septimas as 13. to 7. 26. to 14. 39. to 21. Vndecimas as 17. to 11. 34. to 22. 51. to 33. Decimas tertias as 19. to 13. 38. to 26. 57. to 39. Decimas septimas as 23. to 17. 46. to 34. 60. to 51. Decimas nonas as 25. to 19. 50. to 38. 75. to 57. Vicessimas tertias as 29. to 23. 58. to 46.	which is as much as.	$1\frac{6}{7}$ $1\frac{6}{11}$ $1\frac{6}{13}$ $1\frac{6}{17}$ $1\frac{6}{19}$ $1\frac{6}{23}$

Hitherto

Of Geometricall proportion.

Hitherto of Simplex proportio. **Now** of Multiplex proportio.
Multiplex proportio is when the Antecedent containeth the consequent more then once, as 6. to 2. for 6. containeth 2. three times, which is called Tripla proportio. Also 12. to 5. for 12. comprehendeth 5. twise and ⅖. And this Multiplex proportio is twofolde, that is either exact or not exact.

Multiplex exact, is when the Antecedent containeth the consequet more then once, and nothing remaineth, as 4. to 2. 6. to 3. &c. whereof are infinite kinds, as Dupla, Tripla, and so forth as this Table sheweth.

The kinds of Multiplex exact are these and such like.
{
Dupla as 4. to 2. 6. to 3. 8. to 4.
Tripla as 6. to 2. 9. to 3. 12. to 4.
Quadrupla as 8. to 2. 12. to 3. 16. to 4.
Quintupla as 10. to 2. 15. to 3. 20. to 4.
Sextupla as 12. to 2. 18. to 3. 24. to 4.
Septupla as 14. to 2. 21. to 3. 56. to 8.
Octupla as 16. to 2. 24. to 3. 32. to 4.
Nondupla as 18. to 2. 27. to 3. 36. to 4.
Decupla as 20. to 2. 30. to 3. 40. to 4.
Vndecupla as 22. to 2. 33. to 3. 495. to 45.
} which is as much as.
{
2/1
3/1
4/1
5/1
6/1
7/1
8/1
9/1
10/1
11/1
}

But Multiplex not exact, is when the Antecedent containeth the consequent more then once and some thing remaineth ouer, as 5. to 2. for 5. containeth 2. twise and one remayneth, and this is also twofolde, that is, Multiplex superparticularis, and Multiplex superpartiens

Multiplex superparticularis, is when the Antecedent containeth the consequent more then once & one only remaineth, as 7. to 3. for 7. containeth 3. twise and one only remaineth, whereof are diuers kinds, as Duplex sesquialtera, Duplex sesquitertia, Triplex sesquisexta, & so forth as ye table hereafter following sheweth.

But Multiplex superpartiens, is when the Antecedent containeth the consequent more then once & the remainder is more then 1. as 8. to 3. for 8. containeth 3. twise and 2 thirds ouer, whereof there be many kinds, as Dupla superbipartiens tertias, Dupla supertripartiens quartas, and so forth as this Table following sheweth, which comprehendeth both kinds, that is Multiplex superparticularis, and Multiplex superpartiens

Multiplex

Of Geometricall proportion.

Multiples ouercast, is either

Multiplex Superparticularis or els:
- **Duplex.**
  - Sesquialtera as 5. to 2. 15. to 6.
  - Sesquitertia as 7. to 3. 21. to 9.
  - Sesquiquinta as 11. to 5, 22. to 10.
- **Triplex.**
  - Sesquisexta as 19. to 6. 38. to 12.
  - Sesquiseptima as 22. to 7. 44. to 14.
  - Sesquioctaua as 25. to 8. 50. to 16.
- **Quadruplex.**
  - Sesquinona as 37. to 9. 148. to 36.
  - Sesquidecima as 41. to 10. 82. to 20.
  - Sesquiundecima as 45. to 11. 90. to 22.

which is
- $2\frac{1}{2}$
- $2\frac{1}{3}$
- $2\frac{1}{5}$
- $3\frac{1}{6}$
- $3\frac{1}{7}$
- $3\frac{1}{8}$
- $4\frac{1}{9}$
- $4\frac{1}{10}$
- $4\frac{1}{11}$

Multiplex Superpartiens.
- **Dupla.**
  - Superbipartiens.
    - Tertias as 8. to 3.
    - Quintas as 12. to 5.
  - Supertripartiens.
    - Quartas as 11. to 4.
    - Quintas as 13. to 5.
  - Superquadripartiens.
    - Quintas as 14. to 5.
    - Septimas as 18. to 7.
- **Tripla.**
  - Superquintupartiens.
    - Octauas as 29. to 8.
    - Nonas as 34. to 9.
  - Supersextupartiens.
    - Vndecimas as 39. to 11.
    - Decim tertias as 45. to 13.
  - Superseptupartiens.
    - Octauas as 31. to 8.
    - Nonas as 34. to 9.
- **Quadrupla.**
  - Superbipartiens.
    - Tertias as 14. to 3.
    - Quintas as 22. to 5.
  - Superquadripartiens.
    - Quintas as 24. to 5.
    - Nonas as 40. to 9.
  - Superquintupartiens.
    - Sextas as 29. to 6.
    - Septimas as 33. to 7.

which is
- $2\frac{2}{3}$
- $2\frac{2}{5}$
- $2\frac{3}{4}$
- $2\frac{3}{5}$
- $2\frac{4}{5}$
- $2\frac{4}{7}$
- $3\frac{5}{8}$
- $3\frac{5}{9}$
- $3\frac{6}{11}$
- $3\frac{6}{13}$
- $3\frac{7}{8}$
- $3\frac{7}{9}$
- $4\frac{2}{3}$
- $4\frac{2}{5}$
- $4\frac{4}{5}$
- $4\frac{4}{9}$
- $4\frac{5}{6}$
- $4\frac{5}{7}$

Thus much of proportion of the greater inequalitie: Now wee will speake somewhat of proportion of the lesser inequalitie.

Proportion

Of Musicall proportion.

Cap. 21.

PRoportion of the lesser inequalitie, is when the Antecedent is lesse then the consequent, as 2. to 3. 5. to 7. for if you diuide 2. by 3. the quotient is ⅔. if 5. by 7. then the quotient is 5/7 and this Proportion hath the same names which the Proportion of the greater inequalitie hath, sauing that you must adde to the beginning of euery name, this word Sub. as Subdupla, Subtripla, Subsesquialtera, &c. as 1 to 2 is Subdupla proportio, 3. to 9. is Subtripla proportio, &c.

Of Musicall proportion called in
Latine Harmoniaca proportio.

Cap. 22.

MUsical proportion which requireth 3. numbers at the least, is when the first number hath the same proportion vnto the third, which the difference betwixt the first and the second, hath to the difference which is betwixt the second and the third, as 3. 4. and 6. for looke what proportion 3. hath to 6. which is Subdupla, the same hath the difference betwixt 3. and 4. which is 1. to the difference betwixt 4. and 6. which is 2. for 1. to 2. is Subdupla, and this is called Musicall proportion, because the numbers therein haue the same proportions one to another, which are found to be in Musicall consorts, as 6. 4. and 3. for the proportion which 6. hath to 4. is Sesquialtera, called of the Musitians Diapente, or a fift, and the proportion which 4. hath to 3. is Sesquitertia, called of the Musitians Diatesseron, or a fourth, and the

pro-

Of Musicall proportion.

proportion which 6. hath to 3. is *Dupla*, called a *Diapason*, or an Eight, and of this Musicall proportion there be two kinds, that is simple and compound.

It is said to be simple when it consisteth onely of 3. numbers.

And it is called compound when it consisteth of more then 3. numbers: And of compound there be 2. kinds, that is vnproper and proper.

The vnproper is, when to 3. numbers giuen, 2. other seueral numbers are ioyned, which doe containe the same proportions with the third, which the first 3. numbers haue one to another: As for example, let be giuen these 3. numbers, 3.4.6. vnto which if you ioyne 8. and 12. here 6.8.12. haue the same proportion one to another which 3.4. and 6. haue amongst themselues, for like as here betwixt 6. and 8. the proportion is Subsesquitertia, and betwixt 6. and 12. it is Subdupla, and betwixt 8. and 12. it is Subsesquialtera, so betwixt 3. and 4. is Subsesquitertia, and betwixt 3. and 6. is Subdupla, and betwixt 4. and 6. is Subsesquialtera,

Compound proper, is when diuerse numbers in Musicall proportion standing together, and the first being omitted, the 3. next doe continue still in Musicall proportion. Also when omitting the 2. first numbers the 3. next following are in Musicall proportion. And when omitting the first 3. the next 3. are in Musicall proportion, and so forth how many numbers so euer there be: As for example 10.12.15.20.30. these are numbers belonging to the proper kind of compound Musical proportion, for 10.12.and 15. are in Musicall proportion, then omitting 10. which is the first number, 12. 15. and 20 are still in Musicall proportion, and if you omit the 2. first numbers, that is 10 and 12. then the next 3. that is 15.20. and 30. are still in Musicall proportion, and so forth how many so euer there be of them, so as they be in Musicall proportion: But our Musitians doe make no more but 8. Musicall proportions in all, that is.

Dupla

Of Musicall proportion. 23

Dupla.		Diapason.
Tripla.		Diapason diapente.
Quadrupla.		Bis diapason.
Sesquialtera.	which are	Diapente.
Sesquitertia.	thus na-	Diatesseron.
Sesquiquarta.	med	Diatonus semitonus.
Dupla superbipartiens.		Diapason diatesseron.
Sesquioctaua.		Tonus.

The vse wherof is to be learned at the hands of the Musitians.

Cap. 23.

But now because it is not inough to knowe the foresaid proportions and their names, vnlesse you can both adde, Subtract, Multiply, and diuide them when nœde is, I will therefore briefly here set downe the order thereof, and first shewe how to set them downe in writing, and then how they are to be added, Subtracted, Multiplyed, and diuided.

They are to be set downe in writing in like manner that Fractions are wont to be set downe, sauing that as in Fractions the vpper number is called Numerator, and the inferior number Denominator, so in proportions the vpper number is called Antecedent, and the inferior number Consequent, but whereas in Fractiōs, ther is wont to be drawn a litle line betwixt the Numerator and the Denominator, in proportions no line is drawne betwixt the Antecedent and the Consequent, but the Antecedent is set ouer the Consequent, without any line drawne betwixt them thus.

And looke as Fractions are to be written in as fewe figures as may be, so are proportions to be set downe in so small numbers as may be, which is to be done by the self same rule that Fractions are. Againe as you can neither adde nor Subtract Fractions, hauing diuerse denominations, before that you haue brought them to one selfe denomination: So can you neither adde nor subtract proportions hauing diuerse consequents, vntill you haue brought them to one selfe consequent by

3.|4.|2.|
5.|7.|1.|

the

Of finding out the square

the same rule that Fractions are reduced to one selfe denomination: And looke what order is to be observed in adding, subtracting, multiplying and diuiding of Fractions, the same is to be kept in adding, subtracting, multiplying and diuiding of proportions. Wherefore I wholly referre you to the rules of Fractions before most plainly taught. And thus I end with proportions, the knowledge wherof is very necessarie to such as haue to deale in matters either of Geometry or of Astronomie or with Musicke.

How to finde out the square roote
of any number.

Cap. 24.

What is a square roote? It is any Digit or any other number, which being multiplyed into it self bringeth forth a square number, as 4. being multiplyed in it selfe, in saying 4. times 4. maketh 16. which is a square number, the roote whereof is 4. And to finde out the square roote of any number be it square or not square, you must doe thus. First hauing set downe the number propounded which must consist of 3. figures at the least, set a pricke vnder the first Digit or Cypher of the said number on the right hand, that done, pricke euery other figure thereof towards the left hand, leauing alwaies one voide space betwixt euery 2. pricks as you see here done in this number 475645 and looke how many prickes there be, so many figures shall you haue in the quotient, and this differeth nothing in a manner from Diuision, the order in working is thus, first seeke out a Digit, which being multiplyed in it selfe, may take away that which standeth ouer the last pricke on the left hand, or as much thereof as may be, and set the Digit in the quotient, that done, double the said quotient, then consider whether the double doe consist of one figure or of many, if of one onely, then place that in the next voide space on the right hand, but if the double doe consist of many figures, then take the first figure thereof, that is to say, that figure

Roote of any number. 24

or cipher which standeth in the first place of the double, and place that in the foresaide next voide space towards the right hand, and next to that, place all the rest of the figures of the sayd double orderly towardes the left hand, that done, sæke another Digit, which being multiplyed in it selfe together with the doubled number, may take away all that which standeth right ouer it, or as much thereof as may be, which Digit you must not onely put in the quotient to his fellow, but also set downe the same right vnder the next pricke on the right hand, and then multiply that Digit together with the double whereto it is ioyned, by the sayd last found Digit set downe in the quotient, the product whereof you must subtract out of the vpper number, standing right ouer the foresaid pricke, and if there be any remainder, write it aboue, cancelling the other, and looke how many prickes ther be, so many times must you double the quotient, obseruing alwaies like order of working as before, which you shall more plainely vnderstand by these examples following, and for your better vnderstanding thereof, I will first giue you an example of 3. figures onely as thus, 464. here to finde out the square roote of this number, you must first pricke the said number in such order as is before taught, and then sæke out some one of the 9. Digits, which being multiplyed in it selfe, may take away the first 4. on the left hand, or as much thereof as may be, which you shall finde to be 2 which 2. being set in the quotient, multiply the same in it selfe, in saying 2. times 2. maketh 4. which 4. doth cleane take away the first 4. standing ouer the last pricke on the left hand, and therefore cancell the 4. that done, double the quotient which maketh 4. and set the same in the next voide space on the right hand vnder the figure 6. thus.

And then sæke out some Digit, which being multi- 4 64 (2.
plyed in it selfe together with the doubled, may take   4
away the Somme of 64 that remaineth, which by asking how many times 4. is comprehended in 6. you shall finde to be but one, which you must not onely put in the quotient, but also set it downe vnder the first pricke on the right hand thus.

So as you shall make the lower Somme to be 41. 4 64 (21.
which is to be multiplyed by the last found quotient   41
which is 1. wherefore you must say that 1. times 41.
is 41 which being taken out of 64. there remaineth 2 3. so as the
roote

Of finding out the square

roote is 21. and the remainder 23. as this example sheweth.

And to know whether you haue done well or not, multiply the quotient in it selfe, and if there be any remainder left, adde that vnto the product of the multiplyed quotient, and if you finde the summe thereof to be like to the first, then you haue done well, if it be not like, then you haue erred, the product of this quotient being multiplyed in it selfe amounteth vnto 441. whereto by adding the remainder which is 23. you shall finde the summe to be like vnto the first number that is 464. whereof you sought the square roote. But if such number do consist of many figures, then in working you must double the quotient once, twice or thrice, according as the number doth require, as you shall more plainely perceiue by this example following: as let 50/467.be the number whereof you would know the square roote, here hauing pricked this number in such order as before, first seeke out some Digit which being multiplyed in it selfe may take away the Digit or number standing right ouer the last pricke on the left hand, which you shall finde to be 2. and hauing set downe the saide 2. in the quotient, saye 2. times 2. maketh 4. which being subtracted from 5. there remaineth 1. which you must set ouer 5. and cancell the sayde 5. as this example sheweth.

$$\begin{array}{r} 2\ 3 \\ 4\cancel{6}4 \\ \cancel{4\ 1} \end{array}(21$$

That done, double the quotient which maketh 4. and set downe the 4. in the next space towardes the right hand, vnder the cipher as you see in the former example, then seeke out a Digit, which being multiplyed in it selfe together with the double quotient, may take away all that which standeth ouer the second pricke towards the left hand, or as much thereof as may be, which Digit by asking how many times 4. is comprehended in 10. you shall finde to be 2. then set downe the saide 2. not onely in the quotient, but also vnder the second pricke towardes the right hande, which together with the former 4.maketh 42. as you see in this example.

$$\begin{array}{r} 1 \\ \cancel{5}0467 \\ \cdot\ 4\ \cdot\ \cdot \end{array}(2$$

Then multiply that 42. in a place by it selfe by the figure 2. last set downe in the quotient, and it will make 84. which 84. being taken out of 104. there remaineth 20 which you must set ouer the second pricke and cancell the 104. and also the 42. beneath

$$\begin{array}{r} \cancel{1}20 \\ \cancel{5\cancel{0}}467 \\ \cdot\ \cancel{4\ 2}\ \cdot \\ \hline 84. \end{array}(22$$

Roote of any number. 25

neath, as you sée in the former example: that done, double the whole quotient, which maketh 44. whereof set the first 4. in the next voide space towards the right hand vnder the figure 6. and the other 4. vnder the second pricke towards the left hand, then aske how many times 4. is comprehended in 20. and you shall find 4. which you must not onely set downe in the quotient, but also vnder the next and last pricke on the right hand, and so the nether Summe shall be 444. which being multiplyed by the Digit 4. last set down in ye quotient, will make in all 1776. which being taken out of the vpper number which is 2067. there will remaine 291. so as the square roote of the first number is 224. and the remainder 291. as you may sée in this example.

The square roote whereof being multiplyed in it self amounteth vnto 50176. wherto if you adde the remainder which is 291.
the Summe thereof will bee like vnto the
first number which is 50467. Moreouer you
haue to note that if you finde any number out
of the which your quotient being doubled cannot be subtracted, then you must set downe a

$$\begin{array}{r} x2 \\ x\cancel{2}\cancel{0}91 \\ \cancel{5}\cancel{0}\cancel{4}\cancel{6}\cancel{7}\;(224. \\ 4 \\ 4\;2\;4 \\ 4 \\ \hline 1776. \end{array}$$

Cypher in the quotient, and procéde to the next pricke on your right hand, as in this example following.

In which example the first found Digit set in the quotient is 6. which being multiplyed in it selfe maketh 36. and cleane taketh away the first number standing ouer the last pricke on the left hand, that done, I double the quotient, which maketh 12 and I place the 2. in the next voide space towardes the right hand, and the 1. next to that towards the left hand orderly as in this example.

$$36602\overline{5}\;(6$$

Ouer which 12. there is but 6. remaining, so as I can not take 12. out of 6. and therefore I cancell the 12. and set down a Cypher in the quotient, so as the quotient is now 60. which being doubled maketh 120. the Cypher whereof I set downe in the next voide space towards the right hand, and the other two Digits orderly towards the left hand, as you sée in this example.

$$3\cancel{6}602\overline{5}\;(60.$$
$$\cancel{12}\;\;\cdot$$

C   Then

The vſe of the ſquare

Then I aſke how many times one is contained in 6, and I finde 5. which I ſet downe in the quotient and alſo vnder the laſt pricke on the right hand, ſo as the nether number is 1205. which I multiply aparte by it ſelfe by the 5. laſt ſet downe in the quotient, the product whereof is 6025. which being taken out of the vpper number, which is alſo 6025. there remaineth nothing, as this example ſheweth.

Wherefore I finde the foreſaid number to be a iuſt ſquare number, the roote whereof is 605. which being multiplyed in it ſelfe maketh 366025. a number like vnto the firſt number.

The vſe of the ſquare roote in ſetting of battels.

Cap. 25.

The knowledge of finding out the ſquare roote of any number, is very neceſſary for a Sargent maior in the fielde, that he may the more readily ſet and range his Squadrans of Battel: And therefore I thinke it not amiſſe, to giue them here certaine examples of ſuch manner of ſquares as the Italians were wont to vſe in my time, which is foure manner of waies.

Whereof the firſt teacheth how to ſet the Battel ſquare of ground, the ſecond a ſquare battel of men, the third a long ſquare battel which we call the Hearſe battel, and the fourth teacheth how to ſet a battel of ſo much and a third.

Shew by what rules theſe battels are to be ſet?

Before I ſet downe any rules you haue to vnderſtand that all numbers giuen or ſuppoſed, are not meete for ſuch purpoſe, but onely ſuch as may be diuided into 3. equall parts without leauing any remainder: for otherwiſe all the rules following, take no

place

Roote in setting of battell.

place, and therefore when any number is giuen you to be brought into any one of the foresaid 4. formes of battell, you shall doe well first to deuide that number by 3. and if there be any remainder, to reiect it, and to take the rest for your number, As for example, A Sargent Maior is commanded by his Generall to set a battaile square of ground, appointing him thereunto 1345. men, here the Sargent by diuiding that number by 3. findeth in the remainder one man too many, which being onely reiected, all the rest of the number which is 1344. is fitte for his purpose.

How to set a battell square of ground, called in Italian Battaglia quadra di terreno.

Now shew the order of setting such kinds of battell?

The order is thus, first double the number giuen, then take the square roote thereof, and that roote shall be the Front, that done, diuide the number giuen by the said roote, and the number in the quotient shall be the flanke, now if you multiply the front and the flanke the one into the other, you shall haue your whole number first giuen, vnlesse there be some remainder left in the former Diuision, which remainder must be added to the product to make vp the somme. But if you woulde know how many men are to be put in a rancke, and also how many such ranckes you must haue, then diuide the front into 3. partes, and the quotient will shew you how many men you are to haue in a rancke, that done, multiply the flancke by 3. and the product thereof shall be the number of the ranckes: As for example, take the number before giuen, which was 1345. which being diuided by 3. you finde the remainder to be 1. which you must reiect, and take all the rest which is 1344. and that being doubled maketh 2688. the square roote whereof is 51. by which roote you must diuide the number giuen, Videlicet 1344. and the quotient will be 26. which shall be the flancke, now if you diuide the foresaid front 51. by 3. the quotient will shew how many shall be in a rancke, that is 17. Finally, if you multiply the flancke, which is 26. by 3. the product thereof will be 78. which is the iust number of the rancks which you must haue.

E 2   How

The vſe of the ſquare.

How to ſet a battell ſquare of men called in Italian
Battaglia quadra d'huomini.

TAke the ſquare roote of the number giuen, and that ſhall bee both the front and the flancke, which if you multiply together, the product thereof will make vp the number firſt giuen, and if you would know how many men are to bee put in a rancke, and how many ſuch rancks you muſt haue, then doe as you did before in diuiding the front by 3. and multiplying the flancke by 3.

How to ſet a long ſquare battell which we call
the Hearſe battel, and is called in Italian,
Battaglia dun tanto emezzo.

TO the number giuen, adde the iuſt halfe therof, and the ſquare roote of that Somme ſhall be the front, by which roote, or front you diuide the firſt number giuen, the quotient ſhall be the flancke: As for example, let the number giuen be 6144. the halfe thereof is 3072. which being added to the giuen number 6144. maketh in all 9216. the ſquare roote whereof is 96. which ſhall bee the front, by which roote or front if you diuide the giuen number 6144. you ſhall finde in the quotient 64. and that ſhall bee the flancke, that done, diuide the foreſaide front 96. by 3. and you ſhall finde in the quotient 32. which number ſheweth how many men ſhall be in a rancke, then multiply the foreſaid flancke 64. by 3. and the product thereof ſhall bee 192. which is the number of the ranckes.

How to ſet a battell of ſo much and a third,
called in Italian, Battaglia dun tanto & dun terzo.

TO the number giuen, adde the third part thereof and the ſquare roote of that Somme ſhall be the front, then diuide the firſt number giuen by the foreſaid roote or front, and the number in the quotient ſhall be the flancke: As for example, let the number giuen be 1575. the third part whereof is 525. which being added to 1575. maketh in all 2100. the roote whereof is 45 which muſt be the front, by which roote or front, if you diuide the number giuen which was 1575. the quotient will be 35. which ſhall be your flancke

Roote in setting of battell.

flancke, then by diuiding 45. which is the front by 3. the quotient wil shew you how many men you must haue in a rancke that is 15. Againe by multiplying the flancke which is 35. by 3. the product thereof will be 105. which is the number of your ranckes.

How to finde out the Cubique
roote of any number.

Cap. 26.

S the square roote is said to bee a number which being multiplyed in it selfe, doth make a square superficiall number, hauing onely length and breadth, so as the Cubique roote is a number which being first multiplyed in it selfe, and the product therof being againe multiplyed by the first number, doth make a Cubique or Corporall number hauing both length breadth, and depth: As for example, 2. times 2. maketh 4. and 2. times 4. maketh 8. Againe 3. times 3. maketh 9. and 3. times 9. maketh 27. and so you may deale with all the rest of the 9. Digits, and make thereby a Table containing both the square numbers and Cubique numbers of euery roote, consisting of any one of the 9. Digits, like vnto this table here following.

Then hauing to finde out the Cubique roote of any number that is greater then 1000. (for lesser it cannot be to worke vpon) then set a pricke vnder the first figure on your right hand, and so procéede towards the left hand, omitting alwaies 2. figures betwixt euery two prickes, as in this example 41063625. and looke howe

Rootes	Squares	Cubiques
1.	1.	1.
2.	4.	8.
3.	9.	27.
4.	16.	64.
5.	25	125.
6.	36.	216.
7.	49.	343.
8.	64.	512.
9.	81.	729.

How to finde out the Cubique

many prickes there be, so many figures shall you haue in the quotient: That done, you haue to finde out three seuerall numbers, first such a Cubique number as will cleane take away the number which standeth right ouer the head of the last pricke on the left hand, or so much thereof as may be, which you may easily finde in the foresaid table: The roote of which Cubique number, you must set downe in the quotient, the second number which you haue to find is a number called the Triple, which is easily had by tripling the quotient, and the third number is called the Diuisor which you shall most readily finde by multiplying the quotient into the Triple, both which numbers are to be placed in such order as followeth. First then hauing found the Cubique number as is before said, and taken the same out of the number standing ouer the last pricke on the left hand, write the remainder (if there be any) ouer his head, cancelling that which is vnder it, and then place the roote of the said Cubique number in the quotient: that done, Triple the roote set downe in the quotient, and that shall be the Triple which you must place in the second voide space on the left side of the next pricke which is on your right hand, then multiply that Triple by the quotient, and the product thereof shall be the Diuisor which must be placed right vnder the Triple one figure shorter towards the left hand: that done, drawe a line as you see in the example hereafter following and worke thus: First aske how many times the first figure of the said Diuisor is contained in the number standing right ouer his head: and hauing found an apt Digit for the purpose, put that Digit in the quotient to his fellow, that done, multiply the said Digit into the Diuisor and set the product thereof right vnder the Diuisor: Secondly multiply the said Digit in him selfe, and againe the product thereof into the triple, & set the somme y̆ comes therof right vnder the triple: Thirdly multiply the said Digit in it selfe cubically, and set the product thereof lowest of all, right vnder the next pricke on the right hand: that done, drawe a line, and by Addition bring all the foresaid three products into one somme placing euery figure so as you may easily take the said somme out of the vpper number, whereof you seeke the Cubique roote, and write the remainder (if there be any) ouer the head, and if such somme will not be subtracted out of the vpper number, then you must seeke out a lesser Digit, reseruing still

the

roote of any number.   28

the former triple and Diuisor, and if there bee more pricks in the number then two, then you must for euery pricke finde out a new Triple, and a new Diuisor by tripling the whole quotient, and by working continually in like order as the rule before teacheth, which rule you shall more plainely perceiue by example thus. First then hauing to finde out the Cubique roote of the number before mentioned, which is 4̇1̇063̇625̇ marked with prickes as before is taught: resorte to the Table wherein you shall finde such a Cubique number as will take away as much of the 41. as may bee, which is 27. the roote whereof is 3. for al the other Cubique numbers in the table are either too small or too great, and therefore you must alwaies haue due consideration therof. Then take 27. which is the Cubique number, out of 41. and there remaineth 14. which remainder you must set ouer the last pricke on your left hand cancelling the 41. as in this example.

Then triple the quotient which maketh 9. 1 4
which you must set downe vnder the figure 4̇1̇063̇625̇ (3.
6. next vnto the second pricke which is on your
right hand: Then to find out the Diuisor mul-
tiply the Triple 9. by 3. which is the quotient, and the product therof shal be ÿ Diuisor: which diuisor you must place right vnder the triple one figure shorter towards the left hand: that done, draw a line as you see in this example.

Now hauing orderly set downe
the Triple and the Diuisor, aske how   1 4
many times 2. which is the first figure   4̇1̇063̇625̇ (3
of the Diuisor is contained in 14. stan-   Triple———|9
ding right ouer his head, and remem-   Diuisor——27
ber to make choise of such a Digit as may not cleane take away the whole number at the first, but rather leaue so much as the saide quotient hauing afterward to be multiplyed diuerse waies as shal be said hereafter, may take away the rest or as much therof as may be, as in this example following, you shall finde the aptest Digit for this purpose to be 4. which you must put in the quotient, whereby the quotient shall be now 3 4. that done, you must first multiply this last quotient which is 4. into the Diuisor 27. the product whereof is 108. which you must set downe right vnder the Diuisor. beneath the line already drawne, as you see in this example.

E 4   Secondly

How to finde out the Cubique.

Secondly you must multiply the said 4. in it self quadratly which maketh 16. then multiply that 16. by the triple 9. the product whereof is 144. which is to be set right vnder the triple as in this example.

```
        1 4
      4x063625 (34.
Triple—|  |9.   .
Diuisor—27.
      ─────────
        1 0 8
```

Thirdly you must multiply the said 4. in it selfe cubically, which maketh 64. which is to be set right vnder the next pricke on the right hand beneath the former summes as you see in this other example.

```
        1 4
      4x063625 (34.
Triple—|  |9.   .
Diuisor—27|
      ─────────
        1 0 8|
          1 4 4
```

So as euery one of the foresaide three products doe extend one further then another by one figure towardes the right hand, as you see in ye former example: Now these products being thus placed, drawe another line and bring all the three seuerall products contained betwixt the 2. lines, into one summe by Addition, and you shal finde the totall summe to be 12304. which being subtracted out of the number standing right ouer it which is 14063. there will remaine 1759. which you must set downe aboue as you see in this example.

```
           1 4
         4x063625
Triple    .  9|   .
Diuisor    2 7||
         ─────────
           1 0 8||
             1 4 4|
                 6 4
```

Now to procede with the next, you must finde out a new Triple and a new Diuisor.

How is that done?

```
            1
           x4759
         4x063625 (34
Triple    .  9|   .
Diuisor   27 |  |
         ─────────
           1 0 8| |
             1 4 4|
                 6 4
         ─────────
           1 2 3 0 4
```

Thus, multiply the whole quotient which is 34. by 3. the product whereof is 102. and that is the Triple which is to be placed in the next voide space hard by the next pricke that is on your right hand: Then multiply the whole quotient into the last found Triple 102. and the product thereof which is 3468 shall bee your Diuisor, which is to be placed vnder the Triple one figure shorter towards the left hand as you see in the example following:

which

roote of any number. 29

which two numbers being thus found and rightly placed, draw a line, then aſke how many times the firſt figure of your Diuiſor which is 3. is contained in the number right ouer his head which is 17. and you ſhall finde it to bee 5. times contained therein, wherefore ſet downe 5. in the quotient, that doone, multiply the 5. into the Diuiſor, and place the product thereof which is 17/340. right vnder the Diuiſor beneath the line, as you ſee in this example.

Secondly ſquare the ſame 5 that is to ſay, multiply it in it ſelfe, and that maketh 25. which 25. you muſt multiply againſt into the Triple 102. the product whereof is 2550. which number you muſt ſet right vnder the ſaide triple beneath the line. Thirdly multiply the ſaid 5. in it ſelfe cubically

```
        1
      X 4 7 5 9
     4 X 8 6 3 6 2 5  ( 3 4 5
      Triple  1 0 2
      Diuiſor 3 4 6 8
      ─────────────
              1 7 3 4 0
```

which maketh 125. and place that right vnder the firſt pricke on the right hande, and drawe a line, that done, adde all the foreſayde 3. products together, and you ſhall finde the totall ſumme thereof to bee 1759625. which being ſubtracted out of the vpper number, there remaineth nothing, as this example plainely ſheweth.

Whereby you may conclude that the foreſaid nūber whereof you ſought the roote, is a perfect Cubique number, for if you multiply the whole quotient in it ſelfe cubically, it wil produce the ſelfe ſame number whereof you ſought ye roote. But note that if in making the firſt Subtractiō, the firſt Diuiſor is not to be found in the vpper number, then you muſt ſet a Cypher

```
           X
         X 4 7 5 9
        4 X 8 6 3 6 2 5 ( 3 4 5
         Triple  1 0 2
         Diuiſor 3 4 6 8
         ─────────────
                 1 7 3 4 0
                   2 5 5 0
                       1 2 5
         ─────────────
                 1 7 5 9 / 6 2 5
```

in the quotient, that done, triple the whole quotient, and place the ſaide triple vnder the figure which is next to the next pricke on the right hand, working in ſuch order as before, and to avoide confuſiō, hauing to deale with a multitud of nūbers it ſhal not be amiſſe at ye finding out of euery new triple, & diuiſor to ſet ye remainders in ſeueral places by thēſelues, & to worke then in the ſelf ſame order
that

Of Astronomicall fractions

that is before taught, also note that if you haue to deale with a fewe numbers, and that the Diuisor cannot bee subtracted out of the number standing right ouer his head, then you must set a Cypher in the quotient and so you haue done, as for example, hauing to take the cubicall roote of 8567. here I find 2. to be the quotient, which being cubically multiplyed, doth cleane take away the 8. and now according to my former rule 12. must be my Diuisor, which because I cannot take it out of 5. I set downe therefore a Cypher in the quotient, so as the quotient is now 20. which is the cubicall roote of the foresaide number, for if you multiply 20. cubically and adde therevnto 567. which is the remainder, the summe thereof will be like vnto the first number, as you may sée in this example.

```
                          20
  8567.                   20
        6 Triple (20.    400
   12    Diuisor           2
                        ————
                         8000
                          567
                        ————
                         8567
```

Of Astronomicall Fractions.

Cap. 27.

BEcause the vse of these Fractions is very necessarie for those that haue to calculate the motions of the Starres and the difference of time, I thought good to shew here how the same are to be added, subtracted, multiplyed and diuided, for the measure of time falleth not out alwaies to bee a whole yeare, moneth, day, or hower, nor the mouing of the celestiall bodies are to be measured alwaies by whole circles, signes, or whole degrées, & therfore to haue an exact measure of such things it was thought best by the auncient writers to diuide all whole things called Integra into the lessest parts that might bee, for which purpose no number was thought so méete as 60. for there is no number vn-

Of Astronomicall Fractions.   30

ver 100. that receiueth so many Diuisions as 60 which may bee diuided many sundrie waies, that is by 2.3.4.5.6.10.12.15.20 and by 30. and therefore they diuided euery whole thing that had no vsuall parts into 60. minutes, and euery minute into 60. seconds, and euery second into 60. thirds, and so forth vnto 60. fourthes, fifts, sixts, seuenths, eights, ninthes and tenths, and further if neede were but that seldome chanceth. And you haue to note that minutes are marked with one streeke ouer the head, seconds with two streekes, thirds with three streekes, and so forth thus, $23'.6''.7'''.8''''$. &c. which do signifie 23. minutes 6. seconds 7. thirds, and 8. fourths.

Of Addition.

What is to bee obserued in adding of these kindes of Fractions?

First that you bring Integrums to Integrums, and Fractions to Fractions, that bee of like denomination, beginning alwaies with the least on the right hand, and if the summe of such Addition doe amount any where to the number of 60. or aboue 60 then you must looke how many 60. are comprehended therein, and for euery 60. adde one to the next greater Fraction that is on the left hand, obseruing still that order vntill you come to the Integrums, of which Integrums, it is also necessary to know their value, that is to say, what parts they containe, and what denomination those parts haue: As for example, if you adde common signes such as the twelue signes of the Zodiaque be, then euery signe containeth 30. degrees, so as euery summe exceeding 30 is to be diuided by 30. but if they be Phisicall signes whereof 6. doe make a whole circle, such as be set downe in the table of Alphonsus, then the summe of those degrees is to be diuided by 6. Moreouer so often as the summe of the common signes doe exceede 12. or the summe of the Phisicall signes doe exceede 6. the ouerplus is alwaies to be reiected, and the remainder to be set in the place of the signes, as you may see in this example following wherein seconds are reduced to minutes, minutes to degrees, and finally degrees, to signes.

A2

Of Astronomicall Fractions.

An example of Addition consisting of signes, degrees, minutes and seconds.

Signes	Degrees	Minutes	Seconds
9.	19.	1.	19.
0.	0.	35.	16.
11.	29.	33.	5.
9.	29.	38.	11.
0.	11.	49.	40.
		4.	56.
8.	0.	42.	27.

totall me.

In which example beginning first with the seconds because they are here the least Fractions, you shall finde by Addition that the summe of them amounteth to 147. which being divided by 60. you shall finde in the quotient 2. and the remainder to bee $\frac{27}{}$, which remainder you must set downe under the collum of seconds, keeping the quotient which is 2. in mind to be added to the collum of minutes, the summe whereof is 162. which being divided by 60. you shall find in the quotient 2. degrees, and the remainder to be 42. which is to be set downe under the collum of minutes, and the quotient 2. kept in minde to bee added to the collum of degrees, the summe whereof is 90. degrees, which being divided by 30. (because that 30. degrees doe make one common whole signe) you shall find in the quotient 3. signes and no remainder: wherefore you must set downe a Cypher under the collum of degrees, and adde the 3. signes to the collum of signes, the summe whereof is 32. which if you divide by twice 12. which maketh 24. you shall finde the remainder to be 8. signes, which is to be set downe under the collum of signes, for the quotient here is to be rejected according to the rule before given, so as the totall summe of this Addition is 8. signes, 0. degrees $42'.27''$.

Another example of daies, howers, minutes and seconds to be added together.

Daies	Howres	Minutes	Seconds
21.	14.	32.	11.
16.	16.	19.	41.
8.	16.	30.	30.
46.	23.	22.	22.

n2 to-

In this example one self order as in the other before is to bee observed as touching the secondes and mynutes, for the exceeding

Of Astronomicall Fractions.   31

ding number in both, are diuided by 60. but when you come to the howers you must diuide that number by 24. becaufe that fo many howers doe make a whole day, and hauing fet downe the remainder vnder the collum of howers, adde that one day which was in the quotient, vnto the collum of daies, and fo you fhall find the totall fumme to be as the example aboue fheweth.

Of Subtraction.

VVHat order is to be obferued in Subtraction?

The felfe fame that was before obferued in Addition, fo as you alwaies remember that when you haue to take a greater number of Fractions, as of minutes, feconds, thirds and fuch like, out of a leffer number of Fractions, to borrow 60. and hauing fet downe the remainder to adde the one borrowed, vnto the next collum on the left hand, for there the 60. borrowed, is but one, but if you haue to deale with degrées, which are counted Integrums, then you muft borrow but 30. for fo many degrées doe make one figne, and if you haue to fubtract howers, then you muft borrow 24. for fo many howers doe make one day, as by the example here following you fhall more plainely perceiue.

In this example becaufe you can not take $\frac{53}{}$ out of $\frac{27}{}$, you muft borrowe one minute from the next collum on the left hand, which one my-

Signes	Degrees	Minutes	Secondes
8.	0.	42.	27.
0.	1.	9.	53
7.	29.	32.	34.

nute is $\frac{60}{}$. which being added to 27. doe make in all $\frac{87}{}$. out of. which if you take $\frac{53}{}$. there will remaine $\frac{34}{}$ that done, you muft pay home the minute which you borrowed by adding the fame vnto the next 9. on the left hand, which maketh 10. then fay, take 1'o. out of 42. and there remaineth 32. which being fet down, procéde to the next. But here to take one out of none that cannot be, and therefore borrow one whole figne, from the collum of fignes which is 30. degrées, from whence take one, and there remaineth 29. which being fet downe, take the one which you borrowed out of 8. and there remaineth 7. fo as the whole remainder of this Subtraction

Of Astronomicall Fractions.

traction is 7/ signes, 29/ degrees, 3′2/ and 4″/4. as the former example sheweth.

Of Multiplycation.

Though there bee more difficultie in multiplying and diuiding Astronomicall Fractions then in adding or subtracting them, yet the greatest difficultie thereof chiefly consisteth in the finding out of the true denomination of the products, for first as touching Multiplycation, you must multiply euery number of the multiplyer into all the particular nūbers of the sum that is to be multiplyed, and then seuerally to adde together the products that bee of one selfe denomination, and whatsoeuer in that Addition ariseth to the number of 60. or excædeth 60. it is to be reduced by the sexaginarie Diuision into the greater summe, so shall you collect the whole summe of the Multiplycation. But you haue to note by the way, that if there be any Integrums of diuers denominations in your multiplyer: That then such Integrums must be reduced to one selfe kind of Integrums: As for example, suppose that you would multiply the daily motion of the Moone which according to Alphonsus tables, is 13. degrees, 1′0. ″/. and ‴/. by 29. daies, 12. howers, 4′4. and ″/. here because there be in this multiplyer Integrums of diuers denominations, that is to say daies and howers, you must therefore reduce the same into one selfe denomination, before that you can make your Multiplycation.

How is that to be done?

By multiplying euery number of the said multiplyer by 5. and then by halfing the product thereof, by which halfing you shall reduce mynutes to seconds, and seconds to thirds, and so forth to the smallest Fractions of all, and if any product doe amount to 60. or excæde 60. then you must diuide that product by 60. the emainder whereof is to be let vnder his proper denomination, and you must kéepe the quotient in minde to adde the same to the next greater number as this table sheweth: In the front whereof I haue set downe in seuerall spaces not onely the denominations of the two Integrums, as daies and howers, but also the denominations of so many Fractions as I think méete to serue my turne, vnder which front I place the foresaid multiplyer, and then draw a line as you sée in this example following.

Denomina-

Of Astronomicall Fractions.   32

Denomination.	Daies	Howers	′	″	‴	⁗
The multiplyer to be reduced into one selfe denomination.	29.	12.	44	3		
The products of the reductio.		29.	31	50	7	30

Now beginning first on the right hand with the least Fraction of the saide multiplyer which is ″/3. I multiply 3. by 5. which maketh 15. the one halfe whereof is ‴/7. and halfe a third which is ⁗/30. wherefore I set downe the said ‴/7. and ⁗/30. vnder their proper denominations as you sée in the example aboue, then I multiply 44 by 5. the product whereof is 220. and the halfe thereof is 110. which being diuided by 60. the quotient is 1′ and the remainder ″/50. which remainder I set vnder his proper denomination, kéeping the quotient stil in mind, that done, I multiply 12. howers by 5. the product whereof is 60. and the halfe of that is 30. whereunto by adding the one which I had in mind, I make it 31. and so I set it downe vnder his proper denomination, and because there be no more Fractions to be multiplyed, I set downe on the left hand the Integrum 29. and by this meanes I haue brought the foresaid multiplyer to one selfe denomination and to one kind of Integrum, that is to say to 29. daies 31. ″/50. ‴/7. ⁗/30. which now being the greater number is to be set aboue, and to be Multiplyed by the foresaid daily motion of the Moone, that is 13. degrées, 10 ″/35. and ‴/1. but to the intent that in multiplying these numbers together you may set euery product in his true place, that is to say, vnder his proper denomination, it shall not be amisse in the front of your worke to set downe two rowes of numbers, whereof the first must containe so many denominations or Fractions as you thinke good, as minutes, seconds, thirds, fourths, and so forth, marked with strokes and vulgare numbers, and the second rowe shall be the naturall order of numbers written in Arithmeticall figures as this table sheweth

Integra	′	″	‴	⁗	v.	vi.	vii.	viii.	The denominations.
0.	1.	2.	3.	4.	5.	6.	7.	8.	The naturall numbers.

Vnder which table you must first set the number that is to be multiplyed, and right vnder that the multiplyer in such sorte as

every

Of Astronomicall Fractions.

euery particular product may be placed vnder his proper denomination, and then draw a line as you see in this example following, and when you haue multiplyed 2. numbers the one into the other, and know not where to place the product, then marke vnder which of the naturall numbers in the front, the said two numbers, that is to say the multiplyer and the multiplycand do stand: That done, adde those two naturall numbers of the front together, and the summe thereof will shew you vnder what denomination the product is to be placed, as in this example.

Int gra	′	″	‴	⁗	v	vi	vii	
0	1	2	3	4	5	6	7	Denominations Naturall numbers.
29	31	50	7	30				The multiplycand.
13	10	35	1					The multiplyer
			29	31	50	7	30	The seuerall products.
		1015	1085	1750	245	1050		
	290	310	500	70	300			
377	403	650	91	390				
389	6	24	2	31	12	37	30	The generall product or totall summe.

The remainder.

In which example I first multiply ‴. into ⁗. the product whereof is ⁷⁄ᵥᵢᵢ. which must be set vnder the denomination vii. because the two naturall numbers that is 3. standing ouer 1. the multiplyer, and 4. standing ouer 30. the multiplycand being added together doe make 7. appointing to the product his proper denomination, then multiply againe the same ‴. into ‴. the product whereof is ⁵⁰⁄ᵥᵢ. which must be set vnder the denomination vi. because 3. standeth ouer both their heads, and therefore must bee taken twice, that is to say for each number 3. which being added together doe make 6. appointing thereby to the product his proper place of denomination, that done, multiply the said ‴. into ″. the product whereof is ″⁄ᵥ. which must be set vnder the denomination v. for the 3 which standeth ouer 1‴ and the 2. ouer ″. being added together maketh 5. appointing to his proper place of denomination, then multiply the said ‴. into 3′1. the product whereof is ‴. here the two naturall numbers that is to say 3. standing ouer ‴. and one ouer 3′1. being added together, doe make 4. appointing to the product his proper place of denomination, then multiply the foresaid ‴. into 29. Integrums, the product whereof

is

Of Astronomicall Fractions.   33

is $\frac{27}{9}$. and must be set vnder the denomination ′′′ because 3. standeth ouer ′′. but 29. being an Integrum, hath no natural number standing ouer him but a Cypher, thus hauing gon through out all the numbers of the multiplycand, with the first number of the multiplyer, procede in like order with the second number of the multiplyer which is $\frac{2}{7}$′′. which being multiplyed into $\frac{4}{7}$′′′′. maketh $1\frac{9}{10}\frac{0}{5}$′′. to be set vnder the denomination vi. because 2. standeth ouer $\frac{2}{7}$′′. the multiplyer, and 4. ouer $\frac{4}{7}$′′′′. the multiplycand, which 4. and 2. being added together maketh 6. then multiply $\frac{2}{7}$′′. by $\frac{3}{7}$′′′. which maketh $\frac{6}{24}$′′′′. which you must set downe vnder the denomination v. because 2. and 3. maketh 5. that done, multiply $\frac{2}{7}$′′. by $\frac{2}{7}$′′. which maketh $\frac{4}{25}$′′′′. which you must set downe vnder the denomination ′′′′ because the multiplycand and the multiplyer are both vnder the denomination ′′ which being twice repeated, maketh 4. then multiply $\frac{2}{7}$′′. by 3′1. and that maketh $1\frac{0}{8}\frac{6}{5}$′′. which you must set downe vnder the denomination ′′′ because 1. and 2. maketh 3. finally multiply 29. Integrums, by $\frac{2}{7}$′′. and that maketh $1\frac{0}{1}\frac{3}{5}$′′. thus as you haue gon through with two numbers of your multiplyer, so procede in like order with the other two numbers of the multiplyer which is 10. and 13. and when you haue ended your Multiplycation, and set euery product in his proper place, and so as euery figure may stand one right vnder another, to auoide confusion when you come to Addition, (to which end the spaces of collums had nede to be the larger) then draw a line vnder all the products, and beginning on the right hand, adde all the products contained in euery seuerall collum together, and if the summe of any such particular Addition do arise to the summe of 60. or excede the number of 60. then diuide that summe by 60. and set downe the remainder, kéeping the quotiēt in mind to be added to the product of the next collum on the left hand, so shall you find the total summe of your Multiplycation to be 389. degrees, 6′. $\frac{4}{24}$″. $\frac{2}{2}$‴. $\frac{2}{31}$⁗. $\frac{12}{12}$ᵛ. $\frac{6}{37}$ᵛⁱ. and $\frac{2}{30}$ᵛⁱⁱ. as the former example plainly sheweth. Now if you diuide 389. degrees, by 30. because euery common signe containeth 30. degrees, you shall find your totall summe to be 12. signes, 29. degrees, 6′. $\frac{4}{24}$″. $\frac{2}{2}$‴. $\frac{2}{31}$⁗. $\frac{12}{12}$ᵛ. $\frac{6}{37}$ᵛⁱ. and $\frac{2}{30}$ᵛⁱⁱ. and so much the Moone runneth in the space of 29. daies 12. howers, 4′4 and $\frac{2}{30}$. of an hower, which is her full reuolution betwixt euery two changes, but for as much as it chanceth as wel in this example as in many others like, that

F   Integrums

Of Astronomicall Fractions.

Integrums of two sundrie denominations are propounded in the question, it may be very well doubted with what denomination the product of such multiplycation is to be named, as in this example hauing multiplyed time by motion, a man may aske whether the product shall be named daies or degrees, the resoluing of which doubt dependeth vpon the nature of the question propounded, for in the foresaid example, because time or daies do comprehend any certaine appointed motion, therefore the product of the Multiplycation is to be referred to the degrees of motion which are comprehended vnder time, and not to time which comprehendeth motion, wherfore this product of Integrums videliz. 389 signifieth here degrees and not daies, so likewise when degrees and minutes are multiplyed by myles and minutes, the product of such Multiplycation taketh his name frō myles & not from degrees, because degrees do comprehend myles, for we say in matters of Geography that euery degree of the great circle comprehēdeth 60 myles, thus hauing spoken sufficiently of the Multiplycation of Astronomical Fractions, we wil now proceede to ye Diuision of such Fractions.

Of the Diuision of Fractions Astronomicall.

VVHat is to be obserued therein?

First you must consider whether your Diuisor be compound, or simple, I cal that compound which contayneth Fractions of diuers denominations, and that simple which consisteth of Integrums, or is one whole number of one selfe denomination, wherein there is no dificultie, for then you haue no more to do but to diuide euery particular number contained in the diuidend by ye same Diuisor, and to place the product of euery one vnder such denomination, as the little table of denominations sheweth, & therefore it shall not bee amisse to set the foresaid little table ouer your diuidend euen as you did in Multiplycation: Also the Sexagenary progression is alwaies to be vsed, as well in Diuision as in Multiplycation. Moreouer if your Diuisor be not exactly contained in ye diuidend, then hauing multiplyed the diuidend by 60. you must adde to the product therof the next Fraction following: As for example, knowing by Alphonsus tables that the daily motion of the Moone is 13. degrees, 1′0. ″₃₅. ‴. ″″₇. you would know how much she goeth in the space of an hower, here because that one day containeth

Of Astronomicall Fractions.

caineth 24. howers, the number must be ~~24~~/~~7~~. your Divisor which is simple and not compound, first then set downe in the front of your worke the rowe of denominations onely, and not the natural numbers, because they are not to be vsed in this way of Division, that done, right vnder the rowe of denominations place your dividend, and right vnder ye your Divisor, as you see in this example.

In which example because the Divisor 24. is not contained in 13. therfore I multiply 13. by 60. which maketh 780. where-

degree	′	″	‴	⁗	⁽ᵛ⁾	
13	10	35	1	15		The denominations.
24						The dividend.
						The divisor.
	32	56	27	33	7	The severall summe of every quotient.

vnto by adding the next Fraction on the right hand which is $1'0$. the whole summe is $79'0$. which being divided by 24. the quotient is $3'2$. which because they are minutes, I place them vnder the denomination of minutes, and the remainder is $2'2$. which being multiplyed by 60. maketh $_{120}''$. wherunto I adde the next figure which is 35. and so the whole summe is $_{155}''$. which being divided by 24. the quotient is $_{56}''$. which I place vnder the denomination of seconds, and the remainder of this Division is $_{11}''$. which being multiplyed by 60. maketh $_{660}'''$. whereto I adde the next Fraction which is $_{1}'''$. so that now the whole summe is $_{661}'''$. which being divided by 24. the quotient is $_{27}'''$. which I set downe vnder the denomination of thirds, and the remainder is $_{13}'''$. which being multiplyed by 60. maketh $_{780}''''$. whereunto I adde the next Fraction which is $_{15}''''$. which maketh in all $_{795}''''$. which being divided by 24. the quotient is $_{33}''''$. which I place vnder the denomination of fourths, and the remainder is $_{3}''''$. which being multiplyed by 60. maketh $_{180}^{v}$. whereunto hauing no Fraction to adde, I diuide the same by 24. and so I find in the quotient $_{7}^{v}$. which I set vnder the denomination of fifts, so as I find the howerly motion of the Moone to be $3'2.\,_{56}''\,_{27}'''\,_{33}''''\,_{7}^{v}$. and somewhat more, for I leaue to deale any further with the smaller Fractions that would still grow by multiplying the remainders by 60. thinking this sufficient to shew you in what order you haue to worke, to diuide your dividend by a simple Divisor, into as many small parts as you will: but if your Divisor be compound, then the Division is to be done either by reduction into the smallest Fractions, or without reductiō: which last way

F 2   is

Of Astronomicall Fractions.

is very hard and tedious, and therefore I will onely shew you how to make your diuision whereof the Diuisor is compound by reduction, and that by this one example here following. Suppose then that the Moone according to her owne course which is from West to East, is distant from some fixed Starre 36. degrees. 3'0. $\frac{'''}{24}$. $\frac{''''}{50}$. and $\frac{''''}{75}$. and that you would know in what time she will runne that distaunce, according to her daily moouing which as hath beene said before is 13. degrees, 1'0. $\frac{''}{3}$. $\frac{'''}{5}$. and $\frac{''''}{15}$. here to make this diuision by reduction, you must doe thus. First reduce all the numbers of your diuidend into the smallest Fractions thereof by the Sexagenarie Multiplycation and Addition of the next Fraction vnto the product of that Multiplycation: that done, reduce all the numbers of your Diuisor by like Multiplycation and Addition, into the smallest Fractions, so as the diuidend & the Diuisor may be both of one selfe denomination, and diuide the one by the other, euen as they were Integrums, as in this example you must first multiply 36. degrees, by 60. and it will make 2160. whereto by adding 3'0. you make the whole summe of minutes to be 2190. which being multiplyed againe by 60. doe make 131400. whereto if you adde the $\frac{''}{24}$. the summe of secondes will bee 131424. and so proceeding still with the Sexagenarie Multiplycation and Addition of the next Fraction as you did before, you shall find the diuidend to be 4731129415. Then in like order reduce your Diuisor into the smallest Fraction, and you shall find the totall summe thereof to be 170766075. this reduction being made, diuide the diuidend by the diuisor, so shal you find in the quotient 2. Integrums, that is to say 2. daies, and the remainder to be 1325973265. which remainder if you multiply by 60. and diuide the product by the self same Diuisor, you shal haue in the quotient minutes, then multiply againe that remainder by 60. and diuide the product thereof by the same Diuisor, and you shall haue in the quotient secondes, and so by obseruing still that order you shall bring it into as smal Fractions as you will, thus shall you finde that the Moone according to her daily motion, will runne the foresaide space of distance that was betwixt her and the fixed Starre in 2. daies, 34'6 and $\frac{''}{24}$.

How

Of Astronomicall Fractions.

How to diuide Astronomicall
Fractions when the Diuisor is greater then the diuidend.

Cap. 28.

Though by the last chapter you may learne how to diuide any number in Astronomicall Fractions, whereof the Diuisor is greater then the diuidend, yet I minde once againe, to set downe a generall rule to serue for such purpose, because it commeth often in vse in hauing to deale with Astronomicall tables, and to giue you example therof: First then hauing to diuide any number, whose Diuisor is greater then the diuidend, doe thus: multiply the greatest denomination of the diuidend by 60. and if there be any Fractions annexed thereunto which are of the next inferiour kinde, as minutes are to degrées, or to howers, and seconds to minutes, and thirds to seconds &c. Then adde them to the former product, but if such Fractions be not of the next inferior kind, then let them stand as they are vntill you come to deale with them, and hauing diuided according to the common rule of Diuision the first summe of the diuidend by the Diuisor, multiply the quotient into the whole Diuisor, and subtract that product out of the vpper number if it may bee, if not, then make the quotient lesser and lesser, vntill you can finde such a number as will be subtracted out of the said vpper number, and if there be any remainder left, then multiply that remainder by 60. not leauing to follow the former order of working, vntill you haue found the nearest exact quotient that may be. And you haue to note that the denomination of the first quotient must be of the next inferior kind, to that denomination which the Diuisor hath, and to make this rule the plainer, I will set downe an example vsed by Stadius in the 115 page of his Ephemerides who to know the very instant of the full Moone the second of March 1569. biddeth to diuide the distance of the opposition which was 8. degrées, 4′ 6″. by the diurnall excesse of the Moones

motion

Of Astronomicall Fractions.

motion from the sunne which was then 13. degrees, 4'8. which Divisor because it is greater then the dividend, you must according to the rule before given, worke thus: First multiply the greatest denomination of the dividend which is 8. degrees by 6'0. the product whereof will be 4'80. whereunto by adding the Fraction annexed videliz. 4'6. it maketh in all 52'6. which is to be divided by 13. degrees, 4'8. here in dividing the last product first by 13. degrees, I finde in the quotient 39. which is one too many, considering that I must take out of the foresaid dividend 4'8. as often as I did take out thereof 13. degrees, wherefore I set downe but 38. in the quotient, and then the remainder will be 96. which because I may easily divide by the common Divisor 13. degrees, and 4'8. I divide therefore that 96. first by 13. whereof the quotient is 7. and the remainder is 5. which I reduce into seconds by multiplying that 5. by 60. the product whereof is 3,00. that done, I multiply 4'8. by 7. the product whereof is 336. which though it bee somewhat too great a number to be taken out of 300. yet I let it stand because it approcheth to a very nigh exactnesse, and by this meanes I finde the whole quotient to bee 3'8. and ⅖. and you have to note that if after the first quotient be set downe, there happen any remainder which is lesser then the Divisor, then you must set downe a Cypher in the quotient, and remooue your Divisor one place further, even as you doe in common Division, and then to worke as before.

How to take the square roote of *Astronomicall Fractions*.

Cap. 29.

He greatest difficultie hereof consisteth in finding out the true denomination of the roote, for if the Fraction be seconds, then the roote therof are mynutes, and if the Fraction be fourths, then the roote are seconds, for the Fraction must alwaies haue such denomination as may be halfed, as seconds, fourths, and such like, the one halfe whereof giueth alwaies

name

Of Astronomicall Fractions.

name to the roote, for if the question bee of thirds, you must first reduce them to fourths before you can take the roote, and you must doe the like with any other Fraction, whose denomination is od and not euen. As for example, if you would take the roote of $\frac{'''}{43}$, here by multiplying these $\frac{'''}{43}$, by 60. you shall reduce them into $\frac{''''}{2580}$, the roote whereof is $\frac{''}{50}$. Moreouer the Fractions wherewith you haue to deale, are either simple, or compound, if they bée simple and lesse then minutes, and therewith haue euen denominations and not odde, then you néde to make no further reduction, but to worke as if you had to deale with whole numbers. As in séeking the square roote of $16\overset{''}{00}$, you find it to be iust 4' 0. but if the number be compound, that is to say, consisting of Integrums and Fractions, or of many Fractions hauing diuers denominations, then you must first reduce them all to the smallest Fraction that hath an euen denomination before that you can take the roote: As for example, you would know the roote of 4. degrées, 2' 5. here you must by the Sexagenarie Multiplycation and Addition of the next Fraction, reduce the degrées to minutes, and the minutes to seconds, as you were taught before in Diuision, and then to worke as you were wont to doe in taking the square roote of whole numbers, and in so doing, you shall finde the summe of seconds to be $15\overset{''}{9}00$. the square roote whereof is 12' 6. which if you diuide by 60. it will make 2. degrées, 6'. Another example, as to take the roote of 13. degrées, 4' 2. and $\frac{''}{45}$. here by reduction as before, you shall bring the degrées and minutes to 49 3 6" 5. the square roote whereof is 22' 2. which being diuided by 60. maketh 3. degrées, 4' 2. And thus I end with the Astronomicall Fractions, which kinde of Fractions, though they be very learnedly and orderly taught by Reinoldus in the beginning of his Prutenicall tables, yet in mine opinion not in so plaine order, and so fit for euery mans vnderstanding, as I haue here set them downe according to the doctrine of Gemma Frisius, which being once learned, you shall the soner attaine to the other. And without the knowledge of these Fractions, you can neuer truely calculate any thing out of the Astronomicall tables, and therefore such Fractions are most necessarie to be learned.

F 4   The

The description and vse of the
Sexagenarie table.

His table conſiſteth of two figures, whereof the neather figure hauing foure Angles, is called in Latine Trapezium, marked with the letters A. D. E. B. and the vpper figure is a Triangle, marked with the letters A. B. C. and each figure containeth particular collums of numbers, ſeruing to finde out the products of Aſtrono-
micall Fractions being multiplyed one by another, and alſo the quotients of the like Fractions being diuided one by another, and alſo the ſquare rootes of the ſaid Fractions, for which purpoſe the firſt collum on the left hand, containeth 59. Fractions, counting from 1. ſtanding aboue, and ſo proceeding downe-ward to 59. and are contained betwixt A. and D. and the foote of the ſaid Trape-zium, containeth 30. counting from the left hand towards the right, which are contained betwixt D. and E. and the reſt of the numbers to make vp 59. are to be found in the vppermoſt front of the Triangle, proceeding from C towards A. you haue to note alſo. that in the outermoſt collum of the Trapezium on the right hand, the numbers doe procede downeward from 30. to 59. that is from B. to E. and the numbers in the outermoſt collum of the Triangle on the right hand, doe procede vpward from 31. to 59. contained betwixt B. and C. both which doe ſerue to fill vp the firſt multiplyers, and multiplycands, for when you cannot finde them in the Trapezium, then the outermoſt collum of the Triangle on the right hand; ſerueth to ſupply that want, and when you cannot finde the ſaid numbers in the Triangle, then the outermoſt collum of the Trapezium on the right hand, ſerueth to ſupplie that want, and in ſeeking any multiplyer or multiplycand,

either

The vſe of the Sexagenarie table.   40

either in the Trapezium or in the Triangle, conſider alwaies which way they are moſt readily found out, ſo as they may directly aunſwere one againſt another.

All the reſt of the numbers contained betwixt the two outermoſt collums and are ſet downe in ſquare Angles, called common Angles, do ſignifie either products or dividends or ſquare numbers, according as occaſion ſhall require. And the outermoſt collums do ſignifie ſometime multiplyers, ſometime multiplycands, ſometime quotients, and ſometime rootes. All which thinges you ſhall better vnderſtand by the examples hereafter following.

The vſe of the Table.

By helpe of this table you may more readily multiply and divide Aſtronomicall Fractions, and alſo find out the ſquare roote of ſuch Fractions, then by thoſe rules which I haue heretofore ſet downe according to the doctrine of Gemma Friſius. And firſt I will ſet downe an example of Multiplycation, then another of Diuiſion, and thirdly one example of finding out the ſquare roote of the ſaid Fractions, and let the example of Multiplycation be thus: Suppoſe that you would multiply 29. degrées, $3'\, 1.\, \overset{''}{5^{0}},\, \overset{'''}{7},\, \overset{''''}{3^{0}}$. by 13. degrées, $1'\, 0.\, \overset{''}{3^{5}},\, \overset{'''}{1}$. which you muſt ſet downe in ſuch order as followeth, (that is to ſay) firſt the denominations, then next vnder them the multiplycand, and next vnder that the multiplyer, and vnder them the ſeverall products, and loweſt of all the totall ſumme of the ſaid products.

Integra	De.	′	″	‴	⁗	v	vi	vii	Denominations
29	3	1	50	7	30				The multiplycand.
13	10	35	1						The multiplyer.
			29	3	1	50	7	30	The firſt product.
		17	13	34	14	22	30		The ſecond product.
	4	55	18	21	15	0			The third product.
6	23	53	51	37	30				The fourth product.
6	29	6	24	2	31	12	37	30	The total ſumme

Here beginning with the firſt number of the multiplyer on the right hand which is $\overset{'''}{1}$. ſay thus, one times 30. is 30 which is to beé placed vnder the denomination of 7. becauſe the denomination ouer 1. is 3. and the denomination ouer 30. is 4. which being added together according to the rule before ſet

downe

The vſe of the Sexagenarie table.

downe in the Chapter of Multiplycation of Aſtronomicall Fractions doe make 7. then ſay one times 7. is 7. which you muſt place vnder the ſixt denomination. Againe one times 50. is 50. which is to bee ſet vnder the fift denomination, becauſe 3. and 2. maketh 5, then ſay one times 31. is 31. which is to bee ſet vnder the fourth denomination, for 3. and 1. maketh 4, then one times 29. is 29. which is to be ſet vnder the third denomination, and thus you haue the firſt product: then procéde with the next number of the multiplyer towards your left hand which is 35. and is to be multiplyed into 30. which to doe readily, you muſt enter the table with theſe two numbers, and ſéeking in the firſt Collum of the Trapezium on the left hand, for 35. in the foote of the ſaid Trapezium, for the number of 30. you ſhall finde in the common angle the product to be 17. and 30. whereof you muſt place the 30. vnder the ſixt denomination, and kéeping the number 17. ſtill in minde to be added to the next product, multiply 35. into 7. and you ſhall finde in the Trapezium the product to be 4. and 5. whereunto if you adde the 17. which you had in minde, the product will be 4. and 22. whereof you muſt ſet the 22. vnder the fift denomination, and kéeping the 4. in minde, multiply againe the ſame 35. into 50. the product whereof you ſhall finde in the Triangle to be 29. and 10. whereto if you adde the 4. in mind, it will make 29. and 14. whereof the 14. is to bee placed vnder the fourth denomination, and kéeping the 29. in minde, multiply againe the ſaid 35. vnto 31. and you ſhall finde the product thereof in the Triangle to bee 18. and 5. whereunto if you adde 29. in mind, it will make in all 18. and 34. which 34. is to be ſet vnder the third denomination, then kéeping 18. in mind, multiply the foreſaid 35. into 29. and you ſhall finde the product thereof in the Trapezium to be 16. and 55. whereunto if you adde the 18. in mind, the product will be 17. and 13. which 13. is to be placed vnder the ſecond denomination. Now becauſe you haue gon through all the numbers of the multiplycand with the number 35. you muſt place the 17. which you had in mind vnder the denomination of mynutes, and ſo hauing ended the ſecond product, procéde to the finding out of the third product by multiplying 10. firſt into 30. the product whereof you ſhall find in the Trapezium to bée 5. and 0. whereof you muſt ſet the Cypher vnder the fift denomination

The vſe of the Sexagenarie table. 41

nation, and kæping the 5 in minde, multiply againe 10. into 7. and you ſhall finde the product to be 1. and 10. whereunto if you adde the 5. in mind, the product will be 1. and 15. whereof you muſt ſet the 15. vnder the fourth denomination, and kæping the one in mind, multiply 10. into 50. and you ſhall find the product thereof in the Trapezium to be 8. and 20. whereunto if you adde the 1. in mind, it will make 8. and 21. whereof you muſt ſet the 21. vnder the third denomination, and kæping 8. in mind, multiply 10. into 31. and you ſhall find the product thereof in the Trapezium to bee 5. and 10. vnto which if you adde the 8. in minde, the product will be 5. and 18. whereof ſet downe the 18. vnder the ſecond denomination, and kæping 5. in minde multiply 10. into 29. the product where of you ſhall finde in the Trapezium to bee 4. and 50. whereunto if you adde the 5. in mind, the product will be 4 and 55. Now becauſe you haue gone through all the numbers of the multiplycand, with 10. you muſt ſet downe the 4. which you had in mind vnder the denomination of degrées, and ſo hauing the third product, procéde to the fourth by multiplying 13. into 30. and you ſhall find the product in the Trapezium to be 6. and 30. whereof ſet downe 30. vnder the fourth denomination of Fractions, and kæping 6. in mind multiply 13. into 7. the product whereof you ſhall finde in the Trapezium to be 1. and 31. whereunto if you adde the 6. in minde, the product will be 1. and 37. whereof ſet downe the 37. vnder the third denomination, and kæping the 1. in mind multiply 13. into 50. and you ſhall find the product in the Trapezium to be 10. and 50. whereunto if you adde the one in mind, the product will be 10. and 51. whereof ſet the 51. vnder the ſecond denomination of Fractions, and kæping 10. in mind multiply 13 into 31. the product whereof you ſhall find in the Trapezium to bee 6. and 43. whereunto if you adde the 10. in mind, the product will be 6. and 53. whereof ſet downe 53. vnder the firſt denomination of Fractions, and kæping 6. in mind multiply 13. into 29. the product whereof you ſhall find in the Trapezium to be 6. and 17. wherſuto if you adde ẏ 6. in mind, the product will be 6. and 23. whereof ſet downe 23. vnder the denomination of degrées. Now becauſe you haue gone through all the numbers of the multiplycand with the laſt number of the multiplyer, you muſt ſet downe the 6. which you had in minde vnder

the

The vse of the Sexagenarie table.

the denomination of Integrums, that done, adde all the foure products together, beginning on the right hand, saying thus, 30. and 0 is but 30. which set downe vnder the neathermost line, as you see in the former figure, then say 30. and 7. maketh 37. which you must set downe vnder the neathermost line next vnto 30. then say 22. and 50. maketh 72. out of which by subtracting the Sexagenarie number, the remainder is 12. which is to bee set vnder the line next vnto the 37. kéeping still the 60. in minde, which in this account maketh but one, and is to be added to the next rancke on the left hand, then say 1. in mind and 5. maketh 6. and 4. maketh 10. and 1. maketh 11. here set downe 1. and kéepe the Article in mind, then say 1. in mind and 3. maketh 4. then 1. and 1. and 3. being added to 4. doe make in all 9. out of which 9. you must subtract 6. which is but one 60. and is to be kept in mind and there remaineth 3. which is to be set downe by the 1. then say 1 in mind and 7. is 8. and 1. is 9. and 4. is 13. and 9. is 22. here set downe 2. and kéepe the 2. Articles in minde, then say 2. and 3. maketh 5. and 2. is 7. and 3. is 10. and 2. maketh 12. tennes, which doe make 2. sixties, and are to be kept in mind, then say 2. and 1. maketh 3. and 8. maketh 11. and 3. maketh 14. wherefore set down 4. kéeping the 1. Article in mind, then say 1. in mind and 5. is 6. and 1. is 7. and 1. is 8. then take 6. from 8. and there remaineth 2. which set downe by the 4. and kéepe one sixtie in mind, then say 1. in mind and 3. is 4. and 5. is 9. and 7. is 16. wherefore set downe 6. and kéeping one in mind, say that 1. and 5. is 6. and 5. is 11. and 1. is 12. which maketh two sixties to be kept in minde, then say 2. in mind and 3. is 5. and 4. maketh 9. which set downe, then say 2. is 2. which set downe by the 9. vnder the denomination of degrées, then say 6. and nothing maketh 6. which set downe vnder the denomination of Integrums, for that 6. in this place signifieth 6. sixties, which is in value 360. degrées, and being diuided by 30. because 30. degrées maketh one whole signe, you shal find in the quotient 12. signes, so as the totall summe of the foure products is 12. signes, 29. degrées, $6'.24''.2'''.31''''.12^{v}.37^{vi}.30^{vii}.$ as the former figure sheweth, and this is the very same example which I wrought before when I taught you how to multiply Astronomicall Fractions according to Gemma Frisius his rule, and they both doe wholly agrée in euery condition, sauing that to

worke

The vſe of the Sexagenarie table.   39

worke by the Sexagenarie table is the readier way of the two.

An example how to diuide Aſtronomicall Fracti-
ons by helpe of the Sexagenarie table.

SVppoſe then that you would diuide the former totall ſumme or product found by Multiplycation, which is 6. Integrums, 29. degrées, 6′. 24″. 2‴. 31⁗. 12ᵛ. 37ᵛⁱ. 30ᵛⁱⁱ. by this Diuiſor 29. degrées, 3′ 1″. 50‴. 7⁗. 30ᵛ. which in the former example of Multiplycation was the multiplycand. Now to diuide theſe two numbers the one by the other, you muſt doe thus, firſt you muſt ſet downe the rowe of denominations as you did before in Multiplycation, and next vnder that the diuidend, then right vnder that the diuiſor, and on the right hand behind a crooked line made like a halfe Moone, all the ſeuerall quotients are to be ſet one by another in a right line, as you may ſée by the figures hereafter following.

Héere hauing made your rowe of denominations, and ſet downe your di- uidend, you haue to

Denominatiō	Inte-grum	de	′	″	‴	⁗	ᵛ	ᵛⁱ	ᵛⁱⁱ
The diuidend.	6	29	6	24	2	31	12	37	30
The diuiſor.		29	3	1	50	7	30		

conſider whether the firſt number of your Diuiſor be greater then the firſt number of your diuidend, for if it be, then you muſt place your Diuiſor one ſpace further towardes your right hand, as in this example, becauſe the firſt number of your Diuiſor 29. cannot be taken out of 6. you ſet it vnder the ſecond number of the diui- dend, and ſo all the reſt of the numbers ſucceſſiuely towardes the right hand, as the former example ſheweth, now the order of wor- king is thus: You muſt firſt ſéeke out the firſt number of your Di- uiſor, in the firſt collum of the Trapezium on the left hand which is 29. then in the rowe right againſt that 29. on the right hand, you haue to ſéek out thoſe numbers of the diuidend, which do ſtand right ouer the firſt number of your Diuiſor which is 6. and 29. and if you cannot finde thoſe numbers iuſtly, then ſéeke in the ſelfe ſame rowe a number which is ſomewhat leſſe and neareſt in value vnto it, as in this example, becauſe you cannot finde 6. and 29. you take 6. and 17. the quotient, wherof you ſhall find in the foote of the Trapezium right vnder the ſaid number 6. and 17. to bée

13 degrées

The vse of the Sexagenarie table.

13. degrees, which you must set in the quotient line, and that is your first quotient, hauing his proper denomination ouer his head which are degrees, and are to bee found by the rule before taught. Thus you see here that the number of the Diuisor is to be founde in the outermost collum on the left hand, and the number of the diuidend in that rowe which is right against the said Diuisor, and the quotient in the foote of the Trapezium right vnder the number of the diuidend last found. Then you haue to multiply the whole Diuisor (that is to say) euery particular number thereof by the first quotient 13. which you may doe by helpe of the Table, as you did before in the example of Multiplycation, and is to be set downe in this manner. And you shall finde the whole product to be 6·23·53·51.   29·31·50·7·30.   13.   37·30.

Which being set right vnder the first diuidend, is to be subtracted out of the same, and the remainder to bee written ouer the head of the diuidend as you doe in common Diuision, first then to multiply all the particular numbers of the Diuisor by the quotient 13. and to find euery product thereof, resort to the Table and seeke for 13. in the foote of the Trapezium, and for 30. which is the multiplycand in the outermost collum of the Trapezium on the left hand, and the common angle will shew the product which is 6. and 30. whereof you must set downe 30. vnder 30. and keeping 6. in mind, multiply againe 13. into 7. the product whereof you shall finde to bee 1. and 31. vnto which adde the 6. which you kept in mind, and the product shall be 1. and 37. whereof you must set downe 37. and keeping one in mind,

29	31	50	7	30	The Diuisor.
				13	The first quotient.
6	23	53	51	37	30	The first product.

multiply againe 13. into 50. and you shall find the product to be 10. and 50. whereunto if you adde 1. in mind, the product shall be 10. and 51. whereof you must set downe 51. and keeping 10. in mind, multiply 13. into 31. the product whereof you shall find in the Trapezium to be 6. and 43. whereunto if you adde the 10. which you had in mind, the product shall be 6. and 53. whereof set downe 53. and keeping 6. in mind, multiply 13. into 29. and you shall find the product to be 6. and 17. whereunto if you adde the 6. in mind, the product shall be 6. & 23. whereof set downe 23

vnder

The vse of the Sexagenarie table. 40

vnder the last number of your multiplycand, and because you haue no moe numbers of the Diuisor to be multiplyed, set downe 6. in mind on the left hand, so shall the whole product bee 6.23.53.51. 37.30. as the former example sheweth, which product is to be subtracted out of the first diuidend, & the remainder is to be set downe ouer the head of that diuidend, as you sée in this example next following, wherein the first diuidend is first set downe, and right vnder that the foresaid product which is the first product, and the remainder aboue the diuidend, and the quotient 13. is set in the quotient line which is your first quotient.

The remainder.	5	12	32	25	1			
The first diuidend.	6	29	6	24	2	31	12	37	30
The first product.	6	23	53	51	37	30			

(13.

And remember (in making your Subtraction) to begin with the first number of the foresaide product which is on the right hand, and when you cannot take it out of the number standing right ouer his head, to borrow alwaies 60. of the next nuber on the left hand, and to pay it home againe with 1. for there 60. is but one. This done, remoue your Diuisor one space further towards the right hand (that is to say) set the first number of your Diuisor vnder 12 which is the second nuber of the second diuidend, which together with the first remainder is 5.12.32.25.1.12.37.30. and al other numbers of the Diuisor orderly towards your right hand, as you sée in this example.

Then aske how many times 29. is in 5. & 12. which number you

The second diuidend.	5	12	32	25	1	12	37	30
The Diuisor.		29	31	50	7	30		

must séeke in the Trapezium in the rowe that answereth towards the right hand, to the first number of your Diuisor which is 29. standing in the outermost collum of the Trapezium on the left hand, and because you cannot find 5. and 12. in that rowe, you must take in the same rowe the number which is nighest vnto it, but lesse, which you shall find to be 4. and 50. and right vnder that in the foote of the Trapezium you shall find 10. which must be your second quotient, by which quotient you haue to multiply all the particular numbers of the Diuisor in such order as is before set

downe.

The vse of the Sexagenarie table.

downe and you shall find the product of that Multiplycation to be 4. 55. 18. 21. 15. 0. which product you must place vnder the second diuidend setting 4. vnder 5. and 55. vnder 12. and so forth orderly towards the right hand, that done, subtract the same product out of the numbers of the second diuidend, standing right ouer the said product, and the remainder will be 17. 14. 3. 46. 12. as you see in this example.

The remainder.	17	14	3	46	12	37	30
The second diuidend.	5	12	32	25	1	12	37	30
The second product.	4	55	18	25	15	0		

(13. 10.

Now remoue your Diuisor one space further towards the right hand by setting the first number of your Diuisor which is 29 vnder 14. which is the second number of the third diuidend, and so all the rest orderly towards the right hand, as you see in this example.

Then aske how many times 29. is contained in 17. and 14. which

The third diuidend.	17	14	3	46	12	37	30
The Diuisor.		29	31	50	7	30	

diuidend because you can not find it in the rowe that answereth to that 29. which standeth in the outermost collum of the Trapezium on the left hand (for all those numbers are too little) you must seeke for it in that collum which standeth right vpon 29. in the foote of the Trapezium, neither shall you find it there, but you shall finde 16. and 55. which is somewhat lesser yet in value nighest vnto 17 and 14. and right against that you shall find in the outermost collum on the right hand 35. which must be your third quotient to bee set in the quotient line, whereby you haue againe to multiply the whole Diuisor in such order as before, the product whereof you shall find to be 17. 13. 34. 14. 22. 30. which product being placed vnder the diuidend by setting 17. vnder 17. and 13. vnder 14. and so forth orderly towards the right hand, subtract the said product (beginning at the right hand) out of the numbers which stand right ouer the said product, and the remainder will be 29. 31. 50. 7. 30. which is to be set downe ouer the third diuidend, and the rest to be cancelled as you see in this example.

The

The vſe of the Sexagenarie table.   41

		29	31	50	7	30	
The remainder.							
The third diuidend	17	13	3	46	12	37	30	(13,10,35
The third product.	17	13	34	14	22	30		

And now the remainder of the third diuidend is come to be the **fourth** diuidend, wherefore remoue your Diuiſor one ſpace further towards the right hand by ſetting 29. vnder 29. and the reſt orderly towards the right hand as you ſee in the example following. Then aſke how many times 29. is contained in 29. which being but once, your fourth quotient is 1. and is to bee ſet in the quotient line, whereby the whole Diuiſor being multiplyed, the product will be 29. 31. 50. 7. 30. which you muſt place vnder the fourth diuidend, and being ſubtracted out of the ſame, nothing will remaine, and ſo the whole quotient will be 13. degrees, 1′ 0.″ 35.‴ 1.⁗ as you ſee in this example.

The fourth diuidend.	0	0	0	0	0		degrees ′ ″ ‴
The fourth product.	29	31	50	7	30		(13.10.35.1.
	29	31	50	7	30		

Another example of Diuiſion.

IF the daily motion of the Moone be 13. degrees 3′.″ 53.‴ 50.⁗ 23.ᵛ 58. in what time ſhall ſhe make her whole reuolution allowing 360 degrees to that reuolution, which is otherwiſe called the moneth of Paragration. Heere for ſo much as this example is to bee wrought by Diuiſion, & that your diuidend is a ſimple and whole number (that is to ſay) 360. degrees, without any Fractions of diuers denominations annexed thereunto: you muſt firſt ſet downe the 360. degrees, and next to that towardes your right hand ſet downe a long rowe of Cyphers with ſo many denominations ouer their heads as you ſhall thinke nedful to ſerue your turne, & right vnder your diuidend ſet your Diuiſor as you ſee in this example.

	Integrum	De.	′	″	‴	⁗	V	VI	VII	VIII	IX
The denominations											
The diuidend.		360	0	0	0	0	0	0	0	0	
The Diuiſor.		13	3	53	56	23	58				(27.
The firſt product.		352	45	16	22	47	6				

The vſe of the Sexagenarie table.

Here you muſt firſt aſke how many times 13. is in 360. and by the common rule of Diuiſion you ſhall finde the quotient to be 27. which you muſt ſet downe in the quotient line, and by that quotient you haue to multiply euery particular number of the Diuiſor beginning on the right hand, as you doe in Multiplycation, ſaying that 27. times 58. is 26. and 6, as the Trapezium ſheweth, for by ſéeking 27. in the foote of the Trapezium and for 58. in the outermoſt collum on the right hand, you ſhal find in the common Angle 26. and 6. wherefore ſet downe 6. vnder 58. and kéeping 26. in mind, multiply 23. which is the ſecond number of the Diuiſor by 27. and by the Trapezium you ſhall find the product thereof to be 10. and 21. whereunto if you adde the 26. in mind, it will make 10. and 47. whereof ſet downe 47. vnder 23. and kéeping 10. in mind, multiply 56. which is the third number of the Diuiſor by 27. and in the Trapezium you ſhall finde the product to bee 25. and 12. whereto if you adde the 10. in minde, it will make 25. and 22. whereof ſet downe 22. vnder 56. and kéeping 25. in mind, multiply 53. which is the fourth number of the Diuiſor by 27. the product whereof you ſhall find by the Trapezium to be 23. and 51. whereto if you adde 25. in mind, it wil make 76 which is one 60. and 16. whereof ſet downe 16. vnder 53. and adde the one in mind to 23. and that will make 24. which you muſt kéepe in mind, then multiply 3. which is the fift number of the Diuiſor by 27. the product whereof you ſhall find in the Trapezium to be 1. and 21. whereto if you adde the 24. in mind, it will make 1. and 45. whereof ſet downe 45. vnder 3. and kéeping 1. in mind, multiply 13. which is the laſt number of the Diuiſor by 27. the product whereof you ſhall find in the Trapezium to bee 5. and 51. whereunto if you adde one in mind, it will make 5 and 52. here becauſe the firſt 5. is 5. ſixties, it maketh in all 352 and is to be ſet vnder 360. ſo as the firſt product of this Multiplycation containeth theſe numbers 352.45.16.22.47.6. as you ſée them ſet downe in the former example, and this product is to bee ſubtracted out of 360 which is the firſt diuidend. And to auoide confuſion, it ſhall not bee amyſſe to ſet downe the firſt Diuidend and the firſt product apart by themſelues thus.

The

The vſe of the Sexagenarie table.   42

		7	14	43	37	12	54	0	0	0	
Heere be-ginning on the right hãd ſay thus, take	The firſt remainder	3̶	5̶0̶	0̶	0̶	0̶	0̶	0̶	0̶	0̶	(27
	The firſt diuidend	2̶	5̶2̶	48	1̶6̶	22	47	6			
	The firſt product.										

6. out of nothing which wil not be, wherfore you muſt borrow 60 then by taking 6. out of 60. there will remaine 54. which you muſt ſet aboue that Cypher which ſtandeth right ouer 6. and cancell the 6. kéeping ſtill the one 60. which you borrowed in mind, then ſay 47. and one in mind maketh 48. which will not be taken out of nothing, and therefore you muſt borrow againe one 60. as you did before, ſo ſhall the remainder be 12. which is to be ſet aboue the Cypher which ſtandeth right ouer 47. and cancell 47. and ſo procéede with like order in ſubtracting all the reſt of the numbers of the firſt product out of the firſt diuidend, ſo ſhall the remainder be 7. 14. 43. 37. 12. 54. as you ſée them ſet downe in the former figure. Now hauing to remoue your Diuiſor one ſpace further towards the right hand, you ſhall doe well to make your firſt remainder which is 7. 14. 43. 37. 12. 54. to be your ſecond diuidend, and vnder that to ſet your Diuiſor as you ſée in this example.

	7	14	43	37	12	54	0	0	0	
The ſecond diuidend.										(27.33.
The Diuiſor.		13	3	53	56	23	58			

Then aſke how many times 13. is contained in 7. and 14. & hauing found 13. in the foote of y Trapezium, ſéek in that collum for 7. and 14. and not finding it there, take in the ſelf ſame collum that number which is nigheſt in value vnto it & leſſe, which you ſhal find to be 7. and 9. right againſt which in the outermoſt collum on the right hand is 33. which muſt be your ſecond quotient, and is to be ſet in the quotient line, next vnto 27. and by this quotient you haue to multiply euery number of the Diuiſor as you did before, the product wherof you ſhall find to be 7. 11. 8. 40. 1. 10. 54. wherfore you muſt firſt ſet downe your ſecond diuidend, and then the ſecond product right vnder the ſame as you ſée in this example.

	3	34	57	11	43	6				
The ſecond remainder.										
The ſecond diuidend.	7̶	1̶4̶	4̶3̶	3̶7̶	1̶2̶	5̶4̶	0̶	0	0	(27.33.
The ſecond product.	7	1̶1̶	8	4̶0̶	1̶	1̶0̶	54			

G 2   Which

The vse of the Sexagenarie table.

Which product being subtracted out of the second diuidend, the remainder will be 3. 34. 57. 11. 43. 6. which is to be set aboue the second diuidend, and the product to bee cancelled as you see in the former example. Here hauing to remooue againe your common Diuisor one space further towards the right hand, set downe first the last remainder which now must be your third diuidend, and vnder that set the common Diuisor as you see in this example.

The third diui-dend.	3	34	57	11	43	6	0	0	0	(27. 33. 16.
The Diuisor.		13	3	53	56	23	58			

Then aske how many times 13. is contained in 3. and 34. Here by seeking in the foote of the Trapezium for 13. though you cannot find in that collum 3. and 34. yet you shall find 3. and 28. which is the nighest, right against which in the outermost collum on the left hand you shall find 16. which must be your third quotient, and is to bee set in the quotient line, by which quotient you must multiply the whole Diuisor as before, the product whereof you shall find to be 3. 29. 2. 23. 2. 23. 28. which being the third product you must set downe vnder the third diuidend which was your last remainder and to be subtracted out of the same as you see in this example.

The third diui-dend.		5	54	48	40	42	32	0	0
	3̶	3̶4̶	5̶7̶	1̶1̶	4̶3̶	6̶	0̶	0̶	0̶
The third product.	3	29	2	23	2	23	28		

And so the remainder will be 5. 54. 48. 40. 42. 32. which you must set aboue the third diuidend, and al the inferior numbers are to be cancelled as you see in the former example. Here hauing againe to remoue your Diuisor one space further towardes your right hand, the last remainder must be your fourth diuidend, vnder which the common Diuisor is to be set thus.

The fourth di-uidend.	5	54	48	40	42	32	0	0	daies $'$ $''$ $'''$ (27/33/16/26.
The Diuisor.		13	3	53	56	23	58		

Here asking how many times 13. is contained in 5. and 54. seeke for 13 in the foote of the Trapezium, in whose collum you shal not finde 5. and 54. but 5. and 51. which is nighest vnto it.

The vse of the Sexagenarie table   43

it, and right against that in the outermost collum on the left hand you shall finde the quotient to be 27. by which if you should multiply the whole Diuisor, the whole product thereof would be 5.56 21.24.1.47.6. which is more then the diuidend, and therefore you must make your quotient one lesse, setting downe no more but 26. in the quotient line, by which if you multiply the Diuisor, the product will be 5.26.41.22.25.14.8. which being subtracted out of the fourth diuidend, the remainder will bee 28.7.18.17. 17.52. which remainder if you will, you may make to be a fift diuidend, and then to worke as before, if you would haue your quotient to extend to smaller denominations, which I leaue to doe because I thinke that thirds be small inough. And as often as the product of any particular quotient shall bee greater then the diuidend, remember to take a lesse quotient euen as you doe in common Diuision. But now you haue to note that though this whole quotient here signifieth time, for the first quotient signifieth daies the second quotient minutes of daies, the third quotient seconds, and the fourth quotient thirds, yet for so much as the day is to be counted by 24. howers and not by mynutes, you must therefore reduce all the particular quotients sauing the first, into howers and parts of howers, euen to so small denominations as you shall thinke good your selfe, by helpe of this rule which in diuision of Astronomicall Fractions biddeth to multiply the quotient by 2. and to diuide the product thereof by 5. as here if you multiply the 33. which is the second quotient by 2. the product will bee 66. which being diuided by 5. the quotient will be 13. howers and one sixtie remaining to be kept in minde, wherefore set downe in the place of 3′ 3.13. howers, then multiply the third quotient which is 16. by 2. and that will make 32. whereunto if you adde the one 60. in mind it will make in all 92. which being diuided by 5. the quotient will be 18. and two sixties, which is 120. remaining to be kept in minde, wherefore in stéede of 16″. which was the third quotient, set 1′8 of an hower, and then procéede to the fourth quotient which is ″6. which being multiplyed by 2. maketh 52. wherunto if you adde 120. it will make 172. which if you diuide by 5. the quotient will bee ″$_4$. of an hower, which is to bee set downe in the place of 26. so shall your whole former quotient con-

                     G 3   taine

The vse of the Sexagenarie table

taine 27. daies, 13. howers, 18. of an hower and 3″4. of an hower. And note that as in the Diuision of Astronomicall Fractions to bring the quotients to like denomination, you doe multiply by 2. and diuide by 5. so in Multiplycation to reduce the numbers of the multiplyer being of diuers denominations to one selfe denomination, you must by order reuerse, multiply by 5. and diuide by 2. whereof I haue giuen you an example before, whereas I shew you how to multiply Astronomicall Fractions according to Gemma Frisius, without the helpe of the Sexagenarie table.

An example shewing howto extract the square roote out of Astronomicall Fractions.

Suppose the number giuen to be 17. 12. 33. 4. whereof you haue to take the square roote, here hauing set downe the said numbers with their proper denominations ouer their heads, as you sée in the example following, first set a pricke vnder the last number on the right hand, and then pricke euery other number leauing one voide space betwixt euery two prickes, as you doe when you séeke the square roote of whole numbers or Integrums, as you sée here in this example.

Then resort to the table   The number giuen 17. 12. 33. 4.
& séek amongst the products
which are placed next to the line A. B. as well in the Trapezium as in the Triangle, and sée whether you can find the number standing ouer the first pricke on the left hand which is 17. and 12. but not finding it there, you must take that which is nighest vnto it and lesse, which you shall find in the Triangle towards B. to be 17. 4. the roote whereof you shall finde both in the head, and also in the outward collum of the Triangle on the right hand to be 32. aunswering to the foresaid square roote, which roote you must place behinde the quotient line, then subtract 17. and 4. out of 17. and 12. and there will remaine 8. which is to be set ouer 12. and the 17. and 12. to be cancelled, that done, double the roote 32. which will be 64. that is to say 1. and 4. and setting 1. vnder 8. and 4. vnder 33. aske how many times 1. is in 8. and there is 8. which is the second quotient by which you must multiply 1. and 4.

the

The vse of the Sexagenarie table   44

the product whereof will be 8. and 32. which being subtracted from the vpper number which is 8. and 33. the remainder is 1. which is to be set ouer 33. and the 8. and 33. to be cancelled. Finally multiply the second quotient 8. in it selfe, and the product will be 1. and 4. which being subtracted out of the former remainder of the giuen number nothing remaineth, so as you shall finde 3′ 2″ 8‴. to be the roote of the giuen number, as this example sheweth.

Which roote if you multiply into it selfe squarely, the product will be like vnto the number giuen, and by this meanes you shall find the giuen number to be a iust square number. But you haue to note that if the last denomination standing on the right hand be od & not euen, as thirds or fifts, then you must set downe a Cypher beyond the last denomination towards your right hand, and vnder that Cypher set your first pricke, as in this example 3′. 2″. ‴. ⁗. ⁗′. here because the last denomination on the right hand is od, that is to say fifts, you must therefore set downe next vnto it towards the right hand a Cypher hauing ouer his head the next denomination which is vi. and is euen, wherefore set your first pricke vnder that Cypher and so proceede towards your left hand pricking euery other number, and then worke as followeth.

First then seeke in the table amongst the square numbers nigh vnto the line A. B. for 3. and 2. which should stand in one selfe square Angle next vnto the line A. B. for those onely are square numbers, and not finding it amongst the square numbers, take that square which is nighest vnto it and lesse, which you shall find to be 2. and 49. the roote whereof is 13. and is to be found as well in the foote of the Trapezium right vnder the foresaid square number, as also in the outermost collum of the said Trapezium on the left hand standing right against the same square number, which roote is to be set in the quotient line, that done, subtract the foresaid square number out of the number standing right ouer the last pricke on ye left hand which is 3. and 2. and the remainder will be 13. which is to be set ouer 2.

G 4

The vse of the Sexagenarie table.

and the 3. and 2. to be cancelled as you sée in the former example, that done, double the foresaid found roote 13 which wil make 26. by which you haue to diuide 13. and 9. wherefore resorte to the Trapezium, in the foote whereof you shall find 26. in whose collum not finding 13. and 9. take that which is nighest vnto it and lesse, which is 13. and 0. right against which you shall find in the outermost collum on the right hand 30. which should be your quotient, but because it is too great and that the square thereof cannot bée taken out of the remainder, you must make the quotient one lesse, and set in the quotient line no more but 29. by which 29. if you multiply 26. the product will bee 12. and 34. as the Trapezium sheweth, which being subtracted out of 13. and 9. standing ouer 26. the remainder will be 35. which 35. is to bee set ouer 9. then by multiplying the second quotient which is 29. into it selfe, you shall find by the Trapezium the product to be 14. and 1. which being subtracted out of 35. and 17. the remainder will be 21. and 16. which is to be set ouer 35. and 17. as you sée in this example.

That done, double bothe the quotients or whole roote, and the product will be 26.58. and hauing set 58. vnder 8. and 26. vnder 16. diuide 21.16 and 8 by 26.58. wherefore looke for 26. in the foote of the Trapezi-

```
         2 1
       ✗ 3 ✗ 5  16
    ✗ .✗ .9 .✗ 7 .8 .0 .( 13.29.
    ✗   4 9 26   The duble of the first quotiēt
              14.1. The square of the second
                    quotient.
```

um, and for 21. 16. in that collum, and not finding it there, take that which is nighest vnto it and lesse, which you shall find to be 21 and 14. and right against that in the outermost collum on the right hand standeth 49. which being too great to be your quotient by 2. therefore make your quotient no more but 47. which being set downe in the quotient line, multiply thereby 26.58. and the product will be 21. 7.26. which being subtracted out of 21. 16. and 8. the remainder will be 8. and 42. then setting 42. ouer 8. and 8. ouer 16. cancell 21.16. and 8. then multiply 47. into it selfe squarely, and you shall finde the product thereof in the Triangle to be 36. 49. which being subtracted out of 8. 42. and 0. the remainder will be 8.6.11. as you sée in this example.

And

The vſe of the Sexagenarie table.

And thus you haue found the roote of your giuen number to be 13.29 and 47. and whether you haue done rightly or not, you may know by multiply‐

```
    2′  8   6
 13.38.16.47.11
  3. 2. 9. 17. 8. 0.   (13,29′,47″.
  2.49.26.0.58.49
 12.34.26.26
  14. 7.36
   2′
```

ing the ſaid roote into it ſelfe ſquarely, and by adding to the pro‐
duct thereof the remainder, and if you would finde a more exact
roote, then adde more Cyphers to your giuen number towardes
your right hand, together with their denominations ouer their
heads as they encreaſe ſucceſſiuely, and be ſure alwaies that the
number of the Cyphers bee euen and not odde as 2, 4. 6. and ſo
forth, & the more Cyphers that you adde to the giuen number(ſo
as the number of them be euen) the more exact roote you ſhal haue.

Now to giue to euery number of the roote his proper denomi‐
nation, do thus, take halfe of thoſe denominations that are ouer
euery pricke, and thoſe ſhall be the denominations of the roote, as
in the former example, whereas ſeconds doe ſtand ouer the laſt
pricke on the left hand take halfe of that which is 1. wherefore the
firſt number of your roote muſt be minutes, and ouer the reſt of the
numbers of the roote ſet ſeconds, thirds and ſo forth
ſucceſſiuely vntill you haue brought your
roote to as ſmall denominations
as you deſire.

FINIS.

A BRIEFE DEscription of the tables

of the three speciall right lines belonging to a Circle, called Sines, lines Tangent, and lines Secant.

Not only shewing why they were first *inuented, and defining the proper tearmes thereto belonging:*

But also shewing diuers necessarie vses thereof, written by **Maister** *Blundeuile* 1593.

LONDON.
Printed by *Iohn Windet*, 1594.

THE DESCRIPTION
and vſe of the Table
of Synes.

Ecauſe there is no proportion, compariſon, or likenes betwixt a right line and a crooked, the auntient Philoſophers, as Ptolomey and diuers other, were much troubled in ſéeking to know the meaſures of a Circle or of any portion thereof by his Diameter, and by knowing the Diameter to find out the length of any chorde in a circle, which is alwaies leſſer then the Diameter it ſelfe, and finding that the more parts whereinto the Diameter was diuided, the nearer they approched to the truth: Some of them therefore, as Ptolomey, diuided the Diameter of a circle into 120. parts, and the Semidiameter into 60. parts, and every ſuch part into 60. and every minute into 60. ſeconds &c. And in like manner did Arzahel, an auntient Arabian, who diuided the Diameter into 300. parts and the Semidiameter into 150. and every of thoſe parts into 60. and ſo forth as before, according to which computation they made their tables: but becauſe the working by thoſe tables was very tedious and troubleſome, by reaſon that it was néedfull continually to vſe the art of numbring by Aſtronomicall Fractions: therefore Georgius Purbachius, and Regio montanus his Scholer to auoide that trouble of calculating by Aſtronomicall Fractions, diuided the Diameter of a Circle into a farre greater number of parts, and made ſuch tables as are vſed at this preſent, the deſcription and vſe whereof doth hereafter follow, firſt of thoſe that are ſet downe by Monte Regio in Folio, and then of thoſe that were lately Corrected and made perfect by Clauius the Ieſuite which are Printed in quarto.

And

The description and vse of

And because that the way to find out the proportion which any chord hath to the whole Diameter, was very hard, therefore the said Purbachius and Monte Regio hauing direction from certaine propositions of Euclyd, as from the 47. proposition of his first booke, and from the third proposition of his third booke, and also from the 15. proposition of his fift booke, they made choice of the halfe chord and Semidiameter of the Circle, calling the halfe chord, Sinum rectum, and the Semidiameter Sinum totum. And because that the proportion of any circumference to his Diameter neuer changeth, how great or how little so euer the Circle be: after that they had calculated for one Circle, they made such tables as might serue for all Circles, and though these tables of sines doe suffice to worke thereby all manner of conclusions, as well of Astronomie, as of Geometrie, yet for more ease, our moderne Geometricians haue of late inuented two other right lines belonging to a Circle called lines Tangent, and lines Secant, and haue made like tables for them that were made for sines, and both tables, that is to say as well of the sines, as of the lines Tangent and Secant, haue one selfe manner of working thereby, as shall plainely appeare hereafter when wee come to describe the same. But first we will beginne with the tables of sines, and plainely define euery terme or vocable of Art, belonging thereunto: The termes are these here following: An arch, a Chord, Sinus rectus Sinus versus, Quadrans, Complementum, and sinus Complementi.

The definitions of the foresaid tearmes.

AN Arch is any part or portion of the circumference of a Circle, which in this practise doth not commonly extend beyond 180. degrees which is one halfe of the circumference of any Circle how great or small so euer it be, for euery Circle containeth 360 degrees.

A Chord is a right line drawne from one end of the Arch to the other end thereof, and note that all chordes are alwaies lesser then the Diameter it selfe, for that is the greatest chord in any Circle.

Sinus rectus is the one halfe of a Chord or string of any Arke

which

the tables of Sines.   48

which is double to the Arke that is giuen or supposed, and falleth with right Angles vpon ẏ Semidiameter which diuideth the double Arke into two equall parts.

   Sinus versus that is to say turned the contrary way, is a right line, and that part of the Semidiamiter, which is intercepted betwixt the beginning of the giuen Arke and the right Sine of the same Arke, and this is also called in Latine Sagitta, in English a Shaft or Arrowe, for the Demonstratiue figure thereof hereafter following, is not vnlike to the string of a bowe ready bent hauing a Shaft in the midst thereof.

   Quadrans is the fourth part of a Circle containing 90. degrǽs.

   Complementum arcus, is that portion of the Circle, which sheweth how much the giuen Arke is lesser then the quadrant, if the giuen Arke doe containe fewer degrǽs then the Quadrant, but if it containe more degrǽs then the Quadrant, then the difference betwixt the quarter of the Circle and the saide arch, is the complement of the said giuen Arke.

   Sinus complementi, is the right Sine of that Arch which is the complement of the giuen Arke.

   Sinus totus, is the Semidiamiter of the Circle, & is the greatest Sine that may be in the Quadrant of a Circle, which according to the first tables of Monte Regio containeth 6/000/000. and according to the last tables 10/000/000 parts, for the more parts that the totall Sine hath, the more true and exact shall your worke be, notwithstanding sometime it shall suffice to attribute vnto the totall Sine but 60/000. parts, which number Appian obserueth in teaching the way to find out the distance of two places differing both in Longitude and Latitude by the tables of Sines, and some doe make the totall Sine to containe 100/000 parts, as Wittikindus in his treatise of Dials, and diuers other doe the like. Also Clauius himselfe saith that in the tables set downe by him in quarto, you may sometime make the totall Sine to be but 100/000. so as you cut off the two last figures on the right hand in euery Sine, but you shall better vnderstand euery thing here aboue mentioned, by the figure Demonstratiue here following.

                                                 The

The description and vse of

The figure Demonstratiue.

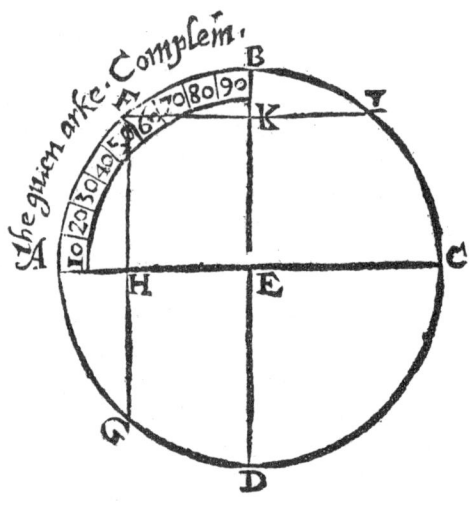

In this figure you ſee firſt a whol Circle drawne vp-on the Centre E. and marked with the letters A.B.C D. which Circle by two croſſe Di-ameters marked with the letters A C, and B, D, & paſ-ſing both through the Centre E. is di-uided into foure Quadrantes or quarters, the vp-per Quadrante whereof on the left hand is marked with the letters A. B. E. in which Quadrant, the right perpendicular line marked with the letters F. H. betokeneth the right Sine of the giuen Arke A. F. which right Sine is the one halfe of the chord or ſtring F. G. and the giuen Arke A. F. is the one halfe of the double Arke or bowe G. A. F. and A. H. is the Shaft called in Latine Sinus verſus: Againe the letters F. B. doe ſhew the complement which together with the giuen Arke A. F. doe make the whole Quadrant A. F. B which is diuided into 9. ſpaces, euery ſpace cōtaining 10. degrees, whereby you may plainely perceiue that in this demonſtration, the giuen Arke A. F. is 50. degrees, and the complement F. B. is 40. degrees, both which being added together doe make vp the whole Quadrant of 90. degrees, marked with the letters A. F. B. Now Sinus complementi is the croſſe line marked with the let-ters F. K. the totall Sine which is the whole Semidiameter and greateſt right Sine, is marked with the letters B. E. But becauſe it is not enough to know the ſignification of the things aboue ſpe-cified to vſe the foreſaid tables when nede is, vnleſſe you know

also

the tables of Sines.

also how to find out those things in the said tables, I thinke it good therefore to shew you the order of the said tables by describing the same as followeth.

You haue then to vnderstand that the tables of Monte Regio printed in Folio, are contained in 18. Pages, and euery Page containeth eleauen partitions, called collums, whereof the first on the left hand containeth 60. minutes, which are to be counted from head to foote, as they stand in order one right vnder another in seuerall places, procéeding from 1. to 60. The second collum containeth Sines. The third containeth onely a portion or part of one second, and from thence forth procéeding towards the right hand all the other collums doe contain in like manner Sines and the portion of one second. And right ouer the head of euery Sine (the first collum of Sines onely excepted, hauing nothing but a Cypher ouer his head) are set downe the degrées of the whole Quadrant called arches, in such order as from the first Page to the last, there are in all 89. degrées, or arches, as by perusing the said tables you may plainely sée. Now to find out in these tables the things aboue mentioned, you must doe as followeth.

First to find out the right Sine of any giuen Arke, you must séeke out the number of the said Arke in the front of the tables, and if the giuen Arke hath no minutes ioyned thereunto, then the first number of Sines right vnder the said Arke, is the right Sine thereof. But if it hath any minutes ioyned thereunto, then you must séeke out in that Page, where you found the giuen Arke, the number of the minutes in the first collum of the said Page, on the left hand, and right against those minutes on the right hand, in the square Angle right vnder the said arch, you shall find the right Sine. As for example, you would find out the right Sine of a giuen Arke containing 8. degrées, and 2'0. here hauing found out in the front of the second Page the figure of 8. standing right ouer the eight collum, séeke in the first collum on the left hand of the said Page, for 20. minutes, and right against the 20. minutes you shall find on the right hand in the common Angle or square 8695 93. which is the right Sine of the foresaid giuen Arke, so as you make 6/000/000. to be the totall Sine: but if you make 60/000. the totall Sine, then you must alwaies reiect the two last figures standing on the right hand of the said right Sine, and the rest of the figures shall be the right Sine.

The description and vse of

Now to find out the complement, there is nothing to be done, but onely to subtract the giuen Arke out of the whole Quadrant which is 90. degrees, and the remainder shall be the complement: as in the former example by subtracting 8. degrees, 2′0. out of 90 degrees, you find that there remaineth 81. degrees, 4′0. which is the complement of that arch. Againe to find out the Sine of the complement you must doe thus, seeke the complement in the front of the tables of Sines, euen as you doe to find out any giuen arke: as in the former example, the complement being 81. degrees 4′0. you must seeke 81. in the front of the 17. Page of the first tables, which being found, seeke out also the 4′0. in the first collum of the said Page on the left hand, and right against those 4′0. in the common Angle right vnder the Arke 81. you shall find 5/936/649. which number is the right Sine of the foresaid complement, so as you make 6/000/000. to be the totall Sine, for if 60/000. be the total Sine, then you must reiect (as I said before) the two last figures on the right hand, and the number remaining shall be the right Sine of the foresaid complement, and therefore in working by these tables, you must alwaies remember what number you make the totall Sine to be.

Sinus versus commeth seldome in vse, notwithstanding if you would know how to find it out, you neede to doe no more but to subtract Sinum complementi of the giuen Arke, out of the totall Sine and the remainder shall be Sinus versus, as in the former example your Sinus complementi was 5/936/649. which being subtracted out of the totall Sine 6/000/000. there remaineth this 63/351. and that number is Sinus versus: for if you adde this remainder to the number which you subtracted, it will make vp the totall Sine 6/000/000. But there is one thing more necessarie to be knowen then this, because it commeth oftner in vse, and that is vpon some diuision made how to find out the Arke of any quotient, which is to be done thus: Enter with the quotient into the body of the tables, and leaue not seeking amongst the squares of the Sines, vntill you haue found out the iust number of the quotient (if it be there) if not, you must take the number of that Sine which is in value most nigh vnto it, whether it be a little more or lesse, it maketh no matter, and hauing found that number, looke in the front of that collum, and you shall find the Arke of your quotient,

the tables of Sines.   50

ent, standing right ouer the head of that collum, and also the mynutes thereof in the first collum of the said Page on the left hand.
As for example, hauing diuided one number by another, I find
the quotient to be 469/012. whereof I would know the arch, now
in seeking this quotient amongst the Sines, I can not find that
iust number, but I find in the first Page, and in the tenth collum
469/015. which is the nighest number vnto it that I can see. In
the front of which collum I find the Arke to be 4. degrees, and directly against that Sine on the left hand, I find 2'9. belonging to
that arch, whereof that quotient is the Sinus, so as I gather hereof that the arch of the foresaid quotient is 4. degrees, 2'9. But you
haue to note by the way that the number of your quotient must neuer be much lesse then 1745. for otherwise it is not to be found in
these tables, vnlesse you make the totall Sine to be but 60/000.
for then by reiecting the last two figures on the right hand, as I
haue said before, the first right Sine of these tables shal be no more
but 17. and by that account a very small quotient may be found in
these tables. And whatsoeuer hath beene said here touching the order that is to be obserued in the first tables of Monte Regio, whose
totall Sine is 6/000/000. the like in all points is to be obserued
in the last tables, whose totall Sine is 10/000/000. Thus much
touching the order of the foresaid tables of Monte Regio Printed
in Folio: but for so much as those tables bee not altogether truely
Printed, and for that they haue beene lately corrected, and made
more perfect by Clauius, who doth set downe the said tables in
quarto and not in folio, whereby they are the more portable, and
the more commodious, as well for that they are more truely Printed, as also for that the complement of euery Arke is set downe in
euery Page at the foote of euery collum, so as you need to spend no
time in subtracting the Arke from 90. I thinke it good therefore
to make a briefe description of those tables, and the rather for
that I haue requested the Printer to print the like here in quarto,
and I doe worke all such conclusions as hereafter follow, by the
said tables, the totall Sine whereof is 10/000/000. according to
the last tables of Monte Regio. But for so much as some may
haue already the tables of Monte Regio Printed in Folio, not
knowing perhaps the vse thereof, I will set downe two conclusions to bee wrought by those tables, and all the rest of the conclusi-

H 2   ons

The description and vse of

ons are to be wrought by those tables which I haue here caused to be Printed in quarto like to those of Clauius: and though the two conclusions next following, which are to shew the vse of the foresaid tables, may be wrought by the tables of Sines in what forme so euer they be truely Printed in Folio, or in quarto, yet because I had appointed them to bee done by the tables of Monte Regio, Printed in folio before that euer I saw Clauius his booke, I mind not now to alter them but to let them stand still as they are.

How to find out by the said tables, the distance betwixt two places differing both in Longitude and Latitude, making the totall Sine to be no more but 60000.

This is done by finding out two numbers, whereof the one is called in Latine Primum inuentum: that is to say, the first found number, and the other is called Secundum inuentum, that is the second found number in such order as followeth.

First then knowing the Longitude of either place, take the differēce of their Longitudes by subtracting the lesser Longitude out of the greater, that done, multiply the right Sine of that difference into the Sine of the complement of the lesser Latitude, and diuide the product of that Multiplycation by the totall Sine, and then seeke out the arch of that quotient accoding to the rule before taught, so shall you haue the first found number: That done, multiply the right Sine of the lesser Latitude by the totall Sine, and hauing diuided the product thereof by the right Sine of the complement of the first found number, subtract the arch of that quotient out of the greater Latitude, and you shall haue the second found number: Then multiply the right Sine of the complement of the first found number into the right Sine of the complement of the second found number, and hauing diuided the product of that Multiplycation by the totall Sine, seeke the Arke of that quotient in the tables, and take that Arke out of the whole Quadrant, and the degrees that doe remaine, are degrees of the great Circle, which if you multiply by 60. the product of y Multiplycation will shew you how many Italian miles the one place is distant from the other, or if you would haue Germane miles, thē multiply the foresaid degrees of the great Circle by 15. or else diuide the product

of

the tables of Sines.   51

of the Italian miles by 4. and you shall haue your desire. As for example, you would know what distance is betwixt Hierusalem and Noremberg a famous towne in Germanie, Hierusalem according to Appian his tables, hath in Longitude 66. degrées o′4. and in Latitude 31. degrées, 4′0. Againe Noremberg hath in Longitude 28. degrées, 2′0. and in Latitude 49. degrées, 2′4. the difference of their Longitudes is 37. degrées, 4′0. the right Sine whereof is 36/664. for in this example Appian maketh 60/000. to be the total Sine, and therefore he reiecteth the two last figures on the right hand found in the first tables of Monte Regio. Now you must multiply 36/664. into the right Sine of the complement of the lesser Latitude which Sine is 51/067. the product of which two Sines being multiplyed the one by the other, amounteth to 1/872/320/488. which if you diuide by the totall Sine. 60/000. you shall find in the quotient 31/205. whose arch is 31. degrées, 2′0. and this shall be your first found number. This done, multiply the right Sine of the lesser Latitude which is 31/498. by the totall Sine 60/000. and the product thereof wil be 1/889/880/000. which summe if you diuide by the Sine of the complement of the first found number which Sine is 51/249. you shall find in the quotient 36876. the Arke whereof is 37. degrées 5′5. which arch being subtracted out of the greater Latitude, there will remaine 11. degrées, 2′9. and that shall be your second found number, then multiply the foresaid Sine of the complement of the first found number which is 51/249. by the Sine of the complement of the second found number which is 58/798. and the product thereof will amount to 3/013/338/702. which if you diuide by the totall Sine, you shall find in the quotient 50/222. the arch whereof is 56. degrées, 5′0. which being subtracted out of the whole Quadrant which is 90. degrées, there will remaine 33. degrées, 1′0. of the greater Circle, which 33. degrées, if you multiply by 60. it will make 1980. miles, whereunto you must adde for the 1′0. 10. miles, so shall you find the distance betwixt the two foresaid places to be 1990. Italian miles, which if you would reduce into Germaine miles, then diuide that number by 4. for 4. Italian miles doe make but one Germaine mile, so shall you haue 497. Germaine miles, and two Italian miles remaining, which is halfe a Germaine mile, which summe agréeth with that which

H 3   Appian

The description and vse of

Appian setteth downe in his Geographie, whereas he vseth the selfe same example, and worketh it in like manner Per tabulas sinuum.

The Altitude of the sunne being knowne how to find out the Longitude of the shadow both right and verse of any body yeelding shadow by helpe of the foresaid tables.

First you haue to vnderstand that euery bodily thing yeelding shadow, is diuided into 12. equall partes, and euery such part into 60. minuts, and euery minute into 60. seconds and so forth: Againe, of shadowes there be 2 kindes, that is, Vmbra recta, and Vmbra versa, Vmbra recta is that which proceedeth from some bodie rightly erected vpon the vpper face of the Horizon, as from some tower or post standing right vp vpon a leauel ground: And that shadowe is called vmbra versa which proceedeth from some right style or pearch being thrust into a wall or post standing right vp and not leaning, in such sort as the sayd style or pearch may be a iust paralell to the vpper face of the Horizon. Now to find out the length of either the foresaid shadowes, you must doe thus.

Multiply the right Sine of the complement of the giuen Solar altitude, by 12. and diuide the product by the right Sine of the said Solar altitude, and you shall haue the Longitude of the right shadow of the said body. Againe if you multiply the right Sine of the foresaid altitude by 12. and diuide the product by the Sine of the complement of the said altitude, you shall haue the Longitude of Vmbra versa, of the said body. As for example, suppose the giuen Solar altitude to be 25. degrees, the complement whereof is 65. and the right Sine of that complement is 54/378. If you make the totall Sine to be 60/000. Then in multiplying the foresaid right Sine by 12. the product will be 652/536. which if you diuide by the Sine of the altitude which is 25/357. you shall find the Longitude of Vmbra recta to be 25. parts, 44'. 4" 4/2 and "'/6. Now if you multiply the Sine of the altitude which is 25/357. by 12. and diuide the product by the Sine of the complement which is 54/378. you shall find the Longitude of Vmbra versa to bee 5. parts, 23' 5. and in saying here parts, I meane alwaies such parts

as

as are the 12. parts, whereinto the body yeelding shadow is diuided: but if you worke this example by the first tables of Sines making the totall Sine 6/000/000. though you finde it true in the parts and minutes, yet shall you not find it so in the seconds and thirds, and if you worke the same by the second tables making the totall Sine 10/000/000. you shal find it to agrée only in the parts, but neither in minutes nor secondes, which maketh me to suspect that the Printer hath committed some error therein, for both the tables were made by one selfe rule.

A briefe description of the tables of Sines Printed in quarto like vnto those which *Clauius* setteth down in his commentaries vpon *Theodasius* his Spheriques.

Hauing heere before plainely described vnto you the tables of Sines made by Monte Regio, which are Printed in folio, and how to vse the same I will now briefly describe the said tables lately corrected by Clauius, and were Printed in quarto at Rome Anno 1586. the totall Sine of which tables according to the last table of Monte Regio, is 10/000/000. by which tables are to be wrought all the conclusions hereafter following

First then you haue to vnderstand that these tables are contained in 36. Pages, in the front whereof are set downe the degrées of the Quadrant procéeding from 1. to 89. but because the whole number of minutes belonging to the said degrées, which is 6'0. can not be all placed in one selfe Page, but onely 3'0. in the left outermost collum of the left hand Page: and other 3'0. in the left outermost collum of the right hand Page: therefore the degrées or arches of the Quadrant are faine to be twice repeated in the front of euery two Pages, as you may plainely sée by viewing the said tables, and euery Page containeth seauen collums, whereof the first on the left hand containeth the minutes belonging to the degrées or arches of the Quadrant, which minutes do procéed downward from 1. to 3'0. and the seauenth collum on the right hand in euery Page containeth the minutes belonging to the complement of euery arch, which minutes doe procéede backward, that is to say from 1. set downe at the lowest end of the last collum of the second Page, and so procéeding vpward to 3 0. which 3'0. is also set

P 4   downe

The description and vse of

downe at the lowest end of the last collum of the left hand Page, and so procædeth vpward to 6′0. which 60. minutes doe helpe to make vp the complement that is answerable to euery arch whereunto no minutes be annexed, for if the arch hath no minutes, then you must adde to the complement thereof 60. minutes, which is one whole degrée, to make vp the complement. As for example, suppose the Arke to be 46. degrées, without any minutes ioyned thereunto, the complement whereof set downe at the foote of the said arch, is but 43. degrées, wherefore you must adde thereunto 6′0. which is one degrée, so shall the complement be 44. degrées, which is the true complement indéede, but if you suppose the foresaid Arke to haue 1′3. ioyned thereunto you shall find the complement to be 43. degrées, 4′7. which is answerable to the foresaid Arke 46. degrées, and 1′3. for if you take 46. degrées, 1′3. out of 90. degrées, the remainder will be 43. degrées, 4′7. which is the complement, so as you néed not to make any Subtraction out of 90. to find the complement of any arch, that hath any minutes annexed thereunto: but when so euer you haue to find out the right Sine of any complement in these tables, you must then make the complement an arch, séeking for the same in the front and not at the foote of the tables, & if the said complement haue any minutes annexed thereunto, you must séeke those minutes in the left outermost collum of euery Page, and not in the outermost right collum belonging to complements, for in this case the complement is an arch and not a complement. The order of working by these tables in all other things differeth not one iotte from that which we haue obserued in working the two former conclusions by the tables of *Monte Regio* Printed in folio, as you shall easily perceiue by the examples here following.

1 How to find out the declination of the Sunne at any time his place in the Zodiaque being giuen, *Per tabulas Sinunm.*

Knowing the place of the Sun for the day, consider how much the said place is distant from the first point of Aries if the place of the Sunne be nigher to Aries then to Libra. But if it be nigher to Libra, then take his distance from the first point of Libra, which distance must not excéede 90. degrées, and séeke that distance amongst

the tables of Sines.

mongst the Arkes in the front of the tables if they bee degrées, if minutes, you shall finde them in the first collum on the left hand, then multiply the Sine of that distance by the Sine of the greatest declination which is 23. degrées, & and divide the product thereof by the totall Sine, and the quotient will shew you the Sine of the declination for that day, the arch whereof, is the very number of the declination it selfe. As for example, you would know the declination of the Sunne the eleauenth of Aprill 1591. when as the Sunne was entred 3'3. into Taurus vnto which you must adde according to the rule before giuen 30. degrées, for that is his distance from the first point of Aries, then seeke out the said 30. degrées, in the front of the last tables among the Arkes and the 3'3 in the first collum on the left hand right against which you shall finde the Sine of the said distance to bee 5082/901. which being multiplyed by 3/982/155. (which is the Sine of the greatest declination) the product thereof will be 20/240/899/631/655. which if you diuide by the totall Sine which is 10/000/000. you shall finde the quotient to be 2/024/089. which quotient must be sought out in the said tables, and if you finde no such number, then take the nearest number thereunto, which is 2025025. and the Arke thereof together with the minutes that stand right against the said quotient in the first collum on the left hand, is the declination of the Sunne for that day which is 11. degrées, 4'1.

2 How to know the right ascention of the Sunne, *Per tabulas Sinuum.*

First knowing the Sunnes place, you shall learne the right ascention thereof thus. First consider how farre his place is from the Equinoctiall point, as is said before in the last proposition, and multiply the Sine of the complement of that distance by the totall Sine, then knowing the declination of the Sunne for that day by the last rule, deuide the former product by the Sine of the complement of the said declination, and the quotient will shew the Sine of that Arke whereof the complement is the right ascention. As for example, the Sunne being in the 3'3. of Taurus and his declination 11 degrées, 4'1. as you found by the last proposition and his distance from the first point of Aries to be in all 30. degrées, 3'3

the

The description and vse of

the complement whereof is 59. degrées, 2′2, the Sine of which complement is 8/611/860. which being multiplyed by the totall Sine which is 10/000/000. the product will be 86/118/600/000/000. which being diuided by the Sine of the complement of the Sunnes declination which is 78. degrées, 19. whose Sine is 9792818. by which Sine if you diuide the former product, the quotient will be 8/773/600. the Arke of which quotient is 61. degrées, and 1′9. and the complement of that, is 28. degrées, and 4′1. which is the right ascention of the place of the Sunne for the foresaid day.

And here note that if the Arke from Aries to the giuen pointe, doe containe iust 90. degrées, the right ascention thereof is also 90. degrées, but if the said Arke be more then 90. degrées, and lesser then 180. degrées, then subtract the same out of 180. and sæke out the right ascention of the remainder as before, whose right ascention if you take out of 180. the remainder shall be the ascention of the propounded Arke. But if the said arch which is comprehended betwixt Aries and the giuen pointe bee greater then 180. degrées, or 6. signes, and lesser then 270. degrées, or 9. signes, then hauing subtracted 180. degrées from the same arch, calculate the right ascention of the arch from the beginning of Libra, as before, and the right ascention thereof being added to 180. degrées, shall be the ascention desired.

Lastly if the arch comprehended betwixt Aries and the pointe giuen, be greater then 270. degrées, or 9. signes, then subtract the same out of 360. degrées, and sæke out the right ascention of that remainder as before, which if you subtract out of 360. degrées, the remainder shall be the right ascention desired.

3 How to finde out the ascentionall difference, *Per tabulas Sinuum.*

Multiply the Sine of the Latitude giuen by the total Sine, and diuide the product by the complement of ÿ said Latitude, that done, multiply the quotient by the Sine of the Sunnes declinatiō, and diuide the product by the Sine of the complement of the declination, and the quotient thereof wil shew the signe of the ascentionall difference: and by working according to this rule, you shal find that

the tables of Sines. 54

that when the Sunne is entred 3′3. into Taurus, at which time his declination is 11. degrees, 4′1. (as hath beene said before) the ascentionall difference will be 15. degrees, 2′1. And here note that the ascentionall differences for one quarter of ye Circle serueth also for all the rest, so that the Latitude be not altered, and that the declination of the pointes in the later quarters be equall to the declination of the points in the first quarter.

4. How to find out the oblique ascention of any pointe of the Ecliptique in any Latitude asigned.

First find the right ascention of the giuen pointe by the second proposition, & also the ascentionall difference thereof by the third proposition: then consider whether the declination of the said point be Northward or Southward, for if it be Northward, then subtract the ascentionall difference out of the right ascention, and the remainder shall be the oblique ascention desired: but if the declination be Southward, then adde the ascentionall difference vnto the right ascention, and the summe shall be the oblique ascention. In working thus for the Latitude 52. and supposing the Sunne to be in ♉ 3′3. of Taurus, & hauing found by the former proposition his right ascention to be 28. degrees, 4′1. and the ascentionall difference to be 15 degrees, 2 1. you shall find his oblique ascention in the foresaid Latitude to be 13. degrees, 2 0.

5. How to finde out the time of the Sunnes rising and set and thereby the length of the artificial day.

First you must know the ascentionall difference, and conuert the same into howers & minutes, then if the Sunne be in any of the Northern signes, adde those howers to 6. howers which is the one halfe of an Equinoctiall day, and the summe of such Addition shall be the one halfe of the artificial day, which being subtracted out of 12. howers, the remainder shal be the hower of the Sunnes rising: As for example, the Sunne being in the 3′3. of Taurus, you found the ascentionall difference to be 15. degrees, and 2′1. which being turned into howers maketh one hower 1/4. and somewhat more, which being added to 6. maketh 7. howers 1/4. and a little more

which

The description and vse of &c

which is the one halfe of the artificiall day which being doubled, maketh in all 14. howers 2′. and 4″ but if the Sunne be in any of the Southerne signes, you must remember to subtract ẏ howers of the ascentionall difference out of 6. howers, and the remainder shall be the one halfe of the artificiall day, and by subtracting the halfe length of the artificiall day out of 12. you shall know the hower of the Sunnes rising, and hauing the time of his rising, you must néeds know the time of his setting.

6 How to find out the Meridian altitude of the Sunne in any day though he doth not shine at all, the eleuation of the Pole being giuen.

Subtract the eleuation of the pole from 90. and the remainder shall be the eleuation of the Equinoctiall, then if the Sunne be in any of the Northern signes, adde the declination of the Sunne, vnto the altitude of the Equinoctiall, or else if he be in any of the Southern signes, subtract the declination, and the summe of the Addition, or remainder of the Subtraction, shal shew the Meridiã altitude. As for example, the second of May 1591. the Sunne being in the 20. degrées, 4′2. of Taurus, and his declination Northward 17. degrées, 5′6. 2″1. here by subtracting 52. degrées, which is our Latitude from 90. you shall find the remainder to be 38. which is the altitude of the Equinoctiall, whereunto if you adde the Sunnes declination for that day which is 17. degrées, 5′6. and 2″1. the summe will bee 55. degrées, 5′6. and 2″1. and that is the Meridian altitude of the Sunne for that day.

  Let these few conclusions serue to shew you the vse of the tables of Sines, for it would make a long booke to set downe so many conclusions as are to be wrought by these tables, and therefore I leaue to trouble you any further therewith, minding now briefly to declare vnto you the vse of the tables of lines Tangent & Secant by one example or two as followeth. But first I thinke it necessary to shew you what those lines be, and whereto they serue.

THE

THE DESCRIPTION
and vſe of the Tables of Tan-
gents and Secants.

Vclid in the ſecond propoſition of his third booke defineth the line Tangent in this ſort. A right line (ſaith hee) is ſaid to touch a Circle when it toucheth it ſo, as being drawne out in length, it would neuer cut the ſaid Circle.

The line Secant is not by him any wher defined, but what theſe two lines are, you ſhal better vnderſtand by this figure Demonſtratiue here following, then by any definition that can be made thereof: for a definition ought to bee plaine and briefe, and not long, intricate or doubtfull, which will be hardly performed in ſhewing the nature of theſe two lines by way of definition, and therefore marke well this figure following.

In this figure you ſée firſt a Circle drawne vpon the Centre C. from which Centre is extended to the circumference of the Circle a right line, called the Semidiameter, marked with the letters A. C. then there is another right line 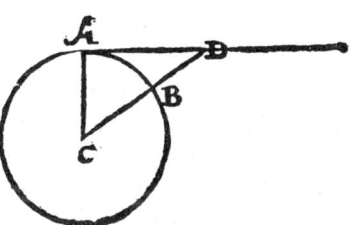 which toucheth the ſaid Circle, and alſo the outermoſt end of the ſaid Semidiameter making therewith a right Angle in the point A. and is called the line Tangent, then there is a third line which procéeding from the Centre C. doth cut the circumference of the Circle in the point B. and alſo méeteth with the line Tangent in the point D. and therefore is called the line Secant, betwixt which two lines, I meane the Tangent and Secant, is intercepted or included a certaine portion or arch of the foreſaide Circle,

The description and vse of

cle lesse then a Quadrant marked with the letters A, B, of which Arke the line A. D. is the Tangent, and the line C. D. is the Secant thereof, which must needs meete with the Tangent in the point D. because that the two Angles C, A, D. and D.C, A, are lesser then two right Angles, for the one is right, and the other sharpe, by reason that the Arke is lesse then a Quadrant. And some doe call the line Tangent the line Ascript, because it is ascribed to the Circle, and they call the line Secant the Hipothenuse, because it subtendeth the right Angle A. & they cal the Semidiameter or total Sine, y base of the rectangle Triangle C. A. D which is called a rectangle Triangle because it containeth one right Angle marked with the letter A. and note y whensoeuer any manner of Angle is propounded by three letters, that the middle letter doth alwaies signifie the Angle propounded, be it right, sharpe, or blunt. Now if you would know to what end the foresaid two lines were inuented, & wherto they serue, you haue to vnderstand that they chiefly serue in calculating the quantitie of Angles and their sides, as well in right lined Triangles, as in sphericall Triangles, for the sides of Triangles are either right or crooked, and if they haue right sides, such Triangles are either right angled Triangles, or oblique angled Triangles, and you haue to note that the quantity of euery Angle is to be measured by the arch of a Circle subtending that Angle, for the point of euery Angle is imagined to be the Centre of a whole Circle, which you may suppose to be so great or little as you will, for euery Circle (be it great or little) is diuided into 360. degrees, and looke how many degrees and minutes the arch subtending that Angle containeth, so much is the quantitie of that Angle, the practise whereof is very well set down by Clauius in his comentaries vpon Theodosius, which I mind (God willing) hereafter to translate into our mother tongue: In the meane time my intention here, is onely to shew you by one example or two, the vse of the tables made for the foresaid lines Tangent and Secant.

The vse of the said tables according to *Clauius*, is thus.

IN seeking out the Tangent or Secant of any Arke giuen, or of the complement of any Arke by either of these tables, you haue to observe

the tables of Tangents and Secants. 56

obserue the selfe same order which you did before in finding out the right Sine of any Arke giuen, or of the complement of any Arke, Per tabulas sinuum. As for example, if you would find out the Tangent of an Arke containing 50. degrées, 2′4. then resorte to the table of Tangents in the front, whereof looke first for the Arke 50. degrées, and then in the first collum on the left hand of the said table, for 2′4. right against which on the right hand vnder the Arke 50. you shall find in the comon Angle the Tangent to be 12/087/923. the totall Sine whereof is 10/000/000. but if you would find out the Secant of the foresaid Arke 50. and 2′4. then you must resort to the table of Secants, and hauing found out the Arke 50. in the front of the said tables, and the 2′4. on the left hand as before, you shall find in the common Angle the Secant to be 156881 44. And if you would haue the Tangent of the complement of the said Arke which is 39. degrées, and 3′6. you shall find the 39. degrées of the coplement in the foote of the table of Tangents right vnder the Arke 50. & the 3′6.. in the outermost collum on the right hand of the said table, with which complement you must enter the table of Tangents, séeking for 39. degrées in the front of the table, and 3′6. in the first collum on the left hand of the said table, right against which in the common Angle you shall find 8272720. to be the Tangent of 39. degrées, 3′6. which is the complement of 50 degrées, 2′4.

And you must work in like manner with the table of Secants: As for example, if you would find the Secant of 72. degrées, 3′6. first then enter the table of Secants, looking for 72. degrees aboue in the front of the table, and 3′6. in the first collum on the left hand of the said table, and in the common Angle you shall find 33/440/240. which is the Secant of 72. degrees, 3′6. But if you would haue the Secant of the complement of the said arch 72. degrees, 3′6. then looking in the foote of the table right vnder 72. degrees, you shall find 17. degrees, and in the outermost collum on the right hand, iust against 3′6. you shall find 2′4. so as you see that 17. degrees 2′4. is the complement of 72. degrees, 3′6. with which complement you must enter the table of Secants, looking for 17. degrees, aboue in the front of the table, and for 2′4. in the first collum on the left hand of the said tables, & in the common Angle you shall find 10/479/542. to bee the Secant of the arch 17.

degrées,

The description and vse of

degrees, 2′4. which is the complement of the arch 72. degrees, 3′6.

The vse of which tables in Astronomi-call matters, I haue here set downe
as followeth.

1. To find out the declination of the Sunne, the place thereof being knowne.

Vltiply the Secant of the complement of the greatest declination by the totall sine, and diuide the product by the sine of the sunnes distance from one of the equinoctiall points, the quotient is the Secant of an arch, whose complement is the declination of the sunne: for example, suppose that the sunne be entering into ♉ to finde the declination therof, first I multiply 25112030 the Secant of 66. degrees 3′2. (which is the complement of 23. degrees 2′8 the greatest declination) by 10000000. the product is 251120300000000. which being diuided by 5000000. (the sine of 30. degrees the sunnes distance from the equinoctial point) the quotient is 50224060. for which number I seeke in the table of Secants, the arch answering vnto it, is 78. degrees 3′2. the complement whereof is 11. degrees 2′9. which is the declination of the sunne.

2 Knowing the declination of the sunne how to finde his distance from the equinoctiall point, and so consequently his place in the Zodiacke.

M Vltiply the Secant of the complement of the greatest declination by the totall sine, and deuide the product by the Secant of the complement of the declination giuen, the quotient is the distance of the sun from the equinoctiall point. As for example, the declination of the sunne is supposed to be 11. degrees 2′9. then to

finde

the tables of Tangents and Secants.   57

finde his distance from the equinoctiall, I multiply 25112030. the Secant of 66. degrées 3'2. (which is the complement of the greatest declination) by the totall sine the product is 251120 300000000. which I diuide by 50224350. the Secant of 78. degrées 3'1 the complement of 11. degrées 2'9. the supposed declination, the quotient is 5000000. the sine whereof is 30. degrées degrées 0' which is the distance of the sunne from the equinoctial: thē for his place you must take the same accoding to ſeaſō of the yeare: for if it be in Aprill, then the sunne is entering into Taurus, but if it be in August it is entering into Virgo, and being in October, it is entering into Scorpio, and being in February it is in the beginning of Pisces.

3   To finde out the right ascention of the sunne.

Multiply the Tangent of the distance of the sunne from the equinoctiall point which is nearest vnto it, by the sine of the complement of the greatest declination, and diuide the product by the totall sine, the quotient is the Tangent of the right ascention of the sunne, for which if you seeke in the table of Tangents, the arch answering vnto it is your desire: For example the sunne being in the first of Taurus, to knowe the right ascention thereof, I multiply 5773502. the Tangent of 30. degrées (for that 30. degrées is the distance of the sunne from the equinoctiall point) by 9172920. the sine of 66. degrées 3'2. which is the complemēt of 23 degrées 2'8. the greatest declination, the product is 52959871965840. which being deuided by 10000000. the totall sine, the quotient is 5295987. which is the Tangent of 27. degrées, 5'4. which is the right ascention of the sunne bǽing entered into Taurus.

4   How to finde out the declination of the sunne knowing onely the right ascention thereof.

Multiply the Tangent of the complement of the greatest declination by the total sine, and deuide the product by the sine of the right ascention giuen, the quotient is the Tangent of the complement of the sunnes declination: for example the right
I   ascention

The defcription and vfe of

afcention of the funne being 27. degrees 5'4. I would knowe the declination thereof, multiplying 23035062. the Tangent of 66. degrees 3'2. the complement of 23. degrees 2'8 by 10000000. the product is 230350620000000. which being deuided by 4679298. the fine of the giuen afcention the quotient is 49224856. the arch of which Tangent is 78. degrees 3'2. which being fubducted out of 90. the remainder is 11. degrees 2'8. and fo much is the declination of the funne.

5 How to finde the place of the funne knowing onely the right afcention thereof.

Subduct the right afcention giuen out of 90. if it bee leffe then 90. but if the fame be more then 90. fubtract the afcention giuen out of 180. & being greater then 180. fubduct the fame out of 270. or being greater then 270. fubduct the fame from 360. and multiply the Tangent of the remainder by the fine of the complement of the greateft declination, and deuide the product by the totall fine, the quotient is the Tangent of the complement of the funs diftance from one of the Equinoctiall points: which diftance being knowne, the place of the funne can not be vnknowne. For example, fuppofing the right afcention of the funne to be 27. degrees 5'4. the complement thereof is 62. degrees 6'. the Tangent whereof is 18886715. which being multiplyed by 9172920. the product is 173245985757800. which being deuided by 10000000. the quotient is 17324598. for which I looke in the table of Tangents, and I finde the arch thereof to be 60. the complement whereof is 30. which is the diftance of the funne from the equinoctiall point, that is, from Aries, for that the right afcention is leffe then 90. I fay then that the funne is in the beginning of Taurus. But if the right afcention had béene more then 90. and leffe then 180. the place of the funne had béene betwixt Cancer and Libra 30. degrees from Libra, fo fhould it haue béene in the firft point of Virgo, but if the right afcention had béene more then 180. the place of the funne fhould be betwixt Libra and Capricorne, that is in the beginning of Scorpio, but being more then 270. the place of the funne fhould be betwixt Capricorne and Aries, that is in the beginning of Pifces.

6   To

the tables of Tangents and Secants.   58

6 To find out the ascentionall difference of the sunne or any starre in the firmament, knowing the declination thereof, and also the latitude of your region.

Multiply the Tangent of the declination of the Sunne or star by the Tangent of the latitude of the place, and deuide the product by the totall sine: the sine of the quotient is the sine of the ascentionall difference, the arch wherof is the desired ascentionall difference. For example let the declination of the sunne, starre, or other point in the firmament be 10. degrees 3'. and suppose the latitude to be 52. degrees, the Tangent of 10. degrees 3'. is 1772268. the tangent of 52. degrees is 12799416. which being multiplyed together, the product will be 22674995395488. which being deuided by 10000000. the totall sine, the quotient is 2267499. for which number I looke amongst the sines, the arch answering thereunto is 13. degrees 6'. which is the ascentionall difference desired.

7 To find out the oblique ascention of the sunne.

Knowing the place of the sunne, finde out the right ascention of the same by the thirde proposition, and finde also the ascentionall difference of the same point, then if the declination of the sunne be North, subduct the ascentionall difference out of the right ascention, the remainder is the oblique ascention. For example, suppose the sunne to be in the beginning of Taurus, now to finde the oblique ascention thereof in the latitude 52. degrees, first by the third proposition I finde out the right ascention of the beginning of Taurus, which I finde to be 27. degrees 5' 4. then by the sixth proposition I finde the ascentionall difference to bee 15. degrees 4'. which being subducted from 27 degrees 5' 4. (for that the declination is North) the right ascention the remainder is 12. degrees 5' 0. and so much is the oblique ascention for the latitude of 52. degrees. But if the sunne be in any of the Southerne sines, and that the declination bee South, then the ascentionall difference is to be added vnto the right ascention before giuen.

I 2   8 To

The description and vse of

8 To finde out the oblique descention of the Sunne at any time.

First finde out the right ascention of the sunne by the third proposition, the same shall be the right descention thereof, then finde the ascentionall difference by the sixt proposition, and if the sunne be in any of the Northern signes, adde the ascentional difference vnto the right ascention, the summe shall be the oblique descention of the sunne: But if it be in any of the Southerne signes subduct the ascentionall difference out of the right ascention, the remainder is the desired descention: as for example the sunne being in the beginning of Taurus, the right ascention thereof by the third proposition is 27. degrees 5′ 4. the ascentional difference thereof by the sixt proposition is 15. degrees 4′. for as much then as Taurus is a Northerne signe, I adde the ascentionall difference vnto the right ascention, the summe is 42. degrees 5′ 8. and so much is the oblique descention of the sunne being in the beginning of Taurus.

9 To finde out the length of the day or night.

Hauing found out the ascentionall difference by the sixth proposition, adde the same vnto 90. if the sunne be in any of the Northerne signes, but if it bee in any of the Southerne signes, subduct the ascentionall difference out of 90. then diuide ye summe of the addition or the remainder of the subtraction by 15. the quotient wil shew the halfe length of the day in houres and minutes, which being doubled you shall haue the whole length of the artificiall day: For example the sunne being in the beginning of Taurus, the ascentionall difference thereof is 15. degrees 4′. which for that the sunne is in a Northerne signe, I adde vnto 90. degrees, the summe is 105. degrees 4′ which being diuided by 15. the quotient is 7. houres 0′. the halfe length of the day which being doubled will be 14. houres the whole length of the artificiall day in the latitude of 52. degrees.

10 To finde the houre of the Sunne his rising or setting in any latitude assigned.

First find out the halfe length of the artificial day by the 9. proposition, and subtract the same from 12. houres, the remainder will

the tables of Tangents and Secants. 59

will shew the houre of the sunnes rising, for example the sunne being in the beginning of Taurus to know the houre of his rising in the latitude of 52.degrees, by the ninth propositiō, the half length of the artificiall day I find to be 7.houres, which I subtract from 12.houres the remainder is 5.which sheweth that the sunne riseth at 5.of the clocke in the morning, but the halfe length of the day it selfe is the houre of the sunnes setting.

11 To find out the length of the planetary houres and to find what Planet raigneth at any houre of the day.

First find out the length of the artificiall day by the ninth proposition, and diuide the same by 12. the quotient is the length of one planetary houre: or thus, hauing an houre of the artificiall day giuen, looke what houre the same is from the sunne rising, and multiply the same by 12. diuide the product by the length of the artificiall day, the quotient is the number of the Planetarye houre. For example the sunne being in the beginning of Taurus, and our latitude being 52. degrees, the length of the artificiall day by the ninth proposition is 14. houres: then doe I diuide 14. by 12. the quotient is $1\frac{1}{6}$. that is 1. houre 1'0. the length of one planetary houre. But if an houre of the artificial day be giuen as ỳ I would know what planetary houre it is at 9.of the clock the sunne being in the first of Taurus in the latitude of 52. degrees, hauing found that the sunne riseth at fiue of the clocke by the tenth proposition, I see that 4.houres of the artificiall day are gone at nine of the clocke, I therefore multiply 12. by 4. the product is 48. which I diuide by 14. the length of the artificiall day, the quotient is $3\frac{3}{7}$. which is the planetary houre at the time set downe: likewise shall you finde the planetary houre of the night, finding the length thereof, and then worke with it as was shewed before for the day.

Thē to know the Planet which raigneth at the appointed time, you must consider the Planet whereof the day taketh his name, for that Planet ruleth the first houre of that day, the next Planet the second houre, and so foorth. But for your ease I haue set down a table whereby to finde the Planet which raigneth at any time.

I 3   Houres

The description and vse of

Houres	1	2	3	4	5	6	7
of the day.	8	9	10	11	12		
Houres	10	11	12			1	2
of the night.	3	4	5	6	7	8	9
Sunday.	☉	♀	☿	☾	♄	♃	♂
Munday.	☾	♄	♃	♂	☉	♀	☿
Tuesday.	♂	☉	♀	☿	☾	♄	♃
Wednesday	☿	☾	♄	♃	♂	☉	♀
Thurseday.	♃	♂	☉	♀	☿	☾	♄
Friday.	♀	☿	☾	♄	♃	♂	☉
Saturday.	♄	♃	♂	☉	♀	☿	☾

The vse whereof is this, hauing founde the number of the Planetary houre, looke for the same in the head of the Table, whether it be in the day or night, and right vnder it iust against the name of the day, you shall haue the Planet which rayneth at that time. For example the sunne being in the beginning of *T* aurus the 11. of April, being Thursday at nine of the clocke in the morning, I find the number of the Planetary houre at that time to be 3 ½. then looking for three amongst the houres of the day, I descend in that collum vntill I come to be iust against thursday, where I sée the Character of Sol to be, I conclude then that the sunne raigneth at that time.

12 To finde the arch of the Equinoctial, comprehended betwixt the Meridian and any Circle of position according vnto Campanus and Gazula.

Multiply the Sine of the complement of your latitude by the Tangent of the distance of the giuen Circle of position from the Zenith, & diuide the product by the totall Sine, the quotient is the tangent of the arch of the Equinoctiall, which is comprehended betwixt the giuen Circle of position, and the Meridian. For example in the latitude of 52. I would knowe what part of the Equinoctiall is comprehended betwixt the Meridian and that Circle of position, which is 30. degrées from the Zenith: the latitude being 52. degrées, the complement thereof is 38. degrées, the

the tables of Tangents and Secants.   60

the Sine whereof is 6156615. and the tangent of 30. degrees, (for that 30. is the distance betwixt the giuen Circle of position and the Zenith) is 5773502. which being multiplyed by 6156615. the product is 35545229015730. which being diuided by 10000000. the totall Sine, the quotient is 3554522. for which I sæke amongst the tangents, and I finde the Arke answering thereunto, to be 19. degrǽs 3'4 the Arch of the Equinoctiall, betwixt the Meridian and the giuen Circle of position.

13 Knowing the Latitude of your Region, and also the eleuation of the Pole aboue any Circle of position, howe to finde the inclination of the saide Circle of position vnto the Meridian, and so consequently the Arch of the Equinoctiall, which is betwixt the said Circle of position & the Meridian.

Multiply the Secant of the complement of the eleuation of the Pole aboue the Circle of position by the Sine of your Latitude, and diuide the product by the totall Sine, the quotient is the Secant of the complement of the inclination of the Circle of position vnto the Meridian, and that is the distance betwixt the Circle of Position and the Zenith, by helpe whereof you shall find the Arch of the Equinoctiall, betwixt the Circle of Position and the Meridian, as in the former proposition. As for example, suppose the eleuation of the Pole aboue a Circle of position, in our Latitude of 52. to be 23. degrǽs 1'2. Now to find out the Inclination of that Circle of Position vnto the Meridian, first I multiply 25384445. the Secant of 66. degrǽs 4'8. (for that is the complement of 23. degrǽs 1'2. the eleuation of the Pole aboue the Circle of position) by 7880108. the Sine of 52. the Latitude of our Region, the product is 200032168120060. which being diuided by 10000000. the quotient is 20003216. for which I looke in the Table of Secants, and the Arch therof is 60. degrǽs 0'. the complement whereof is 30. degrǽs 0. which is the Inclination of the Circle of position vnto the Meridian, or the distance of the Zenith from the sayde Circle, then to find the Arch of the Equinoctiall betwixt the sayd Circle of Position and the Miridian, repeate the worke of the former proposition.

I 4   14 To

The description and vse of

14 To finde out the Eleuation of the Pole aboue any assigned Circle of Position in any giuen Latitude.

Knowing the Inclination of the assigned Circle of Position vnto the Meridian, multiply the Secant of the complement thereof by the totall Sine, and diuide the product by the Sine of your Latitude, the quotient is the Secant of the complement of the Eleuation of the Pole aboue the giuen Circle of Position: as for example, suppose the inclination of a Circle of position to be 30. degrees 0'. Now to finde the eleuation of the Pole aboue the same for the Latitude of 52. degrees 0'. First take the complement of 30. degrees 0'. which is 60. degrees 0'. the Secant whereof is 20000000. which being multiplyed by the totall Sine, the product is 200000000000000. which being diuided by 7880108. the Sine of 52. the assigned Latitude the quotient is 25380362. for which I seeke in the Table of Secants, and the Arch answering thereunto I finde to be 66. degrees 4'8. the complement whereof is, 23. degrees 1'2. and so much is the eleuation of the Pole aboue the assigned Circle of Position in your Latitude.

And thus much shall be sufficient to haue beene spoken for the vse of the lines Tangent and Secant at this present, of whose ample and infinite vse, you shall haue further taste in our Directorie tables and Horologies, whereof a beginning is made, and shal be ended so soone as may be possible: in the meane time I shal desire the Reader fauourably to accept of these, vntill better leasure and more fit opportunitie will be offered. And now followe the Tables themselues.

HERE FOLLOWETH THE TABLE OF RIGHT SINES FOR EVERIE MINVTE OF THE QVADRANT, FIRST CALCVLATED BY *I. REGIO MONTANVS*, BVT NOW EXAMINED AND IN MANY PLACES CORRECTED AND AMENDED BY *CLAVIVS*.

The Table
The degrees of the Quadrant for right Sines

		0	1	2	3	4		
The minutes of the degrees of the Quadrant for right Sines of the Arches of the same Quadrant.	0	0000	174524	348995	523360	697565	60	The minutes of degrees of the Quadrant for right Sines of the cōplement of the Arches of the same Quadrāt.
	1	2909	177433	351902	526265	700467	59	
	2	5818	180341	354809	529170	703369	58	
	3	8727	183250	357716	532075	706270	57	
	4	11636	186158	360623	534980	709172	56	
	5	14544	189066	363530	537884	712073	55	
	6	17453	191975	366437	540789	714975	54	
	7	20362	194883	369344	543694	717876	53	
	8	23271	197792	372251	546598	720777	52	
	9	26180	200700	375158	549503	723678	51	
	10	29088	203608	378064	552407	726579	50	
	11	31997	206517	380971	555312	729480	49	
	12	34906	209425	383878	558216	732381	48	
	13	37815	212333	386785	561120	735282	47	
	14	40724	215241	389692	564024	738183	46	
	15	43632	218149	392598	566928	741084	45	
	16	46541	221057	395505	569832	743985	44	
	17	49450	223965	398412	572736	746886	43	
	18	52359	226873	401318	575640	749787	42	
	19	55268	229781	404225	578544	752688	41	
	20	58177	232689	407131	581448	755588	40	
	21	61086	235597	410038	584352	758489	39	
	22	63995	238505	412944	587256	761389	38	
	23	66904	241413	415851	590160	764290	37	
	24	69813	244321	418757	593064	767190	36	
	25	72721	247229	421663	595967	770090	35	
	26	75630	250137	424570	598871	772991	34	
	27	78539	253045	427476	601775	775891	33	
	28	81448	255953	430382	604678	778791	32	
	29	84357	258861	433288	607582	781691	31	
	30	87265	261769	436194	610485	784591	30	
		89	88	87	86	85		

The degrees of the Quadrant for right Sines

of Sines.

of the Arches of the same Quadrant.

		0	1	2	3	4		
The minutes of the degrees of the Quadrant for right Sines of the Arches of the same Quadrant.	30	87265	261769	436194	610485	784591	30	The minutes of the degrees of the Quadrant for right Sines of the complement of the same Quadrant.
	31	90174	264677	439100	613389	787491	29	
	32	93083	267585	442006	616292	790391	28	
	33	95992	270493	444912	619196	793291	27	
	34	98901	273401	447818	622099	796191	26	
	35	101809	276308	450724	625002	799090	25	
	36	104718	279216	453630	627905	801990	24	
	37	107627	282124	456536	630808	804889	23	
	38	110536	285032	459442	633711	807789	22	
	39	113445	287940	462348	636614	810688	21	
	40	116353	290847	465253	639517	813587	20	
	41	119262	293755	468159	642420	816486	19	
	42	122171	296663	471065	645323	819385	18	
	43	125079	299570	473970	648226	822284	17	
	44	127988	302478	476876	651129	825183	16	
	45	130896	305385	479781	654031	828082	15	
	46	133805	308293	482687	656934	830981	14	
	47	136714	311200	485592	659837	833880	13	
	48	139622	314108	488498	662739	836778	12	
	49	142531	317015	491403	665642	839677	11	
	50	145439	319922	494308	668544	842576	10	
	51	148348	322830	497214	671447	845474	9	
	52	151257	325737	500119	674349	848372	8	
	53	154165	328645	503024	677251	851271	7	
	54	157074	331552	505929	680153	854169	6	
	55	159982	334459	508834	683055	857067	5	
	56	162891	337367	511740	685957	859965	4	
	57	165799	340274	514645	688859	862863	3	
	58	168708	343181	517550	691761	865761	2	
	59	171616	346088	520455	694663	868659	1	
	60	174524	348995	523360	697565	871557	0	
		89	88	87	86	85		

of the complements of the Arches of the same Quadrant.

The Table
The degrees of the Quadrant for right Sines

	5	6	7	8	9	
0	871557	1045285	1218693	1391731	1564345	60
1	874455	1048178	1221580	1394612	1567218	59
2	877353	1051071	1224467	1397492	1570091	58
3	880250	1053964	1227354	1400373	1572964	57
4	883148	1056857	1230241	1403253	1575837	56
5	886045	1059749	1233128	1406133	1578709	55
6	888943	1062642	1236015	1409013	1581581	54
7	891840	1065534	1238901	1411893	1584453	53
8	894737	1068426	1241788	1414772	1587325	52
9	897634	1071318	1244674	1417652	1590197	51
10	900531	1074210	1247560	1420531	1593069	50
11	903428	1077102	1250446	1423410	1595941	49
12	906325	1079994	1253332	1426289	1598812	48
13	909222	1082886	1256218	1429168	1601684	47
14	912119	1085778	1259104	1432047	1604555	46
15	915016	1088669	1261990	1434926	1607426	45
16	917913	1091561	1264876	1437805	1610297	44
17	920809	1094452	1267761	1440684	1613168	43
18	923706	1097344	1270647	1443562	1616038	42
19	926602	1100235	1273532	1446441	1618909	41
20	929498	1103126	1276417	1449319	1621779	40
21	932395	1106017	1279302	1452197	1624649	39
22	935291	1108908	1282187	1455075	1627519	38
23	938187	1111799	1285072	1457953	1630389	37
24	941083	1114690	1287957	1460831	1633259	36
25	943979	1117580	1290841	1463708	1636129	35
26	946875	1120471	1293726	1466586	1638999	34
27	949771	1123361	1296610	1469463	1641868	33
28	952667	1126252	1299495	1472340	1644738	32
29	955563	1129142	1302378	1475217	1647607	31
30	958458	1132032	1305262	1478094	1650476	30
	84	83	82	81	80	

The degrees of the Quadrant for right Sines

of Sines.
of the Arches of the same Quadrant.

	5	6	7	8	9	
30	958458	1132032	1305262	1478094	1650470	30
31	961354	1134922	1308146	1480971	1653345	29
32	964249	1137812	1311030	1483848	1656214	28
33	967144	1140702	1313914	1486724	1659082	27
34	970039	1143592	1316798	1489601	1661951	26
35	972934	1146482	1319681	1492477	1664819	25
36	975829	1149372	1322564	1495353	1667687	24
37	978724	1152261	1325447	1498229	1670555	23
38	981619	1155151	1328330	1501105	1673423	22
39	984514	1158040	1331213	1503981	1676291	21
40	987408	1160929	1334096	1506857	1679159	20
41	990303	1163818	1336979	1509733	1682027	19
42	993198	1166707	1339862	1512608	1684894	18
43	996092	1169596	1342744	1515484	1687761	17
44	998987	1172485	1345627	1518359	1690628	16
45	1001881	1175374	1348509	1521234	1693495	15
46	1004775	1178263	1351392	1524109	1696362	14
47	1007669	1181151	1354274	1526984	1699229	13
48	1010563	1184040	1357156	1529859	1702095	12
49	1013457	1186928	1360038	1532734	1704962	11
50	1016351	1189816	1362920	1535608	1707828	10
51	1019245	1192704	1365802	1538482	1710694	9
52	1022139	1195592	1368683	1541356	1713560	8
53	1025032	1198480	1371564	1544230	1716426	7
54	1027926	1201368	1374446	1547104	1719292	6
55	1030819	1204255	1377327	1549978	1722157	5
56	1033713	1207143	1380208	1552852	1725022	4
57	1036606	1210031	1383089	1555725	1727887	3
58	1039499	1212918	1385970	1558599	1730752	2
59	1042392	1215806	1388851	1561472	1733617	1
60	1045285	1218693	1391731	1564345	1736482	0
	84	83	82	81	80	

of the complements of the Arches of the same Quadrant.

The Table

The degrees of the Quadrant for right Sines

	10	11	12	13	14	
0	1736482	1908090	2079117	2249511	2419219	60
1	1739347	1910945	2081962	2252345	2422041	59
2	1742211	1913800	2084807	2255179	2424863	58
3	1745075	1916655	2087652	2258013	2427685	57
4	1747939	1919510	2090497	2260847	2430507	56
5	1750803	1922365	2093342	2263680	2433329	55
6	1753667	1925220	2096186	2266513	2436150	54
7	1756531	1928074	2099030	2269346	2438971	53
8	1759394	1930928	2101874	2272179	2441792	52
9	1762258	1933782	2104718	2275012	2444613	51
10	1765121	1936636	2107562	2277844	2447434	50
11	1767984	1939490	2110405	2280676	2450254	49
12	1770847	1942344	2113248	2283508	2453074	48
13	1773710	1945197	2116091	2286340	2455894	47
14	1776573	1948050	2118934	2289172	2458714	46
15	1779435	1950903	2121777	2292004	2461533	45
16	1782298	1953756	2124620	2294835	2464352	44
17	1785160	1956609	2127462	2297666	2467171	43
18	1788022	1959462	2130304	2300497	2469990	42
19	1790884	1962314	2133146	2303328	2472809	41
20	1793746	1965166	2135988	2306159	2475628	40
21	1796608	1968018	2138830	2308989	2478446	39
22	1799469	1970870	2141671	2311819	2481264	38
23	1802331	1973722	2144512	2314649	2484082	37
24	1805192	1976574	2147353	2317479	2486900	36
25	1808053	1979425	2150194	2320309	2489717	35
26	1810914	1982276	2153035	2323138	2492534	34
27	1813774	1985127	2155876	2325967	2495351	33
28	1816634	1987978	2158716	2328796	2498168	32
29	1819495	1990829	2161556	2331625	2500984	31
30	1822355	1993679	2164396	2334454	2503800	30
	79	78	77	76	75	

The degrees of the Quadrant for right Sines

of Sines.

of the Arches of the same Quadrant.

	10	11	12	13	14	
30	1822355	1993679	2164396	2334454	2503800	30
31	1825215	1996530	2167236	2337282	2506616	29
32	1828075	1999380	2170076	2340110	2509432	28
33	1830935	2002230	2172916	2342938	2512248	27
34	1833795	2005080	2175755	2345766	2515064	26
35	1836654	2007930	2178594	2348594	2517879	25
36	1839513	2010780	2181433	2351421	2520694	24
37	1842372	2013629	2184272	2354248	2523509	23
38	1845231	2016478	2187111	2357075	2526324	22
39	1848090	2019327	2189949	2359902	2529138	21
40	1850949	2022176	2192787	2362729	2531952	20
41	1853808	2025025	2195625	2365555	2534766	19
42	1856666	2027874	2198463	2368381	2537580	18
43	1859524	2030722	2201300	2371207	2540393	17
44	1862382	2033570	2204137	2374033	2543206	16
45	1865240	2036418	2206974	2376859	2546019	15
46	1868098	2039266	2209811	2379684	2548832	14
47	1870956	2042114	2212648	2382509	2551645	13
48	1873813	2044962	2215485	2385334	2554458	12
49	1876670	2047809	2218322	2388159	2557270	11
50	1879527	2050656	2221158	2390983	2560082	10
51	1882384	2053503	2223994	2393808	2562894	9
52	1885241	2056350	2226830	2396632	2565706	8
53	1888098	2059197	2229666	2399456	2568517	7
54	1890954	2062043	2232502	2402280	2571328	6
55	1893810	2064889	2235337	2405104	2574139	5
56	1896666	2067735	2238172	2407927	2576950	4
57	1899522	2070581	2241007	2410750	2579760	3
58	1902378	2073427	2243842	2413573	2582570	2
59	1905234	2076272	2246677	2416396	2585380	1
60	1908090	2079117	2249511	2419219	2588190	0
	79	78	77	76	75	

of the complements of the Arches of the same Quadrant

The Table
The degrees of the Quadrant for right Sines

	15	16	17	18	19	
0	2588190	2756373	2923717	3090170	3255682	60
1	2591000	2759169	2926499	3092936	3258532	59
2	2593809	2761965	2929280	3095702	3261182	58
3	2596618	2764761	2932061	3098468	3263931	57
4	2599427	2767556	2934842	3101234	3266681	56
5	2602236	2770351	2937623	3103999	3269430	55
6	2605045	2773146	2940403	3106764	3272179	54
7	2607853	2775941	2943183	3109529	3274927	53
8	2610661	2778735	2945963	3112294	3277675	52
9	2613469	2781529	2948743	3115058	3280423	51
10	2616277	2784323	2951523	3117822	3283171	50
11	2619084	2787117	2954302	3120586	3285918	49
12	2621891	2789911	2957081	3123349	3288665	48
13	2624698	2792704	2959860	3126112	3291412	47
14	2627505	2795497	2962638	3128875	3294159	46
15	2630312	2798290	2965416	3131638	3296906	45
16	2633118	2801082	2968194	3134400	3299652	44
17	2635924	2803874	2970972	3137162	3302398	43
18	2638730	2806666	2973750	3139924	3305144	42
19	2641536	2809458	2976527	3142686	3307889	41
20	2644342	2812250	2979304	3145448	3310634	40
21	2647147	2815041	2982081	3148209	3313379	39
22	2649952	2817832	2984857	3150970	3316123	38
23	2652757	2820623	2987633	3153731	3318867	37
24	2655562	2823414	2990409	3156491	3321611	36
25	2658366	2826204	2993185	3159251	3324355	35
26	2661170	2828994	2995960	3162011	3327098	34
27	2663974	2831784	2998735	3164770	3329841	33
28	2666777	2834574	3001510	3167529	3332585	32
29	2669580	2837364	3004284	3170288	3335327	31
30	2672383	2840153	3007058	3173047	3338069	30
	74	73	72	71	70	

The degrees of the Quadrant for right Sines

of Sines.

of the Arches of the same Quadrant.

	15	16	17	18	19	
30	2672383	2840153	3007058	3173047	3338069	30
31	2675186	2842942	3009832	3175805	3340811	29
32	2677989	2845731	3012606	3178563	3343553	28
33	2680792	2848520	3015380	3181321	3346294	27
34	2683595	2851308	3018153	3184079	3349035	26
35	2686397	2854096	3020926	3186837	3351776	25
36	2689199	2856884	3023699	3189594	3354516	24
37	2692001	2859672	3026472	3192351	3357256	23
38	2694802	2862459	3029244	3195108	3359996	22
39	2697603	2865246	3032016	3197864	3362736	21
40	2700404	2868033	3034788	3200620	3365475	20
41	2703205	2870819	3037559	3203375	3368214	19
42	2706005	2873605	3040330	3206130	3370953	18
43	2708805	2876391	3043101	3208885	3373691	17
44	2711605	2879177	3045872	3211640	3376429	16
45	2714405	2881963	3048643	3214395	3379167	15
46	2717204	2884748	3051413	3217150	3381905	14
47	2720003	2887533	3054183	3219904	3384642	13
48	2722802	2890318	3056953	3222658	3387379	12
49	2725601	2893103	3059723	3225412	3390116	11
50	2728400	2895888	3062492	3228165	3392852	10
51	2731198	2898672	3065261	3230918	3395588	9
52	2733996	2901456	3068030	3233671	3398324	8
53	2736794	2904240	3070798	3236423	3401060	7
54	2739592	2907023	3073566	3239175	3403795	6
55	2742389	2909806	3076334	3241927	3406530	5
56	2745186	2912589	3079102	3244679	3409265	4
57	2747983	2915371	3081869	3247430	3411999	3
58	2750780	2918153	3084636	3250181	3414733	2
59	2753577	2920935	3087403	3252932	3417467	1
60	2756373	2923717	3090170	3255682	3420201	0
	74	73	72	71	70	

of the complements of the Arches of the same Quadrant.

The Table
The degrees of the Quadrant for right Sines

	20	21	22	23	24	
0	3420201	3583679	3746066	3907311	4067366	60
1	3422934	3586395	3748763	3909989	4070023	59
2	3425667	3589110	3751460	3912666	4072680	58
3	3428400	3591825	3754156	3915343	4075337	57
4	3431133	3594540	3756852	3918020	4077993	56
5	3433865	3597254	3759548	3920696	4080649	55
6	3436597	3599968	3762243	3923372	4083305	54
7	3439329	3602682	3764938	3926048	4085960	53
8	3442060	3605395	3767633	3928723	4088615	52
9	3444791	3608108	3770327	3931398	4091169	51
10	3447522	3610821	3773021	3934072	4093923	50
11	3450253	3613533	3775715	3936746	4096577	49
12	3452983	3616245	3778408	3939420	4099231	48
13	3455713	3618957	3781101	3942093	4101884	47
14	3458442	3621669	3783794	3944766	4104537	46
15	3461171	3624380	3786486	3947439	4107189	45
16	3463900	3627091	3789178	3950112	4109841	44
17	3466629	3629802	3791870	3952784	4112493	43
18	3469357	3632512	3794562	3955456	4115144	42
19	3472085	3635222	3797253	3958128	4117795	41
20	3474813	3637932	3799944	3960799	4120446	40
21	3477540	3640642	3802635	3963470	4123096	39
22	3480267	3643351	3805325	3966140	4125746	38
23	3482994	3646060	3808015	3968810	4128395	37
24	3485721	3648768	3810704	3971480	4131044	36
25	3488447	3651476	3813393	3974149	4133693	35
26	3491173	3654184	3816082	3976818	4136341	34
27	3493899	3656892	3818771	3979487	4138989	33
28	3496624	3659599	3821459	3982155	4141637	32
29	3499349	3662306	3824147	3984823	4144285	31
30	3502075	3665012	3826834	3987491	4146932	30
	69	68	67	66	65	

The degrees of the Quadrant for right Sines

of Sines.

of the Arches of the same Quadrant.

	20	21	22	23	24	
30	3502075	3665012	3826834	3987491	4146932	30
31	3504799	3667718	3829521	3990159	4149579	29
32	3507523	3670424	3832208	3992826	4152226	28
33	3510247	3673130	3834895	3995493	4154872	27
34	3512971	3675835	3837581	3998159	4157518	26
35	3515694	3678541	3840267	4000825	4160163	25
36	3518417	3681246	3842953	4003491	4162808	24
37	3521140	3683951	3845638	4006156	4165453	23
38	3523862	3686655	3848323	4008821	4168097	22
39	3526584	3689359	3851008	4011486	4170741	21
40	3529306	3692062	3853692	4014150	4173385	20
41	3532027	3694765	3856376	4016814	4176028	19
42	3534748	3697468	3859060	4019478	4178671	18
43	3537469	3700170	3861743	4022141	4181313	17
44	3540190	3702872	3864426	4024804	4183955	16
45	3542910	3705574	3867109	4027467	4186597	15
46	3545630	3708276	3869791	4030130	4189239	14
47	3548350	3710977	3872473	4032792	4191880	13
48	3551070	3713678	3875155	4035454	4194521	12
49	3553789	3716379	3877837	4038115	4197162	11
50	3556508	3719080	3880518	4040776	4199802	10
51	3559227	3721780	3883199	4043437	4202442	9
52	3561945	3724480	3885880	4046097	4205081	8
53	3564663	3727179	3888560	4048757	4207720	7
54	3567380	3729878	3891240	4051416	4210359	6
55	3570097	3732577	3893919	4054075	4212997	5
56	3572814	3735275	3896598	4056734	4215635	4
57	3575531	3737973	3899277	4059392	4218273	3
58	3578247	3740671	3901955	4062050	4220910	2
59	3580963	3743369	3904633	4064708	4223547	1
60	3583679	3746066	3907311	4067366	4226183	0
	69	68	67	66	65	

of the complements of the Arches of the same Quadrant.

K 2

The Table
The degrees of the Quadrant for right Sines

	25	26	27	28	29	
0	4226183	4383712	4539905	4694716	4848096	60
1	4228819	4386326	4542497	4697284	4850640	59
2	4231455	4388940	4545088	4699852	4853184	58
3	4234090	4391554	4547679	4702419	4855727	57
4	4236725	4394167	4550270	4704986	4858270	56
5	4239360	4396780	4552860	4707553	4860812	55
6	4241994	4399392	4555450	4710119	4863354	54
7	4244628	4402004	4558039	4712685	4865895	53
8	4247262	4404616	4560628	4715250	4868436	52
9	4249895	4407227	4563216	4717815	4870977	51
10	4252528	4409838	4565804	4720380	4873517	50
11	4255161	4412449	4568392	4722944	4876057	49
12	4257793	4415059	4570979	4725508	4878596	48
13	4260425	4417669	4573566	4728071	4881135	47
14	4263056	4420278	4576153	4730634	4883674	46
15	4265687	4422887	4578739	4733197	4886212	45
16	4268318	4425496	4581325	4735759	4888750	44
17	4270949	4428104	4583911	4738321	4891287	43
18	4273579	4430712	4586496	4740882	4893824	42
19	4276209	4433320	4589081	4743443	4896361	41
20	4278838	4435927	4591665	4746004	4898897	40
21	4281467	4438534	4594249	4748564	4901433	39
22	4284096	4441140	4596833	4751124	4903968	38
23	4286724	4443746	4599416	4753683	4906503	37
24	4289352	4446352	4601999	4756242	4909037	36
25	4291979	4448957	4604581	4758801	4911571	35
26	4294606	4451562	4607163	4761359	4914105	34
27	4297233	4454167	4609744	4763917	4916638	33
28	4299859	4456771	4612325	4766474	4919171	32
29	4302485	4459375	4614906	4769031	4921703	31
30	4305111	4461978	4617486	4771588	4924235	30
	64	63	62	61	60	

The degrees of the Quadrant for right Sines

of Sines.

of the Arches of the same Quadrant.

	25	26	27	28	29	
30	4305111	4461978	4617486	4771588	4924235	30
31	4307736	4464581	4620066	4774144	4926767	29
32	4310361	4467184	4622646	4776700	4929298	28
33	4312986	4469786	4625225	4779255	4931829	27
34	4315610	4472388	4627804	4781810	4934359	26
35	4318234	4474990	4630382	4784365	4936889	25
36	4320858	4477591	4632960	4786919	4939418	24
37	4323481	4480192	4635538	4789473	4941947	23
38	4326104	4482792	4638115	4792026	4944476	22
39	4328726	4485392	4640692	4794579	4947004	21
40	4331348	4487992	4643268	4797132	4949532	20
41	4333970	4490591	4645844	4799684	4952059	19
42	4336591	4493190	4648420	4802236	4954586	18
43	4339212	4495788	4650995	4804787	4957113	17
44	4341833	4498386	4653570	4807338	4959639	16
45	4344453	4500984	4656145	4809888	4962165	15
46	4347073	4503582	4658719	4812438	4964690	14
47	4349693	4506179	4661293	4814988	4967215	13
48	4352312	4508776	4663866	4817537	4969740	12
49	4354931	4511372	4666439	4820086	4972264	11
50	4357549	4513968	4669012	4822635	4974788	10
51	4360167	4516563	4671584	4825183	4977311	9
52	4362785	4519158	4674156	4827731	4979834	8
53	4365402	4521753	4676727	4830278	4982356	7
54	4368019	4524347	4679298	4832825	4984878	6
55	4370635	4526941	4681869	4835371	4987399	5
56	4373251	4529535	4684439	4837917	4989920	4
57	4375867	4532128	4687009	4840462	4992441	3
58	4378482	4534721	4689578	4843007	4994961	2
59	4381097	4537313	4692147	4845552	4997481	1
60	4383712	4539905	4694716	4848096	5000000	0
	64	63	62	61	60	

of the complements of the Arches of the same Quadrant.

The Table
The degrees of the Quadrant for right Sines

	30	31	32	33	34	
0	5000000	5150381	5299192	5446390	5591929	60
1	5002519	5152874	5301659	5448829	5594340	59
2	5005038	5155367	5304125	5451268	5596751	58
3	5007556	5157859	5306591	5453707	5599161	57
4	5010074	5160351	5309056	5456145	5601571	56
5	5012591	5162843	5311521	5458583	5603981	55
6	5015108	5165334	5313985	5461020	5606390	54
7	5017624	5167825	5316449	5463456	5608798	53
8	5020140	5170315	5318913	5465892	5611206	52
9	5022656	5172805	5321376	5468328	5613614	51
10	5025171	5175294	5323839	5470763	5616021	50
11	5027686	5177783	5326301	5473198	5618427	49
12	5030200	5180271	5328763	5475632	5620833	48
13	5032714	5182759	5331224	5478066	5623239	47
14	5035227	5185246	5333685	5480499	5625644	46
15	5037740	5187733	5336145	5482932	5628049	45
16	5040253	5190220	5338605	5485364	5630453	44
17	5042765	5192706	5341065	5487796	5632857	43
18	5045277	5195192	5343524	5490228	5635260	42
19	5047788	5197677	5345983	5492659	5637663	41
20	5050299	5200162	5348441	5495090	5640066	40
21	5052809	5202646	5350898	5497520	5642468	39
22	5055319	5205130	5353355	5499950	5644869	38
23	5057829	5207614	5355812	5502379	5647270	37
24	5060338	5210097	5358268	5504808	5649670	36
25	5062847	5212580	5360724	5507236	5652070	35
26	5065355	5215062	5363179	5509664	5654469	34
27	5067863	5217544	5365634	5512091	5656868	33
28	5070370	5220025	5368088	5514518	5659266	32
29	5072877	5222506	5370542	5516944	5661664	31
30	5075384	5224986	5372996	5519370	5664062	30
	59	58	57	56	55	

The degrees of the Quadrant for right Sines

of Sines. 68
of the Arches of the same Quadrant.

	30	31	32	33	34	
30	5075384	5224986	5372996	5519370	5664062	30
31	5077890	5227466	5375449	5521795	5666459	29
32	5080396	5229949	5377902	5524220	5668856	28
33	5082901	5232425	5380354	5526645	5671252	27
34	5085406	5234904	5382806	5529069	5673648	26
35	5087911	5237382	5385258	5531493	5676043	25
36	5090415	5239860	5387709	5533916	5678438	24
37	5092619	5242337	5390159	5536338	5680832	23
38	5095422	5244814	5392609	5538760	5683226	22
39	5097925	5147290	5395058	5541182	5685619	21
40	5100427	5249766	5397507	5543603	5688012	20
41	5102929	525224?	5399955	5546024	5690404	19
42	5105430	5254716	5402403	5548444	5692796	18
43	5107931	5257191	5404851	5550864	5695187	17
44	5110431	5259665	5407298	5553283	5697578	16
45	5112931	5262139	5409745	5555702	5699968	15
46	5115431	5264612	5412191	5558120	5702358	14
47	5117930	5267085	5414637	5560538	5704747	13
48	5120429	5269557	5417082	5562956	5707136	12
49	5122927	5272029	5419527	5565373	5709524	11
50	5125425	5274501	5421972	5567790	5711912	10
51	5127922	5276972	5424416	5570206	5714299	9
52	5130419	5279443	5426859	5572622	5716686	8
53	5132916	5281913	5429302	5575037	5719072	7
54	5135412	5284383	5431745	5577452	5721458	6
55	5137908	5286852	5434187	5579866	5723844	5
56	5140402	5289321	5436620	5582280	5726229	4
57	5142898	5291789	5439070	5584693	5728613	3
58	5145393	5294257	5441510	5587106	5730997	2
59	5147887	5296725	5443950	5589518	5733381	1
60	5150381	5299192	5446390	5591929	5735764	0
	59	58	57	56	55	

of the complements of the Arches of the same Quadrant.

K 4

The Table

The degrees of the Quadrant for right Sines

	35	36	37	38	39	
0	5735764	5877852	6018150	6156615	6293204	60
1	5738147	5880205	6020470	6158907	6295464	59
2	5740529	5882558	6022796	6161198	6297724	58
3	5742911	5884910	6025118	6163489	6299983	57
4	5745292	5887262	6027439	6165780	6302242	56
5	5747672	5889613	6029760	6168070	6304501	55
6	5750052	5891964	6032080	6170359	6306759	54
7	5752432	5894314	6034400	6172648	6309016	53
8	5754811	5896664	6036719	6174936	6311273	52
9	5757190	5899013	6039038	6177224	6313529	51
10	5759568	5901361	6041357	6179512	6315784	50
11	5761946	5903709	6043675	6181799	6318039	49
12	5764323	5906056	6045992	6184085	6320293	48
13	5766700	5908403	6048309	6186371	6322547	47
14	5769076	5910750	6050625	6188656	6324800	46
15	5771452	5913096	6052940	6190940	6327053	45
16	5773827	5915442	6055255	6193224	6329305	44
17	5776202	5917787	6057570	6195508	6331557	43
18	5778576	5920132	6059884	6197791	6333808	42
19	5780950	5922476	6062198	6200074	6336059	41
20	5783324	5924820	6064511	6202356	6338310	40
21	5785697	5927163	6066824	6204638	6340560	39
22	5788069	5929505	6069136	6206919	6342809	38
23	5790441	5931847	6071448	6209199	6345058	37
24	5792812	5934189	6073759	6211479	6347306	36
25	5795183	5936530	6076069	6213758	6349553	35
26	5797553	5938871	6078379	6216037	6351800	34
27	5799923	5941211	6080688	6218315	6354046	33
28	5802292	5943551	6082997	6220593	6356292	32
29	5804661	5945890	6085306	6222870	6358537	31
30	5807030	5948228	6087614	6225146	6360782	30
	54	53	52	51	50	

The degrees of the Quadrant for right Sines

of Sines.   69
of the Arches of the same Quadrant.

	35	36	37	38	39	
30	5807030	5948228	6087614	6225146	6360782	30
31	5809398	5950566	6089922	6227422	6363026	29
32	5811766	5952904	6092229	6229698	6365270	28
33	5814133	5955241	6094536	6231973	6367513	27
34	5816499	5957578	6096842	6234248	6369756	26
35	5818865	5959914	6099147	6236522	6371999	25
36	5821230	5962250	6101452	9238796	6374241	24
37	5823595	5964585	6103756	6241069	6376482	23
38	5825959	5966919	6106060	6243342	6378722	22
39	5828323	5969253	6108364	6245614	6380962	21
40	5830687	5971586	6110667	6247885	6383201	20
41	5833050	5973919	6112970	6250156	6385440	19
42	5835412	5976251	6115272	6252426	6387678	18
43	5837774	5978583	6117573	6254696	6389916	17
44	5840136	5980915	6119873	6256966	6392153	16
45	5842497	5983146	6122173	6259235	6394390	15
46	5844858	5985577	6124473	6261503	6396626	14
47	5847218	5987907	6126772	6263771	6398862	13
48	5849578	5990237	6129071	6266038	6401097	12
49	5851937	5992566	6131369	6268305	6403332	11
50	5854295	5994894	6133667	6270572	6405566	10
51	5856653	5997222	6135964	6272838	6407799	9
52	5859010	5999549	6138261	6275103	6410032	8
53	5861367	6001876	6140557	6277368	6412264	7
54	5863724	6004202	6142853	6279632	6414496	6
55	5866080	6006528	6145148	6281895	6416728	5
56	5868436	6008853	6147442	6284158	6418959	4
57	5870791	6011178	6149736	6286420	6421189	3
58	5873145	6013502	6152030	6288682	6423419	2
59	5875499	6015826	6154323	6290943	6425648	1
60	5877852	6018150	6156615	6293204	6427876	0
	54	53	52	51	50	

of the complements of the Arches of the same Quadrant.

The Table

The degrees of the Quadrant for right Sines

	40	41	42	43	44	
0	642787	656059	669130	681998	694658	60
1	643010	656278	669346	682211	694867	59
2	643233	656497	669562	682423	695076	58
3	643455	656717	969778	682636	695285	57
4	643678	656936	669994	682848	695494	56
5	643901	657156	670210	683061	695703	55
6	644123	657375	670426	683273	695912	54
7	644346	657594	670642	683486	696121	53
8	644560	657813	670858	683698	696330	52
9	644790	658032	671073	683910	696539	51
10	645013	658251	671289	684122	696747	50
11	645235	658470	671505	684335	696956	49
12	645457	658689	671720	684547	697165	48
13	645679	658908	671936	684759	697373	47
14	645902	659127	672151	684971	697582	46
15	646124	659345	672366	685183	697790	45
16	646346	659564	672582	685394	697998	44
17	646567	659783	672797	685606	698207	43
18	646789	660016	673012	685818	698415	42
19	647011	660220	673227	686030	698623	41
20	647233	660438	673442	686241	698831	40
21	647455	660657	673657	686453	699039	39
22	647676	660875	673872	686664	699247	38
23	647898	661093	674087	686876	699455	37
24	648119	661311	674302	687087	699663	36
25	648341	661530	674517	687298	699871	35
26	648562	661748	674731	687510	700078	34
27	648784	661966	674946	687721	700286	33
28	649005	662184	675161	687932	700494	32
29	649226	662402	675375	688143	700701	31
30	649448	662620	675590	688354	700909	30
	49	48	47	46	45	

The degrees of the Quadrant for right Sines

of Sines. 70
of the Arches of the same Quadrant.

	40	41	42	43	44	
30	6494480	6626200	6755902	6883546	7009093	30
31	6496692	6628379	6758047	6885656	7011167	29
32	6498903	6630557	6760191	6887765	7013241	28
33	6501114	6632734	6762334	6389874	7015314	27
34	6503324	6634911	6764477	6891982	7017387	26
35	6505533	6637087	6766619	6894089	7019459	25
36	6507742	6639263	6768760	6896196	7021530	24
37	6509950	6641438	6770901	6898302	7023601	23
38	6512158	6643012	6773041	6900408	7025671	22
39	6514365	6645786	6775181	6902513	7027741	21
40	6516572	6647959	6777320	6904617	7029810	20
41	6518778	6650132	6779459	6906721	7031879	19
42	6520984	6652304	6781597	6908824	7033947	18
43	6523189	6654470	6783734	6910927	7036014	17
44	6525394	6656647	6785871	6913029	7038081	16
45	6527598	6658817	6788007	6915131	7040147	15
46	6529801	6660987	6790143	6917232	7042213	14
47	6532004	6663156	6792278	6919332	7044278	13
48	6534206	6665325	6794413	6921432	7046342	12
49	6536408	6667493	6796547	6923531	7048406	11
50	6538609	6669661	6798681	6925630	7050469	10
51	6540809	6671828	6800814	6927728	7052432	9
52	6543009	6673994	6802946	6929725	7054594	8
53	6545208	6676160	6805078	6931922	7056655	7
54	6547407	6678326	6807209	6934018	7058716	6
55	6549606	6680491	6809340	6936114	7060776	5
56	6551804	6682655	6811470	6938209	7062836	4
57	6554001	6684818	6813599	6940303	7064895	3
58	6556198	6686981	6815728	6942397	7066953	2
59	6558394	6689144	6817856	6944491	7069011	1
60	6560590	6691306	6819984	6946584	7071068	0
	49	48	47	46	45	

of the complements of the Arches of the same Quadrant.

The Table
The degrees of the Quadrant for right Sines

	45	46	47	48	49	
0	7071068	7193398	7313537	7431448	7547096	60
1	7073125	7195418	7315521	7433394	7549004	59
2	7075181	7197438	7317504	7435339	7550911	58
3	7077236	7199457	7319486	7437284	7552818	57
4	7079291	7201476	7321468	7439229	7554724	56
5	7081345	7203494	7323449	7441173	7556630	55
6	7083399	7205511	7325429	7443116	7558535	54
7	7085452	7207527	7327409	7445058	7560439	53
8	7087504	7209543	7329388	7447000	7562343	52
9	7089556	7211559	7331367	7448941	7564246	51
10	7091607	7213574	7333345	7450882	7566148	50
11	7093658	7215588	7335322	7452822	7568050	49
12	7095708	7217601	7337298	7454761	7569951	48
13	7097757	7219614	7339274	7456690	7571851	47
14	7099806	7221627	7341250	7458637	7573751	46
15	7101854	7223639	7343225	7460574	7575650	45
16	7103902	7225661	7345199	7462511	7577548	44
17	7105949	7227662	7347173	7464447	7579446	43
18	7107995	7229672	7349146	7466382	7581343	42
19	7110041	7231681	7351118	7468317	7583240	41
20	7112086	7233689	7353090	7470251	7585136	40
21	7114131	7235697	7355061	7472184	7587031	39
22	7116175	7237704	7357031	7474117	7588925	38
23	7118218	7239711	7359001	7476049	7590819	37
24	7120261	7241718	7360970	7477981	7592714	36
25	7122303	7243724	7362939	7479912	7594606	35
26	7124344	7245729	7364907	7481842	7596498	34
27	7126385	7247733	7366874	7483771	7598389	33
28	7128425	7249737	7368841	7485700	7600280	32
29	7130465	7251741	7370807	7487629	7602170	31
30	7132504	7253744	7372773	7489557	7604060	30
	44	43	42	41	40	

The degrees of the Quadrant for right Sines

of Sines.
of the Arches of the same Quadrant.

	45	46	47	48	49	
30	7132504	7253744	7372773	7489557	7604960	30
31	7134543	7255746	7374738	7491484	7605949	29
32	7136581	7257747	7376702	7493410	7607837	28
33	7138618	7259748	7378666	7495336	7609725	27
34	7140655	7261749	7380629	7497262	7611612	26
35	7142691	7263749	7382592	7499187	7613498	25
36	7144727	7265748	7384554	7501111	7615384	24
37	7146762	7267746	7386515	7503034	7617269	23
38	7148796	7269744	7388475	7504957	7619153	22
39	7150830	7271741	7390435	7506879	7621037	21
40	7152863	7273737	7392394	7508801	7622920	20
41	7154895	7275733	7394353	7510722	7624802	19
42	7156927	7277728	7396311	7512642	7626683	18
43	7158958	7279722	7398268	7514561	7628564	17
44	7160989	7281716	7400225	7516480	7630445	16
45	7163019	7283710	7402181	7518398	7632325	15
46	7165049	7285703	7404137	7520316	7634204	14
47	7167078	7287695	7406092	7522233	7636082	13
48	7169106	7289687	7408046	7524149	7637960	12
49	7171134	7291678	7410000	7526065	7639838	11
50	7173161	7293668	7411953	7527980	7641715	10
51	7175187	7295658	7413905	7529894	7643591	9
52	7177213	7297647	7415856	7531808	7645466	8
53	7179238	7299635	7417807	7533721	7647341	7
54	7181263	7301623	7419758	7535634	7649215	6
55	7183287	7303610	7421708	7537546	7651088	5
56	7185310	7305597	7423657	7539457	7652961	4
57	7187333	7307583	7425605	7541367	7654833	3
58	7189355	7309568	7427552	7543277	7656704	2
59	7191377	7311553	7429501	7545187	7658575	1
60	7193398	7313537	7431448	7547096	7660445	0
	44	43	42	41	40	

of the complements of the Arches of the same Quadrant.

The Table
The degrees of the Quadrant for right Sines

	50	51	52	53	54	
0	7660445	7771460	7880108	7986355	8090170	60
1	7662314	7773290	7881898	7988105	8091879	59
2	7664183	7775120	7883688	7989855	8093588	58
3	7666051	7776949	7885477	7991604	8095296	57
4	7667919	7778777	7887266	7993352	8097004	56
5	7669786	7780605	7889054	7995100	8098711	55
6	7671652	7782432	7890841	7996847	8100417	54
7	7673517	7784258	7892927	7998593	8102122	53
8	7675382	7786084	7894413	8000339	8103827	52
9	7677246	7787909	7896198	8002084	8105531	51
10	7679110	7789833	7897983	8003828	8107234	50
11	7680973	7791557	7899767	8005571	8108936	49
12	7682835	7793380	7901550	8007314	8110638	48
13	7684697	7795202	7903332	8009056	8112339	47
14	7686558	7797024	7905114	8010797	8114040	46
15	7688418	7798845	7906895	8012538	8115740	45
16	7690278	7800665	7908676	8014278	8117439	44
17	7692137	7802485	7910456	8016017	8119137	43
18	7693995	7804304	7912235	8017756	8120835	42
19	7695853	7806123	7914014	8019494	8122532	41
20	7697710	7807941	7915792	8021232	8124229	40
21	7699566	7809758	7917569	8022969	8125925	39
22	7701422	7811574	7919345	8024705	8127620	38
23	7703277	7813390	7921121	8026440	8129314	37
24	7705132	7815205	7922896	8028175	8131008	36
25	7706986	7817020	7924671	8029909	8132701	35
26	7708839	7818834	7926445	8021642	8134393	34
27	7710692	7820647	7928218	8033375	8136084	33
28	7712544	7822459	7929990	8035107	8137775	32
29	7714395	7824271	7931762	8036838	8139465	31
30	7716246	7826082	7933533	8038569	8141155	30
	39	38	37	36	35	

The degrees of the Quadrant for right Sines

of Sines.

of the Arches of the same Quadrant.

	50	51	52	53	54	
30	7716246	7826082	7933533	8038569	8141155	30
31	7718096	7827892	7935303	8040299	8142844	29
32	7719945	7829702	7937073	8042028	8144532	28
33	7721794	7831511	7938842	8043757	8146220	27
34	7723642	7833320	7940611	8045485	8147907	26
35	7725490	7835128	7942379	8047212	8149593	25
36	7727337	7836935	7944146	8048938	8151278	24
37	7729183	7838741	7945912	8050664	8152963	23
38	7731028	7840547	7947678	8052389	8154647	22
39	7732872	7842352	7949443	8054114	8156330	21
40	7734716	7844157	7951208	8055838	8158013	20
41	7736559	7845961	7952972	8057561	8159695	19
42	7738402	7847764	7954735	8059283	8161376	18
43	7740244	7849566	7956497	8061005	8163057	17
44	7742085	7851368	7958259	8062726	8164737	16
45	7743926	7853169	7960020	8064446	8166416	15
46	7745766	7854970	7961780	8066166	8168094	14
47	7747606	7856770	7963540	8067885	8169772	13
48	7749445	7858569	7965299	8069603	8171449	12
49	7751283	7860368	7967057	8071321	8173126	11
50	7753121	7862166	7968815	8073038	8174802	10
51	7754958	7863963	7970572	8074754	8176477	9
52	7756794	7865759	7972328	8076470	8178151	8
53	7758630	7867555	7974084	8078185	8179825	7
54	7760465	7869350	7975839	8079899	8181498	6
55	7762299	7871145	7977593	8081613	8183170	5
56	7764132	7872939	7979347	8083326	8184841	4
57	7765965	7874732	7981100	8085038	8186512	3
58	7767797	7876525	7982852	8086749	8188182	2
59	7769629	7878317	7984604	8088460	8189851	1
60	7771460	7880108	7986355	8090170	8191520	0
	39	38	37	36	35	

of the complements of the Arches of the same Quadrant.

The minutes of the degrees of the Quadrant for right Sines of the Arches of the same Quadrant.

The minutes of degrees of the Quadrant for right Sines of the complements of the same Quadrant.

72

The Table
The degrees of the Quadrant for right Sines

	55	56	57	58	59	
0	8191520	8290376	8386706	8480481	8571673	60
1	8193188	8292002	8388290	8482022	8573171	59
2	8194855	8293628	8389873	8483562	8574668	58
3	8196522	8295253	8391456	8485102	8576164	57
4	8198188	8296877	8393038	8486641	8577660	56
5	8199854	8298501	8394619	8488180	8579155	55
6	8201519	8300124	8396199	8489718	8580649	54
7	8203183	8301746	8397778	8491255	8582142	53
8	8204846	8303367	8399357	8492791	8583635	52
9	8206508	8304987	8400935	8494326	8585127	51
10	8208170	8306607	8402513	8495860	8586619	50
11	8209831	8308226	8404090	8497394	8588110	49
12	8211491	8309844	8405666	8498927	8589600	48
13	8213151	8311462	8407241	8500459	8591089	47
14	8214810	8313079	8408816	8501991	8592577	46
15	8216469	8314696	8410390	8503522	8594064	45
16	8218127	8316312	8411963	8505052	8595551	44
17	8219784	8317927	8413536	8506582	8597037	43
18	8221440	8319541	8415108	8508111	8598523	42
19	8223096	8321155	8416679	8509639	8600008	41
20	8224751	8322768	8418250	8511167	8601492	40
21	8226405	8324380	8419820	8512694	8602975	39
22	8228058	8325991	8421389	8514220	8604457	38
23	8229711	8327602	8422957	8515745	8605939	37
24	8231363	8329212	8424525	8517270	8607420	36
25	8233015	8330822	8426092	8518794	8608901	35
26	8234666	8332431	8427658	8520317	8610381	34
27	8236316	8334039	8429223	8521839	8611860	33
28	8237965	8335646	8430788	8523361	8613338	32
29	8239614	8337252	8432352	8524882	8614815	31
30	8241262	8338858	8433915	8526402	8616292	30
	34	33	32	31	30	

The degrees of the Quadrant for right Sines

of Sines.

of the Arches of the same Quadrant.

	55	56	57	58	59	
30	8241262	8338858	8433915	8526402	8616292	30
31	8242909	8340463	8435477	8527921	8617768	29
32	8244556	8342067	8437039	8529440	8619243	28
33	8246202	8343671	8438600	8530958	8620718	27
34	8247847	8345247	8440161	8532476	8622192	26
35	8249492	8346877	8441721	8533993	8623665	25
36	8251136	8348479	8443280	8535509	8625137	24
37	8252779	8350080	8444838	8537024	8626608	23
38	8254421	8351680	8446396	8538538	8628079	22
39	8256062	8353279	8447953	8540052	8629549	21
40	8257703	8354878	8449509	8541565	8631019	20
41	8259343	8356476	8451064	8543077	8632488	19
42	8260982	8358073	8452618	8544588	8633956	18
43	8262621	8359670	8454172	8546099	8635423	17
44	8264259	8361266	8455725	8547609	8636889	16
45	8265897	8362862	8457278	8549119	8638355	15
46	8267534	8364457	8458830	8550628	8639820	14
47	8269170	8366051	8460381	8552136	8641284	13
48	8270806	8367644	8461932	8553643	8642748	12
49	8272441	8369236	8463482	8555149	8644211	11
50	8274075	8370828	8465031	8556655	8645673	10
51	8275708	8372419	8466579	8558160	8647134	9
52	8277340	8374009	8468126	8559664	8648595	8
53	8278972	8375599	8469673	9561168	8650055	7
54	8280603	8377188	8471219	8562671	8651514	6
55	8282234	8378776	8472765	8564173	8652973	5
56	8283864	8380363	8474310	8565675	8654431	4
57	8285493	8381950	8475854	8567176	8655888	3
58	8287121	8383536	8477397	8568676	8657344	2
59	8288749	8385121	8478939	8570175	8658799	1
60	8290376	8386706	8480481	8571673	8660254	0
	34	33	32	31	30	

of the complements of the Arches of the same Quadrant.

L

The Table

The degrees of the Quadrant for right Sines

	60	61	62	63	64	
0	8660254	8746197	8829476	8910065	8987940	60
1	8661708	8747607	8830841	8911385	8989215	59
2	8663162	8749016	8832205	8912704	8990489	58
3	8664615	8750425	8833569	8914023	8991762	57
4	8666067	8751833	8834932	8915341	8993035	56
5	8667518	8753240	8836295	8916659	8994307	55
6	8668968	8754646	8837657	8917976	8995578	54
7	8670417	8756051	8839018	8919292	8996848	53
8	8671866	8757456	8840378	8920607	8998117	52
9	8673314	8758860	8841737	8921921	8999386	51
10	8674762	8760263	8843095	8923234	9000654	50
11	8676209	8761665	8844452	8924546	9001921	49
12	8677655	8763067	8845809	8925858	9003187	48
13	8679100	8764468	8847165	8927169	9004453	47
14	8680544	8765868	8848521	8928479	9005718	46
15	8681988	8767268	8849876	8929789	9006982	45
16	8683431	8768667	8851230	8931098	9008245	44
17	8684874	8770065	8852583	8932406	9009508	43
18	8686316	8771462	8853936	8933714	9010770	42
19	8687757	8772859	8855288	8935021	9012031	41
20	8689197	8774255	8856639	8936327	9013292	40
21	8690636	8775650	8857989	8937632	9014552	39
22	8692074	8777044	8859338	8938936	9015811	38
23	8693512	8778437	8860687	8940240	9017069	37
24	8694949	8779830	8862035	8941543	9018326	36
25	8696386	8781222	8863383	8942845	9019582	35
26	8697822	8782613	8864730	8944146	9020838	34
27	8699257	8784003	8866076	8945446	9022093	33
28	8700691	8785393	8867421	8946746	9023347	32
29	8702124	8786782	8868765	8948045	9024600	31
30	8703557	8788171	8870108	8949344	9025853	30
	29	28	27	26	25	

The degrees of the Quadrant for right Sines

of Sines. 74

of the Arches of the same Quadrant.

	60	61	62	63	64	
30	8703557	8788171	8870108	8949344	9025853	30
31	8704989	8789559	8871451	8950642	9027105	29
32	8706420	8790946	8872793	8951939	9028356	28
33	8707851	8792332	8874134	8953235	9029606	27
34	8709281	8793717	8875475	8954530	9030856	26
35	8710710	8795102	8876815	8955824	9032105	25
36	8712138	8796486	8878154	8957117	9033353	24
37	8713565	8797869	8879492	8958410	9034600	23
38	8714992	8799251	8880830	8959702	9035847	22
39	8716418	8800633	8882167	8960994	9037093	21
40	8717844	8802014	8883503	8962285	9038338	20
41	8719269	8803394	8884838	8963575	9039582	19
42	8720693	8804773	8886172	8964864	9040825	18
43	8722116	8806152	8887506	8966152	9042068	17
44	8723538	8807530	8888839	8967440	9043310	16
45	8724960	8808907	8890171	8968727	9044551	15
46	8726381	8810283	8891502	8970013	9045791	14
47	8727801	8811659	8892833	8971299	9047031	13
48	8729221	8813034	8894163	8972584	9048270	12
49	8730640	8814408	8895492	8973868	9049508	11
50	8732058	8815782	8896821	8975151	9050746	10
51	8733475	8817155	8898149	8976433	9051983	9
52	8734891	8818527	8899476	8977715	9053219	8
53	8736307	8819898	8900802	8978996	9054454	7
54	8737722	8821268	8902127	8980276	9055688	6
55	8739137	8822638	8903452	8981555	9056922	5
56	8740551	8824007	8904776	8982833	9058155	4
57	8741964	8825375	8906099	8984111	9059387	3
58	8743376	8826743	8907422	8985388	9060618	2
59	8744787	8828110	8908744	8986664	9061848	1
60	8746197	8829476	8910065	8987940	9063078	0
	25	26	27	28	29	

of the complements of the Arches of the same Quadrant.
L 2

The Table

The degrees of the Quadrant for right Sines

	65	66	67	68	69	
0	9063078	9135455	9205049	9271839	9335804	60
1	9064307	9136638	9206185	9272928	9336846	59
2	9065535	9137820	9207321	9274017	9337887	58
3	9066763	9139001	9208456	9275105	9338928	57
4	9067990	9140181	9209590	9276192	9339968	56
5	9069216	9141361	9210723	9277278	9341007	55
6	9070441	9142540	9211855	9278363	9342045	54
7	9071665	9143718	9212986	9279448	9343082	53
8	9072889	9144895	9214117	9280532	9344119	52
9	9074112	9146072	9215247	9281615	9345155	51
10	9075334	9147248	9216376	9282697	9346190	50
11	9076555	9148423	9217504	9283778	9347224	49
12	9077775	9149597	9218631	9284859	9348257	48
13	9078995	9150770	9219758	9285939	9349289	47
14	9080214	9151943	9220884	9287018	9350321	46
15	9081432	9153115	9222010	9288096	9351352	45
16	9082649	9154286	9223135	9289173	9352382	44
17	9083866	9155457	9224259	9290250	9353411	43
18	9085082	9156627	9225382	9291326	9354440	42
19	9086297	9157796	9226504	9292401	9355468	41
20	9087512	9158964	9227625	9293476	9356495	40
21	9088726	9160131	9228746	9294550	9357521	39
22	9089939	9161297	9229866	9295623	9358546	38
23	9091151	9162463	9230985	9296695	9359571	37
24	9092362	9163628	9232103	9297766	9360595	36
25	9093572	9165792	9233220	9298836	9361618	35
26	9094781	9165955	9234337	9299905	9362640	34
27	9095990	9167117	9235453	9300974	9363662	33
28	9097198	9168279	9236568	9302042	9364683	32
29	9098406	9169440	9237682	9303109	9365703	31
30	9099613	9170601	9238795	9304176	9366722	30
	24	23	22	21	20	

The degrees of the Quadrant for right Sines

of Sines.

of the Arches of the same Quadrant.

	65	66	67	68	69	
30	9099613	9170601	9238795	9304176	9366722	30
31	9100819	9171761	9239908	9305242	9367740	29
32	9102024	9172920	9241020	9306307	9368758	28
33	9103228	9174078	9242131	9307371	9369775	27
34	9104432	9175235	9243242	9308434	9370791	26
35	9105636	9176391	9244352	9309497	9371806	25
36	9106837	9177547	9245461	9310559	9372820	24
37	9108038	9178702	9246569	9311620	9373834	23
38	9109238	9179856	9247676	9312680	9374847	22
39	9110438	9181009	9248782	9313739	9375859	21
40	9111637	9182161	9249888	9314798	9376870	20
41	9112835	9183313	9250993	9315856	9377880	19
42	9114032	9184464	9252097	9316913	9378889	18
43	9115229	9185614	9253200	6317969	9379898	17
44	9116425	9186763	9254303	9319024	9380906	16
45	9117620	9187912	9255405	9320079	9381913	15
46	9118814	9189060	9256506	9321133	9382919	14
47	9120007	9190207	9257606	9322186	9383925	13
48	9121200	9191353	9258706	9323238	9384930	12
49	9122392	9192499	9259805	9324290	9385934	11
50	9123584	9193644	9260963	9325241	9386937	10
51	9124775	9194788	9262000	9326391	9387939	9
52	9125965	9195931	9263096	9327440	9388941	8
53	9127154	9197073	9264192	9328488	9389942	7
54	9128342	9198215	9265287	9329535	9390942	6
55	9129529	9199356	9266381	9330582	9391941	5
56	9130716	9200496	9267474	9331628	9392940	4
57	9131902	9201635	9268566	9332673	9393938	3
58	9133087	9202774	9269658	9333717	9394935	2
59	9134271	9203912	9270749	9334761	9395931	1
60	9135455	9205049	9271839	9335804	9396926	0
	24	23	22	21	20	

of the complements of the Arches of the same Quadrant.

The Table
The degrees of the Quadrant for right Sines

	70	71	72	73	74	
0	9396926	9455186	9510565	9563048	9612617	60
1	9397921	9456133	9511464	9563898	9613418	59
2	9398915	9457079	9512362	9564747	9614219	58
3	9399908	9458024	9513259	9565596	9615019	57
4	9400900	9458968	9514155	9566444	9615818	56
5	9401891	9459911	9515050	9567291	9616616	55
6	9402882	9460854	9515944	9568137	9617413	54
7	9403872	9461796	9516838	9568982	9618209	53
8	9404861	9462737	9517731	9569826	9619005	52
9	9405849	9463677	9518623	9570670	9619800	51
10	9406836	9464616	9519514	9571513	9620594	50
11	9407822	9465555	9520404	9572355	9621387	49
12	9408808	9466493	9521294	9573196	9622179	48
13	9409793	9467430	9522183	9574036	9622971	47
14	9410777	9468366	9523071	9574875	9623762	46
15	9411760	9469301	9523958	9575714	9624552	45
16	9412742	9470236	9524844	9576552	9625341	44
17	9413724	9471170	9525730	9577389	9626129	43
18	9414705	9472103	9526615	9578225	9626917	42
19	9415685	9473035	9527499	9579061	9627704	41
20	9416665	9473967	9528382	9579896	9628490	40
21	9417644	9474898	9529264	9580730	9629275	39
22	9418622	9475828	9530146	9581563	9630059	38
23	9419599	9476757	9531027	9582395	9630843	37
24	9420575	9477685	9531907	9583226	9631626	36
25	9421550	9478612	9532786	9584057	9632408	35
26	9422525	9479539	9533664	9584887	9633189	34
27	9423499	9480465	9534541	9585716	9633969	33
28	9425472	9481390	9535418	9586544	9634748	32
29	9425444	9482314	9536294	9587371	9635527	31
30	9426415	9483237	9537169	9588197	9636305	30
	19	18	17	16	15	

The degrees of the Quadrant for right Sines

of Sines.

of the Arches of the same Quadrant.

	70	71	72	73	74	
30	9426415	9483237	9537169	9588197	9636305	30
31	9427386	9484160	9538043	9589023	9637082	29
32	9428356	9485082	9538917	9589848	9637858	28
33	9429325	9486003	9539790	9590672	9638633	27
34	9430293	9486923	9540662	9591495	9639408	26
35	9431260	9487842	9541533	9592318	9640182	25
36	9432227	9488761	9542403	9593140	9640955	24
37	9433193	9489679	9543272	9593961	9641727	23
38	9434158	9490596	9544141	9594781	9642498	22
39	9435122	9491512	9545009	9595600	9643268	21
40	9436085	9492427	9545876	9596419	9644038	20
41	9437048	9493341	9546742	9597237	9644807	19
42	9438010	9494255	9547607	9598054	9645575	18
43	9438971	9495168	9548472	9598870	9646342	17
44	9439931	9496080	9549336	9599685	9647108	16
45	9440890	9496991	9550199	9600499	9647873	15
46	9441849	9497902	9551061	9601313	9648638	14
47	9442807	9498812	9551922	9602126	9649402	13
48	9443764	9499721	9552783	9602938	9650165	12
49	9444720	9500629	9553643	9603749	9650927	11
50	9445676	9501536	9554502	9604559	9651689	10
51	9446631	9502443	9555360	9605368	9652450	9
52	9447585	9503349	9556217	9606177	9653210	8
53	9448538	9504254	9557074	9606985	9653969	7
54	9449490	9505158	9557930	9607792	9654727	6
55	9450441	9506061	9558785	9608598	9655484	5
56	9451392	9506963	9559639	9609403	9656240	4
57	9452342	9507865	9560492	9610208	9656996	3
58	9453291	9508766	9561345	9611012	9657751	2
59	9454239	9509666	9562197	9611815	9658505	1
60	9455186	9510565	9563048	9612617	9659258	0
	19	18	17	16	15	

of the complements of the Arches of the same Quadrant.

The Table
The degrees of the Quadrant for right Sines

	75	76	77	78	79	
0	9659258	9702957	9743600	9781476	9816272	60
1	9660011	9703660	9744355	9782080	9816827	59
2	9660163	9704363	9745008	9782684	9817381	58
3	9661514	9705065	9745660	9783281	9817934	57
4	9662264	9705766	9746312	9783889	9818486	56
5	9663013	9706466	9746963	9784490	9819037	55
6	9663761	9707165	9747613	9785090	9819587	54
7	9664508	9707863	9748262	9785689	9820137	53
8	9665255	9708561	9748910	9786288	9820686	52
9	9666001	9709258	9749557	9786886	9821234	51
10	9666746	9709954	9750203	9787483	9821781	50
11	9667490	9710649	9750849	9788079	9822327	49
12	9668233	9711343	9751494	9788674	9822872	48
13	9668976	9712036	9752138	9789268	9823417	47
14	9669718	9712729	9752781	9789862	9823961	46
15	9670459	9713421	9753424	9790455	9824504	45
16	9671199	9714112	9754065	9791047	9825046	44
17	9671938	9714802	9754706	9791638	9825587	43
18	9672677	9715491	9755346	9792228	9826128	42
19	9673415	9716180	9755985	9792818	9826668	41
20	9674152	9716868	9756623	9793407	9827207	40
21	9674888	9717555	9757260	9793995	9827745	39
22	9675623	9718241	9757897	9794582	9828282	38
23	9676357	9718926	9758533	9795168	9828818	37
24	9677091	9719610	9759168	9795753	9829354	36
25	9677824	9720294	9759802	9796337	9829889	35
26	9678556	9720977	9760435	9796921	9830423	34
27	9679287	9721659	9761067	9797504	9830956	33
28	9680017	9722340	9761699	9798086	9831488	32
29	9680747	9723020	9762330	9798667	9832019	31
30	9681476	9723699	9762960	9799247	9832549	30
	14	13	12	11	10	

The degrees of the Quadrant for right Sines

of Sines.

of the Arches of the same Quadrant.

	75	76	77	78	79	
30	9681476	9723699	9762960	9799247	9832549	30
31	9682804	9724378	9763589	9799827	9833079	29
32	9682931	9725056	9764217	9800406	9833608	28
33	9683657	9725733	9764845	9800984	9834136	27
34	9684383	9726409	9765472	9801561	9834663	26
35	9685108	9727085	9766098	9802137	9835189	25
36	9685832	9727760	9766723	9802712	9835714	24
37	9686555	9728434	9767347	9803287	9836239	23
38	9687277	9729107	9767970	9803861	9836763	22
39	9687998	9729779	9768593	9804434	9837286	21
40	9688719	9730450	9769215	9805006	9837808	20
41	9689439	9731120	9769836	9805577	9838329	19
42	9690158	9731789	9770456	9806147	9838850	18
43	9690879	9732458	9771075	9806716	9839370	17
44	9691593	9733126	9771693	9807285	9839889	16
45	9692309	9733793	9772311	9807853	9840407	15
46	9693025	9734459	9772928	9808420	9840924	14
47	9693740	9735124	9773544	9808986	9841440	13
48	9694454	9735789	9774159	9809551	9841956	12
49	9695167	9736453	9774773	9810116	9842471	11
50	9695879	9737116	9775387	9810680	9842985	10
51	9696590	9737778	9776000	9811243	9843498	9
52	9697301	9738439	9776612	9811805	9844010	8
53	9698011	9739099	9777223	9812366	9844521	7
54	9698720	9739759	9777833	9812926	9845032	6
55	9699428	9740418	9778442	9813486	9845542	5
56	9700135	9741076	9779050	9814045	9846051	4
57	9700842	9741733	9779658	9814603	9846559	3
58	9701548	9742389	9780265	9815160	9847066	2
59	9702253	9743045	9780871	9815716	9847572	1
60	9702957	9743700	9781476	9816272	9848078	0
	14	13	12	11	10	

of the complements of the Arches of the same Quadrant.

The Table

The degrees of the Quadrant for right Sines

	80	81	82	83	84	
0	9848078	9876883	9902681	9925461	9945219	60
1	9848583	9877338	9903085	9925816	9945523	59
2	9849087	9877792	9903489	9926169	9945826	58
3	9849590	9878245	9903892	9926521	9946128	57
4	9850092	9878697	9904294	9926873	9946429	56
5	9850593	9879148	9904695	9927224	9946729	55
6	9851093	9879598	9905095	9927574	9927028	54
7	9851593	9880048	9905494	9927923	9947337	53
8	9852092	9880497	9905893	9928271	9947625	52
9	9852590	9880945	9906291	9928618	9947622	51
10	9853087	9881392	9906688	9928965	9948218	50
11	9853583	9881838	9907084	9929311	9948513	49
12	9854079	9882283	9907479	9929656	9948807	48
13	9854574	9882728	9907873	9930000	9949100	47
14	9855068	9883172	9908266	9930343	9949393	46
15	9855561	9883615	9908659	9930685	9949685	45
16	9856053	9884057	9909051	9931026	9949976	44
17	9856544	9884498	9909442	9931367	9950266	43
18	9857035	9884938	9909832	9931707	9950555	42
19	9857525	9885378	9910221	9932046	9950844	41
20	9858014	9885817	9910610	9932384	9951132	40
21	9858502	9886255	9910998	9932721	9951419	39
22	9858989	9886692	9911385	9933057	9951705	38
23	9859475	9887128	9911771	9933393	9951990	37
24	9859961	9887564	9912156	9933728	9952274	36
25	9860446	9887999	9912540	9934062	9952557	35
26	9860930	9888433	9912923	9934395	9952840	34
27	9861413	9888866	9913306	9934727	9953122	33
28	9861895	9889298	9913688	9935058	9953403	32
29	9862376	9889729	9914069	9935389	9953683	31
30	9862856	9890159	9914449	9935719	9953962	30
	9	8	7	6	5	

The degrees of the Quadrant for right Sines

of Sines.

of the Arches of the same Quadrant.

	80	81	82	83	84	
30	9862856	9890159	9914449	9935719	9953962	30
31	9863336	9890588	9914828	9936048	9954240	29
32	9863815	9891017	9915206	9936376	9954518	28
33	9864293	9891445	9915584	9936703	9954795	27
34	9864770	9891872	9915961	9937029	9955071	26
35	9865246	9892298	9916337	9937355	9955346	25
36	9865722	9892723	9916712	9937680	9955620	24
37	9866197	9893147	9917086	9938004	9955893	23
38	9866671	9893571	9917459	9938327	9956165	22
39	9867144	9893994	9917832	9938649	9956437	21
40	9867616	9894416	9918204	9938970	9956708	20
41	9868087	9894837	9918575	9939290	9956978	19
42	9868557	9895257	9918945	9939609	9957247	18
43	9869027	9895677	9919314	9939928	9957515	17
44	9869496	9896096	9919682	9940246	9957782	16
45	9869964	9896514	9920049	9940563	9958049	15
46	9870431	9896931	9920416	9940879	9958315	14
47	9880897	9897347	9920782	9941194	9958580	13
48	9871362	9897762	9921147	9941509	9958844	12
49	9871827	9898177	9921511	9941823	9959307	11
50	9872291	9898591	9921874	9942136	9959370	10
51	9872754	9899004	9922236	9942448	9959632	9
52	9873216	9899416	9922598	9942759	9959893	8
53	9873677	9899827	9922959	9943069	9960153	7
54	9874137	9900237	9923319	9943379	9960412	6
55	9874597	9900646	9923678	9943688	9960670	5
56	9875056	9901055	9924036	9943996	9960927	4
57	9875514	9901463	9924393	9944303	9961183	3
58	9875971	9901870	9924750	9944609	9961438	2
59	9876427	9902276	9925106	9944914	9961693	1
60	9876883	9902681	9925461	9945219	9961947	0
	9	8	7	6	5	

of the complements of the Arches of the same Quadrant.

The Table

The degrees of the Quadrant for right Sines

	85	86	87	88	89	
0	9961947	9975640	9986295	9993908	9998477	60
1	9962200	9975843	9986447	9994009	9998527	59
2	9962452	9976045	9986598	9994109	9998576	58
3	9962703	9976246	9986748	9994208	9998625	57
4	9962954	9976446	9986897	9994307	9998673	56
5	9963204	9976645	9987045	9994405	9998720	55
6	9963453	9976843	9987193	9994502	9998766	54
7	9963701	9977040	9987340	9994598	9998811	53
8	9963948	9977237	9987486	9994693	9998855	52
9	9964194	9977433	9987631	9994787	9998899	51
10	9964440	9977628	9987775	9994881	9998942	50
11	9964685	9977822	9987918	9994974	9998984	49
12	9964929	9978015	9988061	9995066	9999025	48
13	9965172	9978207	9988203	9995157	9999065	47
14	9965414	9978398	9988344	9995247	9999104	46
15	9965655	9978589	9988484	9995336	9999143	45
16	9965895	9978779	9988623	9995424	9999181	44
17	9966235	9978968	9988761	9995512	9999218	43
18	9966374	9979156	9988899	9995599	9999254	42
19	9966612	9979343	9989036	9995685	9999289	41
20	9966849	9979530	9989172	9995770	9999323	40
21	9967085	9979716	9989307	9995854	9999356	39
22	9967320	9979901	9989441	9995937	9999389	38
23	9967555	9980085	9989574	9996019	9999421	37
24	9967789	9980268	9989706	9996101	9999452	36
25	9968022	9980450	9989837	9996182	9999482	35
26	9968254	9980631	9989968	9996262	9999511	34
27	9968485	9980811	9990098	9996341	9999539	33
28	9968715	9980992	9990227	9996419	9999566	32
29	9968944	9981170	9990355	9996496	9999593	31
30	9969173	9981348	9990482	9996573	9999619	30
	4	3	2	1	0	

The degrees of the Quadrant for right Sines

of Sines. 79
of the Arches of the same Quadrant.

	85	86	87	88	89	
30	9969173	9981348	9990482	9996573	9999616	30
31	9969401	9981525	9990608	9996649	9999644	29
32	9969628	9981701	9990734	9996724	9999668	28
33	9969854	9981877	9990859	9996798	9999691	27
34	9970079	9982052	9990983	9996871	9999713	26
35	9970304	9982226	9991106	9996943	9999735	25
36	9970528	9982399	9991228	9997014	9999756	24
37	9970751	9982571	9991349	9997085	9999776	23
38	9970973	9982742	9991470	9997155	9999795	22
39	9971194	9982912	9991590	9997224	9999813	21
40	9971414	9983082	9991709	9997292	9999830	20
41	9971633	9983251	9991827	9997359	9999846	19
42	9971851	9983419	9991944	9997425	9999862	18
43	9972069	9983586	9992060	9997491	9999877	17
44	9972286	9983752	9992175	9997556	9999891	16
45	9972502	9983917	9992290	9997620	9999904	15
46	9972717	9984081	9992404	9997683	9999916	14
47	9972931	9984245	9992517	9997745	9999927	13
48	9973145	9984408	9992629	9997806	9999938	12
49	9973358	9984570	9992740	9997867	9999948	11
50	9973570	9984731	9992850	9997927	9999957	10
51	9973781	9984891	9992960	9997986	9999965	9
52	9973991	9985050	9993069	9998044	9999972	8
53	9974200	9985209	9993177	9998101	9999978	7
54	9974408	9985367	9993284	9998157	9999984	6
55	9974615	9985524	9993390	9998212	9999989	5
56	9974822	9985680	9993495	9998267	9999993	4
57	9975028	9985835	9993599	9998321	9999996	3
58	9975233	9985989	9993703	9998374	9999998	2
59	9975437	9986143	9993806	9998426	9999999	1
60	9975640	9986295	9993908	9998477	10000000	0
	4	3	2	1	0	

of the complements of the Arches of the same Quadrant.

THE TABLE
OF TANGENTS,
OTHERWISE CALLED
THE FRVITFVLL
TABLE.

The Table

The degrees of the Quadrant for the Tangents

	0	1	2	3	4	
0	0000	174550	349207	524078	699269	60
1	2909	177459	352120	526995	702193	59
2	5818	180369	355033	529911	705116	58
3	8727	183279	357945	532828	708039	57
4	11636	186189	360858	535745	710962	56
5	14544	189100	363770	538663	713886	55
6	17452	192010	366683	541580	716809	54
7	20361	194920	369596	544498	719733	53
8	23270	197830	372508	547415	722657	52
9	26179	200740	375421	550333	725580	51
10	29088	203650	378334	553251	728504	50
11	31996	206561	381247	556169	731428	49
12	34905	209471	384160	559087	734353	48
13	37814	212381	387073	562005	737277	47
14	40723	215291	389987	564923	740202	46
15	43632	218201	392900	567841	743127	45
16	46541	221111	395814	570759	746052	44
17	49450	224022	398727	573678	748978	43
18	52359	226932	401641	576596	751903	42
19	55268	229842	404554	579514	754829	41
20	58177	232752	407468	582433	757754	40
21	61086	235663	410382	585352	760680	39
22	63995	238574	413295	588270	763606	38
23	66904	241485	416209	591189	766532	37
24	69813	244395	419123	594108	769459	36
25	72722	247306	422037	597028	772385	35
26	75631	250217	424951	599947	775311	34
27	78540	253128	427866	602866	778238	33
28	81450	256038	430780	605786	781164	32
29	84359	258949	433694	608705	784091	31
30	87268	261859	436609	611625	787017	30
	89	88	87	86	85	

The degrees of the Quadrant for the Tangents

of Tangents.

of the Arches of the same Quadrant.

	0	1	2	3	4	
30	87268	261859	436609	611625	787017	30
31	90177	264770	439523	614544	789944	29
32	93086	267681	442438	617464	792871	28
33	95995	270592	445353	620384	795799	27
34	98904	273503	448267	623304	798726	26
35	101814	276414	451182	626225	801653	25
36	104723	279325	454097	629145	804581	24
37	107632	282237	457012	632066	807509	23
38	110541	285148	459927	634986	810437	22
39	113450	288059	462842	637907	813365	21
40	116360	290970	465757	640828	816293	20
41	119269	293882	468672	643749	819221	19
42	122178	296794	471588	646671	822150	18
43	125088	299705	474503	649592	825079	17
44	127997	302617	477419	652514	828008	16
45	130906	305528	480335	655435	830937	15
46	133816	308439	483251	658357	833866	14
47	136725	311351	486166	661278	836795	13
48	139635	314262	489082	664200	839724	12
49	142544	317174	491997	667121	842653	11
50	145454	320085	494913	670043	845583	10
51	148363	322997	497829	672965	848513	9
52	151273	325909	500745	675888	851443	8
53	154182	328821	503662	678810	854374	7
54	159092	331733	506578	681733	857304	6
55	160001	334645	509495	684656	860234	5
56	162911	337558	512411	687578	863164	4
57	165820	340470	515328	690501	866025	3
58	168730	343382	518244	693423	869025	2
59	171640	346295	521161	696346	871956	1
60	174550	349207	524078	699269	874886	0
	89	88	87	86	85	

of the complements of the Arches of the same Quadrant.

The Table
The degrees of the Quadrant for the Tangents

		5	6	7	8	9		
The minutes of the degrees of the Quadrant for the Tangents of the Arches of the same Quadrant.	0	874886	1051042	1227846	1405408	1583844	60	The minutes of degrees of the Quadrant for the Tangents of the complements of the Arches of the same Quadrant.
	1	877817	1053983	1230798	1408374	1586826	59	
	2	880748	1056924	1233751	1411341	1589808	58	
	3	883680	1059866	1236704	1414308	1592791	57	
	4	886611	1062808	1239658	1417275	1595774	56	
	5	889543	1065750	1242612	1420242	1598757	55	
	6	892475	1068692	1245566	1423210	1601740	54	
	7	895407	1071634	1248520	1426178	1604723	53	
	8	898339	1074576	1251474	1429146	1607707	52	
	9	901271	1077518	1254428	1432115	1610691	51	
	10	904204	1080461	1257383	1435084	1613675	50	
	11	907137	1083404	1260338	1438053	1616660	49	
	12	910070	1086347	1263293	1441022	1619645	48	
	13	913003	1089291	1266249	1443992	1622630	47	
	14	915936	1092234	1269205	1446961	1625615	46	
	15	918870	1095178	1272161	1449931	1627601	45	
	16	921804	1098122	1275117	1452901	1631587	44	
	17	924738	1101066	1278073	1455871	1634573	43	
	18	927771	1104010	1281029	1458842	1637560	42	
	19	930605	1106954	1283986	1461813	1640547	41	
	20	933539	1109899	1286943	1464784	1643534	40	
	21	936473	1112844	1289900	1467755	1646522	39	
	22	939407	1115789	1292857	1470727	1649510	38	
	23	942342	1118734	1295815	1473699	1652499	37	
	24	945277	1121680	1298773	1476671	1655488	36	
	25	948212	1124625	1301731	1479644	1658477	35	
	26	951147	1127571	1304689	1482617	1661466	34	
	27	954083	1130517	1307648	1485590	1664456	33	
	28	957019	1133463	1310607	1488563	1667446	32	
	29	959954	1136409	1313566	1491536	1670436	31	
	30	962890	1139355	1316525	1494510	1673426	30	
		84	83	82	81	80		

The degrees of the Quadrant for the Tangents

of Tangents.

of the Arches of the same Quadrant.

	5	6	7	8	9	
30	962890	1139355	1316525	1494510	1673426	30
31	965826	1142302	1319485	1497484	1676417	29
32	968763	1145249	1322445	1500458	1679408	28
33	971699	1148196	1325405	1503433	1682399	27
34	974636	1151144	1328365	1506408	1685390	26
35	977573	1154092	1331325	1509383	1688382	25
36	980509	1157040	1334285	1512358	1691374	24
37	983446	1159988	1337246	1515334	1694366	23
38	986383	1162936	1340207	1518310	1697358	22
39	989320	1165884	1343168	1521286	1700351	21
40	992257	1168832	1346129	1524262	1703344	20
41	995195	1171781	1349091	1527239	1706337	19
42	998133	1174730	1352053	1530216	1709331	18
43	1001072	1177679	1355015	1533193	1712325	17
44	1004010	1180628	1357977	1536170	1715319	16
45	1006949	1183577	1360940	1539148	1718313	15
46	1009887	1186527	1363903	1542126	1721308	14
47	1012825	1189477	1366866	1545104	1724304	13
48	1015763	1192427	1369830	1548082	1727300	12
49	1018702	1195377	1372793	1551061	1730296	11
50	1021641	1198328	1375757	1554040	1733292	10
51	1024580	1201279	1378721	1557019	1736287	9
52	1027519	1204230	1381686	1559999	1739284	8
53	1030459	1207181	1384650	1562979	1742281	7
54	1033399	1210132	1387615	1565959	1745278	6
55	1036339	1213084	1390580	1568939	1748275	5
56	1039279	1216036	1393545	1571920	1751273	4
57	1042219	1218988	1396510	1574901	1754271	3
58	1045160	1221940	1399476	1577882	1757270	2
59	1048101	1224892	1402442	1580863	1760269	1
60	1051042	1227845	1405408	1583844	1763268	0
	84	83	82	81	80	

of the complements of the Arches of the same Quadrant.

M 2

The Table

The degrees of the Quadrant for the Tangents

	10	11	12	13	14	
0	1763268	1943803	2125565	2308682	2493280	60
1	1766268	1946822	2128605	2311746	2496370	59
2	1769268	1949841	2131646	2314810	2499411	58
3	1772268	1952861	2134687	2317875	2502552	57
4	1775269	1955881	2137729	2320940	2505643	56
5	1778270	1958901	2140771	2324006	2508735	55
6	1781271	1961922	2143814	2327072	2511827	54
7	1784272	1964943	2146857	2330139	2514920	53
8	1787274	1967964	2149900	2333206	2518013	52
9	1790276	1970985	2152944	2336273	2521106	51
10	1793278	1974007	2155988	2339341	2524200	50
11	1796281	1977029	2159032	2342419	2527294	49
12	1799284	1980052	2162077	2345478	2530389	48
13	1802287	1983075	2165122	2348547	2533484	47
14	1805291	1986098	2168167	2351616	2536580	46
15	1808295	1989122	2171213	2354686	2539676	45
16	1811299	1992146	2174259	2357757	2542773	44
17	1814303	1995171	2177306	2360828	2545870	43
18	1817308	1998196	2180352	2363899	2548968	42
19	1820313	2001221	2183400	2366971	2552066	41
20	1823318	2004247	2186448	2370043	2555165	40
21	1826324	2007273	2189496	2373116	2558264	39
22	1829329	2010299	2192544	2376189	2561364	38
23	1832335	2013326	2192544	2379263	2564464	37
24	1835342	2016353	2195593	2382337	2567564	36
25	1838349	2019380	2201692	2385411	2570665	35
26	1841357	2022408	2204742	2388486	2573766	34
27	1844365	2025436	2207792	2391561	2576868	33
28	1847373	2028464	2210843	2394636	2579970	32
29	1850382	2031493	2213894	2397712	2583073	31
30	1853391	2034522	2216946	2400788	2586176	30
	79	78	77	76	75	

The degrees of the Quadrant for the Tangents

of Tangents.
of the Arches of the same Quadrant.

	10	11	12	13	14	
30	1853301	2034522	2216946	2400788	2586176	30
31	1856400	2037552	2219998	2403865	2589280	29
32	1859409	2040582	2223051	2406942	2592384	28
33	1862419	2043612	2226104	2410020	2595489	27
34	1865429	2046643	2229157	2413098	2598594	26
35	1868439	2049674	2232211	2416176	2601700	25
36	1871449	2052705	2235265	2419255	2604806	24
37	1874460	2055737	2238319	2422334	2607912	23
38	1877471	2058769	2241374	2425414	2611019	22
39	1880482	2061801	2244429	2428494	2614126	21
40	1883494	2064834	2247485	2431574	2617234	20
41	1886506	2067867	2250541	2434655	2620342	19
42	1889516	2070900	2253597	2437736	2623451	18
43	1892531	2073934	2256654	2440818	2626560	17
44	1895544	2076968	2259711	2443900	2629670	16
45	1898558	2080002	2262769	2446983	2632780	15
46	1901572	2083037	2265827	2450066	2635891	14
47	1904586	2086073	2268885	2453150	2639002	13
48	1907601	2089109	2271944	2456234	2642114	12
49	1910616	2092145	2275003	2459319	2645226	11
50	1913632	2095182	2278063	2462404	2648339	10
51	1916648	2098219	2281123	2465490	2651452	9
52	1999664	2101256	2284183	2468576	2654566	8
53	1922680	2104293	2287244	2471662	2657680	7
54	1925697	2107331	2290305	2474749	2660795	6
55	1928714	2110369	2293367	2477836	2663910	5
56	1931731	2113407	2296429	2480924	2667026	4
57	1934749	2116446	2299492	2484012	2670142	3
58	1937767	2119485	2302555	2487101	2673258	2
59	1940785	2122525	2305618	2490191	2676375	1
60	1943803	2125565	2308682	2493280	2679492	0
	79	78	77	76	75	

of the complements of the Arches of the same Quadrant.

M 3

The Table
The degrees of the Quadrant for the Tangents

	15	16	17	18	19	
0	2679492	2867453	3057307	3249197	3443276	60
1	2682610	2870601	3060487	3252413	3446530	59
2	2685728	2873749	3063669	3255630	3449785	58
3	2688847	2876898	3066851	3258848	3453040	57
4	2691966	2880048	3070034	3262066	3456296	56
5	2695086	2883198	3073218	3265285	3459553	55
6	2698206	2886349	3076402	3268504	3462810	54
7	2701327	2889501	3079587	3271724	3466068	53
8	2704448	2892653	3082772	3274944	3469326	52
9	2707570	2895806	3085958	3278165	3472585	51
10	2710693	2898960	3085144	3281387	3475845	50
11	2713816	2902114	3092331	3284609	3479105	49
12	2716940	2905268	3095518	3287832	3482366	48
13	2720064	2908423	3198706	3291055	3485628	47
14	2723189	2911578	3101895	3294280	3488891	46
15	2726314	2914734	3105084	3297505	3492154	45
16	2729439	2917890	3108274	3300731	3495418	44
17	2732565	2921047	3111464	3303957	3498683	43
18	2735691	2924204	3114655	3307184	3501949	42
19	2738818	2927362	3117846	3310411	3505215	41
20	2741945	2930520	3121038	3313639	3508482	40
21	2745073	2933679	3124230	3316868	3511749	39
22	2748201	2936839	3127423	3320097	3515017	38
23	2751330	2939999	3130617	3323327	3518286	37
24	2754459	2943160	3133811	3326558	3521555	36
25	2757589	2946321	3137006	3329789	3524825	35
26	2760729	2946483	3140201	3333020	3528096	34
27	2763850	2952645	3143397	3336252	3531368	33
28	2766981	2955808	3146594	3339485	3534640	32
29	2770113	2958971	3149791	3342719	3537913	31
30	2773245	2962135	3152989	3345953	3541186	30
	74	73	72	71	70	

The degrees of the Quadrant for the Tangents

of Tangents.

of the Arches of the same Quadrant.

	15	16	17	18	19	
30	2773245	2962135	3152989	3345953	3541186	30
31	2776378	2965299	3156187	3349188	3544460	29
32	2779511	2968464	3159386	3352423	3547735	28
33	2782645	2971629	3162585	3355659	3551010	27
34	2785779	2974795	3165785	3358896	3554286	26
35	2788914	2977962	3168986	3362133	3557563	25
36	2792050	2981129	3172187	3365371	3560840	24
37	2795186	2984297	3175389	3368610	3564118	23
38	2798323	2987465	3178591	3371850	3567397	22
39	2801460	2990634	3181794	3375090	3570676	21
40	2804597	2993804	3184998	3378331	3573956	20
41	2807735	2996973	3188202	3381572	3577237	19
42	2810873	3000143	3191407	3384814	3580519	18
43	2814012	3003314	3194613	3388057	3583801	17
44	2817151	3006486	3197819	3391300	3587084	16
45	2820291	3009658	3201026	3394544	3590367	15
46	2823432	3012831	3204233	3397798	3593651	14
47	2826573	3016004	3207441	3401033	3596936	13
48	2829714	3019178	3210649	3404279	3600221	12
49	2832856	3022353	3213858	3407525	3603507	11
50	2835999	3025528	3217067	3410772	3606794	10
51	2839142	3028703	3220277	3414020	3610082	9
52	2842286	3031879	3223488	3417268	3613370	8
53	2845430	3035055	3226699	3420517	3616659	7
54	2848575	3038232	3229911	3423766	3619949	6
55	2851720	3041410	3233124	3427016	3623239	5
56	2854866	3044588	3236337	3430167	3626530	4
57	2858012	3047767	3239551	3433518	3629822	3
58	2861159	3050946	3242766	3436770	3633115	2
59	2864306	3054126	3245981	3440023	3636408	1
60	2867453	3057307	3249197	3443276	3639702	0
	74	73	72	71	70	

of the complements of the Arches of the same Quadrant.

M 4

The Table
The degrees of the Quadrant for the Tangents

	20	21	22	23	24	
0	3639702	3838640	4040262	4244748	4452286	60
1	3642997	3841978	4043647	4248182	4455772	59
2	3646293	3845316	4047031	4251617	4459259	58
3	3649589	3848655	4050416	4255052	4462747	57
4	3652886	3851995	4053802	4258488	4466236	56
5	3656183	3855336	4057189	4261925	4469726	55
6	3659481	3858678	4060577	4265363	4473216	54
7	3662780	3862020	4063966	4268801	4476707	53
8	3666079	3865363	4067356	4265363	4480199	52
9	3669379	3868707	4070747	4268801	4483692	51
10	3672680	3872052	4074139	4272240	4487186	50
11	3675982	3875397	4077531	4275680	4490681	49
12	3679284	3878743	4080924	4279121	4494177	48
13	3682587	3882090	4084318	4282563	4497674	47
14	3685891	3885438	4087713	4286006	4501172	46
15	3689195	3888787	4091109	4289450	4504671	45
16	3692500	3892136	4094506	4292895	4508171	44
17	3695806	3895486	4097903	4296340	4511672	43
18	3699113	3898837	4101301	4299786	4515173	42
19	3702420	3902188	4104699	4303233	4518675	41
20	3705728	3905540	4108097	4306681	4522178	40
21	3709037	3908893	4111497	4310130	4525682	39
22	3712347	3912247	4114898	4313580	4529187	38
23	3715657	3915601	4118300	4317031	4532693	37
24	3718968	3918956	4121703	4327387	4536200	36
25	3722279	3922312	4125107	4330841	4539708	35
26	3725591	3925669	4128511	4334296	4543217	34
27	3728904	3929027	4131916	4337752	4546727	33
28	3732218	3932385	4135322	4341209	4550238	32
29	3735533	3935744	4138728	4344666	4553750	31
30	3738848	3939104	4142135	4348124	4557264	30
	69	68	67	66	65	

The degrees of the Quadrant for the Tangents

of Tangents.

of the Arches of the same Quadrant.

	20	21	22	23	24	
30	3738848	3939104	4142135	4348124	4557264	30
31	3742164	3942465	4145544	4351583	4560778	29
32	3745480	3945826	4148953	4355043	4564293	28
33	3748797	3949188	4152363	4358504	4567809	27
34	3752115	3952551	4155773	4361966	4571326	26
35	3755434	3955915	4159184	4365429	4574843	25
36	3758753	3959280	4162596	4368893	4578361	24
37	3762073	3962646	4166009	4372357	4581880	23
38	3765394	3966012	4169423	4375822	4585400	22
39	3768716	3969379	4172838	4379288	4588921	21
40	3772038	3972746	4176255	4382755	4592443	20
41	3775361	3976114	4179672	4386223	4595966	19
42	3778685	3979483	4183090	4389692	4599490	18
43	3782010	3982853	4186509	4393162	4603015	17
44	3785335	3986224	4189928	4396633	4606541	16
45	3788661	3989596	4193348	4400105	4610068	15
46	3791988	3992969	4196769	4403578	4613596	14
47	3795315	3996342	4200191	4407051	4617125	13
48	3798643	3999716	4203613	4410525	4620654	12
49	3801972	4003090	4207036	4414000	4624184	11
50	3805302	4006465	4210460	4417476	4627715	10
51	3808632	4009841	4213885	4420953	4631247	9
52	3811963	4013217	4217311	4424432	4634780	8
53	3815295	4016594	4220738	4427910	4638314	7
54	3818628	4019972	4224165	4431390	4641849	6
55	3821961	4023351	4227593	4434871	4645385	5
56	3825295	4026731	4231022	4438352	4648922	4
57	3828630	4030112	4234452	4441834	4652460	3
58	3831966	4033494	4237883	4445317	4655999	2
59	3835303	4036877	4241315	4448801	4659540	1
60	3838640	4040262	4244748	4452286	4663081	0
	69	68	67	66	65	

of the complements of the Arches of the same Quadrant.

The Table

The degrees of the Quadrant for Tangents

	25	26	27	28	29	
0	4663081	4877328	5095254	5317024	5543090	60
1	4666623	4880930	5098919	5320826	5546893	59
2	4670166	4884533	5102585	5324559	5550697	58
3	4673710	4888137	5106252	5328293	5554503	57
4	4677255	4891742	5109920	5332028	5558310	56
5	4680801	4895347	5113589	5335765	5562118	55
6	4684348	4898953	5117259	5339503	5565927	54
7	4687896	4902560	5120930	5343242	5569738	53
8	4691444	4906168	5124602	5346982	5573550	52
9	4694993	4909777	5128275	5350723	5577363	51
10	4698543	4913387	5131949	5354465	5581177	50
11	4702094	4916998	5135625	5358209	5584993	49
12	4705646	4920610	5139302	5361954	5588810	48
13	4709199	4924223	5142980	5365700	5592628	47
14	4712753	4927838	5146659	5369447	5596447	46
15	4716308	4931454	5150339	5373195	5600268	45
16	4719864	4935071	5154020	5376944	5604090	44
17	4723422	4938689	5157702	5380694	5607913	43
18	4726981	4942308	5161385	5384445	5611737	42
19	4730541	4945928	5165069	5388198	5615562	41
20	4734102	4949549	5168755	5391952	5619388	40
21	4737664	4953171	5172442	5395707	5623216	39
22	4741227	4956794	5176130	5399463	5627045	38
23	4744790	4960418	5179819	5403221	5630875	37
24	4748354	4964043	5183509	5406980	5634707	36
25	4751919	4967669	5187200	5410740	5638540	35
26	4755485	4971296	5190892	5414501	5642374	34
27	4759052	4974924	5194585	5418263	5646210	33
28	4762620	4978553	5198279	5422026	5650047	32
29	4766189	4982184	5201974	5425791	5653885	31
30	4769759	4985816	5205670	5429557	5657725	30
	64	63	62	61	60	

The degrees of the Quadrant for the Tangents

of Tangents. 86
of the Arches of the same Quadrant.

	25	26	27	28	29	
30	4769759	4985816	5205670	5429557	5657725	30
31	4773330	4989448	5209368	5433324	5661566	29
32	4776902	4993081	5213067	5437092	5665408	28
33	4780475	4996716	5216767	5440861	5669251	27
34	4784049	5000352	5220468	5444632	5673096	26
35	4787624	5003989	5224170	5448404	5676942	25
36	4791200	5007627	5227873	5452177	5680789	24
37	4794777	5011266	5231577	5455951	5684637	23
38	4798355	5014906	5235283	5459726	5688486	22
39	4801934	5018547	5238990	5463503	5692337	21
40	4805515	5022189	5242698	5467281	5696189	20
41	4809096	5025832	5246407	5471060	5700043	19
42	4812678	5029476	5250117	5474840	5703898	18
43	4816261	5033121	5253828	5478621	5707754	17
44	4819845	5036767	5257540	5482404	5711611	16
45	4823430	5040414	5261254	5486188	5715469	15
46	4827016	5044062	5264969	5489973	5719329	14
47	4830603	5047712	5268685	5493759	5723190	13
48	4834191	5051363	5272402	5497546	5727052	12
49	4837780	5055015	5276120	5501335	5730916	11
50	4841371	5058668	5279839	5505125	5734781	10
51	4844962	5062322	5283959	5508916	5738647	9
52	4848554	5065977	5287280	5512708	5742515	8
53	4852147	5069633	5291003	5516501	5746384	7
54	4855741	5073290	5294727	5520296	5750254	6
55	4859336	5076948	5298452	5524092	5754125	5
56	4862932	5080607	5302178	5527889	5757998	4
57	4866529	5084267	5305905	5531687	5761872	3
58	4870127	5087928	5309633	5535487	5765747	2
59	4873727	5091590	5313363	5539288	5769624	1
60	4877328	5095254	5317094	5543090	5773502	0
	64	63	62	61	60	

of the complements of the Arches of the same Quadrant.

The Table

The degrees of the Quadrant for Tangents

	30	31	32	33	34	
0	5773502	6008606	6248693	6494076	6745085	60
1	5777381	6012566	6252738	6498212	6749318	59
2	5781262	6016528	6256785	6502350	6753553	58
3	5785144	6020491	6260834	6506489	6757789	57
4	5789027	6024455	6264884	6510630	6762027	56
5	5792911	6028420	6268935	6514773	6766267	55
6	5796797	6032387	6272988	6518917	6770508	54
7	5800684	6036355	6277042	6523063	6774751	53
8	5804572	6040324	6281098	6527200	6778996	52
9	5808462	6044295	6285155	6531359	6783243	51
10	5812353	6048267	6289214	6535510	6787491	50
11	5816245	6052241	6293274	6539662	6791741	49
12	5820139	6056216	6297336	6543816	6795993	48
13	5824034	6060193	6301399	6547971	6800246	47
14	5827930	6064171	6305464	6552128	6804501	46
15	5831828	6068150	6309530	6556287	6808758	45
16	5835727	6072131	6313598	6560447	6813016	44
17	5839627	6076113	6317667	6564609	6817276	43
18	5843528	6080096	6321738	6568772	6821538	42
19	5847431	6084081	6325810	6572937	6825801	41
20	5851335	6088067	6329883	6577103	6830066	40
21	5855241	6092055	6333958	6581271	6834333	39
22	5859148	6096044	6338034	6585440	6838602	38
23	5863056	6100035	6342112	6589611	6842872	37
24	5866966	6104027	6346191	6593784	6847144	36
25	5870877	6108020	6350272	6597958	6851417	35
26	5874489	6112015	6354355	6602134	6855692	34
27	5878702	6116011	6358439	6606312	6859969	33
28	5882617	6120009	6362525	6610491	6864247	32
29	5886533	6124008	6366613	6614672	6868527	31
30	5890450	6128008	6370702	6618855	6872809	30
	59	58	57	56	55	

The degrees of the Quadrant for the Tangents

of Tangents.

of the Arches of the same Quadrant.

	30	31	32	33	34	
30	5890450	6128008	6370702	6618855	6872809	30
31	5894369	6132010	6374792	6623039	6877093	29
32	5898289	6136013	6378884	6627225	6881379	28
33	5902211	6140018	6382977	6631413	6885666	27
34	5906134	6144024	6387072	6635603	6889955	26
35	5910058	6148032	6391169	6639792	6894246	25
36	5913984	6152041	6395267	6643984	6898539	24
37	5917911	6156052	6399366	6648178	6902833	23
38	5921839	6160064	6403467	6652373	6907129	22
39	5925769	6164077	6407569	6656570	6911426	21
40	5929700	6168092	6411673	6660768	6915725	20
41	5933633	6172108	6415779	6664968	6920026	19
42	5937567	6176126	6419886	6669170	6924329	18
43	5941502	6180147	6423995	6673373	6928634	17
44	5945438	6184168	6428105	6677578	6932940	16
45	5949376	6188190	6432216	6681785	6937248	15
46	5953315	6192213	6436329	6685994	6941558	14
47	5957255	6196237	6440444	6690204	6945869	13
48	5961197	6200263	6444560	6694416	6950182	12
49	5965140	6204290	6458678	6698630	6954497	11
50	5969084	6208319	6452798	6702845	6958813	10
51	5973030	6212350	6456919	6707062	6963131	9
52	5976977	6216382	6461042	6711281	6967451	8
53	5980926	6220416	6465166	6715501	6971773	7
54	5984876	6224451	6469292	6719723	6976097	6
55	5988827	6228488	6473419	6723946	6980423	5
56	5992780	6232526	6477548	6728171	6984750	4
57	5996734	6246566	6481678	6732397	6989079	3
58	6000690	6240607	6485809	6736625	6993409	2
59	6004647	6244649	6489942	6740854	6997741	1
60	6008606	6248693	6494076	6745085	7002075	0
	95	58	57	56	55	

of the complements of the Arches of the same Quadrant.

The Table
The degrees of the Quadrant for Tangents

	35	36	37	38	39	
0	7002075	7265424	7535541	7812856	8097840	60
1	7006411	7269869	7540103	7817542	8102658	59
2	7010749	7274316	7544667	7822230	8107478	58
3	7015088	7278765	7549233	7826920	8112300	57
4	7019429	7283216	7553801	7831612	8117124	56
5	7023772	7287669	7558371	7836306	8121951	55
6	7028117	7292124	7562943	7841002	8126780	54
7	7032463	7296581	7567517	7845700	8131611	53
8	7036811	7301040	7572093	7850400	8136444	52
9	7041161	7305501	7576670	7855102	8141280	51
10	7045513	7309563	7581249	7859807	8146118	50
11	7049867	7314427	7585830	7864514	8150958	49
12	7054223	7318893	7590413	7869223	8155801	48
13	7058581	7323361	7594999	7873934	8160646	47
14	7062940	7327831	7599587	7878647	8165493	46
15	7067301	7332303	7604177	7883363	8170343	45
16	7071664	7336777	7608769	7888081	8175195	44
17	7076029	7341253	7613363	7892801	8180049	43
18	7070395	7345731	7617959	7897523	8184905	42
19	7084763	7350219	7622557	7902247	8189764	41
20	7089133	7354691	7627157	7906973	8194625	40
21	7093505	7359174	7631759	7911702	8199488	39
22	7097879	7363659	7636363	7916433	8204354	38
23	7102254	7368146	7640969	7921166	8209222	37
24	7106631	7372635	7645577	7925901	8214092	36
25	7111010	7377126	7650187	7930638	8218965	35
26	7115391	7381619	7654799	7935378	8223840	34
27	7119773	7386114	7659413	7940120	8228717	33
28	7124167	7390611	7664030	7944864	8233597	32
29	7128543	7395110	7668649	7949610	8238479	31
30	7132931	7399610	7663270	7954358	8243363	30
	54	53	52	51	50	

The degrees of the Quadrant for the Tangents

of Tangents.

of the Arches of the same Quadrant.

	35	36	37	38	39	
30	7132931	7399610	7673270	7954358	8243363	30
31	7137321	7404112	7677893	7959109	8248250	29
32	7141713	7408616	7682518	7963862	8253139	28
33	7146106	7413122	7687145	7968617	8258031	27
34	7150501	7417630	7691774	7973374	8262925	26
35	7154898	7422140	7696405	7978133	8267821	25
36	7159297	7426652	7701038	7982895	8272720	24
37	7163698	7431167	7705673	7987659	8277621	23
38	7168100	7435684	7710310	7992425	8282524	22
39	7172504	7440203	7714949	7997193	8287429	21
40	7176910	7444724	7719590	8001963	8292337	20
41	7181318	7449246	7724233	8006736	8297247	19
42	7185728	7453770	7728878	8011511	8302160	18
43	7190140	7458296	7733525	8016288	8307075	17
44	7194554	7462824	7738175	8021067	8311992	16
45	7198970	7467354	7742827	8025849	8316912	15
46	7203387	7471886	7747481	8030633	8321834	14
47	7207806	7476420	7752137	8035419	8326759	13
48	7212227	7480956	7756795	8040207	8331686	12
49	7216650	7485494	7761455	8044997	8336615	11
50	7221075	7490033	7766117	8049790	8341547	10
51	7225502	7494574	7770781	8054585	8346481	9
52	7229931	7499117	7775447	8059382	8351418	8
53	7234362	7503663	7780116	8064181	8356357	7
54	7238794	7508211	7784787	8068983	8361298	6
55	7243228	7512761	7789460	8073787	8366242	5
56	7247664	7517313	7794135	8078593	8371188	4
57	7252102	7521867	7798812	8083401	8376136	3
58	7256541	7526423	7803491	8088212	8381087	2
59	7260982	7530981	7808172	8093025	8386040	1
60	7265424	7535541	7812856	8097840	8390996	0
	54	53	52	51	50	

of the complements of the Arches of the same Quadrant.

The Table
The degrees of the Quadrant for Tangents

	40	41	42	43	44	
0	8390996	8692867	9004040	9325151	9656888	60
1	8395954	8697975	9009308	9330591	9662511	59
2	8400915	8703085	9014579	9336034	9668137	58
3	8405878	8708198	9019853	9341480	9673766	57
4	8410844	8713344	9025130	9346929	9679398	56
5	8415812	8718433	9030410	9352381	9685034	55
6	8420782	8723555	9035693	9357835	9690674	54
7	8425754	8728679	9040978	9363292	9696315	53
8	8430729	8733806	9046266	9368752	9701960	52
9	8435706	8738935	9051557	9374215	9707609	51
10	8440686	8744067	9056850	9379682	9713261	50
11	8445668	8749201	9062146	9385152	9718916	49
12	8450653	8754338	9067445	9390625	9724574	48
13	8455640	8759478	9072747	9396101	9730235	47
14	8460630	8764620	9078052	9401580	9735900	46
15	8465622	8769764	9083360	9407062	9741568	45
16	8470617	8774911	9088670	9412547	9747239	44
17	8475614	8780061	9093983	9418034	9752913	43
18	8480614	8785214	9099299	9423524	9758591	42
19	8485617	8790369	9104618	9429017	9764272	41
20	8490622	8795527	9109940	9434513	9769956	40
21	8495629	8800688	9115265	9440012	9775643	39
22	8500639	8805851	9120593	9445514	9781334	38
23	8505651	8811017	9125923	9451019	9787028	37
24	8510666	8816186	9131256	9456528	9792725	36
25	8515683	8821357	9136592	9462040	9798425	35
26	8520703	8826531	9141930	9467555	9804128	34
27	8525725	8831708	9147271	9473072	9809835	33
28	8530750	8836887	9152615	9478594	9815545	32
29	8535777	8842069	9157962	9484118	9821258	31
30	8540806	8847353	9163312	9489645	9826974	30
	49	48	74	46	45	

The degrees of the Quadrant for the Tangents

of Tangents.

of the Arches of the same Quadrant.

	40	41	42	43	44	
30	8540806	8847253	9163312	9489645	9826974	30
31	8545838	8852440	9168665	9495175	9832694	29
32	8550872	8857630	9174021	9400708	9838417	28
33	8555909	8862822	9179380	9506244	9844143	27
34	8560949	8868017	9184741	9511783	9849872	26
35	8565991	8873015	9190105	9517325	9855605	25
36	8571036	8878415	9195472	9522870	9861341	24
37	8576083	8883628	9200842	9528419	9867180	23
38	8581133	8888824	9206215	9533971	9872922	22
39	8586185	8899033	9211590	9539526	9878668	21
40	8591239	8899244	9216968	9545084	9884317	20
41	8596296	8904458	9222349	9550645	9890070	19
42	8601355	8909675	9227733	9556209	9895826	18
43	8606417	8914894	9233120	9561776	9901585	17
44	8611482	8920116	9238510	9567346	9907347	16
45	8616549	8925341	9243903	9572919	9913113	15
46	8621619	8930568	9249299	9578495	9918882	14
47	8626692	8935798	9254698	9584074	9924654	13
48	8631767	8941031	9260100	9589656	9930430	12
49	8636845	8946267	9265505	9595241	9936209	11
50	8641926	8951506	9270913	9600830	9941991	10
51	8647009	8956747	9276324	9606422	9947777	9
52	8652095	8961991	9281738	9612017	9953566	8
53	8657683	8967238	9287155	9617615	9959359	7
54	8662273	8972487	9292574	9623216	9965155	6
55	8667366	8977739	9297996	9628820	9970954	5
56	8672461	8982994	9303421	9634427	9976756	4
57	8677559	8988252	9308849	9640037	9982562	3
58	8682659	8993512	9314280	9645651	9988371	2
59	8687762	8998775	9319714	9651268	9994184	1
60	8692867	9004040	9325151	9656888	10000000	0
	49	48	47	46	45	

of the complements of the Arches of the same Quadrant.

N

The Table
The degrees of the Quadrant for the Tangents

	45	46	47	48	
0	10000000	10355302	10723686	11106124	60
1	10005820	10361332	10729942	11112623	59
2	10011643	10367365	10736202	11119126	58
3	10017469	10373402	10742466	11125634	57
4	10023299	10379443	10748734	11132146	56
5	10029132	10385487	10755006	11138662	55
6	10034968	10391535	10761282	11145182	54
7	10040808	10397587	10767562	11151706	53
8	10046651	10403643	10773845	11158235	52
9	10052497	10409702	10780132	11164768	51
10	10058347	10415765	10786423	11171305	50
11	10064201	10421832	10792718	11177846	49
12	10070058	10427902	10799017	11184392	48
13	10075918	10433976	10805320	11190942	47
14	10081782	10340054	10811627	11197496	46
15	10087649	10446135	10817938	11204054	45
16	10093520	10452220	10824253	11210617	44
17	10099394	10458309	10830572	11217184	43
18	10105272	10464401	10836895	11223755	42
19	10111153	10470497	10843222	11230330	41
20	10117038	10476597	10849554	11236910	40
21	10122926	10482701	10855889	11243494	39
22	10128818	10488808	10862228	11250082	38
23	10134713	10494919	10868571	11256675	37
24	10140611	10501034	10874918	11263272	36
25	10146513	10507153	10881269	11269873	35
26	10152418	10513275	10887624	11276478	34
27	10158327	10519401	10893983	11283088	33
28	10164239	10525531	10900346	11289702	32
29	10170154	10531664	10906713	11296321	31
30	10176073	10537801	10913084	11302944	30
	44	43	42	41	

The degrees of the Quadrant for the Tangents

of Tangents.
of the Arches of the same Quadrant.

	45	46	47	48	
30	10176073	10537801	10913084	11302944	30
31	10181996	10543942	10919459	11309571	29
32	10187922	10550087	10925838	11316203	28
33	10193852	10556235	10932221	11322899	27
34	10199785	10562387	10938608	11329480	26
35	10205722	10568543	10945000	11336125	25
36	10211663	10574703	10951396	11342774	24
37	10217607	10580867	10957796	11349428	23
38	10223555	10587034	10964200	11356086	22
39	10229506	10593205	10970608	11362748	21
40	10235460	10599280	10977020	11369415	20
41	10241418	10605559	10983436	11376086	19
42	10247380	10611742	10989856	11382762	18
43	10253345	10617929	10996280	11389442	17
44	10259314	10624119	11002708	11396126	16
45	10265286	10630313	11009140	11402815	15
46	10271262	10636511	11015577	11409508	14
47	10277242	10642713	11022028	11416206	13
48	10283225	10648919	11028463	11422908	12
49	10289212	10655128	11034912	11429615	11
50	10295202	10661341	11041365	11436326	10
51	10301196	10667558	11047832	11443042	9
52	10307193	10673779	11054283	11449762	8
53	10313194	10680004	11060748	11456487	7
54	10319199	10686233	11067218	11463216	6
55	10325207	10692466	11073692	11469950	5
56	10331219	10698702	11080170	11476688	4
57	10337234	10704942	11086652	11483431	3
58	10343253	10711186	11093138	11490178	2
59	10349276	10717434	11099629	11496929	1
60	10355302	10723686	11106124	11503684	0
	44	43	42	41	

of the complements of the Arches of the same Quadrant.

The Table
The degrees of the Quadrant for the Tangents

	49	50	51	52	
0	11503684	11917537	12348972	12799416	60
1	11510444	11924580	12356320	12807093	59
2	11517208	11931628	12363673	12814776	58
3	11523977	11938680	12371032	12822465	57
4	11530751	11945737	12378394	12830159	56
5	11537529	11952799	12385762	12837859	55
6	11544312	11959866	12393136	12845565	54
7	11551100	11966938	12400515	12853277	53
8	11557893	11974015	12407999	12860994	52
9	11564691	11981097	12415288	12868717	51
10	11571494	11988183	12422683	12876445	50
11	11578301	11995274	12430083	12884179	49
12	11585112	12002370	12437489	12891919	48
13	11591928	12009471	12444900	12899665	47
14	11598748	12016578	12452317	12907417	46
15	11605572	12023690	12459739	12915175	45
16	11612401	12030807	12467167	12922939	44
17	11619234	12037929	12474600	12930709	43
18	11626072	12045056	12482029	12938485	42
19	11632915	12052188	12489484	12946267	41
20	11639763	12059325	12496934	12954055	40
21	11646615	12066467	12504389	12961848	39
22	11653472	12073614	12511850	12969647	38
23	11660334	12080766	12519316	12977457	37
24	11667200	12087923	12526787	12985263	36
25	11674071	12095085	12534264	12993080	35
26	11680947	12102252	12541746	13000903	34
27	11687827	12109424	12549233	13008732	33
28	11694712	12116601	12556725	13016567	32
29	11701602	12123783	12564222	13024407	31
30	11708497	12130970	12571724	13032253	30
	40	39	38	37	

The degrees of the Quadrant for the Tangents

of Tangents.

of the Arches of the same Quadrant.

	49	50	51	52	
30	11708497	12130970	12571724	13032253	30
31	11715396	12138162	12579232	13040105	29
32	11722300	12145359	12586746	13047963	28
33	11729208	12152561	12594265	13055827	27
34	11736121	12159768	12601790	13063697	26
35	11743039	12166981	12609321	13071573	25
36	11749962	12174199	12616858	13079455	24
37	11756989	12181422	12624400	13087343	23
38	11763821	12188650	12631948	13095237	22
39	11770758	12195883	12639501	13103138	21
40	11777700	12203121	12647060	13111045	20
41	11784646	12210364	12654624	13118958	19
42	11791597	12217613	12662194	13126877	18
43	11798553	12224867	12669769	13134802	17
44	11805514	12232126	12677350	13142732	16
45	11812479	12239390	12684937	13150668	15
46	11819449	12246659	12692530	13158610	14
47	11826424	12253933	12700128	13166558	13
48	11833404	12261212	12707732	13174512	12
49	11840388	12268496	12715341	13182472	11
50	11847377	12275786	12722956	13190438	10
51	11854371	12283081	12730577	13198411	9
52	11861370	12290381	12738203	13206390	8
53	11868374	12297687	12745835	13214375	7
54	11875383	12304998	12753473	13222367	6
55	11882397	12312314	12761116	13230365	5
56	11889417	12319635	12768765	13238369	4
57	11896438	12326961	12776420	13246379	3
58	11903466	12334293	12784080	13254396	2
59	11910499	12341630	12791745	13262419	1
60	11917537	12348972	12799416	13270448	0
	40	39	38	38	

of the complements of the Arches of the same Quadrant.

The Table

The degrees of the Quadrant for the Tangents

	53	54	55	56	
0	13270448	13763820	14281480	14825610	60
1	13278483	13772243	14290325	14834491	59
2	13286524	13780673	14299177	14844230	58
3	13294571	13789109	14308037	14853553	57
4	13302624	13797552	14316905	14862884	56
5	13310683	13806002	14325780	14872223	55
6	13318749	13814452	14334662	14881570	54
7	13326821	13822922	14343552	14890925	53
8	13334899	13831392	14352451	14909288	52
9	13342984	13839869	14361354	14909659	51
10	13351075	13848352	14370266	14919038	50
11	13359172	13856842	14379186	14928426	49
12	13367276	13865339	14388113	14937822	48
13	13375386	13873843	14397048	14947226	47
14	13383502	13882354	14405990	14956638	46
15	13391624	13890872	14414939	14966058	45
16	13399753	13899397	14423896	14975486	44
17	13407888	13907930	14432861	14984923	43
18	13416029	13916470	14441833	14994368	42
19	13424177	13925017	14450812	15003821	41
20	13432331	13933571	14459799	15013283	40
21	13440492	13942131	14468794	15022753	39
22	13448659	13950698	14477797	15032231	38
23	13456832	13959272	14486807	15041717	37
24	13465011	13967853	14495825	15051211	36
25	13473197	13976441	14504850	15060714	35
26	13481390	13985035	14513883	15070225	34
27	13489589	13993636	14522924	15079744	33
28	13497794	14002244	14531972	15089271	32
29	13506006	14010859	14541028	15078807	31
30	13514224	14019481	14550091	15108351	30
	36	35	34	33	

The degrees of the Quadrant for the Tangents

of Tangents.
of the Arches of the same Quadrant.

	53	54	55	56	
30	13514224	14019481	14550091	15108351	30
31	13522449	14028110	14559162	15117903	29
32	13530680	14036746	14568241	15127464	28
33	13538918	14045389	14577327	15137034	27
34	13547162	14054040	14586421	15146612	26
35	13555413	14062698	14595523	15156199	25
36	13563670	14071363	14604633	15165794	24
37	13571834	14080035	14613750	15175398	23
38	13580104	14088715	14622875	15185011	22
39	13588381	14097402	14632007	15194632	21
40	13596764	14106097	14641146	15204261	20
41	13605054	14114798	14650293	15213899	19
42	13613350	14123506	14659449	15223545	18
43	13621653	14132221	14668613	15233200	17
44	13629963	14140923	14677785	15242863	16
45	13638279	14149672	14686965	15252535	15
46	13646602	14158409	14696153	15262216	14
47	13654932	14167153	14705349	15271905	13
48	13663268	14175904	14714553	15281603	12
49	13671610	14184663	14723765	15291309	11
50	13679959	14193429	14732985	15301024	10
51	13688315	14202202	14742212	15310748	9
52	13696677	14210982	14751447	15320481	8
53	13705046	14219769	14760690	15330222	7
54	13713422	14228563	14769941	15339972	6
55	13721805	14237365	14779200	15349730	5
56	13730194	14246174	14788466	15359497	4
57	13738590	14254990	14797740	15369273	3
58	13746993	14263813	14807022	15379057	2
59	13755403	14272643	14816312	15388850	1
60	13763820	14281480	14825610	15398651	0
	36	35	34	33	

of the complements of the Arches of the same Quadrant.

The Table
The degrees of the Quadrant for Tangents

	57	58	59	60	
0	15398651	16003347	16642794	17320508	60
1	15408461	16013710	16653766	17332150	59
2	15418280	16024083	16664749	17343804	58
3	15428108	16034466	16675742	17355469	57
4	15437945	16044859	16686746	17367146	56
5	15447791	16055261	16697760	17378834	55
6	15457646	16065673	16708785	17390534	54
7	15467510	16076095	16719820	17402246	53
8	15477382	16086527	16730866	17413969	52
9	15487263	16096968	16741922	17425704	51
10	15497153	16107419	16752989	17437451	50
11	15507052	16117880	16764067	17449210	49
12	15516960	16128351	16775156	17460981	48
13	15526877	16138832	16786256	17472764	47
14	15536803	16149322	16797367	17484559	46
15	15546738	16159822	16808489	17496366	45
16	15556682	16170332	16819621	17508185	44
17	15566636	16180852	16830764	17520026	43
18	15576599	16191381	16841918	17531869	42
19	15586571	16201920	16853083	17543724	41
20	15596552	16212469	16864259	17555591	40
21	15606542	16224028	16875446	17567470	39
22	15616541	16233597	16886644	17579362	38
23	15626549	16244176	16897853	17591266	37
24	15636566	16254766	16909074	17603182	36
25	15646592	16265366	16920306	17615111	35
26	15656627	16275976	16931549	17627052	34
27	15666671	16286596	16942803	17639006	33
28	15676724	16297226	16954068	17650972	32
29	15686786	16307866	16965344	17662951	31
30	15696857	16318516	16976631	17674941	30
	32	31	30	29	

The degrees of the Quadrant for the Tangents

of Tangents.

of the Arches of the same Quadrant.

	57	58	59	60	
30	15696857	16318516	16976631	17674941	30
31	15706938	16329176	16987929	17686945	29
32	15717028	16339847	16999239	17698960	28
33	15727127	16350528	17010560	17710987	27
34	15737235	16361219	17021892	17723027	26
35	15747353	16371920	17033236	17735079	25
36	15757480	16382631	17044591	17747143	24
37	15767616	16393352	17055957	17759220	23
38	15777761	16404083	17067325	17771309	22
39	15787915	16414824	17078714	17783410	21
40	15798078	16425575	17090115	17795524	20
41	15808251	16436337	17101527	17808651	19
42	15818433	16447109	17112950	17819790	18
43	15828625	16457892	17124384	17831942	17
44	15838827	16468685	17135829	17844107	16
45	15849038	16479488	17147285	17856285	15
46	15859259	16490302	17158752	17868475	14
47	15869489	16501126	17170231	17880678	13
48	15879729	16511960	17181721	17892894	12
49	15889979	16522805	17193222	17905123	11
50	15900238	16533660	17204734	17917364	10
51	15910507	16544526	17216258	17929618	9
52	15920785	16555402	17227794	17941885	8
53	15931073	16566289	17239342	17954164	7
54	15941370	16577186	17250902	17966456	6
55	15951676	16588094	17262473	17978761	5
56	15961992	16599013	17274056	17991079	4
57	15972317	16609942	17285651	18003410	3
58	15982651	16620882	17297258	18015753	2
59	15992994	16631833	17308877	18028109	1
60	16003347	16642794	17320508	18040478	0
	32	31	30	29	

of the complements of the Arches of the same Quadrant.

The Table

The degrees of the Quadrant for Tangents

	61	62	63	64	
0	18040478	18807265	19626104	20503034	60
1	18052860	18820471	19640225	20518180	59
2	18065255	18833691	19654362	20533344	58
3	18077663	18846925	19668516	20548526	57
4	18090084	18860174	19682686	20563726	56
5	18102518	18873437	19696872	20578945	55
6	18114966	18886715	19711074	20594182	54
7	18127427	18900007	19725293	20609437	53
8	18139901	18913314	19739528	20624711	52
9	18152388	18926636	19753780	20640003	51
10	18164889	18939972	19768048	20655313	50
11	18177403	18953323	19782333	20670642	49
12	18189930	18966689	19796634	20685989	48
13	18202470	18980070	19810951	20701355	47
14	18215024	18993466	19825285	20716739	46
15	18227591	19006876	19839635	20732142	45
16	18240171	19020301	19854002	20747564	44
17	18252765	19033741	19868386	20763004	43
18	18265372	19047196	19882786	20778463	42
19	18277992	19060665	19897203	20793941	41
20	18290626	19074149	19911637	20809438	40
21	18303273	19087648	19926088	20824953	39
22	18315934	19101162	19940555	20840487	38
23	18328608	19114691	19955039	20856040	37
24	18341296	19128235	19669540	26871612	36
25	18353997	19141795	19984057	20887202	35
26	18366712	19155370	19998591	20902811	34
27	18379440	19168960	20013142	20918439	33
28	18392182	19182565	20027709	20934086	32
29	18404938	19196185	20042297	20949752	31
30	18417707	19209821	20056898	20965436	30
	28	27	26	25	

The degrees of the Quadrant for the Tangents

of Tangents.
of the Arches of the same Quadrant.

	61	62	63	64	
30	18417707	19209821	20056898	20965436	30
31	18430490	19223472	20071516	20981140	29
32	18443287	19237138	20086152	20996863	28
33	18456098	19250819	20100805	21012605	27
34	18468922	19264516	20115475	21028367	26
35	18481760	19278228	20130163	21044148	25
36	18494612	19291955	20144868	21059949	24
37	18507478	19305698	20159590	21075769	23
38	18520357	19319456	20174329	21091609	22
39	18533250	19333230	20189086	21107468	21
40	18546157	19347019	20203860	21123347	20
41	18559078	19360824	20218651	21139246	19
42	18572013	19374644	20233460	21155164	18
43	18584962	19388480	20248286	21171102	17
44	18597925	19402331	20263130	21187059	16
45	18610902	19416198	20277991	21203036	15
46	18623894	19430081	20292870	21219032	14
47	18636900	19443980	20307767	21235048	13
48	18649920	19457894	20322681	21251083	12
49	18662954	19471824	20337613	21267138	11
50	18676002	19485770	20352563	21283213	10
51	18689064	19499732	20367531	21299308	9
52	18702140	19513710	20382516	21315423	8
53	18715231	19527704	20397519	21331558	7
54	18728335	19541714	20412539	21347713	6
55	18741454	19555739	20427577	21363888	5
56	18754587	19569780	20442633	21380083	4
57	18767735	19583837	20457706	21396298	3
58	18780897	19597910	20472797	21412534	2
59	18794074	19611999	20487906	21428790	1
60	18807265	19626104	20503034	21445067	0
	28	27	26	25	

of the complements of the Arches of the same Quadrant.

The Table
The degrees of the Quadrant for the Tangents

	65	66	67	68	
0	21445067	22460371	23558529	24750869	60
1	21461364	22477965	23577595	24771613	59
2	21477681	22495582	23596687	24792387	58
3	21494019	22513222	23615805	24813191	57
4	21510377	22530885	23634950	24834024	56
5	21526756	22548571	23654121	24854887	55
6	21543155	22566281	23673318	24875780	54
7	21559575	22584014	23692542	24896704	53
8	21576015	22601771	23711793	24917659	52
9	21592475	22619551	23731071	24938644	51
10	21608956	22637355	23750375	24959659	50
11	21625458	22655183	23769706	24980705	49
12	21641981	22673034	23789064	25001782	48
13	21658525	22690909	23808448	25022890	47
14	21675090	22708808	23827859	25044029	46
15	21691676	22726730	23847297	25065198	45
16	21708283	22744676	23866762	25086398	44
17	21724911	22762646	23886254	25107629	43
18	21741559	22780639	23905773	25128991	42
19	21758228	22798656	23925320	25150183	41
20	21774918	22816696	23944895	25171506	40
21	21791629	22834760	23964496	25192861	39
22	21808362	22852848	23984124	25214248	38
23	21825116	22870960	24003771	25235666	37
24	21841892	22889096	24023462	25257116	36
25	21858689	22907256	24043172	25278597	35
26	21875508	22925441	24062910	25300115	34
27	21892348	22943650	24082675	25321655	33
28	21909210	22961883	24102468	25343232	32
29	21926094	22980141	24122289	25364841	31
30	21943000	22998424	24142137	25386482	30
	24	23	22	21	

The degrees of the Quadrant for the Tangents

of Tangents.

of the Arches of the same Quadrant.

	65	66	67	68	
30	21944000	22998424	24142137	25386482	30
31	21959926	23016731	24162013	25408154	29
32	21976874	23035062	24181917	25429858	28
33	21993843	23053418	24201849	25451594	27
34	22010834	23071798	24221809	25473362	26
35	22027846	23090203	24241798	25495162	25
36	22044879	23108632	24261815	25516995	24
37	22061934	23127086	24281860	25538860	23
38	22079011	23145565	24301934	25560758	22
39	22096109	23164068	24322037	25582688	21
40	22113229	23182597	24342169	25604651	20
41	22130372	23201151	24362329	25626647	19
42	22147537	23219730	24382518	25648675	18
43	22164725	23238335	24402735	25670736	17
44	22181935	23256965	24422981	25692830	16
45	22199168	23275621	24443256	25714957	15
46	22216424	23294302	24463559	25737218	14
47	22233703	23313008	24483891	25759312	13
48	22251004	23331740	24504252	25781540	12
49	22268328	23350498	24524641	25803801	11
50	22285675	23369281	24545061	25826096	10
51	22303044	23388092	24565509	25848424	9
52	22320435	23406927	24585986	25870786	8
53	22337848	23425788	24606492	25893181	7
54	22355284	23444674	24627028	25915610	6
55	22372742	23463586	24647594	25938073	5
56	22390223	23482523	24668189	25960569	4
57	22407726	23501486	24688814	25983099	3
58	22425252	23520475	24709469	26005663	2
59	22442800	23539489	24730154	26028261	1
60	22460371	23558529	24750869	26050893	0
	24	23	22	21	

of the complements of the Arches of the same Quadrant.

The Table

The degrees of the Quadrant for Tangents

	69	70	71	72	
0	26050893	27474777	29042105	30776834	60
1	26073559	27499665	29069569	30807323	59
2	26096260	27524592	29097080	30837866	58
3	26118996	27549559	29124638	30868465	57
4	26141766	27574565	29152243	30899119	56
5	26164571	27599612	29179895	30929828	55
6	26187411	27624699	29207595	30960593	54
7	26210286	27649827	29235343	30991413	53
8	26233196	27674995	29263139	31022289	52
9	26256141	27700204	29290982	31053221	51
10	26279120	27725453	29318873	31084208	50
11	26302135	27750742	29346811	31115252	49
12	26325185	27776072	29374797	31146352	48
13	26348270	27802443	29402831	31177508	47
14	26371390	27826855	29430913	31208720	46
15	26394546	27852308	29459043	31239989	45
16	26417738	27877803	29487221	31271315	44
17	26440966	27903339	29515446	31302698	43
18	26464229	27928917	29543719	31334138	42
19	26487528	27954536	29572041	31365636	41
20	26510863	27980196	29600411	31397191	40
21	26534234	28005898	29628831	31428805	39
22	26557641	28031642	29657301	31460476	38
23	26581084	28057429	29685820	31492205	37
24	26604563	28083258	29714388	31523992	36
25	26628079	28109129	29743006	31555838	35
26	26651631	28135043	29771674	31587742	34
27	26675220	28160999	29800392	31619705	33
28	26698845	28186998	29829160	31651727	32
29	26722507	28213040	29857978	31683807	31
30	26746206	28239125	29886847	31715946	30
	20	19	18	17	

The degrees of the Quadrant for the Tangents

of Tangents.

of the Arches of the same Quadrant.

	69	70	71	72	
30	26746206	28239125	29886847	31715946	30
31	26769942	28265253	29915765	31748144	29
32	26793716	28291424	29944734	31780401	28
33	26816527	28317638	29973753	31812717	27
34	26841375	28343895	30002823	31845093	26
35	26865260	28379195	30031943	31877528	25
36	26889183	28396539	30061113	31910 24	24
37	26913143	28412926	30090334	31942580	23
38	26937141	28449357	30119605	31975197	22
39	26961177	28475832	30148927	32007875	21
40	26985251	28502350	30178299	32040613	20
41	27009362	28528913	30207723	32073413	19
42	27033511	28555520	30237200	32106275	18
43	27057698	28582172	30266730	32139200	17
44	27081922	28608868	30296312	32172187	16
45	27106184	28635608	30325947	32205237	15
46	27130484	28662393	30355635	32238349	14
47	27154823	28689222	30385375	32271524	13
48	27179200	28716096	30415169	32304762	12
49	27203616	28743015	30445015	32338064	11
50	27228070	28769979	30474915	32371430	10
51	27252563	28796987	30504867	32404858	9
52	27272025	28824040	30534872	32438348	8
53	27301667	28851139	30564930	32471901	7
54	27326278	28878283	30595041	32505517	6
55	27350929	28905472	30625205	32539196	5
56	27375620	28932707	30655423	32572937	4
57	27400350	28959988	30685695	32606741	3
58	27425120	28987315	30716020	32640907	2
59	27449929	29014687	30746400	32674536	1
60	27474777	29042105	30776834	32708528	0
	20	19	18	17	

of the complements of the Arches of the same Quadrant.

The Table
The degrees of the Quadrant for Tangents

	73	74	75	76	
0	32708528	34874151	37320517	40107808	60
1	32745286	34912477	37363987	40157569	59
2	32776709	34950881	37407551	40207446	58
3	32810898	34989364	37451210	40257440	57
4	32845153	35027925	37494964	40307552	56
5	32879747	35066565	37538814	40357781	55
6	32913862	35105283	37582760	40408129	54
7	32948317	35144080	37626803	40458596	53
8	32982839	35182956	37670943	40509183	52
9	33017427	35221911	37715180	40559890	51
10	33052082	35260945	37759515	40610718	50
11	33086802	35300059	37803948	40661665	49
12	33121588	35339253	37848479	40712731	48
13	33156441	35378528	37893109	40763917	47
14	33191362	35417883	37937838	40815224	46
15	33226351	35457320	37982666	40866652	45
16	33261408	35496838	38027592	40918201	44
17	33296534	35536438	38072616	40969871	43
18	33331728	35576121	38117740	41021663	42
19	33366990	35615888	38162963	41073577	41
20	33402321	35655739	38208285	41125614	40
21	33437720	35695672	38253708	41177775	39
22	33473188	35735689	38299232	41230062	38
23	33508725	35775789	38344857	41282475	37
24	33544330	35815973	38390584	41335015	36
25	33580005	35856241	38436414	41387683	35
26	33615750	35896593	38482347	41440480	34
27	33651566	35937029	38528384	41493407	33
28	33687453	35977550	38574525	41546464	32
29	33723410	36018156	38620772	41599653	31
30	33759438	36058848	38667125	41652974	30
	16	15	14	13	

The degrees of the Quadrant for the Tangents

of Tangents.
of the Arches of the same Quadrant.

	73	74	75	76	
30	33759438	36058848	38667125	41652974	30
31	33795535	36099623	38713580	41706424	29
32	33831703	36140483	38760139	41760003	28
33	33867942	36181427	38806801	41813712	27
34	33904252	36222456	38853567	41867550	26
35	33940634	36263570	38900438	41921518	25
36	33977088	36304771	38947416	41975617	24
37	34013615	36346060	38994501	42029848	23
38	34050215	36387437	39041695	42084211	22
39	34086888	36428903	39088998	42138706	21
40	34123634	36470459	39136409	42193334	20
41	34160453	36512103	39183929	42248096	19
42	34197345	36553836	39231557	42302993	18
43	34234310	36595659	39279294	42358025	17
44	34271348	36637572	39327139	42413193	16
45	34308459	36679574	39375094	42468497	15
46	34345644	36721666	39423158	42523937	14
47	34382903	36763849	39471331	42579514	13
48	34420237	36806121	39519614	42635228	12
49	34457647	36848483	39568006	42691080	11
50	34495132	36890936	39616509	42747070	10
51	34532692	36933479	39665124	42803199	9
52	34570327	36976114	39713852	42859468	8
53	34608038	37018840	39762695	42915878	7
54	34645824	37061659	39811654	42972429	6
55	34683686	37104570	39860729	43029122	5
56	34721625	37147574	39909917	43085958	4
57	34759640	37190670	39959218	43142937	3
58	34797733	37233859	40008633	43200060	2
59	34835903	37277141	40058103	43257328	1
60	34874151	37320517	40107808	43314742	0
	16	15	14	13	

of the complements of the Arches of the same Quadrant.

O

The Table
The degrees of the Quadrant for the Tangents

	77	78	79	80	
0	43314742	47046295	51445543	56712854	60
1	43372301	47113680	51525561	56809480	59
2	43430006	47181249	51605820	56906425	58
3	43487857	47249003	51686321	57003690	57
4	43545855	47316942	51767065	57101277	56
5	43604000	47385067	51848053	57199188	55
6	43662293	47453380	51929285	57297425	54
7	43720733	47521882	52010762	57395990	53
8	43779321	47590575	52092485	57494885	52
9	43838057	47659460	52174455	57594111	51
10	43896942	47728538	52256673	57693670	50
11	43955977	47797809	52339140	57793564	49
12	44015163	47867274	52421857	57893795	48
13	44074501	47936934	52504826	57994366	47
14	44133992	48006790	52588048	58095279	46
15	44193637	48076841	52671525	58196536	45
16	44253435	48147088	52755259	58298138	44
17	44313387	48217531	52839251	58400087	43
18	44373494	48288171	52923503	58502385	42
19	44433756	48359008	53008016	58605934	41
20	44494174	48430043	53092792	58708035	40
21	44554749	48501278	53177831	58811388	39
22	44615481	48572714	53263134	58915095	38
23	44676371	48644352	53348702	59019157	37
24	44737419	48716193	53434536	59123576	36
25	44798626	48788238	53520637	59228353	35
26	44859993	48860488	53607006	59333490	34
27	44921521	48932945	53693644	59438989	33
28	44983211	49005610	53780552	59544852	32
29	45045065	49078483	53867731	59651081	31
30	45107083	49151565	53955183	59757678	30
	12	11	10	9	

The degrees of the Quadrant for the Tangents

of Tangents.

of the Arches of the same Quadrant.

	77	78	79	80	
30	45107083	49151565	53955183	59757678	30
31	45169263	49224856	54042909	59864646	29
32	45231607	49298357	54130911	59971987	28
33	45294114	49372069	54219190	60079703	27
34	45356785	49445993	54307748	60187796	26
35	45419621	49520130	54396586	60296268	25
36	45482623	49594481	54485705	60405121	24
37	45545790	49669047	54575107	60514358	23
38	45609123	49743829	54664793	60623981	22
39	45672623	49818827	54754764	60733992	21
40	45736291	49894042	54845022	60844392	20
41	45800128	49969475	54935569	60955184	19
42	45864135	50045127	55029406	61066370	18
43	45928314	50120999	55117535	61177952	17
44	45992666	50197092	55208958	61289930	16
45	46057192	50273407	55300676	61402307	15
46	46121892	50349935	55392692	61515085	14
47	46186767	50246707	55485007	61628267	13
48	46251817	50503605	55577622	61741856	12
49	46318043	50580910	55670539	61855854	11
50	46382445	50658353	55763759	61970263	10
51	46448023	50736025	55857283	62085085	9
52	46513778	50813927	55951112	62200323	8
53	46579711	50892060	56045247	62315979	7
54	46645823	50970425	56139689	62432056	6
55	46712115	51049023	56234439	62548556	5
56	46778587	51127855	56329498	62665481	4
57	46845240	51206922	56424868	62782833	3
58	46912075	51286225	56520550	62900615	2
59	46979093	51365765	56616545	63018829	1
60	47046295	51445543	56712854	63137478	0
	12	11	10	9	

of the complements of the Arches of the same Quadrant.

O 2

The Table

The degrees of the Quadrant for the Tangents

	81	82	83	84	
0	63137478	71153707	81443502	95143611	60
1	63256564	71304198	81639821	95410585	59
2	63376089	71455313	81837074	95679034	58
3	63496056	71607058	82035268	95948971	57
4	63616468	71759440	82234410	96220411	56
5	63737327	71912459	82434508	96493467	55
6	63858635	72066117	82635570	96767939	54
7	63980394	72220422	82837603	97044063	53
8	64102607	72375376	83040614	97321646	52
9	64225276	72530983	83244610	97600890	51
10	64348404	72687247	83449598	97881716	50
11	64471994	72844173	83655585	98164135	49
12	64596049	73001766	83862572	98448162	48
13	64720571	73160031	84070565	98733810	47
14	64845563	73318972	84279571	99021104	46
15	64971028	73478593	84489598	99310047	45
16	65096969	73638898	84700687	99600655	44
17	65223388	73799892	84912817	99893042	43
18	65350287	73961579	85125995	100187022	42
19	65477669	74123964	85340229	100482822	41
20	65605537	74287052	85555525	100780346	40
21	65733894	74450847	85771891	101079507	39
22	65862743	74615354	85989335	101380525	38
23	65992087	74780577	86207866	101683314	37
24	66121928	74946521	86427493	101987889	36
25	66252268	75113189	86648225	102294266	35
26	66383110	75280586	86870072	102602473	34
27	66514457	75448716	87093043	102912514	33
28	66646313	75617584	87317150	103224405	32
29	66778681	75787195	87542404	103538166	31
30	66911564	75957554	87768816	103853919	30
	8	7	6	5	

The degrees of the Quadrant for the Tangents

of Tangents.

of the Arches of the same Quadrant.

	81	82	83	84	
30	66911564	75957554	87768816	103853919	30
31	67044965	76128666	87996394	104171468	29
32	67178887	76300536	88225146	104491055	28
33	67313334	76473170	88455079	104812581	27
34	67448309	76646573	88686196	105136063	26
35	67583815	76820751	88918508	105461519	25
36	67719855	76995710	89152021	105788969	24
37	67856423	77171455	89386745	106118428	23
38	67993549	77347991	89622688	106449917	22
39	68131209	77525324	89859858	106783466	21
40	68269416	77703459	90098268	107119198	20
41	68408173	77882402	90337927	107456902	19
42	68547438	78062159	90578848	107796712	18
43	68687350	78242737	90821043	108138767	17
44	68827777	78424142	91064526	108482852	16
45	68968769	78606379	91309309	108829233	15
46	69110326	78789454	91555401	109177805	14
47	69252455	78973371	91802810	109528589	13
48	69395158	79158136	92051546	109881598	12
49	69538439	79343754	92301618	110236864	11
50	69682302	79530231	92553036	110594415	10
51	69826751	79717572	92805759	110954264	9
52	69971789	79905783	93059875	111316432	8
53	70117419	80094869	93315361	111680940	7
54	70263645	80284835	93572238	112047814	6
55	70410470	80475688	93830595	112417202	5
56	70557898	80667435	94090270	112788878	4
57	70705932	80860083	94351448	113163656	3
58	70854576	81053639	94614055	113539681	2
59	71003833	81248110	94878103	113918875	1
60	71153706	81443502	95143611	114300579	0
	8	7	6	5	

of the complements of the Arches of the same Quadrant.

The Table
The degrees of the Quadrant for the Tangents

	85	86	87	
0	114300579	143006601	190811200	60
1	114684819	143606943	191879163	59
2	115071619	144212307	192959095	58
3	115461005	144822757	194051200	57
4	115853017	145438358	195155685	56
5	116247668	146059175	196273146	55
6	116644985	146685275	197403054	54
7	117044995	147316726	198545993	53
8	117447864	147953611	199702191	52
9	117853346	148595987	200871878	51
10	118261757	149244148	202055705	50
11	118672834	149897753	203253093	49
12	119086890	150557233	204464726	48
13	119503669	151222301	205691260	47
14	119923488	151893462	206932111	46
15	120346233	152570581	208188402	45
16	120771937	153253487	209459545	44
17	121200643	153942729	210746693	43
18	121632370	154638158	212049271	42
19	122067151	155339855	213368214	41
20	122505017	156047923	214704085	40
21	122946003	156762433	216056022	39
22	123390142	157483474	217425507	38
23	123837634	158211136	218812405	37
24	124288195	158945509	220217049	36
25	124742169	159686753	221639784	35
26	125199280	160434770	223080983	34
27	125659878	161189849	224540987	33
28	126123842	161952305	226020167	32
29	126521211	162721698	227518902	31
30	127062036	163498660	229037584	30
	4	3	2	

The degrees of the Quadrant for the Tangents

of Tangents.

of the Arches of the same Quadrant.

	85	86	87	
30	127062036	163498660	229037584	30
31	127536341	164282764	230576614	29
32	128014165	165074651	232136427	28
33	128495548	165873906	233717425	27
34	128980521	166681172	235320041	26
35	129469305	167496287	236945285	25
36	129961652	168319085	238592501	24
37	130457692	169150247	240262714	23
38	130957670	169989613	241957021	22
39	131461286	170837304	243674732	21
40	131968930	171693461	245417543	20
41	132480297	172558198	247184785	19
42	132995769	173431641	248978216	18
43	133515636	174313925	250797165	17
44	134038804	175205183	252643455	16
45	134566419	176105555	254517088	15
46	135098153	177015180	256417991	14
47	135634096	177934219	258348100	13
48	136174272	178862806	260307416	12
49	136718731	179801085	262296605	11
50	137267523	180749537	264316358	10
51	137820702	181707670	266366704	9
52	138378319	182676299	268449755	8
53	138940429	183654941	270565570	7
54	139507087	184644417	272714927	6
55	140078545	185644562	274898633	5
56	140654481	186655202	277117516	4
57	141235334	187677207	279372435	3
58	141820765	188710414	281664304	2
59	142411234	189755028	283994009	1
60	143006601	190811200	286362498	0
	4	3	2	

of the complements of the Arches of the same Quadrant.

The Table
The degrees of the Quadrant for the Tangents

	88		89	
0	286362498		572899830	60
1	288770746		582610421	59
2	291219764		592655713	58
3	293710598		603057015	57
4	296244357		613825994	56
5	298823024		624990311	55
6	301445987		636564040	54
7	304115322		648591509	53
8	306833212		661050728	52
9	309599077		674016435	51
10	312416191		687500725	50
11	315283945		701531474	49
12	318204757		716149676	48
13	321181137		731385593	47
14	324212583		747289264	46
15	327302782		763899813	45
16	330451272		781259259	44
17	333661982		799432199	43
18	336934467		818463792	42
19	340272744		838430438	41
20	343677949		859395374	40
21	347150587		881427652	39
22	350695255		904627361	38
23	354312962		929081086	37
24	358006024		954893332	36
25	361776788		982180553	35
26	365626388		1011062679	34
27	369560062		1041705454	33
28	373579199		1074263399	32
29	377686614		1108922084	31
30	381885288		1145891136	30
	1		0	

The degrees of the Quadrant for the Tangents

of Tangents.
of the Arches of the same Quadrant.

	88		89	
30	381885288		1145891136	30
31	386178258		1185395877	29
32	390568737		1227736470	28
33	395060088		1273213435	27
34	399655828		1322188681	26
35	404359642		1375082163	25
36	409175388		1432363027	24
37	414111295		1494645462	23
38	419159137		1562590046	22
39	424335793		1637005697	21
40	429641796		1718863124	20
41	435082056		1809337410	19
42	440661780		1909864971	18
43	446386310		2022219818	17
44	452261453		2148619711	16
45	458293185		2291873854	15
46	464487853		2455533838	14
47	470852152		2644433955	13
48	477393195		2864819229	12
49	484118353		3125276745	11
50	491038024		3437829002	10
51	498155754		3819696333	9
52	505482730		4297181900	8
53	513030946		4911098124	7
54	520805157		5729633839	6
55	528821258		6875680006	5
56	537085003		8594003953	4
57	545610968		11457529506	3
58	554414914		17188033688	2
59	563504309		34376070815	1
60	572899830		Infinita.	0
	1		0	

of the complements of the Arches of the same Quadrant.

THE TABLE
OF SECANTS OTHER-
WISE CALLED THE
BENEFICIALL
TABLE.

The Table
The degrees of the Quadrant for Secants.

	0	1	2	3	
0	10000000	10001524	10006095	10013723	60
1	10000001	10001574	10006198	10013875	59
2	10000002	10001626	10006301	10014029	58
3	10000004	10001679	10006405	10014184	57
4	10000008	10001733	10006509	10014339	56
5	10000010	10001788	10006615	10014495	55
6	10000014	10001844	10006721	10014653	54
7	10000020	10001900	10006828	10014811	53
8	10000027	10001957	10006936	10014970	52
9	10000034	10002015	10007045	10015130	51
10	10000042	10002074	10007155	10015291	50
11	10000051	10002134	10007265	10015453	49
12	10000060	10002195	10007376	10015615	48
13	10000071	10002256	10007488	10015778	47
14	10000083	10002318	10007601	10015942	46
15	10000095	10002381	10007716	10016107	45
16	10000108	10002445	10007831	10016273	44
17	10000122	10002510	10007946	10016440	43
18	10000137	10002576	10008062	10016608	42
19	10000152	10002642	10008179	10016777	41
20	10000168	10002709	10008298	10016946	40
21	10000186	10002777	10008417	10017116	39
22	10000204	10002846	10008537	10017287	38
23	10000223	10002916	10008658	10017459	37
24	10000243	10002987	10008779	10017632	36
25	10000264	10003058	10008902	10017806	35
26	10000285	10003130	10009025	10017981	34
27	10000308	10003203	10009149	10018157	33
28	10000332	10003277	10009274	10018333	32
29	10000357	10003352	10009400	10018510	31
30	10000381	10003428	10009527	10018687	30
	89	88	87	86	

The degrees of the Quadrant for the Secants

The minutes of the degrees of the Quadrant for the Secants of the Arches of the same Quadrant.

The minutes of degrees of the Quadrant for the Secants of the cōplements of the Arches of the same Quadrāt.

of Secants.

of the Arches of the same Quadrant.

	0	1	2	3	
30	10000381	10003428	10009527	10018687	30
31	10000407	10003505	10009655	10018865	29
32	10000433	10003582	10009783	10019044	28
33	10000461	10003660	10009912	10019224	27
34	10000489	10003739	10010043	10019405	26
35	10000518	10003819	10010174	10019587	25
36	10000548	10003900	10010306	10019770	24
37	10000579	10003982	10010439	10019954	23
38	10000611	10004060	10010572	10020138	22
39	10000643	10004148	10010706	10020324	21
40	10000677	10004232	10010841	10020510	20
41	10000711	10004317	10010977	10020698	19
42	10000746	10004403	10011114	10020886	18
43	10000782	10004490	10011252	10021086	17
44	10000819	10004578	10011390	10021266	16
45	10000857	10004666	10011529	10021456	15
46	10000895	10004755	10011670	10021649	14
47	10000934	10004845	10011811	10021842	13
48	10000975	10004936	10011952	10022035	12
49	10001016	10005028	10012098	10022239	11
50	10001058	10005122	10012238	10022424	10
51	10001100	10005216	10012383	10022620	9
52	10001144	10005310	10012528	10022817	8
53	10001188	10005405	10012674	10023015	7
54	10001233	10005501	10012822	10023213	6
55	10001280	10005598	10012970	10023412	5
56	10001327	10005696	10013119	10023612	4
57	10001375	10005795	10013269	10023813	3
58	10001423	10005894	10013419	10024014	2
59	10001473	10005994	10013570	10024217	1
60	10001524	10006095	10013723	10024420	0
	89	88	87	86	

of the complements of the Arches of the same Quadrant.

The Table

The degrees of the Quadrant for Secants.

	4	5	6	7	
0	10024420	10038198	10055082	10075098	60
1	10024625	10038454	10055390	10075459	59
2	10024830	10038710	10055699	10075820	58
3	10025036	10038968	10056009	10076182	57
4	10025242	10039226	10056320	10076545	56
5	10025450	10039486	10056632	10076909	55
6	10025658	10039747	10056944	10077274	54
7	10025868	10040008	10057256	10077639	53
8	10026078	10040269	10057570	10078005	52
9	10026289	10040532	10057884	10078372	51
10	10026500	10040796	10058200	10078740	50
11	10026713	10041061	10058517	10079009	49
12	10026927	10041326	10058834	10079479	48
13	10027141	10041592	10059153	10079850	47
14	10027357	10041859	10059472	10080222	46
15	10027573	10042128	10059792	10080595	45
16	10027790	10042397	10060113	10080968	44
17	10028009	10042667	10060435	10081332	43
18	10028227	10042936	10060757	10081717	42
19	10028447	10043207	10061080	10082093	41
20	10028667	10043479	10061405	10082470	40
21	10028889	10043752	10061730	10082848	39
22	10029111	10044025	10062056	10083226	38
23	10029334	10044300	10062383	10083606	37
24	10029559	10044576	10062711	10083987	36
25	10029784	10044853	10063039	10084368	35
26	10030009	10045130	10063369	10084750	34
27	10030236	10045409	10063700	10085134	33
28	10030463	10045689	10064031	10085518	32
29	10030692	10045969	10064364	10085903	31
30	10030920	10046250	10064696	10086289	30
	85	84	83	82	

The degrees of the Quadrant for the Secants

of Secants.

of the Arches of the same Quadrant.

	4	5	6	7	
30	10030920	10046250	10064696	10086289	30
31	10031150	10046532	10065035	10086677	29
32	10031381	10046815	10065365	10087065	28
33	10031614	10047098	10065701	10087454	27
34	10031846	10047383	10066038	10087843	26
35	10032079	10047669	10066376	10088243	25
36	10032314	10047954	10066715	10088623	24
37	10032550	10048241	10067054	10089015	23
38	10032786	10048529	10067394	10089408	22
39	10033023	10048818	10067735	10089802	21
40	10033261	10049107	10068076	10090196	20
41	10033500	10049398	10068419	10090592	19
42	10033740	10049690	10068763	10090988	18
43	10033981	10049983	10069107	10091385	17
44	10034223	10050276	10069452	10091783	16
45	10034465	10050571	10069808	10092182	15
46	10034708	10050865	10070155	10092582	14
47	10034952	10051160	10070493	10092983	13
48	10035196	10051456	10070842	10093385	12
49	10035441	10051753	10071192	10093787	11
50	10035688	10052051	10071543	10094190	10
51	10035936	10052350	10071895	10094624	9
52	10036184	10052649	10072247	10095030	8
53	10036434	10052951	10072600	10095406	7
54	10036684	10053252	10072954	10095813	6
55	10036934	10053555	10073310	10096221	5
56	10037185	10053858	10073666	10096630	4
57	10037438	10054162	10074023	10097040	3
58	10037690	10054468	10074380	10097451	2
59	10037944	10054775	10074737	10097863	1
60	10038198	10055082	10075098	10098275	0
	85	84	83	82	

of the complements of the Arches of the same Quadrant.

The Table
The degrees of the Quadrant for the Secants.

	8	9	10	11	
0	10098275	10124650	10154264	10187166	60
1	10098698	10125117	10154786	10187743	59
2	10099103	10125585	10155308	10188320	58
3	10099518	10126054	10155831	10188899	57
4	10099934	10126524	10156356	10189478	56
5	10100351	10126994	10156881	10190058	55
6	10100769	10127465	10157407	10190639	54
7	10101188	10127947	10157934	10191221	53
8	10101607	10128410	10158462	10191804	52
9	10102028	10128684	10158991	10192387	51
10	10102450	10129358	10159520	10192972	50
11	10102872	10129634	10160051	10193557	49
12	10103295	10130311	10160582	10194144	48
13	10103720	10130788	10161114	10194732	47
14	10104144	10131266	10161648	10195320	46
15	10104570	10131746	10162182	10195910	45
16	10104996	10132226	10162707	10196500	44
17	10105423	10132707	10163252	10197092	43
18	10105851	10133189	10163789	10197684	42
19	10106286	10133672	10164327	10198277	41
20	10106710	10134156	10165865	10198872	40
21	10107140	10134641	10165495	10199467	39
22	10107572	10135127	10165944	10200063	38
23	10108005	10135614	10166485	10200060	37
24	10108438	10136102	10167028	10201258	36
25	10108873	10136591	10167571	10201857	35
26	10109309	10137080	10168116	10202457	34
27	10109755	10137571	10168661	10203058	33
28	10110182	10138163	10169207	10203659	32
29	10110620	10138555	10169765	10204262	31
30	10111059	10139048	10170303	10204867	30
	81	80	79	78	

The degrees of the Quadrant for the Secants

of Secants.
of the Arches of the same Quadrant.

	8	9	10	11	
30	10111059	10139048	10170303	10204867	30
31	10111509	10139543	10170852	10205470	29
32	10111940	10140038	10171401	10206075	28
33	10112482	10140534	10171952	10206681	27
34	10112825	10141036	10172504	10207289	26
35	10113279	10141528	10173056	10207897	25
36	10113713	10142027	10173609	10208506	24
37	10114159	10142526	10174163	10209116	23
38	10114606	10143026	10174718	10209727	22
39	10115053	10143528	10175274	10210339	21
40	10115501	10144030	10175831	10210952	20
41	10115951	10144533	10176389	10211566	19
42	10116401	10145037	10176947	10211180	18
43	10116852	10145542	10177507	10212796	17
44	10117303	10146048	10178068	10213412	16
45	10117754	10146554	10178630	10214030	15
46	10118209	10147062	10179193	10214668	14
47	10118663	10147572	10179756	10215268	13
48	10119118	10148082	10180321	10215889	12
49	10119574	10148593	10180886	10216510	11
50	10120031	10149104	10181453	19217113	10
51	10120489	10149615	10182021	10217756	9
52	10120948	10150128	10182589	10218380	8
53	10121408	10150642	10183158	10219015	7
54	10121868	10151156	10183728	10219631	6
55	10122330	10151672	10184299	10220258	5
56	10122792	10152188	10184870	10220885	4
57	10123256	10152705	10185443	10221514	3
58	10123720	10153224	10186017	10222143	2
59	10124275	10153744	10186591	10222774	1
60	10124650	10154264	10187166	10223405	0
	81	80	79	78	

of the complements of the Arches of the same Quadrant.

The Table
The degrees of the Quadrant for the Secants.

	12	13	14	15	
0	10223405	10263040	10306136	10352762	60
1	10224037	10263730	10306884	10353569	59
2	10224671	10264420	10307633	10354377	58
3	10225305	10265112	10308383	10355186	57
4	10225941	10265804	10309134	10355996	56
5	10226577	10266498	10309886	10356807	55
6	10227215	10267192	10310639	10357619	54
7	10227854	10267888	10311393	10358433	53
8	10228493	10268584	10312148	10359247	52
9	10229134	10269281	10312903	10360063	51
10	10229775	10269979	10313660	10360880	50
11	10230417	10270688	10314417	10361698	49
12	10231060	10271379	10315176	10362517	48
13	10231644	10272080	10315935	10363337	47
14	10232288	10272782	10316696	10364158	46
15	10232994	10273485	10317457	10364980	45
16	10233641	10274190	10318220	10365802	44
17	10234289	10274895	10318984	10366626	43
18	10234938	10275601	10319749	10367450	42
19	10235587	10276318	10320525	10368276	41
20	10236238	10277016	10321282	10369102	40
21	10236889	10277726	10322050	10369930	39
22	10237541	10278436	10322819	10370758	38
23	10238195	10279148	10323589	10371588	37
24	10238849	10279860	10324359	10372418	36
25	10239505	10280573	10325131	10373250	35
26	10240161	10281287	10325903	10374092	34
27	10240818	10282002	10326677	10374916	33
28	10241476	10282717	10327451	10375750	32
29	10242135	10283434	10328127	10376586	31
30	10242795	10284151	10329003	10377422	30
	77	76	75	74	

The degrees of the Quadrant for the Secants

of Secants.
of the Arches of the same Quadrant.

	12	13	14	15	
30	10242795	10284151	10329003	10377422	30
31	10243456	10284870	10329781	10378260	29
32	10244118	10285589	10330559	10379098	28
33	10245782	10286310	10331339	10379938	27
34	10245445	10287032	10332119	10380778	26
35	10246110	10287754	10332902	10381620	25
36	10246776	10288478	10333684	10382463	24
37	10247442	10289202	10334467	10383307	23
38	10248110	10289928	10335252	10384153	22
39	10248778	10290654	10336037	10384999	21
40	10249448	10291381	10336824	10385846	20
41	10250119	10292119	10337612	10386694	19
42	10250790	10292838	10338400	10387543	18
43	10251461	10293569	10339189	10388393	17
44	10252136	10294300	10339980	10389244	16
45	10252811	10295043	10340771	10390096	15
46	10253482	10295766	10341564	10390949	14
47	10254162	10296501	10342347	10391803	13
48	10254839	10297237	10343152	10392657	12
49	10255517	10297973	10343947	10393513	11
50	10256196	10298710	10344743	10394370	10
51	10256876	10299449	10345541	10395228	9
52	10257557	10300188	10346340	10396087	8
53	10258239	10300928	10347139	10396947	7
54	10258922	10301669	10347940	10397808	6
55	10259606	10302411	10348741	10398670	5
56	10260291	10303154	10349544	10899533	4
57	10260977	10303898	10350347	10400397	3
58	10261661	10304643	10351151	10401262	2
59	10262351	10305390	10351956	10402128	1
60	10263040	10306136	10352762	10402994	0
	77	76	75	74	

of the complements of the Arches of the same Quadrant.
P 2

The Table
The degrees of the Quadrant for the Secants.

	16	17	18	19	
0	10402994	10456917	10514621	10576207	60
1	10403862	10457847	10515616	10577267	59
2	10404730	10458779	10516612	10578328	58
3	10405590	10459711	10517609	10579400	57
4	10406471	10460645	10518607	10580463	56
5	10407343	10461580	10519606	10581518	55
6	10408216	10462516	10520606	10582583	54
7	10409091	10463453	10521607	10583650	53
8	10409966	10464391	10522608	10584717	52
9	10410843	10465330	10523611	10585795	51
10	10411721	10466270	10524615	10586855	50
11	10412600	10467211	10525620	10587925	49
12	10413479	10468153	10526626	10588997	48
13	10414360	10469096	10527633	10590070	47
14	10415241	10470041	10528642	10591145	46
15	10416124	10470986	10529651	10592220	45
16	10417007	10471933	10530662	10593297	44
17	10417892	10472880	10531673	10594375	43
18	10418778	10473829	10532686	10595455	42
19	10419665	10474778	10533699	10596534	41
20	10420553	10475729	10534714	10597615	40
21	10421442	10476680	10535730	10598697	39
22	10422333	10477633	10536747	10599780	38
23	10423224	10478587	10537765	10600865	37
24	10424116	10479542	10538785	10601950	36
25	10425009	10480498	10539805	10603037	35
26	10425903	10481454	10540826	10604125	34
27	10426798	10482412	10541848	10605214	33
28	10427694	10483371	10542872	10606304	32
29	10428591	10484331	10543897	10607395	31
30	10429489	10485292	10544923	10608487	30
	73	72	71	70	

The degrees of the Quadrant for the Secants

of Secants.

of the Arches of the same Quadrant.

	16	17	18	19	
30	10429489	10485292	10544923	10608487	30
31	10430388	10486254	10545950	10609580	29
32	10431288	10487217	10546977	10610675	28
33	10432189	10488181	10548006	10611770	27
34	10433091	10489146	10549036	10612867	26
35	10433995	10490113	10550067	10613964	25
36	10434899	10491080	10551099	10615063	24
37	10435805	10492049	10552133	10616163	23
38	10436711	10493018	10553168	10617264	22
39	10437619	10493989	10554204	10618366	21
40	10438528	10494961	10555241	10619469	20
41	10439436	10494934	10556279	10620574	19
42	10440346	10496908	10557318	10621680	18
43	10441257	10497883	10558359	10622787	17
44	10442170	10498059	10559400	10623895	16
45	10443083	10499836	10560443	10625004	15
46	10443998	10500814	10561496	10626114	14
47	10444913	10501793	10562531	10627226	13
48	10445830	10502773	10563577	10628338	12
49	10446749	10503754	10564623	10629451	11
50	10447668	10504736	10565670	10630566	10
51	10448588	10505719	10566719	10631682	9
52	10449509	10506704	10567769	10632799	8
53	10450431	10507689	10568820	10633917	7
54	10451354	10508676	10569872	10635037	6
55	10452279	10509664	10570925	10636157	5
56	10453204	10510653	10571980	10637279	4
57	10454131	10511643	10573034	10638402	3
58	10455058	10512635	10574091	10639526	2
59	10455987	10513627	10575149	10640651	1
60	10456917	10514621	10576207	10641777	0
	73	72	71	70	

of the complements of the Arches of the same Quadrant.

The Table
The degrees of the Quadrant for the Secants.

	20	21	22	23	
0	10641777	10711449	10785347	10863603	60
1	10642905	10712646	10786616	10864945	59
2	10644034	10713888	10787885	10866628	58
3	10645164	10715042	10789155	10867633	57
4	10646295	10716242	10790427	10868979	56
5	10647427	10717444	10791700	10870326	55
6	10648560	10718647	10792974	10871675	54
7	10649694	10719850	10794250	10873024	53
8	10650829	10721056	10795527	10874374	52
9	10651965	10722261	10796805	10875626	51
10	10653103	10723469	10798085	10877079	50
11	10654242	10724677	10799365	10878434	49
12	10655381	10725887	10800647	10879790	48
13	10656522	10727098	10803214	10881147	47
14	10657664	10728310	10803214	10882506	46
15	10658807	10729524	10804500	10883865	45
16	10659951	10730738	10805787	10885226	44
17	10661097	10731953	10807074	10886588	43
18	10662244	10733170	10808363	10887952	42
19	10663392	10734387	10809652	10889317	41
20	10664541	10735606	10810942	10890683	40
21	10665692	10736826	10812234	15892051	39
22	10666844	10738048	10813528	10893417	38
23	10667996	10739270	10814823	10894788	37
24	10669150	10740494	10816119	10896159	36
25	10670304	10741719	10817417	10897531	35
26	10671460	10742945	10818715	10898905	34
27	10672617	10744173	10820015	10900280	33
28	10673776	10745401	10821316	10901656	32
29	10674936	10746631	10822617	10903033	31
30	10676096	10747864	10823920	10904413	30
	69	68	67	66	

The degrees of the Quadrant for the Secants

of Secants.
of the Arches of the same Quadrant.

	20	21	22	23	
30	10676096	10747864	10823920	10904413	30
31	10677258	10749094	10825225	10905790	29
32	10678420	10750327	10826531	10907171	28
33	10679584	10751561	10827838	10908553	27
34	10680749	10752797	10829146	10909936	26
35	10681915	10754034	10830455	10911322	25
36	10683082	10755273	10831766	10912709	24
37	10684250	10756513	10833078	10914096	23
38	10685420	10757753	10834391	10915484	22
39	10686591	10758995	10835706	10916874	21
40	10687763	10760237	10837023	10918265	20
41	10688936	10761481	10838341	10919657	19
42	10690111	10762726	10839660	10921051	18
43	10691287	10763972	10840980	10922436	17
44	10692464	10765220	10842301	10923833	16
45	10693642	10766469	10843623	10925241	15
46	10694821	10767720	10844947	10926641	14
47	10696001	10768971	10846272	10928041	13
48	10697182	10770224	10847597	10929442	12
49	10698364	10771477	10848924	10930846	11
50	10699548	10772732	10850252	10932249	10
51	10700732	10773988	10851583	10933654	9
52	10701918	10775244	10852914	10935061	8
53	10703105	10776502	10854246	10936469	7
54	10704294	10777761	10855578	10937879	6
55	10705483	10779022	10856912	10939290	5
56	10706674	10780284	10858247	10940702	4
57	10707866	10781547	10859584	10942115	3
58	10709059	10782802	10860922	10943527	2
59	10710254	10784078	10862262	10944945	1
60	10711449	10785347	10863603	10946362	0
	69	68	67	66	

of the complements of the Arches of the same Quadrant.

P 4

The Table

The degrees of the Quadrant for the Secants.

	24	25	26	27	
0	1094636	1103378	1112602	1122326	60
1	1094778	1103528	1112760	1122492	59
2	1094920	1103677	1112918	1122659	58
3	1095062	1103827	1113076	1122826	57
4	1095204	1103978	1113234	1122992	56
5	1095346	1104128	1113393	1123159	55
6	1095489	1104278	1113551	1123327	54
7	1095632	1104429	1113710	1123494	53
8	1095774	1104579	1113869	1123661	52
9	1095917	1104730	1114028	1123829	51
10	1096060	1104881	1114187	1123996	50
11	1096203	1105032	1114346	1124164	49
12	1096346	1105183	1114506	1124332	48
13	1096490	1105335	1114665	1124501	47
14	1096633	1105486	1114825	1124669	46
15	1096777	1105638	1114985	1124837	45
16	1096921	1105789	1115145	1125006	44
17	1097065	1105942	1115305	1125175	43
18	1097209	1106093	1115465	1125344	42
19	1097353	1106246	1115626	1125513	41
20	1097497	1106398	1115786	1125682	40
21	1097642	1106551	1115947	1125851	39
22	1097786	1106703	1116108	1126021	38
23	1097931	1106856	1116269	1126190	37
24	1098076	1107009	1116430	1126360	36
25	1098221	1107162	1116591	1126530	35
26	1098366	1107315	1116753	1126700	34
27	1098511	1107468	1116914	1126870	33
28	1098656	1107621	1117076	1127041	32
29	1098802	1107775	1117238	1127211	31
30	1098948	1107928	1117400	1127382	30
	65	64	63	62	

The degrees of the Quadrant for the Secants

of Secants.   109
of the Arches of the same Quadrant.

	24	25	26	27	
30	10989480	11079289	11174006	11273820	30
31	10990938	11080827	11175627	11275528	29
32	10992398	11082366	11177249	11277238	28
33	10993859	11083906	11178873	11278949	27
34	10995321	11085448	11180499	11280661	26
35	10996783	11086990	11182125	11282374	25
36	10998247	11088536	11183753	11284089	24
37	10999712	11090082	11185383	11285805	23
38	11001179	11091629	11187014	11287524	22
39	11002647	11093178	11188647	11289244	21
40	11004116	11094729	11190281	11290965	20
41	11005587	11096280	11191916	11292688	19
42	11007059	11097833	11193553	11294412	18
43	11008533	11099387	11195191	11296132	17
44	11010008	11100943	11196831	11297864	16
45	11011484	11102500	11198472	11299593	15
46	11019262	11104058	11200114	11301324	14
47	11014441	11105618	11201758	11303056	13
48	11015921	11107179	11203404	11304789	12
49	11017402	11108741	11205051	11306523	11
50	11018884	11110306	11206700	11308259	10
51	11020367	11111871	11208550	11309996	9
52	11021852	11113438	11210001	11311735	8
53	11023338	11115006	11211654	11313476	7
54	11024826	11116575	11213308	11315218	6
55	11026315	11118145	11214963	11316961	5
56	11027806	11119717	11216620	11318706	4
57	11029298	11121290	11218278	11319452	3
58	11030791	11122865	11219938	11322199	2
59	11032287	11124442	11221599	11323949	1
60	11033783	11126021	11223262	11325700	0
	65	64	63	62	

of the complements of the Arches of the same Quadrant.

The Table
The degrees of the Quadrant for the Secants.

	28	29	30	31	
0	11325700	11433540	11547004	11666331	60
1	11327452	11435384	11548944	11668371	59
2	11329206	11437230	11550886	11670413	58
3	11330961	11439078	11552829	11672457	57
4	11332718	11440927	11554774	11674502	56
5	11334479	11442777	11556720	11676548	55
6	11336237	11444629	11558669	11678597	54
7	11337999	11446483	11560619	11680647	53
8	11339762	11448339	11562570	11682698	52
9	11341526	11450196	11564523	11684752	51
10	11343292	11452054	11566480	11686807	50
11	11345060	11453915	11568434	11688864	49
12	11346830	11455776	11570393	11690923	48
13	11348601	11457639	11572353	11692984	47
14	11350373	11459503	11574314	11695046	46
15	11352149	11461370	11576277	11697110	45
16	11353923	11463238	11578242	11699176	44
17	11355698	11465107	11580208	11701243	43
18	11357475	11466978	11582175	11703312	42
19	11359255	11468850	11584145	11705383	41
20	11361036	11470723	11586116	11707455	40
21	11362819	11472599	11588089	11709530	39
22	11364603	11474483	11590064	11711606	38
23	11366389	11476354	11592040	11713684	37
24	11368177	11478235	11594018	11715764	36
25	11369966	11480117	11595998	11717845	35
26	11371756	11482001	11597979	11719928	34
27	11373548	11483887	11599961	11722012	33
28	11375341	11485774	11601946	11724099	32
29	11377136	11487662	11603932	11726187	31
30	11378933	11489553	11605919	11728276	30
	61	60	59	58	

The degrees of the Quadrant for the Secants

The minutes of the degrees of the Quadrant for the Secants of the Arches of the same Quadrant.

The minutes of degrees of the Quadrant for the Secants of the complements of the Arches of the same Quadrant.

of Secants.

of the Arches of the same Quadrant.

	28	29	30	31	
30	11378933	11489353	11605919	11728276	30
31	11380731	11491445	11607909	11730367	29
32	11382530	11493338	11609900	11732460	28
33	11384331	11495233	11611893	11734555	27
34	11386134	11497140	11613888	11736652	26
35	11387938	11499028	11615876	11738751	25
36	11389744	11500928	11617882	11740851	24
37	11391551	11502829	11619881	11742953	23
38	11393359	15504731	11621882	11745057	22
39	11395169	11506626	11623885	11747162	21
40	11396981	11508532	11625889	11749269	20
41	11398794	11510450	11627996	11751378	19
42	11400609	11512360	11629904	11753489	18
43	11402425	11514271	11631913	11755603	17
44	11404243	11516183	11633924	11757718	16
45	11406063	11518097	11635937	11759834	15
46	11407884	11520013	11637952	11761951	14
47	11409706	11521930	11639968	11764069	13
48	11411530	11523849	11641986	11766190	12
49	11413356	11525770	11644005	11768312	11
50	11415183	11527692	11646026	11770437	10
51	11417012	11529616	11648049	11772564	9
52	11418842	11531542	11650075	11774696	8
53	11420673	11533469	11652099	11776822	7
54	11422507	11535398	11654127	11778954	6
55	11424342	11537328	11656156	11781088	5
56	11426178	11539260	11658188	11783223	4
57	11428016	11541193	11660221	11785361	3
58	11429856	11543128	11662256	11787500	2
59	11431689	11545065	11664292	11789640	1
60	11433540	11547004	11666331	11791783	0
	61	60	59	58	

of the complements of the Arches of the same Quadrant.

The Table
The degrees of the Quadrant for the Secants.

	32	33	34	35	
0	11791783	11923633	12062179	12207745	60
1	11793927	11925886	12064546	12210233	59
2	11796073	11928141	12066916	12212723	58
3	11798221	11930397	12069286	12215214	57
4	11800371	11932656	12071660	12217708	56
5	11802522	11934917	12074036	12220204	55
6	11804675	11937180	12076413	12222702	54
7	11806830	11939445	12078792	12225201	53
8	11808987	11941701	12081174	12227703	52
9	11811145	11943979	12083558	12230207	51
10	11813306	11946250	12085943	12232713	50
11	11815468	11948522	12088330	12235221	49
12	11817632	11950796	12090720	12237732	48
13	11819797	11953071	12093111	12240245	47
14	11821965	11955349	12095504	12242759	46
15	11824134	11957629	12097899	12245275	45
16	11826306	11959910	12100296	12247794	44
17	11828479	11962194	12102696	12250315	43
18	11830654	11964479	12105097	12252837	42
19	11832830	11966766	12107500	12255361	41
20	11835008	11969055	12109905	12257888	40
21	11837188	11971346	12112312	12260417	39
22	11839369	11973638	12114722	12262948	38
23	11841552	11975932	12117133	12265481	37
24	11843737	11978229	12119546	12268016	36
25	11845924	11980527	12121960	12270553	35
26	11848114	11982828	12124377	12273093	34
27	11850305	11985131	12126796	12275634	33
28	11852498	11987435	12129216	12278187	32
29	11854693	11989741	12131638	12280722	31
30	11856890	11992050	12134063	12283270	30
	57	56	55	54	

The degrees of the Quadrant for the Secants

of Secants.
of the Arches of the same Quadrant.

	32	33	34	35	
30	11856890	11992050	12134063	12283270	30
31	11859088	11994360	12136490	12285820	29
32	11861288	11996672	12138919	12288372	28
33	11863489	11998986	12141350	12290925	27
34	11865693	12001303	12143783	12293481	26
35	11867899	12003619	12146218	12296039	25
36	11870107	12005938	12148656	12298599	24
37	11872316	12008259	12150095	12301161	23
38	11874527	12010582	12153536	12303725	22
39	11876739	12012907	12155978	12306291	21
40	11878954	12015233	12158423	12308859	20
41	11881171	12017562	12160870	12311430	19
42	11883389	12019893	12163319	12314003	18
43	11885609	12022226	12165770	12316578	17
44	11887831	12024560	12168223	12319156	16
45	11890054	12026897	12170677	12321736	15
46	11892280	12029236	12173135	12324317	14
47	11894508	12031576	12175594	12326900	13
48	11896737	12033919	12178055	12329486	12
49	11898968	12036264	12180518	12332074	11
50	11901202	12038610	12182983	12334664	10
51	11903437	12040958	12185450	12337256	9
52	11905674	12043309	12187719	12339851	8
53	11907912	12045661	12190390	12342448	7
54	11910153	12048016	12192864	12345046	6
55	11912395	12050372	12195340	12347646	5
56	11914640	12052730	12197817	12350249	4
57	11916886	12055089	12200296	12352854	3
58	11919133	12057451	12202777	12355460	2
59	11921382	12059814	12205260	12358068	1
60	11923633	12063179	12207745	12360678	0
	57	56	55	54	

of the complements of the Arches of the same Quadrant.

The Table

The degrees of the Quadrant for the Secants.

	36	37	38	39	
0	12360678	12521357	12690184	12867599	60
1	12363290	12524103	12693070	12870632	59
2	12365906	12526851	12695957	12873667	58
3	12368524	12529601	12698847	12876704	57
4	12371144	12532354	12701739	12879744	56
5	12373766	12535110	12704634	12882787	55
6	12376391	12537867	12707531	12885832	54
7	12379018	12540627	12710430	12888879	53
8	12381647	12543389	12713332	12891929	52
9	12384278	12546152	12716236	12894982	51
10	12386911	12548918	12719143	12898035	50
11	12389546	12551686	12722052	12901094	49
12	12392183	12554456	12724964	12904155	48
13	12394822	12557229	12727878	12907218	47
14	12397464	12560005	12730794	12910283	46
15	12400108	12562783	12733713	12913351	45
16	12402754	12565563	12736635	12916422	44
17	12405402	12568345	12739559	12919494	43
18	12408053	12571130	12742485	12922569	42
19	12410705	12573917	12745413	12925647	41
20	12413359	12576706	12748344	12928727	40
21	12416015	12579597	12751277	12931809	39
22	12418674	12582912	12754213	12934895	38
23	12421335	12585087	12757151	12937983	37
24	12423998	12587885	12760092	12941073	36
25	12426663	12590685	12763035	12944166	35
26	12429331	12593488	12765981	12947262	34
27	12432001	12596293	12768929	12950360	33
28	12434673	12599101	12771880	12953461	32
29	12437348	12601911	12774833	12956565	31
30	12440024	12604724	12777788	12959671	30
	53	52	51	50	

The degrees of the Quadrant for the Secants

of Secants.
of the Arches of the same Quadrant.

	36	37	38	39	
30	12440024	12604724	12777788	12959671	30
31	12442702	12607539	12780746	12962780	29
32	12445383	12610356	12783707	12965892	28
33	12448066	12613175	12786670	12969007	27
34	12450751	12615997	12789635	12972124	26
35	12453438	12618821	12792602	12975243	25
36	12456128	12621648	12795573	12978366	24
37	12458821	12624477	12798546	12981491	23
38	12461516	12627308	12801521	12984618	22
39	12464213	12630141	12804498	12987747	21
40	12466913	12632977	12807478	12990880	20
41	12469614	12635815	12810460	12994015	19
42	12472317	12638655	12813445	12997153	18
43	12475022	12641597	12816432	13000293	17
44	12477730	12644343	12819422	13003436	16
45	12480440	12646191	12822415	13006582	15
46	12483152	12650041	12825410	13009730	14
47	12485866	12652893	12828407	13012881	13
48	12488583	12655748	12831407	13016034	12
49	12491302	12658605	12834409	13019189	11
50	12494022	12661464	12837414	13022348	10
51	12496743	12664325	12840421	13025509	9
52	12499469	12667189	12843431	13028673	8
53	12502197	12670055	12846443	13031839	7
54	12504927	12672924	12849458	13035008	6
55	12507659	12675795	12852475	13038180	5
56	12510394	12678668	12855495	13041354	4
57	12513132	12681543	12858517	13044530	3
58	12515871	12684421	12861542	13047710	2
59	12518613	12687301	12864569	13050892	1
60	12521357	12690184	12867599	13054077	0
	53	52	51	50	

of the complements of the Arches of the same Quadrant.

The Table

The degrees of the Quadrant for the Secants.

	40	41	42	43	
0	13054077	13250131	13456326	13673275	60
1	13057264	13253482	13459851	13676986	59
2	13060455	13256835	13463380	13680700	58
3	13063646	13260192	13466912	13684417	57
4	13066843	13263582	13470447	13688138	56
5	13070041	13266915	13473985	13691861	55
6	13073242	13270282	13477527	13695587	54
7	13076445	13273651	13481071	13699316	53
8	13079651	13277023	13484618	13703048	52
9	13082859	13280397	13488168	13706783	51
10	13086071	13283775	13491721	13710523	50
11	13089285	13287155	13495276	13714266	49
12	13092502	13290538	13498835	13718012	48
13	13095721	13293924	13502397	13721761	47
14	13098944	13297313	13505962	13725514	46
15	13102169	13300704	13509530	13729270	45
16	13105397	13304098	13513101	13733029	44
17	13108627	13307495	13516675	13736790	43
18	13111861	13310896	13520252	13740555	42
19	13114098	13314299	13523832	13744322	41
20	13118337	13317705	13527416	13748092	40
21	13121578	13321114	13531003	13751867	39
22	13124823	13324526	13534593	13755644	38
23	13128070	13327941	13538185	13759424	37
24	13131320	13331359	13541781	13763209	36
25	13134572	13334779	13545380	13766997	35
26	13134828	13338203	13548981	13770788	34
27	13141085	13341629	13552585	13774582	33
28	13144346	13345058	13556193	13778380	32
29	13147509	13348490	13559803	13782181	31
30	13150874	13351924	13563417	13785985	30
	49	48	47	46	

The degrees of the Quadrant for the Secants

of Secants.

of the Arches of the same Quadrant.

The minutes of the degrees of the Quadrant for the Secants of the Arches of the same Quadrant.

	40	41	42	43	
30	13150874	13351924	13563417	13785985	30
31	13154142	13355361	13567034	13789791	29
32	13157413	13358802	13570654	13793603	28
33	13160687	13362245	13574277	13797416	27
34	13163964	13365691	13577903	13801233	26
35	13167243	13369140	13581532	13805053	25
36	13170526	13372592	13585164	13808876	24
37	13173811	13376057	13588799	13812703	23
38	13177099	13379505	13592438	13816534	22
39	13180389	13382966	13596079	13820368	21
40	13183682	13386430	13599723	13824205	20
41	13186978	13389897	13603370	13828045	19
42	13190276	13393367	13607021	13831889	18
43	13193577	13396839	13610975	13835736	17
44	13196882	13400315	13614332	13839586	16
45	13200189	13403794	13617992	13843439	15
46	13203499	13407275	13621656	13847296	14
47	13206812	13410759	13625323	13851156	13
48	13210128	13414247	13628993	13855019	12
49	13213447	13417738	13632666	13858885	11
50	13216769	13421232	13636342	13862755	10
51	13220093	13424728	13640021	13866628	9
52	13223421	13428227	13643704	13870505	8
53	13226750	13431729	13647390	13874385	7
54	13230082	13435234	13651078	13878268	6
55	13233417	13438742	13654769	13882154	5
56	13236754	13442253	13658464	13886044	4
57	13240094	13445767	13662162	13889636	3
58	13243437	13449284	13665863	13893833	2
59	13246783	13452804	13669567	13897733	1
60	13250131	13456326	13673275	13901636	0
	49	48	47	46	

The minutes of degrees of the Quadrant for the Secants of the complements of the Arches of the same Quadrant.

of the complements of the Arches of the same Quadrant.

Q

The Table
The degrees of the Quadrant for the Secants.

	44	45	46	47	
0	13901636	14142135	14395564	14662790	60
1	13905542	14146251	14399901	14667366	59
2	13909451	14150371	14404242	14671946	58
3	13913365	14154494	14408587	14676530	57
4	13917281	14158621	14412937	14681119	56
5	13921201	14162751	14417290	14685712	55
6	13925126	14166884	14421647	14690309	54
7	13929052	14171021	14426008	14694910	53
8	13932982	14175162	14430374	14699514	52
9	13936916	14179306	14434743	14704122	51
10	13940854	14183454	14439116	14708735	50
11	13944795	14187606	14443493	14713352	49
12	13948739	14191761	14447874	14717973	48
13	13952686	14195919	14452259	14722598	47
14	13956638	14200082	14456648	14727228	46
15	13960592	14204248	14461040	14731862	45
16	13964550	14208418	14465437	14736500	44
17	13968511	14212591	14469838	14741142	43
18	13972476	14216769	14474242	14745788	42
19	13976444	14220950	14478650	14750438	41
20	13980416	14225135	14483062	14755094	40
21	13984391	14229324	14487478	14759753	39
22	13988370	14233517	14491898	14764416	38
23	13992352	14237713	14496322	14769083	37
24	13996338	14241912	14500750	14773755	36
25	14000327	14246115	14505182	14778430	35
26	14004319	14250321	14509617	14783110	34
27	14008315	14254531	14514056	14787794	33
28	14012314	14258745	14518500	14792482	32
29	14016316	14262961	14522946	14797174	31
30	14020322	14267182	14527397	14801871	30
	45	44	43	42	

The degrees of the Quadrant for the Secants

of Secants.

of the Arches of the same Quadrant.

	44	45	46	47	
30	14020322	14267182	14527397	14801871	30
31	14024332	14271407	14531852	14806571	29
32	14028345	14275635	14536311	14811276	28
33	14032361	14279867	14540773	14815985	27
34	14036381	14284103	14545250	14820698	26
35	14040404	14288343	14549711	14825416	25
36	14044431	14292587	14554186	14830139	24
37	14048461	14296834	14558665	14834866	23
38	14052494	14301086	14563148	14839597	22
39	14056531	14305331	14567635	14844332	21
40	14060572	14309599	14572126	14849072	20
41	14064616	14313861	14576621	14853815	19
42	14068664	14318127	14581120	14858563	18
43	14072715	14322396	14585624	14863315	17
44	14076770	14326670	14590131	14868071	16
45	14080829	14330947	14594642	14872831	15
46	14084891	14335228	14599157	14877597	14
47	14088956	14339513	14603676	14882377	13
48	14093026	14343802	14608199	14887141	12
49	14097099	14348095	14612725	14891919	11
50	14101175	14352391	14617256	14896701	10
51	14105255	14356691	14621791	14901487	9
52	14109339	14360995	14626330	14906278	8
53	14113427	14365303	14630873	14911073	7
54	14117518	14369615	14635421	14915873	6
55	14121612	14373930	14639973	14920677	5
56	14125709	14378350	14644528	14925486	4
57	14129810	14382573	14649087	14930299	3
58	14133915	14386900	14653651	14935116	2
59	14138023	14391230	14658218	14939938	1
60	14142135	14395564	14662790	14944764	0
	45	44	43	42	

of the complements of the Arches of the same Quadrant.

Q 2

The Table
The degrees of the Quadrant for the Secants.

	48	49	50	51	
0	14944764	15242532	15557239	15890158	60
1	14949594	15247634	15562635	15895869	59
2	14954429	15252741	15568036	15901586	58
3	14959268	15257852	15573441	15907307	57
4	14964112	15262969	15578852	15913034	56
5	14968960	15268090	15584267	15918766	55
6	14973812	15273216	15589688	15924504	54
7	14978668	15278347	15595114	15930247	53
8	14983530	15283484	15600545	15936095	52
9	14988396	15288626	15605981	15941748	51
10	14993266	15293773	15611422	15947508	50
11	14998104	15298924	15616868	15953273	49
12	15003020	15304080	15622319	15959044	48
13	15007903	15309240	15627775	15964820	47
14	15012791	15314405	15633237	15970603	46
15	15017683	15319574	15639704	15976390	45
16	15022580	15324748	15644177	15982184	44
17	15027481	15329926	15649655	15987983	43
18	15032387	15335109	15655138	15993788	42
19	15037297	15340297	15660626	15999599	41
20	15042212	15345491	15666119	16005416	40
21	15047131	15350689	15671617	16011237	39
22	15052054	15355892	15677121	16017065	38
23	15056982	15361100	15682630	16022898	37
24	15061915	15366313	15688144	16028736	36
25	15066852	15371530	15693663	16034579	35
26	15071791	15376753	15699188	16040429	34
27	15076739	15381980	15704717	16046283	33
28	15081690	15387212	15710252	16052143	32
29	15086645	15392449	15715791	16058008	31
30	15091605	15397692	15721337	16063878	30
	41	40	39	38	

The degrees of the Quadrant for the Secants

of Secants.
of the Arches of the same Quadrant.

	48	49	50	51	
30	15091605	15397692	15721337	16063878	30
31	15096569	15402939	15726887	16069754	29
32	15101538	15408191	15732443	16075637	28
33	15106571	15413447	15738003	16081524	27
34	15111490	15418708	15743569	16087418	26
35	15116472	15423974	15749141	16093318	25
36	15121459	15429246	15754718	16099224	24
37	15126451	15434522	15760300	16105135	23
38	15131447	15439803	15765887	16111053	22
39	15136447	15445089	15771479	16116976	21
40	15141453	15450380	15777077	16122905	20
41	15146463	15455675	15782680	16128839	19
42	15151478	15460976	15788289	16134779	18
43	15156497	15466182	15793903	16140724	17
44	15161520	15471593	15799523	16146676	16
45	15166548	15476908	15805147	16152634	15
46	15171581	15482229	15810777	16158598	14
47	15176619	15487554	15816412	16164567	13
48	15181661	15492885	15822052	16170542	12
49	15186708	15498220	15827697	16176522	11
50	15191760	15503560	15833349	16182509	10
51	15196816	15508905	15839005	16188501	9
52	15201877	15514256	15844667	16194499	8
53	15206943	15519611	15850335	16200503	7
54	15212013	15524972	15856008	16206513	6
55	15217088	15530338	15861676	16212528	5
56	15222168	15535710	15867370	16218550	4
57	15227253	15541083	15873058	16224577	3
58	15232342	15546463	15878753	16230610	2
59	15237435	15551848	15884453	16236648	1
60	15242532	15557239	15890158	16242692	0
	41	40	39	38	

of the complements of the Arches of the same Quadrant.

Q 3

The Table
The degrees of the Quadrant for the Secants.

	52	53	54	55	
0	16242692	16616401	17013017	17434469	30
1	16248742	16622819	17019832	17441715	29
2	16254799	16629243	17026654	17448968	28
3	16260861	16635673	17033482	17456229	27
4	16266929	16642109	17040318	17463499	26
5	16273003	16648551	17047160	17470775	25
6	16279083	16655001	17054010	17478059	24
7	16285169	16661457	17060866	17485351	23
8	16291261	16667919	17067729	17492650	22
9	16297358	16674408	17074599	17499957	21
10	16303461	16680864	17081476	17507272	20
11	16309570	16687345	17088359	17514594	19
12	16315685	16693834	17095250	17521924	18
13	16321806	16700328	17102148	17529262	17
14	16327934	16706829	17109053	17536607	16
15	16334067	16713336	17115965	17543959	15
16	16340197	16719850	17122885	17551319	14
17	16346353	16726362	17129812	17558687	13
18	16352505	16732877	17136747	17566063	12
19	16358663	16739430	17143689	17573446	11
20	16364827	16745970	17150638	17580837	10
21	16370996	16752517	17157593	17588236	9
22	16377172	16759070	17164556	17595643	8
23	16383359	16765629	17171525	17603057	7
24	16389542	16772195	17178502	17610480	6
25	16395736	16778767	17185485	17617909	5
26	16401936	16785347	17192476	17625347	4
27	16408152	16791933	17199472	17632793	3
28	16414365	16798525	17206477	17640246	2
29	16420573	16805124	17213488	17647707	1
30	16426798	16811729	17220507	17655174	0
	37	36	35	34	

The degrees of the Quadrant for the Secants

of Secants.

of the Arches of the same Quadrant.

	52	53	54	55	
30	16426798	16811729	17220507	17655175	60
31	16433027	16818341	17227532	17662651	59
32	16439263	16824960	17234565	17670136	58
33	16445505	16831585	17241605	17677627	57
34	16451754	16838217	17248653	17685127	56
35	16458008	16844856	17255708	17692635	55
36	16464269	16851502	17262770	17700151	54
37	16470536	16858154	17269839	17707674	53
38	16476809	16864813	17276917	17715206	52
39	16483089	16871479	17284002	17722744	51
40	16489385	16878151	17291095	17730290	50
41	16495668	16884830	17298194	17737844	49
42	16501967	16891515	17305300	17745407	48
43	16508272	16898207	17312413	17752978	47
44	16514582	16904907	17319514	17760555	46
45	16520898	16911613	17326662	17768142	45
46	16527220	16918326	17333798	17775740	44
47	16533548	16925046	17340941	17783343	43
48	16539883	16931772	17348091	17790955	42
49	16546224	16948504	17355249	17798575	41
50	16552571	16945244	17362415	17806203	40
51	16558925	16951990	17369587	17813838	39
52	16565286	16958743	17376767	17821481	38
53	16571642	16965495	17383954	17829132	37
54	16578026	16972270	17391148	17836792	36
55	16584406	16979044	17398350	17844460	35
56	16590792	16985824	17405560	17852135	34
57	16597184	16992611	17412776	17859818	33
58	16603584	16999406	17420000	17867509	32
59	16609989	17006208	17427231	17875209	31
60	16616401	17013017	17434469	17882917	30
	37	36	35	34	

of the complements of the Arches of the same Quadrant.

Q 4

The Table

The degrees of the Quadrant for the Secants.

	56	57	58	59	
0	17882917	18360816	18870800	19416039	60
1	17890632	18369014	18879589	19425445	59
2	17898356	18377251	18888389	19434862	58
3	17906089	18385497	18897196	19444290	57
4	17913830	18393753	18906018	19453727	56
5	17921579	18402017	18914846	19463175	55
6	17929337	18410291	18923685	19472635	54
7	17937102	18418574	18932534	19482114	53
8	17944876	18426865	18941393	19491595	52
9	17952658	18435165	18950261	19501076	51
10	17960448	18443454	18959139	19510578	50
11	17968247	18451792	18968027	19520091	49
12	17976054	18460120	18976926	19529615	48
13	17983869	18468456	18985834	19539150	47
14	17991693	18476802	18994752	19548697	46
15	17999525	18485157	19003680	19558254	45
16	18007365	18493521	19012618	19567822	44
17	18015214	18501895	19021516	19577401	43
18	18023071	18510278	19030523	19586991	42
19	18030936	18518670	19039491	19596592	41
20	18038811	18527072	19048468	19606204	40
21	18046693	18535483	19057455	19615827	39
22	18054584	18543903	19066453	19625462	38
23	18062482	18552332	19075461	19635107	37
24	18070389	18560770	19084480	19644765	36
25	18078305	18569217	19093509	19654434	35
26	18086229	18577674	19102549	19664114	34
27	18094161	18586139	19111598	19673805	33
28	18102102	18594614	19120658	19683507	32
29	18110051	18603098	19129727	19693220	31
30	18118009	18611591	19138807	19702945	30
	33	32	31	30	

The degrees of the Quadrant for the Secants

The minutes of the degrees of the Quadrant for the Secants of the Arches of the same Quadrant.

The minutes of degrees of the Quadrant for the Secants of the complements of the Arches of the same Quadrant.

of Secants.
of the Arches of the same Quadrant.

	56	57	58	59	
30	18118009	18611591	19128807	19702945	30
31	18125975	18620094	19147897	19712680	29
32	18133950	18629606	19156998	19722428	28
33	18141934	18637127	19166109	19732186	27
34	18149926	18645658	19175231	19741956	26
35	18157927	18654198	19184362	19751738	25
36	18165937	18662748	19193504	19761531	24
37	18173956	18671507	19202656	19771335	23
38	18181984	18679875	19211818	19781141	22
39	18190021	18688452	19220990	19790968	21
40	18198065	18697038	19230172	19800808	20
41	18206118	18705634	19239365	19810658	19
42	18214179	18714239	19248569	19820320	18
43	18222246	18722854	19257783	19830393	17
44	18230328	18731480	19267008	19840277	16
45	18238416	18740115	19276242	19850172	15
46	18246513	18748760	19285488	19860079	14
47	18254618	18757414	19294744	19869997	13
48	18262732	18766078	19304010	19879927	12
49	18270854	18774752	19313287	19889068	11
50	18278986	18783436	19322574	19899820	10
51	18287126	18792130	19331872	19909784	9
52	18295276	18800833	19341181	19919760	8
53	18303434	18809546	19350501	19929748	7
54	18211601	18818268	19359831	19939749	6
55	18319776	18826999	19369172	19949760	5
56	18327961	18835741	19378524	19959784	4
57	18337154	18844492	19387886	19966820	3
58	18344356	18853252	19397260	19979868	2
59	18352567	18862021	19406644	19989928	1
60	18360816	18870800	19416939	20000000	0
	33	32	31	30	

of the complements of the Arches of the same Quadrant.

The Table
The degrees of the Quadrant for the Secants.

	60	61	62	63	
0	20000000	20626654	21300545	22026892	60
1	20010083	20637484	21312206	22039475	59
2	20020179	20648338	21323882	22052074	58
3	20030285	20659184	21335570	22064690	57
4	20040404	20670054	21347275	22077322	56
5	20050534	20680937	21358993	22089970	55
6	20060676	20691834	21370727	22102635	54
7	20070832	20702744	21382475	22115316	53
8	20080995	20713667	21394238	22128014	52
9	20091172	20724603	21407016	22140728	51
10	20101361	20735554	21417808	22153459	50
11	20111562	20746517	21429615	22166204	49
12	20121776	20757494	21441438	22178971	48
13	20132001	20768484	21453275	22191751	47
14	20142239	20779488	21465128	22204548	46
15	20152489	20790505	21476995	22217361	45
16	20162751	20801535	21488877	22230191	44
17	20173035	20812579	21500774	22243038	43
18	20183321	20823636	21512686	22255902	42
19	20193619	20834706	21524612	22268782	41
20	20203930	20845791	21536553	22281680	40
21	20214252	20856888	21548509	22294595	39
22	20224588	20868000	21560481	22307526	38
23	20234936	20879125	21572467	22320474	37
24	20245296	20890264	21584469	22333439	36
25	20255669	20901416	21596487	22346420	35
26	20266054	20912582	21608520	22359419	34
27	20276452	20923761	21620568	22372434	33
28	20286863	20934955	21632631	22385456	32
29	20297286	20946162	21644710	22398418	31
30	20307721	20957383	21656804	22411584	30
	29	28	27	26	

The degrees of the Quadrant for the Secants

of Secants.

of the Arches of the same Quadrant.

	60	61	62	63	
30	20307721	20957383	21656804	22411584	30
31	20318170	20968618	21668913	22424667	29
32	20328630	20979867	21681038	22437768	28
33	20339102	20991130	21693178	22450886	27
34	20349587	21002406	21705334	22464022	26
35	20360084	21013696	21717505	22477175	25
36	20370594	21025001	21729691	22490346	24
37	20381116	21036319	21741893	22503543	23
38	20391751	21047651	21754111	22516748	22
39	20402198	21058997	21766344	22529965	21
40	20412758	21070357	21778593	22543201	20
41	20423331	21081731	21790858	22556358	19
42	20433916	21093119	21803138	22569723	18
43	20444514	21104522	21815434	22583025	17
44	20455126	21115938	21827745	22596336	16
45	20465750	21127368	21840071	22609663	15
46	10476387	21138814	21852415	22623009	14
47	20487037	21150273	21864774	22636372	13
48	20497700	21161747	21877149	22649754	12
49	20508376	21173235	21889539	22663152	11
50	20519064	21184737	21901946	22676569	10
51	20529765	21196253	21914369	22690004	9
52	20540479	21207783	21926808	22703456	8
53	20551205	21219328	21939263	22716924	7
54	20561945	21230887	21951734	22730414	6
55	30572697	21242460	21964220	22743919	5
56	20583463	21254048	21976722	22757443	4
57	20594142	21265650	21989242	22770984	3
58	20605033	21277267	22001775	22784543	2
59	20615837	21288899	22014325	22798120	1
60	20626654	11300545	22026892	22811726	0
	29	28	27	26	

of the complements of the Arches of the same Quadrant.

The Table

The degrees of the Quadrant for the Secants.

	64	65	66	67	
0	22811726	23662013	24585936	25593051	60
1	22825329	23676784	24602010	25610602	59
2	22838962	23691575	24618107	25628180	58
3	22852612	23706387	24634227	25645783	57
4	22866281	23721220	24650370	25663414	56
5	22879968	23736073	24666536	25681071	55
6	22893674	23750947	24682727	25698754	54
7	22907387	23765842	24698940	25716464	53
8	22921140	23780757	24715178	25734201	52
9	22934901	23795692	24731439	25751965	51
10	22948680	23810648	24747724	25769755	50
11	22962478	23825625	24764033	25787582	49
12	22976294	23840623	24780365	25805417	48
13	22990129	23855642	24796721	25823287	47
14	23003983	23870683	24813101	25841185	46
15	23017855	23885844	24829504	25859104	45
16	23031747	23900827	24845932	25877061	44
17	23045657	23915931	24862383	25895040	43
18	23059586	23931055	24879958	25913046	42
19	23073534	23946200	24895356	25931080	41
20	23087501	23961366	24911878	25949142	40
21	23101486	23976553	24928423	25967230	39
22	23115490	23991762	24944993	25985345	38
23	23129513	24006992	24961587	26003487	37
24	23143556	24022245	24978205	26021658	36
25	23157616	24037518	24994847	26039855	35
26	23171696	24052814	25011514	26058081	34
27	23185795	24068130	25028205	26076333	33
28	23199913	24083469	25044920	26094614	32
29	23214050	24098850	25061660	26112923	31
30	23228205	24114213	25078426	26131259	30
	25	24	23	22	

The degrees of the Quadrant for the Secants

of Secants.

of the Arches of the same Quadrant.

	64	65	66	67	
30	23228205	24114213	25078426	26131259	30
31	23242380	24129616	25095216	26149623	29
32	23256574	24145041	25112030	26168015	28
33	23270797	24160487	25128869	26186436	27
34	23285021	24175956	25145732	26204884	26
35	23299273	24191445	25162620	26223361	25
36	23313546	24206956	25179532	26241867	24
37	23327838	24222488	25196469	26260400	23
38	23342150	24238043	25213432	26278963	22
39	23356481	24253619	25230418	26297555	21
40	23370832	24269217	25247431	26316176	20
41	23385203	24284838	25264468	26334825	19
42	23399593	24300481	25281531	26353503	18
43	23414003	24316147	25298620	26372209	17
44	23428433	24331835	25315734	26390945	16
45	23442882	24347546	25332874	26409709	15
46	23457351	24363281	25350039	26428502	14
47	23471840	24379038	25367229	26447323	13
48	23486348	24324818	25384445	26366174	12
49	23500876	24410620	25401687	26485053	11
50	23515424	24426446	25418956	26503962	10
51	23529922	24442294	25436250	26522890	9
52	23544580	24458164	25453570	26541867	8
53	23559188	24474056	25470915	26560863	7
54	23573817	24489973	25488286	26579889	6
55	23588565	24505908	25505683	26598945	5
56	23603134	24521869	25523005	26618030	4
57	23617822	24537851	25540553	26637145	3
58	23632532	24553857	25558072	26656251	2
59	23647262	24569885	25575526	26675466	1
60	23662013	24585936	25593051	26694672	0
	25	24	23	22	

of the complements of the Arches of the same Quadrant.

The Table
The degrees of the Quadrant for the Secants.

	68	69	70	71	
0	26694672	27904284	29238045	30715531	60
1	26713907	27925445	29261433	30741500	59
2	26733172	27946642	29284861	30767516	58
3	26752467	27967873	29308328	30793579	57
4	26771791	27989139	29331835	30819689	56
5	26791145	28010440	29355382	30845584	55
6	26810529	28031776	29378970	30872051	54
7	26819942	28053147	29402599	30898304	53
8	26849390	28074553	29426268	30924605	52
9	26868867	28095994	29449978	30950953	51
10	26888373	28117469	29473728	30977350	50
11	26907910	28138980	29497519	31003793	49
12	26927479	28160527	29521350	31030285	48
13	26947078	28182108	29545222	31056824	47
14	26966709	28203725	29569136	31083412	46
15	29986370	28225378	29593090	31110047	45
16	27006062	28247067	29617087	31136731	44
17	27025785	28268793	29641124	31163462	43
18	27045539	28290553	29665204	31190241	42
19	27065323	28312349	29689326	31217019	41
20	27085138	28334181	29713488	31243945	40
21	27104985	28356049	29737692	31270871	39
22	27124864	28377954	29761938	31297848	38
23	27144774	28399894	29786227	31324873	37
24	27164717	28421871	29810558	31351948	36
25	27184690	28443884	29834931	31379072	35
26	27204686	28465934	29859347	31406247	34
27	27224734	28488021	29883705	31433472	33
28	27244804	28510144	29908306	31460747	32
29	27264906	28532304	29932850	31488072	31
30	27285040	28554501	29957438	31515448	30
	21	20	19	18	

The degrees of the Quadrant for the Secants

of Secants. 120

of the Arches of the same Quadrant.

	68	69	70	71	
30	27285040	28554501	29957438	31515448	30
31	27305205	28576735	29982069	31542873	29
32	27325402	28599007	30006743	31570349	28
33	27345631	28621316	30031460	31597875	27
34	27365893	28643662	30056220	31625453	26
35	27386186	28666045	30081023	31653080	25
36	27406513	28688467	30105870	31680758	24
37	27426872	28710925	30130760	31708486	23
38	27447264	28733422	30155714	31736265	22
39	27467688	28755956	30180672	31764094	21
40	27488145	28778549	30205694	31791974	20
41	27508635	28801139	30230760	31819906	19
42	27529157	28823787	30255871	31847891	18
43	27549722	28846473	30281026	31875929	17
44	27570301	28869196	30306226	31904019	16
45	27590922	28891957	30331460	31932164	15
46	27611578	28914756	30356759	31960358	14
47	27632266	28937594	30382092	31988606	13
48	27952989	28960471	30407470	32016909	12
49	27673745	28983386	30432893	32045263	11
50	27694535	29006340	30458361	32073672	10
51	27715358	29029332	30483873	32102132	9
52	27736215	29052363	30509430	32130649	8
53	27757105	29075435	30535033	32159212	7
54	27778029	29098546	30560682	32187832	6
55	27798987	29121697	30586375	32216504	5
56	27819978	29144888	30612115	32245231	4
57	27841003	29168118	30637890	32274012	3
58	27862060	29191388	30663732	32302846	2
59	27883156	29214697	30689608	32331735	1
60	27904284	29238045	30715531	32360678	0
	21	20	19	18	

of the complements of the Arches of the same Quadrant.

The Table

The degrees of the Quadrant for the Secants.

	72	73	74	75	
0	32360678	34203038	36279559	38637042	60
1	32389676	34235609	36316402	38679033	59
2	32418726	34268245	36353333	38721117	58
3	32447837	34300947	36390323	38763296	57
4	32477001	34333716	36427401	38805571	56
5	32506219	34366553	36464558	38847941	55
6	32535494	34399452	36501793	38890408	54
7	32564823	34432420	36539107	38932971	53
8	32594209	34465456	36570511	38975632	52
9	32623651	34498557	36613973	39018390	51
10	32653148	34531726	36651525	39061246	50
11	32682701	34564959	36689156	39104200	49
12	32712311	34598259	36726868	39147252	48
13	32741977	34631626	36764660	39190423	47
14	32771699	34665061	36802533	39233653	46
15	32801478	34698564	36840488	39277002	45
16	32831314	34732135	36878524	39320449	44
17	32861207	34765775	36916641	39363994	43
18	32891157	34799483	36954842	39407640	42
19	32921165	34833259	36993127	39451384	41
20	32951231	34867105	37031496	39495228	40
21	32981355	34901024	37069947	39539172	39
22	33011537	34935005	37108482	39583218	38
23	33041776	34966052	37147101	39627364	37
24	33072074	35003172	37185803	39671613	36
25	33102431	35037361	37224589	39715965	35
26	33131846	35071621	37263459	39760420	34
27	33163320	35105952	37302413	39804979	33
28	33193853	35140354	37341453	39849642	32
29	33224444	35174826	37380577	39894411	31
30	33255094	35209369	37419788	39939286	30
	17	16	15	14	

The degrees of the Quadrant for the Secants

of Secants.

of the Arches of the same Quadrant.

	72	73	74	75	
30	33255094	35209369	37419788	39939226	30
31	33285803	35243981	37459081	39984263	29
32	33316571	35278664	37498460	40029344	28
33	33347398	35313418	37537923	40074528	27
34	33378286	35348244	37577471	40119816	26
35	33409132	35383140	37617104	40165289	25
36	33440240	35418110	37656824	40210709	24
37	33471307	35453152	37696632	40256316	23
38	33502436	35488268	37736518	40302033	22
39	33533625	35523456	37776513	40347858	21
40	33564875	35558718	37816588	40393792	20
41	33596187	35594052	37856751	40439834	19
42	33627561	35629460	37897004	40485985	18
43	33658998	35664940	37937146	40532245	17
44	33690497	35700494	37977779	40578613	16
45	33722059	35736121	38018300	40625091	15
46	33753683	35771822	38058912	40671678	14
47	33785370	35807597	38099614	40718374	13
48	33817120	35843447	38140406	40765180	12
49	33848934	35879373	38181288	40812093	11
50	33880813	35915374	38222261	40859121	10
51	33912753	35951451	38263324	40906259	9
52	33944756	35987602	38304479	40953510	8
53	33976821	36023829	38345725	41004876	7
54	34008950	36060132	38387064	41048358	6
55	34041141	36096510	38428495	41095957	5
56	34073395	36132966	38470019	41143668	4
57	34105712	36169497	38511635	41191492	3
58	34138091	36206107	38553344	41239431	2
59	34170523	36242794	38595146	41287425	1
60	34203038	36279559	38637042	41335654	0
	17	16	15	14	

of the complements of the Arches of the same Quadrant.

The Table
The degrees of the Quadrant for the Secants

	76	77	78	79	
0	41335654	44454097	48097335	52408433	60
1	41383937	44510183	48163151	52486983	59
2	41432338	44566415	48229350	52565774	58
3	41480856	44622793	48295633	52644807	57
4	41529492	44679318	48362102	52724084	56
5	41578245	44735990	48428756	52803604	55
6	41627117	44792810	48495599	52883368	54
7	41676108	44849777	48562631	52963377	53
8	41725219	44906892	48629854	53043632	52
9	41774450	44964155	48697269	53124134	51
10	41823802	45021567	48764877	53204885	50
11	41873273	45079129	48832678	53285884	49
12	41922863	45136843	48900673	53367134	48
13	41972573	45194707	48968853	53448635	47
14	42022405	45252726	49037249	53530390	46
15	42072357	45310898	49105830	53612399	45
16	42122431	45369223	49174607	53694666	44
17	42172625	45427703	49243590	53777191	43
18	42222942	45486338	49312751	53859976	42
19	42273380	45545127	49382118	53943022	41
20	42323942	45604073	49451684	54026331	40
21	42374627	45663175	49521449	54109903	39
22	42425439	45722435	49591416	54193739	38
23	42476377	45781853	49661584	54277840	37
24	42527442	45841429	49731956	54362207	36
25	42578635	45901164	49802532	54446842	35
26	42629957	45961059	49873313	54531744	34
27	42681409	46021115	49944301	54616915	33
28	42732991	46081333	50015497	54702356	32
29	42784705	46141715	50086901	54788068	31
30	42836551	46202261	50158514	54874053	30
	13	12	11	10	

The degrees of the Quadrant for the Secants

of Secants.

of the Arches of the same Quadrant.

	76	77	78	79	
30	42836551	46202261	50158514	54874053	30
31	42888527	46262969	50230335	54960312	29
32	42940631	46323841	50302367	55046847	28
33	42992865	46384877	50374610	55133659	27
34	43045229	46446076	50447065	55220751	26
35	43097722	46507440	50519732	55308122	25
36	43150347	46568970	50592614	55395775	24
37	43203103	46630665	50665711	55483710	23
38	43255992	46692527	50739024	55571930	22
39	43309012	46754555	50812553	55660434	21
40	43362166	46816752	50886299	55749226	20
41	43415454	46879117	50960263	55838300	19
42	43468877	46941653	51034447	55927677	18
43	43522435	47004361	51108850	56017340	17
44	43576129	47067242	51183475	56107297	16
45	43629959	47130297	51258321	56197549	15
46	43683925	47193526	51333391	56288099	14
47	43738728	47256930	51408684	56378948	13
48	43792268	47320509	51484204	56470097	12
49	43846646	47384264	51559951	56561548	11
50	43901162	47448195	51635936	56653302	10
51	43955817	47512302	51712129	56745360	9
52	44000612	47576586	51788563	56837723	8
53	44065548	47641048	51865227	56930392	7
54	44120625	47705689	51942124	57023369	6
55	44175844	47770510	52019254	57116653	5
56	44231207	47835511	52096618	57210246	4
57	44286712	47900693	52174216	57304150	3
58	44342362	47966058	52252051	57398367	2
59	44398156	48031605	52330123	57492896	1
60	44454097	48097335	52408433	57587740	0
	13	12	11	10	

of the complements of the Arches of the same Quadrant.

R. 2

The Table

The degrees of the Quadrant for the Secants

	80	81	82	83	
0	5758774o	63924495	71852975	82055127	60
1	57682901	64042118	72002006	82249986	59
2	57778381	64160180	72151659	82445779	58
3	57874180	64278683	72301942	82642513	57
4	57970302	64397632	72452863	82840196	56
5	58066748	64517028	72604421	83038833	55
6	58163520	64636873	72756618	83238436	54
7	58260619	64757168	72909461	83439009	53
8	58358049	64877918	73062954	83640561	52
9	58455810	64999124	73217100	83843097	51
10	58553904	65120789	73371903	84046626	50
11	58652333	65242916	73527367	84251153	49
12	58751099	65365508	73683499	84456680	48
13	58850205	65488566	73840302	84663213	47
14	58949653	65612095	73997782	84870760	46
15	59049444	65736097	74155942	85079327	45
16	59149581	65859675	74314786	85288957	44
17	59250065	65985531	74474318	85499628	43
18	59350898	66110967	74634544	85711347	42
19	59452082	66246886	74795468	85924121	41
20	59553618	66363291	74957095	86137958	40
21	59655506	66490185	75119429	86352864	39
22	59757728	66617572	75282475	86568849	38
23	59860346	66745453	75446238	86785921	37
24	59963291	66873831	75610721	87004089	36
25	60066612	67002708	75775928	87223362	35
26	60170285	67132088	75941864	87443750	34
27	60274319	67261972	76108533	87665261	33
28	60378718	67392365	76275941	87887909	32
29	60483482	67523270	76444091	88111704	31
30	60588615	67654691	76612989	88336657	30
	9	8	7	6	

The degrees of the Quadrant for the Secants

of Secants.

of the Arches of the same Quadrant.

	80	81	82	83	
30	60588615	67654691	76612989	88336657	30
31	60694118	67786629	76782641	88562776	29
32	60799995	67919089	76953050	88790069	28
33	60906246	68052073	77124223	89018543	27
34	61012875	68185585	77296165	89248201	26
35	61119882	68319630	77468882	89479054	25
36	61227271	68454208	77642381	89711108	24
37	61335043	68589313	77816665	89944373	23
38	61443202	68724977	77991740	90178856	22
39	61551749	68861175	78167612	90414568	21
40	61660686	68997920	78344287	90651519	20
41	61770013	69135315	78521769	90889717	19
42	61879735	69273018	78700066	91129181	18
43	61989853	69411469	78879183	91369917	17
44	62100367	69550434	79059128	91611941	16
45	62211280	69689963	79239905	91855265	15
46	62322594	69830059	79421520	92099899	14
47	62434312	69970726	79603976	92345849	13
48	62546437	70111967	79787381	92592126	12
49	62658971	70253786	79971439	92841739	11
50	62771918	70396188	80156456	93091699	10
51	62885274	70539174	80342336	93342963	9
52	62999049	70682751	80529087	93595620	8
53	63113241	70826919	80716713	93849647	7
54	63227855	70971684	80905219	94105066	6
55	63342890	71117047	81094612	94361964	5
56	63458352	71263014	81284899	94620181	4
57	63574240	71409586	81476087	94879901	3
58	63690559	71556760	81668183	95141050	2
59	63807309	71704564	81861195	95403639	1
60	63924495	71852975	82055127	95667689	0
	9	8	7	6	

of the complements of the Arches of the same Quadrant.

R 3

The Table
The degrees of the Quadrant for the Secants

	84	85	86	
0	95667689	114737188	143355808	60
1	95933204	115119970	143954694	59
2	96200195	115505313	144558602	58
3	96468673	115893242	145167595	57
4	96738655	116283797	145781740	56
5	97010253	116676991	146401101	55
6	97283267	117072851	147025745	54
7	97557932	117471403	147655740	53
8	97834057	117872815	148291169	52
9	98111843	118276840	148932108	51
10	98391211	118683794	149578791	50
11	98672171	119093414	150230942	49
12	98954738	119506013	150888966	48
13	99236930	119921335	151552578	47
14	99524766	120339695	152222283	46
15	99812250	120760985	152897946	45
16	100101400	121185232	153579394	44
17	100392329	121612482	154267179	43
18	100684851	122042752	154961155	42
19	100979193	122476076	155661396	41
20	101275259	122912485	156368008	40
21	101572962	123352014	157081063	39
22	101872522	123794696	157800648	38
23	102173854	124240732	158526854	37
24	102476971	124689836	159259771	36
25	102781890	125142353	159999560	35
26	103088639	125598007	160746121	34
27	103397202	126057149	161499724	33
28	103707656	126519656	162260744	32
29	104019959	126985568	163028671	31
30	104334254	127454936	163804188	30
	5	4	3	

The degrees of the Quadrant for the Secants

of Secants.
of the Arches of the same Quadrant.

	84	85	86	
30	104334254	127454936	163804188	30
31	104650345	127927785	164586836	29
32	104968474	128404152	165377268	28
33	105288542	128884078	166175067	27
34	105610566	129367604	166980877	26
35	105934564	129854921	167794536	25
36	106260557	130345812	168615879	24
37	106588558	130840395	169445585	23
38	106918589	131338917	170283495	22
39	107250680	131841076	171129820	21
40	107584955	132347264	171984431	20
41	107921201	132857174	172847712	19
42	108259554	133371390	173719700	18
43	108600151	133889600	174600528	17
44	108942779	134411312	175490331	16
45	109287702	134937471	176389247	15
46	109634817	135467749	177297417	14
47	109984143	136002235	178215000	13
48	110335695	136540955	179142131	12
49	110689503	137083887	180078954	11
50	111045597	137631223	181025951	10
51	111403988	138183016	181982628	9
52	111764699	138739177	182949802	8
53	112127750	139299830	183926988	7
54	112493167	139865032	184915009	6
55	112861097	140435034	185913698	5
56	113231316	141009514	186922883	4
57	113604036	141588910	187943432	3
58	113979204	142172885	188975184	2
59	114356941	142761897	190018342	1
60	114737188	143355808	191073059	0
	5	4	3	

of the complements of the Arches of the same Quadrant.

The Table
The degrees of the Quadrant for the Secants

		87	88	89		
The minutes of the degrees of the Quadrant for the Secants of the Arches of the same Quadrant.	0	191073059	286537048	572987098	60	The minutes of degrees of the Quadrant for the Secants of the complements of the Arches of the same Quadrant.
	1	192139567	288943841	582696234	59	
	2	193218044	291391404	592740072	58	
	3	194308693	293880683	603139919	57	
	4	195411723	296413087	613907444	56	
	5	196527729	298990299	625070305	55	
	6	197656182	301611807	636642580	54	
	7	198797665	304279687	648655621	53	
	8	199952408	306996123	661126359	52	
	9	201120639	309760533	674090521	51	
	10	202303011	312576192	687573461	50	
	11	203498943	315544491	701612741	49	
	12	204709121	318361849	716229489	48	
	13	205934200	321336774	731453951	47	
	14	207173596	324366765	747356168	46	
	15	208428431	327455509	763965262	45	
	16	209698119	330602545	781323254	44	
	17	210983811	333811800	799494739	43	
	18	212284914	337082830	818524878	42	
	19	213602421	340419652	838490069	41	
	20	214936837	343823403	859453551	40	
	21	216287319	347294586	881484374	39	
	22	217655350	350837799	904682629	38	
	23	219040792	354454051	929134899	37	
	24	220443981	358145679	954945691	36	
	25	221865261	361914968	982231437	35	
	26	223305005	365763113	1011112129	34	
	27	224763453	369695332	1041753449	33	
	28	226241278	373713015	1074309940	32	
	29	227738558	377818975	1108967170	31	
	30	229255785	382016194	1145934768	30	
		2	1	0		

The degrees of the Quadrant for the Secants

of Secants.
of the Arches of the same Quadrant.

		87	88	89		
The minutes of the degrees of the Quadrant for the Secants of the Arches of the same Quadrant.	30	229255785	382016194	1145934768	30	The minutes of the côplements of the Arches of the same Quadrant for the Secants of the Quadrant.
	31	230793360	386307709	1185438054	29	
	32	232351718	390696734	1227777193	28	
	33	233931261	395186630	1273252703	27	
	34	235532422	399780916	1322226495	26	
	35	237156211	404483275	1375118522	25	
	36	238801972	409397566	1432297932	24	
	37	240470730	414227875	1494678912	23	
	38	242163582	419278406	1562622042	22	
	39	243879838	424453607	1637036239	21	
	40	245621193	429758156	1718892212	20	
	41	247386980	435196961	1809365043	19	
	42	249178956	440775230	1909891150	18	
	43	250996450	446498305	2022234532	17	
	44	252841285	452371994	2148642981	16	
	45	254713463	458402271	2291895669	15	
	46	256612911	464595485	2455554199	14	
	47	258541565	470958329	2644450861	13	
	48	260499426	477497828	2864894681	12	
	49	262487160	484221619	3125282743	11	
	50	264505458	491139838	3437843546	10	
	51	266554348	498256113	3819709423	9	
	52	268635944	505581634	4297193536	8	
	53	270750304	513128395	4911255640	7	
	54	272898206	520901152	5729642566	6	
	55	275080457	528915798	6875687278	5	
	56	277297985	537178089	8594018365	4	
	57	279551349	545702599	11458691197	3	
	58	281841763	554505091	17188036598	2	
	59	284170013	563593031	34376072269	1	
	60	286537048	572987098	INFINITA	0	
		2	1	0		

of the complements of the Arches of the same Quadrant.

A plaine Treatise of the first principles of Cosmographie, and specially of the Spheare, representing the shape of the whole world:

Together with all the chiefest and most necessarie vses thereof, written by M. *Blundevill* of Newton Flotman, *Anno Dom.* 1594

The heauens declare the glory of God, and the firmament sheweth his handy worke. Psal.19.

LONDON
Printed by *John Windet.* 1594.

The expofition of certaine
termes or principles of
Geometrie.

Inding to treate of the principles of Cosmographie, and especially of the Sphere, I thinke it good first to expound vnto you certaine termes of Geometry, without the which the vnlearned shall hardly vnderstand the Contents of this Treatise, which tearmes are certaine principles of Geometrie, called definitions: For there are but three kinds of principles, whereon the demonstration of all Geometricall conclusions dependeth: that is, definitions, petitions, and maxims, but I mind here to deale onely with certaine definitions, Whereof I will set downe as many as I thinke needful for this purpose, in such order as followeth.

1. A point called in Latine punctus, is a thing supposed to be indiuisible, hauing neither length, breadth, nor deapth, as the point or pricke a.

2. A line called in Lataine linea, is a supposed length, hauing neither breadth nor thicknesse, as the line a. b.

3. The ends or bounds of a line, are two points, and of lines some be right, and some be crooked.

4. A

The first Booke

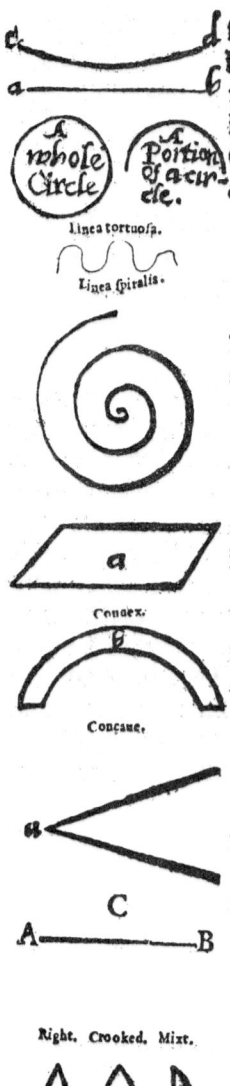

Linea tortuosa.

Linea spiralis.

Conuex.

Concaue.

Right. Crooked. Mixt.

4 A right line is that which goeth right from one point to another, and not bowingly as doth the line c. d. but so straight as is possible, as the line a. b. Againe the crooked line is either a whole Circle, or portion of a Circle, or else goeth winding in and out, as a serpent called in Latine Linea tortuosa, or else winding about like the shell of a Snaile, called of the Latines, linea Spiralis, as these figures here doe shew.

5 Superficies or vpperface, is that which onely hath length and breadth, without deapth, which is twofould, that is to say, plaine and crooked.

6 The bounds of superficies are lines.

7 A plaine superficies is that which lyeth straight betwixt his lines, as the figure a, & a crooked superficies is that which goeth bowing and lyeth not straight betwixt his lines as the figure b. Againe superficies being considered in an hollow body, as a barrell, tunne, or vault, may be deuided into two other kinds, that is Conuexe, and Concaue, for the vpper part of such vault is said to be conuex, and yinward part cancaue as the figure b. sheweth.

8 A plaine Angle is when two lines being drawne vpon a plaine superficies not directly one against another, but so as by meeting or touching one an other in one selfe point, they may make a plaine Angle, as the Angle a. For if two lines be drawne directly one against another, though they meete in one point, yet they make no Angle, but rather one selfe line as the two lines a. b. meeting in the point c.

9 Of plaine Angles, some are called right line angles, because that both the lines whereof it consisteth are right, and some are called

of the Sphere.

called crooked line angles, because that both lines are crooked, & some are said to be mixt, because the one line is crooked and the other right, as you may perceiue by the thrée sundry shapes thereof made in the Margent.

10 If one right line standing vpon another right line do make two equall angles, that is to say, of ech side one, then eyther of those Angles is a right Angle, and the line so standing vpon his fellow, is called the perpendicular or plumbe line, as the line a. b. standing vpon the line c. d.

11 A blunt Angle, is that which is greater then a right Angle, as the Angle e.

12 A sharpe Angle, is that which is lesser then a right Angle, as f. so as there be in all thrée Angles, that is right, blunt, and sharpe, in Latine rectus, obtusus, & acutus: And besides these Angles, there be also certaine Sphericall, that is to say, round Angles, which consist of two circular lines, drawne vppon a Sphericall superficies, which doe crosse one another in some point, eyther rightly, or obliquely: if rightly, then they make right Angles, on ech side of the point where they crosse, as the figure E. sheweth, if obliquely then they make sharpe Angles of the inside, and blunt Angles on the out side of the point, where they crosse as the figure F. doth partly shew, for such Angles cannot be so well described in Plano, as vpon the superficies of some Sphericall body.

13 A terme called in Latine terminus, is the bound or limite of any thing, as points are the bound of lines, & lines the bounds of superficies, and superficies ÿ bounds of a body, which is that which hath imaginatiuely, but not materially, both length, bredth, and depth, and if such body haue many faces

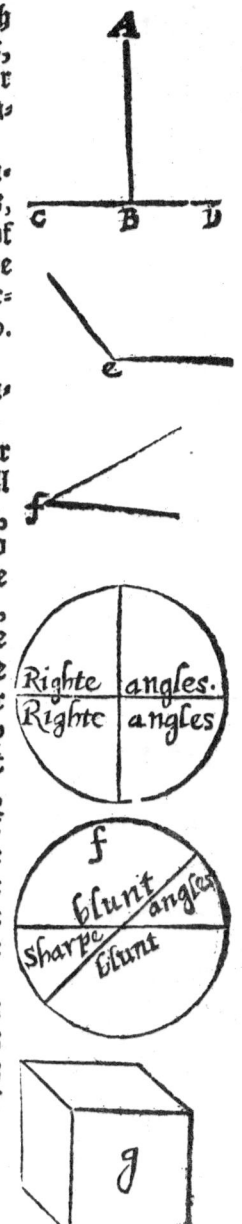

The first Booke

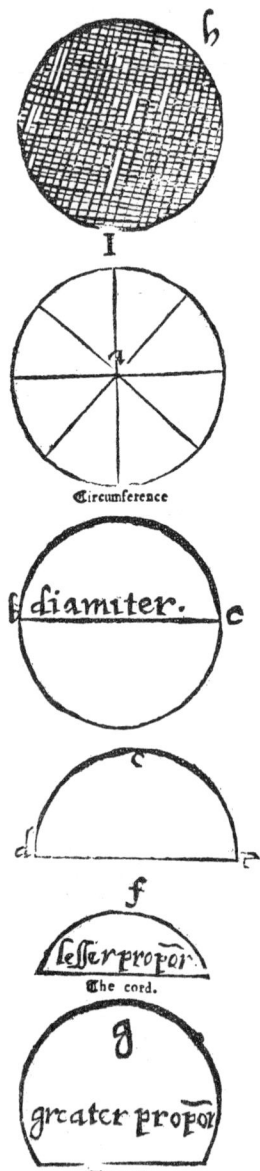

or sides, then it is bounded with many superficieses, as the figure g. made like a sixe square dye. But if such body be round as a bowle, Sphere, or Globe, then it is bounded or couered with one superficies onely, as the figure H. doth shewe.

14 A figure is that which is comprehended within one bound, or many bounds.

15 A Circle is a plaine figure bounded with one circular line, which is called the circumference, vnto the which as many right lines as are drawne from the point standing in the midst, are all equall one to the other, that is to say, one is as long as another, as the figure I. here doth shew.

16 And the middle point of a Circle is called the Center, as the point A. in the figure I.

17 The Diameter of a Circle, is a right line passing through the Center, from one side of the Circumference to the other, which deuideth the Circle into two equall parts, as the line B. C.

18 A Semicircle is a figure, contayned within the Diameter and halfe the Circumference of a Circle as the figure D.C.E.

19 The portion of a Circle, is a part of a Circle, greater or lesser than the Semicircle, as the figures F. and G. doe shewe, the right line in eyther of which figures is called in Latine chorda, in English the string, and the Circular line in Latine is called arcus in English the bowe, but the Diameter is alwayes the greatest cord in any Circle.

20 Right line figures, are those which are bounded with right lines.

21 Trian-

of the Spheare.

21 Triangles or three cornerd figures, are those which are bounded with three right lines as the figure H.

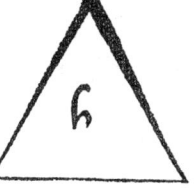

22 Foure square figures, are those which are bounded with foure right lines, as the figure I.

23 Many square figures, are those which are bounded with more right lines then foure as K.

24 Of Triangles or three cornerd figures, there be sixe kindes, whereof the first is called Isoplurus, hauing three equall sides and three equal Angles as A.

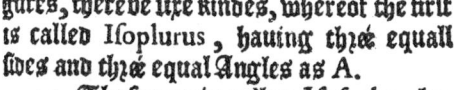

25 The second is called Isosceles, hauing but two equall sides and angles as B.

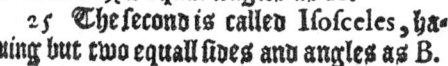

26 The thirde called scalenos hauing no side equall, one with the other, but one shorter or longer then another as C.

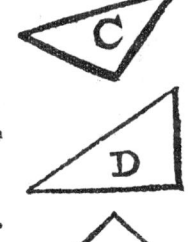

27 The fourth is called Orthogonius, hauing one right Angle, as D.

28 The fift is called ampligonius, because it hath one blunt Angle, as E.

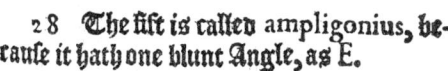

29 The sixt is called oxigonius, because it hath three sharpe Angles, but not equall sides as F.

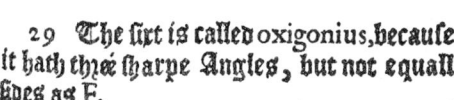

The first Booke of the Spheare.

30 Of foure square figures there is one iust foure square, hauing foure equall sides, and is right angled as G.

31 Another is called a long Square, which is right angled, but not hauing foure equall sides, as H.

32 There is another called rombus, that is to say a Turbut, in shape like to a Diamant, which hath foure equall sides, but is not right angled, as I.

33 There is another called romboides, which though it hath sides and Angles one right against another, yet hath it neither foure equall sides, nor yet is right angled, as K.

34 All other kinds of foure squares are called trapezia, as L. or such like.

35 Paralels are two right lines equally distant one from another, which béeing drawne foorth infinitely, would neuer touch or méete one another in any part, as the two lines M. N. And though the lines bee circular, yet if they be equally distant in all places one from another, they are also called paralels, as the figure P. sheweth, yea though the lines be winding in and out as Serpents, yet if they be equally distant in al places one from another, they are paralels as well as if they were right lines as the figure Q. sheweth.

Right line paralels.

Circuler paralels.

Serpentine paralels.

The order and contents of this Treatise, touching the first principles of Cosmography, and specially of the Spheare.

Auing set downe the exposition of certaine termes of Geometry, for the better vnderstanding of this Treatise, I doe first define what Cosmography is, and therewith doe briefely showe what Sciences it comprehendeth, and who were first inuentors thereof: That done, because the Spheare doth represent the shape of the whole world, I do define what a Spheare is, and then I doe deuide the world into two essentiall parts, that is, the Celestiall and Elementall part, according to which two parts I do also deuide this Treatise into two Bookes, the first whereof treateth of the Celestiall part, and the second of the Elementall part of the world, with what order the Chapters of ech Booke hereafter following, do plainely shew.

The Chapters and Contents
of the first Booke.

OF Cosmography, what it is, and what kinde of Sciences it comprehendeth, and who were first inuentors thereof, Chap. 1.

The definition of a Spheare, and of the vnitie, roundnesse, and capacitie of the worlde, also of the Poles and Axletree thereof, Chap. 2.

Of the diuision of the world, and of the two Essentiall parts thereof, and what things ech part conteyneth. Chap. 3.

A figure of the whole world, wherein are set forth the two Essentiall parts before mentioned, that is to say, the eleauen heauens and the foure Elements.

A demonstration to prooue the pluralitie of the heauens, Chap. 4.

The Contents.

Of the highest Spheare or heauen, called the Imperial heauen. Chap. 5.

Of the tenth Spheare or heauē called in Latine *Primum mobile*, that is, the first moueable, & what motion it hath. Chap. 6.

Of the ninth heauen, what motions & names it hath, & whether there be any waters aboue the firmament or not. Chap. 7.

Of the eight heauen what motions it hath, and what circles are imagined by the Astronomers to be in that heauen, and to what vse and purpose they serue: Also in what time euery one of the seuen Planets maketh his reuolution, & of what thicknes their Spheares be. Chap. 8.

Of the circles whereof a materiall Spheare consisteth, and of their diuers diuisions. Chap. 9.

Of the Equinoctiall line, why it is so called, and of the diuerse vses thereof. Chap. 10.

A figure shewing the Equinoctiall line, the two poles and the Axeltree of the world.

Of the Zodiaque, why it is so called, and of the 12. signes therein conteined. Also of the Latitude, Longitude and declination thereof. Chap. 11.

How much the Zodiaque declineth from the Equinoctiall towards either of the Poles, and of the greatest declination of the sunne, what it is at this present, and what it hath beene in times past. Chap. 12.

How to know the quantitie of the sunnes declination, be it Northward or Southward, euery day throughout the yeare, aswell by a Table as by helpe of a materiall Spheare or globe. Chap. 13.

An instrument to knowe thereby in what signe or degree the sunne is euery day throughout the yeare.

A Table shewing the declination of the sunne euery day throughout the yeare lately calculated.

Vpon what poles the Zodiaque turneth about, also of the Ecliptique line and of the diuerse vses thereof. Chap. 14.

Of the Eclipses both Solar and Lunar, & of the head & taile of the Dragon with certaine figures shewing the same. cap. 15.

Of the two Colures, why they are so named, and whereto they serue: also of the foure Cardinall points, that is, the two

Equi-

The Contents.

Equinoctiall, and the two Solsticiall points, and of the entrance of the sunne into any of those points, or into any other signe. Chap. 16.

Of the Horizon both right and oblique, making thereby three kinds of Spheares, that is, the right, the paralell, and the oblique Spheare. Chap. 17.

A figure shewing the Latitude of any place to be equall vnto the eleuation of the Pole.

Of the Meridian, and of the vses thereof. Chap. 18.

Of the verticall circle and vses thereof, whereof no mention is commonly made by those that write of the Spheare. Chap. 19.

Of the foure lesser circles, that is to say, the circle Artique, the circle Antartique, the Tropique of Cancer, and the Tropique of Capricorne, and also of the fiue Zones, that is to say, two cold, two temperate, and one extremely hoat. Chap. 20.

A figure shewing the fiue foresaid Zones.

A Table shewing how many minutes are requisite to make one degree, in euery lesser circle answerable to one degree of the Equinoctiall.

Of the starres and celestiall bodies contained in the firmament, and first of their substance. Chap. 21.

Of the moouing and shape of the starres. Chap. 22.

Of the number of the starres, and of their magnitude and greatnes, and into how many Images they are deuided, & how many starres euery Image contayneth. Chap. 23.

Of the xij. Images or signes of the Zodiaque. Chap. 24.

Of the xxj. Northerne Images. Chap. 25.

Of the 15. Southerne Images. Chap. 26.

Of the longitude of the fixed starres, and of the precession of the vernall Equinoctiall point, and what it is. Chap. 27.

Of the Latitude of the fixed starres. Chap. 28.

Of the Declination of the fixed starres. Chap. 29.

Of the ascention and discention, that is the rising and setting of the starres, aswell according to the Astronomers, as according to the Poets. Chap. 30.

Of the Astronomicall ascention and discention in generall both right, meane, and oblique, & what a giuen ark is. Cap. 31.

The Contents.

Of the right, oblique, and meane ascention in particular, and of the chiefe causes of such diuersitie of ascentions. cap. 32.

How to know the diuersities of the ascentions and discentions, as well in the right as oblique Spheare. Chap. 33.

Of the ascentionall difference and vses thereof. Chap. 34.

Of the threefold Poeticall rising and setting of the starres, Chap. 35.

Of time, what it is, and into what parts it is deuided. chap. 36.

Of the yeare, and of his diuerse kindes, and of the diuerse computations had thereof in diuerse ages, and amongst diuers Nations. Chap. 37.

Of the sunnes yeare called in Latine *annus solaris*, and of the diuerse kinds thereof, and first of the Tropicall yeare, both equall and vnequall. Chap. 38.

Of the Syderall yeare, & how much it containeth. chap. 39.

Of the Politicall yeare, and diuerse kinds thereof. chap. 40.

Of the Iulian yeare, why it is so called, and of the 2. kinds thereof, that is, the common yeare, and the bisextile yeare, otherwise called the leape yeare. Chap. 41.

Of the Egyptian yeare, and how many daies it containeth. Chap. 42.

How many Moones the Iewes yeare, and the Athenian yeare doe containe. Chap. 43.

Of the yeare Lunar, and of the kinds thereof. Chap. 44.

Of the diuers kinds of monthes, and into what parts euery Solar moneth is deuided according to the Romanes, that is, into Kallends, Nones, and Ides. Chap. 45.

Of the diuerse kindes of moneths Lunar. Chap. 46.

Of a weeke. Chap. 47.

Of dayes and nights both naturall and artificiall, as well in the right Spheare, as in the oblique Spheare. Chap. 48.

Two figures shewing the right and oblique Spheare, together with the arckes of the dayes and nightes in ech Spheare. Chap. 49.

How to find out by the materiall Spheare or Globe, and by helpe of the ascentionall difference before defined, the increase & decrease of euery day throughout the yeare in euery seueral latitude, & at what houre the sun riseth & setteth. c. 50

How

The Contents.

How to know by the materiall Spheare or Globe, in what part of the Horizon the sunne riseth and setteth euery day, and thereby the length of the day. Also how to know the Meridian altitude of the sunne euery day throughout the yeare, & being at his Meridian altitude, to know how farre distant he is from the Zenith euery day, Chap. 51.

Of houres as well equall as vnequall, and into what partes they are deuided. Chap. 52.

How and in what manner the Iewes doe deuide the artificiall day and night, ech of them into foure quarters. chap. 53.

How to knowe what Planet raigneth in euery houre of the day or night artificiall, as well by helpe of a table, as by a rule contained in one verse. chap. 54.

Here ende the Contents of the first Booke.

The Chapters and Contents of the second
Booke, containing the Elementall part of the world.

OF the Elementall part of the world. Chap. 1.
Of the fire, his nature and motion, Chap: 2.
Of the Aire, and into howe many Regions it is deuided, Chap. 3.
A figure shewing the three seuerall Regions of the aire.
Of the water, and whether it be round or not. Chap. 4.
A figure shewing that the water is round.
Of the Earth, and whether it be in all parts round or not. Chap. 5.
Of the compasse of the earth, and of the diuersitie of measures, according to diuerse countries. chap. 6.
Of the Longitude and Latitude of the earth. chap. 7.
A figure shewing the Longitude and Latitude of the world.
How to knowe the Latitude of any place, as well in the day as in the night. chap. 8.
How to knowe the true Longitude of any place. chap. 9.
A readie way to finde out the Longitude of anie place inuented by *Gemma Phrizius*. chap. 10.

The Contents.

Another way taught by *Appian* to find out the Longitude of any place with the Croſſe-ſtaffe, by knowing the diſtance vnto the Eclipticque line. chap. 11.

How to know the diſtance of places, that is to ſay, how many miles one place is diſtant from another, and howe many wayes one place is ſayd to differ in diſtance one from another. Chap. 12.

A Table ſhewing how many miles are anſwereable to one degree of euery ſeuerall Latitude.

How to knowe by the helpe of the foreſaide Table, the diſtance of two places differing only in Longitude. chap. 13.

How to finde out the true diſtance of two places differing both in Longitude and Latitude, by the Arithmeticall way. chap. 14.

Alſo how to find out the ſame by helpe of a Demicircle diuided into 180. parts, without any Arithmetique.

How to finde out the diſtance of places by the Geometricall way. chap. 15.

Of the fiue Zones. chap. 16.

A figure of the foreſaid fiue Zones, that is to ſay, two cold, two temperate, and one extremely hoat.

Of Paralels. chap. 17.

A figure ſhewing the 21. Paralels.

The Table of Paralels, ſhewing how many degrees and minutes euery one is diſtant from the Equinoctiall, made according to the rule of Ptolomie.

Of Climes both old and new. chap. 18:

A figure ſhewing the order of the ſeuen auncient common clymes whereunto *Appian* addeth two more, making in all nine clymes.

A Table ſhewing the degrees of Latitude in the beginning, midſt, and end, of euery one of the foreſaid nine clymes.

A Table ſhewing the longeſt day in euery degree of Latitude, proceeding orderly from the Equinoctiall to the North pole, by whole degrees without minutes from one degree to 90.

A Table ſhewing how many miles euery one of the ſeuen clymes

The Contents.

clymes doe containe, afwell in breadth, as in length.

Of the diuerfe feafons and fhadowes incident to diuerfe climes and paralels, and firft what feafons and fhadowes they haue that dwell right vnder the Equinoctiall. chap. 19.

Of the feafons & fhadowes which they haue that dwell betwixt the Equinoctiall and the Tropique of Cancer. chap. 20.

Of the feafons and fhadowes which they haue that dwell right vnder the Tropique of Cancer. chap. 21:

Of the feafons and fhadowes which they haue that dwell betwixt the Tropique of Cancer, and the circle Artique. chap. 22.

Of the feafons and fhadowes which they haue that dwell right vnder the circle Artique, and howe long their day is. chap. 23.

What feafons, fhadowes, and length of day they haue that dwell betwixt the circle Artique, and the pole Artique. chap. 24.

Of thofe that dwell right vnder the Pole. chap. 25.

By what names certaine inhabitants of the earth are called, according to the diuerfitie or likenes, as well of their fhadow, as of their fituation. chap. 26.

By what names certaine partes of the earth are called by reafon of their diuerfe fhapes. chap. 27:

Of the wind, what it is, what motion it hath, and of the diuerfe names and diuifions thereof. chap. 28.

Of the nature and qualitie of the auncient twelue windes. chap. 29.

Of the moderne diuifion of the windes. chap. 30.

A figure of the 32. windes reprefenting the Marriners compafle.

The first part of the Spheare.

Of Cosmography, what it is, and what kinds of Sciences it comprehendeth, and who were first inuentors thereof.

Chap. 1.

Hat is Cosmography?

Cosmography is the description of the whole world, that is to say, of heauen and earth, and all that is contained therein.

What speciall kindes of knowledge are comprehended vnder this Science.

These foure, Astronomie, Astrologie, Geographie, and Chorographie.

What is Astronomy?

Astronomy is a Science, which considereth and describeth the magnitudes and motions of the celestiall or superiour bodies.

Which call you superiour bodies?

The Spheares or Heauens, and the Starres as well fixed as moueable, which we shall define hereafter in their proper places.

What is Astrologie?

It is a Science which by considering the motions, aspects, and influences of the starres, doth foresee and prognosticate things to come.

What is Geographie?

It is a knowledge teaching to describe the whole earth, and all the places contained therein, whereby vniuersall Maps and Cardes of the earth and sea are made.

What is Chorographie?

It is the description of some particular place, as Region, Ile, Citie, or such like portion of ye earth seuered by it selfe frō the rest.

Who

The first Booke

Who were first inuentors of these Sciences.

Some say that Atlas was the first inuentor, whom the Poets faine to beare vp the heauens with his shoulders, hauing his head placed in the North Pole, and his fœte in the South Pole, and his right hand bearing the East part, and his left hande the West part of the world. Albeit that some apply this fiction of the Poets to an high mountaine in Affrica called Atlas, which for his great height surmounting the cloudes, is sayde to beare vp the skyes. But some say that Adam our first Parent, was the first inuentor thereof: And some others affirme that Abraham was the first inuentor thereof. But whosoeuer was the first inuentor, it well appeareth by Ptolomy his booke, called Almagesti, that he hath bæne no small furtherer thereof, and since his time in these latter dayes Georgius Purbacchius, Iohannes de Monte Regio, Copernicus, and diuerse others haue learnedly treated thereof. But leauing to speake of the first inuentors, or of the furtherers of these sciences, I will speake of the shape, capacitie, and vnitie of the world, and because the shape thereof is likned to a round bodie called a Spheare, I will first define what a Spheare is.

The definition of a Spheare, and of the vnitie, roundnesse, and capacitie of the worlde, also of the poles and axeltre thereof.

Cap. 2.

Hat is a Spheare?
A Spheare is a round body, contained or couered with one only superficies or vpperface, in the midst of which bodie is a point or pricke, called the Centre, from which all right lines being drawne to any part of the superficies or vpper face therof, are of equal length, as you may perceiue by this figure here following, the Centre whereof is marked with the letter A.

And

of the Spheare.

And this Spheare before defined repre-senteth the frame of the whole world, con-taining all thinges, and is contepned of none.

How prooue you that there is but one world?

By the authoritie of Aristotle, who sai-eth, that if there were any other world out of this, then the earth of that worlde would moue towardes the Centre of this world, which cannot be, vnlesse it should first ascend from his own Centre to the superficies or vpper face of the same, which is against nature, for euery heauy thing naturally tendeth downeward, and euery light thing vpward, which common expe-rience sheweth: for a stone being cast vpwards, will fall downe-wardes, and the flame of a candle being turned downeward, will fly vpward.

How proue you the frame of the world to be round?

By three reasons, First by comparison, for the likenes which it hath with the chiefe Idea or shape of Gods minde, which hath neither beginning nor ending, and therefore is compared to a Circle. Secondly by aptnesse aswell of mouing as of containing, for if it were not round of shape, it should not be so apt to turne about as it continually doth, nor to containe so much as it doth, for the round figure is of greatest capacitie and containeth most. Thirdly, necessitie proueth it to be round, for if it were with an-gles or corners, it should not be so apt to turne about, and in turning about, it should leaue voide and emptie places, which nature abhorreth, for no place by nature can be without a bodie, nor a body without a place.

Whereupon is this Spheare or great round frame turned?

Upon two most firme and immoueable hookes called in Latin Cardines mundi, and in Græke Poli, deriued of the verbe Polo, which is asmuch to say, as to turne, for as a doore turneth vppon his hookes, so the world turneth vpon his two Poles, whereof the one is fixed in the North, and the other in the South, & the North pole is called in Latine Polus arcticus, and the South pole is called Antarcticus, through which Poles from the one to the o-ther passeth a right immaginatiue line called of the Astronomers

the

The first Booke

the Axeltrée of the world, about the which the worlde continually turneth like a Cart wheéle, which you may sée liuely expressed in a materiall Spheare made for the purpose, for in Plano it cannot be so well described.

Of the diuision of the world, and of the two essentiall parts thereof, and what things ech part contayneth.

Cap. 3.

How is the world deuided?

Most writers do make a twofold diuision therof, that is according to substance, and according to accidentes, which accidents because they are none other thing, but points, lines, and circles, and so incident and necessarily belonging (by position though not by nature) to the substantiall parts, as the one can not be described without the other, neither the subiect without the accidents, nor the accidents without their subiect, I will therefore make but one diuision of the world, diuiding the same into two essentiall parts, that is, the celestiall part, and the Elementall part, in the description of which parts, it shal also plainely appeare what those lines and Circles be, and to what end they serue.

What doth the celestiall part containe?

The eleuen Heauens and Spheares.

Make you no difference betwixt these Spheares, and the great Spheare which you last defined to be couered with one superficies or vpperface.

Yes verily, for these haue two superficies or vpper faces, that is, the Conuexe and Concaue.

Which call you Conuexe and which Concaue?

The Conuexe is the outermost vpper face of any thing that is round, and therewith hollow, as of an ouen or vault, and the Concaue is the innermost or hollow superficies of the saide ouen or vault, or of any such like rounde bodie hauing concauitie or hollownesse.

Which

of the Spheare.

Which be those eleuen Spheares or Heauens whereof you did last speake?

In ascending orderly vpwardes from the Elements they bee these, The first is the Spheare of the Moone, The seconde the Spheare of Mercury, The third the Spheare of Venus, The fourth the Spheare of the Sunne, The fift the Spheare of Mars, The sixt the Spheare of Iupiter, The seuenth the Spheare of Saturne, The eight the Spheare of the fixed Starres, commonly called the firmament. The ninth is called the second moueable or Christall heauen, The tenth is called the first moueable, and the eleuenth is called the Emperiall heauen, where God and his Angels are said to dwell.

What doth the Elementall part containe?

It containeth the foure Elements.

Which be those?

The Element of the fire which is next to the Spheare of the Moone, and next to that more downeward is the Element of the aire, and next to that is the Element of the water, and next to that is the earth, which is the lowest of all. All which things you may see here plainly set foorth in the figure following on the other side.

The firſt Booke

A figure of the whole world, wherein are ſet forth the two eſſentiall partes before mentioned, that is to ſay, the eleuen heauens and the foure Elements.

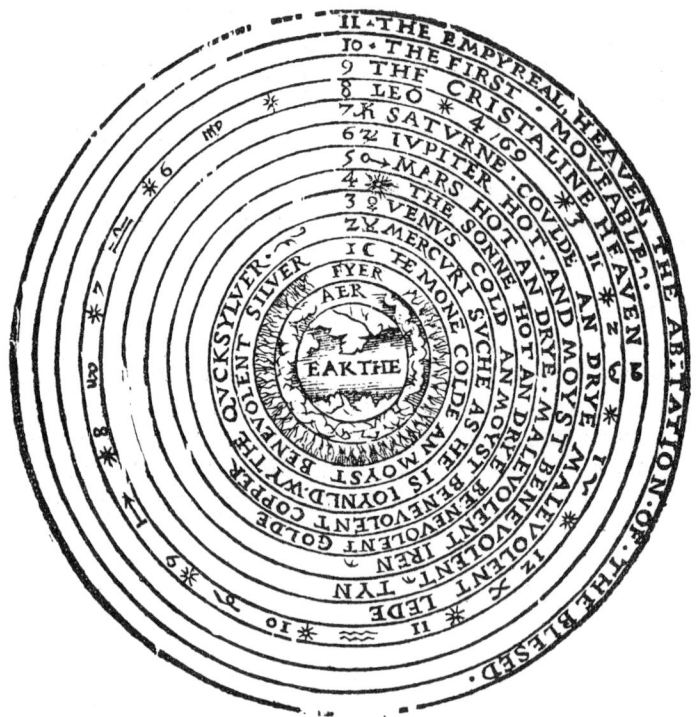

A demonſtration to prooue the pluralitie of the heauens.

Chap. 4.

HOw is it to be prooued that there is ſuch multiplicitie of heauens, ſith to the eye it ſeemeth one whole bodie?

If there were not diuerſe heauens and diuerſe motions, there ſhould neither bee generation nor corruption of any thing, for all things then ſhould be as one, and euer

euer in one estate, but besides that reason men haue found out by experience in obseruing the diuerse rising and setting of diuerse wandring starres called the Planets, that euery such Planet hath a seuerall Spheare or heauen by it selfe, for otherwise they should continually kéepe their places as the fixed starres doe, which do moue altogether according to the moouing of the viii. heauen, wherein they are placed.

I graunt that there may be diuerse heauens or Spheares. But how prooue you that there be tenne moueable heauens, besides the imperiall heauen before mentioned, and specially sith of the olde writers, some affirme that there bee but nine, and some haue saide that there be no more then eight.

The latter writers haue found by good obseruation (as they say) that the firmament or eight heauen hath thrée seuerall motions wherewith it is mooued and turned about, which could not be, vnlesse there were two mooueable heauens higher then that, as shall be declared when we come to shewe the thréefold motion of the saide eight heauen, for here I mind briefely to describe vnto you all the heauens beginning at the highest, and so procéede to the lowest.

Of the highest Spheare or Heauen, called the imperiall heauen.

Chap. 5.

He imperiall heauen (as our auncient Deuines affirme) is immoueable and being created by God the first day that he began the first creation of the world, was by him immediatly replenished with his ministers the holy Angels, and this heauen being the foundation of the world, is most fine and pure in substance, most round of shape, most great in quantitie, most cleare in qualitie, and most high in place, where God and his Angels are said to dwell.

The first Booke

Of the tenth Spheare or heauen, called in Latine *primum mobile*, that is, the first moueable, and what motion it hath.

Chap. 6.

This heauen is also of a most pure and cleare substance and without starres, and it continually moueth with an equall gate from East to West, making his reuolution in 24. houres, which kind of mouing is otherwise called the diurnall or daily mouing, & by reason of the swiftnesse thereof, it violently caryeth & turneth about all the other heauens that are beneath it from East to West, in the selfe same space of 24. houres, whether they will or not, so as they are forced to make their owne proper reuolutions, which is contrary from West to East, euery one in longer or shorter time, according as they be far or neare placed to the same: yet some writers doe not admit any violence in the heauens, but that euer the heauen hath his proper and naturall mouing without any violence.

I doe not well vnderstande by what reason the first moueable should haue such force ouer the rest.

For your better vnderstanding how this may be, suppose that you turne with your hande from East to West a Grindstone, or some other turning wheele, whereon is placed a Fly or some other creeping worme which cometh towardes you from West to East, & you shal easily conceiue that by your swift turning of the wheele, you shal turne about the said Fly or Worme many times contrary to her owne course, whether she will or not, before she can get about the said stone or wheele.

That is true, proceede therefore to the rest, and tell the nature and mouing of the ninth heauen.

Of the ninth heauen, what motions and names it hath, and whether there be any waters aboue the firmament or not.

Chap. 7.

The ninth heauen is also cleare of substance, and without starres, hauing two moouinges, the one from East to West vpon the poles of the world, according to the daily mouing of the first moueable, and the other

of the Spheare.

other from West to East vpon his owne poles, according to the succession of the signes of the Zodiaque, which is in the first moueable, turning so slowly about as it maketh but one degree in 100. yeares, and accomplisheth his full reuolution in 36000. yeares, or as Alfonsus saith in 49000. yeares.

Is there then in the first moueable any Zodiaque, that is to say, a broad Circle carying the 12. signes and yet no starres?

Yea though there be no stars, yet many superstitious fooles doe imagine that there be diuerse Characters and liniaments not to be seéne, but with a most sharpe and subtill sight (by vertue of which Characters & fained constellations) they imagine that they can worke wonders and strange effects, inough to deceiue themselues and others too.

What is the cause that the ninth heauen is so long in making his proper reuolution.

Because as I sayd euen now, this heauen is placed next to the first moueable, which carieth him about cōtrary to his owne course with such violence as he cannot make his owne proper reuolution so soone as the other heauens which are placed further off.

If the ninth heauen be so long in making his course, it will neuer be compleat whilst the world lasteth, for the whole age of the world according to some, is but 6000. yeares?

Yet Plato was of an other opinion, and therefore this reuolution was called, magnus annus Platonis, that is to say, the great yeare of Plato, because hee affirmed that when this reuolution was once complet, all things should be in the same estate wherein they were at the first, and that he should then stand reading to his Schollers in the selfe same chaire wherein he stood at that present, which fond opinion S. Augustine confuteth in his 12. Booke de Ciuitate Dei, speaking in this maner, God forbid (saith he) that we should credit these things, for Christ dyed once for our sinnes, and being risen againe from the dead, he is no more to die, neither can death haue any more power ouer him, by vertue of whose resurrection, we also that beleéue shall rise againe, and dwell with God for euer.

With what names is this ninth heauen called?

Some do call it the Chrystalline heauen, because of the clearenes thereof, and some the watrie heauen, because the Scripture

T 2 affirmeth

affirmeth that there be waters aboue the firmament as we reade in Gen. cap. 1. let the firmament be made, and deuide the waters from waters. Againe we read in the Psalmes, all ye waters that be aboue the firmament, blesse ye the Lord.

The naturall Philosophers allow no waters to dwell aboue the firmament.

That is true, yet notwithstanding if the holy Scriptures doe manifestly affirme that there be waters aboue the firmament, it behooueth a Christian to beléeue it, but question perhaps may bée mooued, what manner of waters they are that are aboue the firmament, whether such as bréd raine, or whether they are onely to be referred to the Christalline heauen, because it is of a watrie substance, and therefore of some is called the watry heauen, affirming it to be placed next to the primum mobile, or first moueable, to the intent that with the coldnesse thereof, it might asswage and represse the extreame heat of the same primum mobile, which otherwise (as some affirm) with his swift & violent moouing, would set all the heauens on fire, and yet no raine is bred therein, for the great raine that drowned the world in Noahes time, did not fall from aboue the firmament, but from the Aire, which in the holy Scripture is many times signified by this Latine word Coelum, that is to say, Heauen, as when the Scripture saith, the flood gates of heauen were opened, is as much to say as the flood-gates of the Aire were opened, &c. but leauing this question, let vs procéde to the eight heauen.

Of the eight heauen what motions it hath, and what circles are imagined by the Astronomers to be in that heauen, and to what vse and purpose they serue: Also in what time euery one of the seuen Planets maketh his reuolution, & of what thicknes their Spheares be.

Chap. 8.

The eight heauen otherwise called the firmament, is a most glorious heauen adorned with all the fixed starres.

Why are they called fixed?

Because they are fastned in this heauen, like knots in a knotty board, hauing no moouing of them-

themselues, but are moued according to the moouing of this eight Spheare or heauen wherein they are fixed.

How manyfold is the moouing of this heauen?

The moouing of this heauen (as hath béene said before) is thréefold, for first it turneth about euery day from East to West in 24. houres, according to the moouing of the first mooueable, otherwise called the diurnall moouing. Secondly it moueth from West to East, according to the moouing of the ninth heauen which maketh but one degrée in a 100. yeares, and this moouing is called motus angium Stellarum fixarum.

What are Anges?

They be certaine imagined points in the heauen, notifying the furthest distance of any Orbe or Spheare from the Centre of the world: Thirdly it moueth sometime towardes the South, and sometime towards the North, by vertue of his owne proper moouing, called in Latine motus trepidationis, that is to say, the trembling moouing, whereby it is moued vpon two little Circles, the poles whereof are in the beginning of that Aries and that Libra which are imagined to be in the ninth heauen, the Semidiameter of which little Circles is 4. degrées 1'8. and 4"3. & maketh his whole reuolution in 7000. yeares, and this moouing is otherwise called motus accessus & recessus, that is to say, the moouing of approching and retiring, and is onely proper to the 8. Spheare: for euery naturall body by order of nature, can haue but one proper moouing, and by reason of this moouing the fixed stars be not alwayes equally distant from the immoueable poles of the tenth Spheare, nor alwayes vnder ỹ Ecliptique of the said 10. Sphere, but oftentimes cleane without it. Nor the fixed stars are in equall times equally distant from the beginning of that Aries & Libra, which are imagined to be in the tenth heauen, but they séeme to moue sometimes towards the East, and sometimes towards the West, now more swiftly, and now more slowly. For Ptolomy in his time found them to be moued in 100. yeares one degrée. And the latter obseruers haue found them when they be in their swift motion to be moued one degrée in 63. yeares, all which thrée moouings are easie (as some affirme) to be demonstrated by an instrument made of purpose, shewing euery seuerall moouing of this heauen by it selfe, but such demonstrations as some others thinke be

not

The first Booke

not altogether necessary conclusions: for it may be that the daily moouing is comon to the whole frame, and that the moouing of the Angles of the fixed starres, is proper to the eight Spheare, hauing some other moouer, and the like obiection may be made touching the trembling moouing, called motus trepidationis. But now you haue to vnderstande that for the better describing, deuiding, and measuring of this heauen, and all the apparances thereof, the Astronomers with the Pensill of imagination, or rather of most necessarie inuention, doe (as it were) trace the same with certaine Circles both greater and lesser, whereof wee shall speake so soone as we haue briefely shewed the mouing of the rest of the heauens, contained in the celestiall part of the world, in such order as they follow one another. For next to the firmament is the heauen of Saturne, which maketh his reuolution from West to East in 30. yeares, next to him is the heauen of Iupiter, who maketh his reuolution from West to East in 12. yeares, next to him is the heauen of Mars, which maketh his reuolution also from West to East in 2. yeares, next to him is the heauen of the Sunne, which maketh his reuolution from West to East in 365. dayes, and 6. houres lacking certaine minutes, next to him is the heauen of Venus, which maketh her reuolution from West to East in like time as doth the Sunne, next to Venus is the heauen of Mercury, which maketh his reuolution from West to East in the like space that Venus doth: And next to Mercury is the heauen of the Moone, which maketh her reuolution from West to East in 28. dayes and certaine minutes. Thus hauing briefely described the eleuen heauens, I will now treat of the Circles that are imagined (as I sayd before) by the Astronomers to be in the eight heauen or firmament, to the intent that the measures and distances of the starres, images, signes, and other apparances therein contayned, might be the better demonstrated.

Before you deale with these Circles, I would gladly vnderstand some reason why all these seuerall heauens seeme to the eye but as one entire bodie.

That is because they are all cleare and transparent like fine birall glasse, or Christall, through the which the sight doth easily pearce, though there were neuer so many coates of such cleare substance, couering one another like the skales of an Onion, for so the heauens doe couer and enclose one another, and euery one is of an excæding great thicknesse.

Why how thicke is euery such heauen?

The heauen of the Moone containeth in thicknesse.	$105/222\frac{2}{33}$ miles.
The heauen of Mercury containeth	$253372\frac{1}{3}$.
The heauen of Venus	$3274494\frac{6}{11}$.
The heauen of Sol,	$343996\frac{4}{11}$.
The heauen of Mars,	$26/308/800$.
The heauen of Iupiter,	$1899654\frac{6}{11}$.
The heauen of Saturne,	$19604454\frac{6}{11}$.

What neede haue the heauens to be so thicke?

Because otherwise they could not containe ech one his starre or starres, for there is no fixed starre so little, but that it is farre greater in compasse then the earth, neither is there any wandring starre, but that it is bigger then the earth, the Moone, Venus, and Mercury excepted.

For the Sunne contayneth the earth	166. times.
Saturne containeth it	95. times.
Iupiter containeth it	91. times.
Mars containeth it	2. times.
Venus is lesse then the earth	39. times.
Mercury is least of all and is contained of the earth,	3143. times.

You haue well satisfied mee touching these matters, and therefore nowe proceede with the Circles if you thinke so good.

Before that I describe vnto you the Circles whereof a Sphere is made, or declare the vses thereof: I thinke it not amisse here to set downe the shape of a Sphere, together with the names of euery Circle written vpon the same, to the intent that you may be acquainted with all the partes of a Sphere before you come to vse the same: which shape or figure was first drawne by Master Blagraue, and is set downe in his booke called the Mathematicall Iewell.

The firſt Booke

Of the Circles whereof a materiall Sphere conſiſteth, and of their diuerſe diuiſions.

Chap. 9.

F Circles which are imagined to be in the firmament, and whereof a Sphere repreſenting the ſhape of the world is commonly made, be in all 10. that is to ſay, the Equinoctiall, the Zodiaque, the two Colures, the Horizon, the Meridian, the two Tropiques and the two polar Circles, of which Circles ſome be greater and ſome leſſer.

Which

of the Spheare. 141

Which call you greater and which lesser?

Those are called the greater Circles, which passing through the Centre or midst of the firmament, do deuide the whole circuit thereof into two equall parts, and of such Circles there be sixe, that is to say, the Equinoctiall, the Zodiaque, the Colure of the Solstices, the Colure of the Equinoxes, the Horizon and the Meridian. The lesser Circles are the foure last mentioned, that is the two Tropiques, and the two polar Circles, which are called lesser, because they do not deuide the world into two equall parts, but into vnequall portions.

What other diuisions doe the Astronomers make of the said Circles?

Diuerse, for some are said to be paralels, some right, some oblique, some moueable, and some immoueable.

Which are said to be paralels?

The two polar Circles, and the two Tropiques, and the Equinoctiall which is in the very midst of them all.

Which are said to be right and which oblique?

The right be the two Colures, the right Horizon, & the Meridian, because they cut the Spheare or Globe with right Angles: And the oblique are these two, that is, the Zodiaque, and the oblique Horizon, which are said to be oblique because they cut the Spheare or globe with oblique Angles.

Which are said to be moueable, and which immoueable?

The moueable are these, the Equinoctiall, the Zodiaque, the two Tropiques, & the 2. polar Circles which are said to be moueable, because they continually moue with y firmament & are like in al places. And these also hauing respect to a materiall Sphere, are said to be intrinsicall or inward. The immoueable are these 2. that is, the Horizon, & the Meridian, which are said to be immoueable because in the turning of the Spheare they remaine vnmoued, for though we change both Meridian and Horizon by going from one habitation to another vpon the earth, yet euery place hath still his owne Meridian and Horizon, which doe remaine immoueable, and these two Circles hauing respect to a materiall Spheare, are said to be extrinsicall or outward, because they do inclose on the outside all the other Circles of a materiall Spheare. Thus hauing sette downe the diuisions of the sayde Circles,

I will

The first Booke

I will now describe them all in order, and shew to what vses they serue, beginning first with the great Circles, and so procéede to the lesser, and first I will speake of the Equinoctiall.

Of the Equinoctiall line, why it is so called, and of the diuerse vses thereof.

Chap. 10.

WHat is the Equinoctiall?

The Equinoctiall is a great Circle, which being in euery part equally distant from the two poles of the world, deuideth the Spheare in the very midst thereof, into two equall parts, and therefore it is called of some, the girdle of the world.

Wherefore is it called by the name of the Equinoctiall?

Because that when the Sunne toucheth this Circle, which is twice in the yeare, the day and night is of equall length throughout the whole world.

At what time of the yeare is it Equinox?

In the spring of the yeare, about the 11. day of March when as the Sunne entereth into the first degrée of Aries, and againe in Autumne about the 13. of September when as the Sun entereth into the first degrée of Libra, And by reason that this Circle deuideth the world in the very midst, those that dwell right vnder it, are said to haue no Latitude either Northward or Southward, to whom the dayes and nights are alwayes equall. But if they dwell any thing distant from the Equinoctiall, towardes any of the poles, then they are said to haue Latitude more or lesse, eyther Southward or Northward, as shall be declared hereafter more at large, when we come to treat of the Longitude and Latitude of the earth, both which are to be knowne by helpe of the Equinoctiall.

To what other vses serueth this Circle?

This Circle hath many most necessary vses, for first it sheweth the daily moouing of the first moueable, which maketh his reuolution

of the Spheare.

lution in 24. houres, which houres are equall, and are to be measured by the degrées of the Equinoctiall, by allowing 15. degrées thereof to an houre. And therefore the degrées of the Equinoctiall are commonly called in Latine of the Astronomers tempora, of which degrées it containeth 360. which being deuided by 15. do make 24. houres, which is a naturall day, contayning both day and night: Moreouer it sheweth the declination of the fixed starres and their right ascentions, whereof we shall speake when we come to treat of the starres, and by his equall motion, all the inequalities of the Zodiaque, & of all the signes conteyned therein, are measured: And to be short, this Circle hath many other good vses not to be declared vntill you haue some knowledge of the rest of the other Circles hereafter described.

A figure shewing the Equinoctiall line, the two Poles and Axletree of the world.

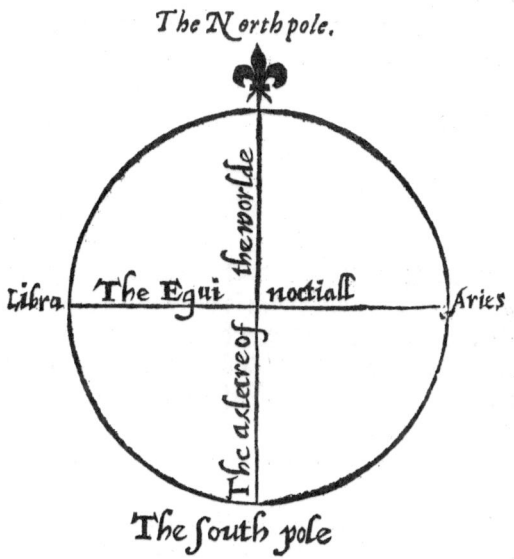

Of the Zodiaque why it is so called, and of the 12. signes therein contained: Also of the Latitude, Longitude, and declination thereof.

Chap.

The first Booke

Chap. 11.

WHat is the Zodiaque?

It is a broad, oblique, or slope Circle, hauing a circular line in the midst thereof, called the Ecliptique line, and deuideth the Sphere into two equall parts, by crossing the Equinoctiall with oblique Angles in two points, that is, in the beginning of Aries, and in the beginning of Libra, so as the one halfe of this broad Circle declineth towards the North, and the other halfe towards the South: in which Circle many thinges are to be considered, first the name, then what breadth it hath, and why it hath such breadth, and into what parts it is deuided aswell touching the breadth or Latitude, as circuit or Longitude thereof, also into how many parts the firmament is deuided by the spaces described in the Zodiaque, and appointed to the 12. signes. Also how much it declineth from the Equinoctiall towards the South or North, and vpon what poles it turneth about, and why the line in the midst is called the Ecliptique line, and many other necessary points, which I mind here briefely to touch.

Why is this Circle named the Zodiaque?

It is named the Zodiaque either of this Græke worde zoe, which is asmuch to say as life, because the sunne being moued vnder this Circle, giueth life to the inferiour bodies, or else of this Græke word Zodion, which is as much to say as a beast, because that 12. Images of stars, otherwise called the 12. signes, named by the name of certaine beasts, are formed in this Circle: and therefore the Latines doe call this Circle Signifer because it beareth the 12. signes.

How are these signes called, and with what Characters are they marked.

Their names and Characters this table here following doth shew, and also which be opposite one to another, as Aries to Libra, Taurus to Scorpio, and so forth.

The

The sixe Northerne signes:		The sixe Southerne signes.	
Aries	♈	Libra	♎
Taurus	♉	Scorpio	♏
Gemini	♊	Sagittarius	♐
Cancer	♋	Capricornus	♑
Leo	♌	Aquarius	♒
Virgo	♍	Pisces	♓

Of which signes the first sixe on the left hande are called the Northerne signes, because they are conteyned in that halfe of the Zodiaque, which declineth towardes the North. And the other sixe on the right hand being right opposite to the first 6. are called the Southerne signes, because they are conteyned in the other halfe of the Zodiaque, declyning towardes the South: And against euery signe is set his proper Character.

What diuisions doe Astronomers make of the 12 signes?

Diuerse, as followeth: for some are said to be ascendent, and some descendent.

Ascendent are those that rise from the South towards our Zenith, tending Northward, which be these, Capricornus, Aquarius, Pisces, Aries, Taurus, and Gemini.

The descendent are these, Cancer, Leo, Virgo, Libra, Scorpius, and Sagittarius.

Againe some are saide to be vernall, as Aries, Taurus, and Gemini.

Some Estiuall as Cancer, Leo, Virgo.

Some Autumnall as, Libra, Scorpio and Sagittarius.

And some Hiemall or Brumall, as Capricornus, Aquarius, and Pisces, signifying the foure seasons of the yeare, that is to say, the Spring, Sommer, Autumne, and Winter.

And some doe make diuerse other diuisions, which because they appertaine to Astrologie rather then to the Treatise of a Spheare, I willingly omit them.

Now tell what breadth the Zodiaque hath, and why it is imagined to haue such breadth?

It hath (according to the ancient writers) 12. degrees in breadth, that is to say, 6. degrees on the one side of the Ecliptique line, & 6. degrees on the other side of the Ecliptique line, but according to

the

the moderne writers, it hath 16. degrées in breadth, that is, eight degrées on ech side of the Ecliptique line, which breadth was necessarily imagined, first to the intent that by the measures and degrées thereof it might be knowne, how much the Planets (otherwise called the wandring starres, whose course is to passe continually vnder this broad Circle) doe wander at any time on either side of the Ecliptique line, for they all wander, but some more, some lesse, the Sunne onely excepted, which neuer swarueth from that line, but alwayes goeth right vnder the same, and therefore the saide line is called of some writers the way of the Sunne: And secondly it hath such breadth to the intent it may containe within the same, the 12. signes aforesaid, by meanes of which signes the whole circuit or longitude of the saide Circle is deuided into 12. equall parts, and euery such part is deuided into 30. degrées, and euery degrée into 60. minutes, and euery minute into 60. seconds, &c. so as the whole Longitude thereof contayneth 360. degrées, according vnto which diuision, all the rest of the Circles both greater and lesser described in the Spheare, are made to containe the like number of degrées, and euery halfe Circle to containe 180. degrées, and euery quarter of a Circle to containe 90. degrées, and by this diuision aswell of the breadth as of the length of the Zodiaque, it appeareth that euery one of the 12. Signes hath 30. degrées in length, and 12. degrées in breadth, and thereof the Planets, Starres, and all other Celestiall bodies are said to haue both Longitude and Latitude, the sunne onely excepted.

How is such Longitude and Latitude to be vnderstood?

The Longitude of any Planet or Starre is to be counted in the Ecliptique line containing in circuit 360. degrées, reckoning from the first point of Aries, and so to Taurus, Gemini, and Cancer, & so foorth according to the succession of the signes, vntill you come againe vnto the first point of Aries, at which point such Longitude both endeth & beginneth. The Latitude is counted frō the said Ecliptique line towards any of ỹ Poles of the Zodiaque. And hereof looke how many degrées any fixed starre or Planet is distant from the Ecliptiqueline towards any of the said poles, so much Latitude it is said to haue either Northerne or Southerne: moreouer, by this diuision of the signes the whole firmament is

deuided

deuided into 12. parts by reason of 6. Circles imagined to passe through the poles of the Zodiaque, and also through the beginning of euery signe, wherby we knowe vnder what signe euery star is situated though it be cleane out of the Zodiaque as this figure here plainely sheweth, marked with these letters A.B.C.D. A signifieth the North pole of the worlde, B. the North pole of the Zodiaque, C. the South pole of the world, D. the South pole of the Zodiaque.

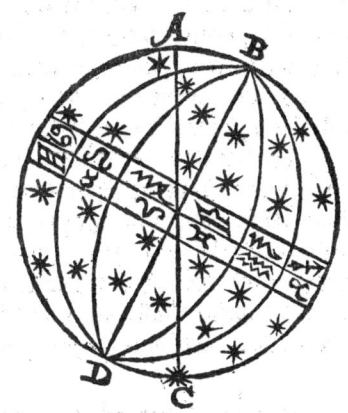

How much the Zodiaque declineth from the Equinoctiall towards either of the Poles, and of the greatest declination of the Sunne, what it is at this present, and what it hath beene in times past.

Chap. 12.

You haue to vnderstande that the Zodiaque or rather the Ecliptique line, declineth from the Equinoctiall towardes eyther of the poles, 23. degrées 2′8. and that is said in these dayes to be the greatest declination of the sunne, which declination is twofold, that is Northerne and Southerne, for like as the Sunne entring into the first point of Aries, beginneth then to decline from the Equinoctiall Northward, to the quantitie of 23. degrées and 2′8. so entring into the first point of Libra, he declineth againe from the Equinoctiall as much Southward. And note by the way that by reason of his slowe motion, when he is in the Northerne signes, he spendeth 7. dayes, and ⅓. of a day more in making his North declination then in making his South declina=

declination, becauſe he is then in his ſwift motion, and the time hath béene that he hath ſpent aboue ten dayes more in making his North declination, then in making his South declination: neither is the greateſt declination of the Sunne in all ages of like quantitie. For in Ptolomies time it was 23. degrées, 51. and 2″ o. euer ſince whoſe time it hath alwaies continually decreaſed vntill this preſent, ſo as now the greateſt declination is no more but 23. degrées and 2′8. And Copernicus maketh the declination of the ſunne in reſpect of quantitie to be twofold, that is, greateſt and leaſt, affirming the greateſt to be 23. degrées and 5′2. and the leaſt to be 23. degrées and 2′8. as it is now accounted, the difference whereof is 2′4. and whilſt the Ecliptique departeth from the Equinoctiall, and turneth againe towardes the Equinoctiall, there do runne (as he ſayth) 3434. yeares.

How to know the quantitie of the Sunnes declination be it Northward or Southward, euery day throughout the yeare, as well by a Table as by helpe of the Spheare.

Chap. 13.

This is chiefely to be knowne by Tables calculated of purpoſe, which Tables moſt commonly are eyther made to anſwere euery day of the moneth, or elſe to the degrée of the ſigne wherein the ſunne is euery day, which kind of Table is contained in leſſer roome then the other, but to worke by ſuch a Table, you muſt firſt knowe in what ſigne and degrée the ſunne is euery day.

How is that to be done?

It is to be knowne moſt truely by the Ephemerides or ſuch like Table calculated of purpoſe, ſhewing not onely the degrée of the ſigne, but alſo the very minute wherein the ſunne is euery day, and for want of ſuch a Table, you may without conſideration of the minutes, know it by ſuch an inſtrument or figure as this following, which conſiſteth of diuerſe Circles, whereof the outermoſt conteyneth the degrées of the 12. ſignes, together with the

of the Spheare. 145

the names of the said signes, & the next the dayes of ech moneth, together with the names of the said moneths, much like the backside of an Astrolabe, in the center or midst of which instrument or figure is a thred, which if you lay vpon the day of the month which you seek, it wil straight direct you to the degrée of the signe wherein the Sun is that day, as for example, if you would know in what signe and degrée the Sun is the 4. of May, then by drawing the thred right vpon the said day ouer and beyond the outermost circle, you shal finde that it will fall right vpon the 23. degrée of Taurus.

An instrument to know thereby in what signe and degree the Sunne is euery day throughout the yeare.

Then

The first Booke

Then hauing found the degree of the Sun you must resort therewith to this Table following, made for the declination of the sun.

A Table shewing the declination of the sun euery day throughout the year

Degrees of the Signes.	♈ ♎			♉ ♍			♊ ♐			Degrees of the signes.
	D	M	S	D	M	S	D	M	S	
1	0	23	53	11	50	6	20	22	57	29
2	0	47	46	12	10	56	20	35	7	28
3	1	11	39	12	31	34	20	40	55	27
4	1	35	30	12	51	59	20	58	20	26
5	1	59	20	13	12	12	21	9	21	25
6	2	23	8	13	32	12	21	19	59	24
7	2	46	54	13	51	58	21	30	13	23
8	3	10	37	14	11	30	21	40	3	22
9	3	34	18	14	30	48	21	49	29	21
10	3	57	54	14	49	51	21	58	29	20
11	4	21	28	15	8	40	22	7	6	19
12	4	44	57	15	27	13	22	15	17	18
13	5	8	22	15	45	30	22	23	3	17
14	5	31	42	16	3	32	22	30	24	16
15	5	54	57	16	21	17	22	37	19	15
16	6	18	6	16	38	44	22	43	48	14
17	6	41	9	16	55	55	22	49	50	13
18	7	4	6	17	12	48	22	55	27	12
19	7	26	57	17	29	23	23	0	38	11
20	7	49	40	17	45	40	23	5	22	10
21	8	12	16	18	1	39	23	9	39	9
22	8	34	45	18	17	18	23	13	29	8
23	8	57	5	18	32	37	23	16	53	7
24	9	16	16	18	47	38	23	19	50	6
25	9	41	19	19	2	18	23	22	19	5
26	10	3	12	19	16	37	23	24	22	4
27	10	24	56	19	30	36	23	25	57	3
28	10	46	30	19	44	14	23	27	5	2
29	11	7	53	19	57	30	23	27	46	1
30	11	29	5	20	10	25	23	28	9	0
D	D	M	S	D	M	S	D	M	S	D
	♍	♓		♌	♒		♋	♑		

The description and vſe of the Table.

This Table conſiſteth of 5. collums, whereof the firſt on the left hand, and the laſt on the right hand do containe the degrées of the 12. ſignes of the Zodiaque, counting from one to 30. And the thrée middle collums do contain the degrées minutes, and ſeconds of declination, ouer the head of which thrée middle collums are ſet downe the characters of theſe 6. oppoſit ſignes, Aries and Libra, Taurus and Scorpio, Gemini and Sagittarius, and at the foot of the ſaid midlde colums are ſet down the characters of the other 6. oppoſit ſignes, that is, Virgo and Piſces, Leo & Aquarius, Cancer and Capricornus, for euery 2. oppoſit ſignes, aſwell aboue as beneath, haue like declination, the vſe of which Table is thus: firſt hauing found out the degrée of the Sun by the inſtrument before deſcribed, or rather by ſome true Ephemerides, you muſt ſéeke out the number of the ſaid degrée, either in the firſt or laſt collum, according as the character of the ſigne is placed. For if the ſigne or character be aboue, then you muſt ſéeke for the ſaid number in the firſt collum on the left hand, which numbers do diſcend from 1. to 30. but if the ſigne be beneath, then you muſt find it out in the the outermoſt colum on the right hand, the numbers whereof do aſcend from 1. to 30. and the common angle or ſquare, ſtanding right againſt the ſaid number will ſhew you the degrée, minutes and ſeconds of the declination, as for example, hauing found by the former inſtrument that the 4. day of May the Sun is in the 23. degrée of Taurus, I ſéeing the character of Taurus to ſtand aboue, doe ſéeke my foreſaid number of 23. degrées in the firſt collum on the left hand, and in the common angle or ſquare right againſt that number, and vnder the ſigne Taurus, I find the declination of the Sun to be 18. degrées 32′. and 37″. But this Table cannot ſerue alwayes: yea rather ſuch tables are to bee renewed as our Aſtronomers ſay euery 30. yeares. Alſo you may know the daylie declination of the Sun, by helpe of a materiall Spheare or globe: thus hauing ſet the Spheare at your latitude, bring the degrée of the ſigne wherein the Sunne is that preſent day vnto the mouable Meridian, and ſtaying it there, marke whether it falleth on the South ſide or on the North ſide of the Equinoctiall:

for if it be in any of the Northern signes, it will fall on the North side of the Equinoctial, and if it be in any of the Southern signes, it will fall on the south side of the Equinoctiall, and by counting the degrees vpon the Meridian, contained betwixt the degree of the Sun and the Equinoctial, you shal know what declination the Sun hath that day, as for example in the latitude 52. in the yeare 1590. the fift day of May, I find the Sunne by the Ephemerides to be in the 23. degree and 48′. of Taurus, which point I bring to the moouable Meridian, and there staying it, I find that point to be distant from the Equinoctiall northward 18. degrees & certaine minutes, and so much of North declination I conclude the Sun to haue that day.

Vpon what Poles the Zodiaque turneth about, also of the Ecliptique line and of the diuers vses thereof.

Chap. 14.

The Zodiaque turneth about vpon his own proper Poles from West to East, whereof the one being placed in the Colure of the Solstices towards the North, is distant from the pole Arctique 23. degrees and 28′. and the other is placed in the said Colure towards the South, being of like distance from the pole Antarctique, whereof the Astronomers haue a generall rule, affirming that the distance of the two Poles of the world from the Poles of the Zodiaque, is alwaies equal to the greatest declination of the sun, which as hath bene said before, is 23. degrees & 28′. as you may plainly see by the Sphear. And note that these 2. Poles are otherwise called the Poles of the Ecliptique, for in considering the declination of the Sun or of the Zodiaque from the Equinoctiall, you must haue respect onely to the Ecliptique line, which is in the midst of the Zodiaque, and not to any other part of the Zodiaque. And as the Equinoctiall line sheweth the moouing of the first moouable, which is from East to West, so the Ecliptique line sheweth the moouing of the second moouable, which is from West to East, clean contrary to the first moouable, the causes whereof haue bene before declared.

What other vses hath this line more than you haue already declared?

It hath diuerse, for in this line or circle are noted the degrees, wherewith any starre riseth or goeth downe, either rightly or obliquely, for all the apparances of the heauens are chiefly referred to this circle. Againe, by this circle the chiefest distinctions and parts of times, as yeares and moneths are knowne, and also the foure seasons of the yeare, as Spring, Summer, Autumne, and Winter: Moreouer, the obliquitie of this circle vnder which the Sun continually walketh, is cause that the dayes both naturall and artificiall are vnequall. Finally, this circle doth shew the places and times of the Eclipses both Solar and Lunar, from whence this line taketh his name, of which Eclipses we minde here briefly to treate.

Of the Eclipses both Solar and Lunar, and of the head and taile of the Dragon, with certaine figures shewing the same. Chap. 15.

WHat signifieth this word Eclipsis?

It is as much to say, as to want light, & to be darkened or hidden from our sight, as when the Sun & Moon are both at one self time right vnder the Ecliptique line, the one of these 2. lights most commonly is eclipsed and darkned: for there be two Eclipses, the one of the Sunne, and the other of the Moone, but sith that neither the eclipse of the Sun or of the Moone doth chance, but when they meet either in the head or tail of the Dragon, I think it good to shew first what is meant by the head and tayle of the Dragon. The Dragon then signifieth none other thing but the intersection of 2. circles, that is to say, of the Ecliptique, & of the circle that carieth the Moon, called her Defferent, cutting one another in 2. pointes, whereof that intersection which is westward when as the Moon goeth towards the North, is called the head, and that which is Eastwardes when the Moone goeth towards the South is called the taile, marked with such characters as you see in the figure following, and that part towards the South is called of some the belly of the Dragon. And note that the Defferent of the Moone is at no time distant from the Ecliptique aboue 5. degrees at the most.

The first Booke
The figure of the Dragon.

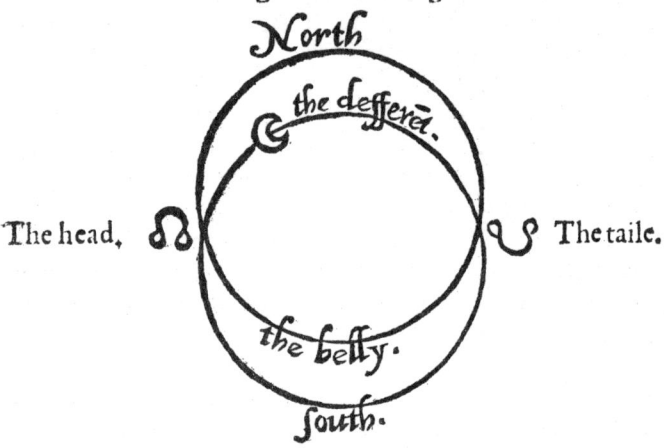

This being presupposed, I will speake first of the Eclipse of the Moone, and then of the Sunne, both which may bee eclipsed either totally or in part.

When is the Moone said to be totally eclipsed?

When the Sunne and Moone are opposite one to the other diametrallie, and the earth in the verie midst betwéen them both, for the bodie of the earth being thick and not transparent, casting his shadow to that point which is opposite to the place of the Sun, will not suffer the Moone to receiue any light from the Sun, from whome she alwaies borroweth her light.

At what time is the Moone said to be diametrallie opposit to the Sunne?

When a right line drawne from the center of the Sun to the center of the Moone, passeth through the center of the earth: & note that euery time that she is at the ful, she is opposit to the Sun, and yet the earth is not at euery such full diametrally betwixt her and the Sun, for then she should be eclypsed at euery ful, which indéed cannot be but when she is either in the head or tail of the Dragon.

When is the Moone said to be eclipsed in part?

When the Sunne, the earth, and the Moone be met in one selfe Diametrall line, but the Moone is declining either on the one side or on the other, as you may plainlie sée by this figure following.

But

of the Spheare. 148

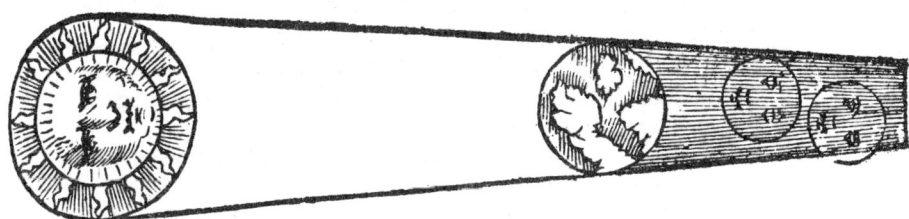

But note that the Eclipses of the Moone may be vniuersall, because the Earth is farre bigger than the Moone, and thereby able to shadow her whole bodie.

When is the Sunne saide to be eclipsed?

When the Moone is betwixt the Sunne and the Earth, which chanceth in a Coniunction, and yet not in euery coniunction, but when it falleth either in the head or taile of the Dragon, which may chance, as I saide before, either totally or in part: totallie I say, in respect of those parts of the earth whereon the shadowe directly falleth. For sith the Moone is farre lesser than the Earth, she cannot shadow all the Earth, and therefore the Eclipse of the Sunne cannot be vniuersall, but yet to some partes of the Earth totally, and to some partlie, and to other some nothing at all, as you may plainlie see by this figure following.

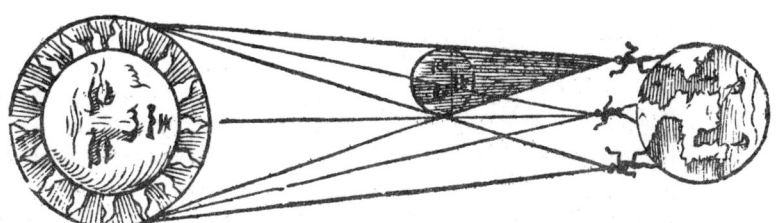

Yet all the histories doe affirme that the Eclipse of the *Sun* was vniuersall at the death of Christ.

Yea, that was miraculous, and also it was then at the full of the Moone, which was also as miraculous: and therefore Dionisius a Senator of Athens, beholding that Eclipse cryed out, saying

V 4

ing these words, Either God this day suffereth, or els the world must needs perish for euer: which Dionisius was the first that conuerted the Frenchmen to the faith of Christ, doing there great miracles: in honour of whome was erected the rich Abbey of S. Denise, not far from Paris, whereas the Kinges and Princes of France were woont to be buried.

How is it to be prooued that the Eclipse at Christ his death was at the full of the Moone?

Aswell by ancient historie, as by S. Augustine, who saith that the Iewes were woont to kéepe their feast of Passouer (at which time God suffered) alway at the full of the Moone.

If the Sun and Moone be eclipsed but in part, how are such partes to be accounted?

By the parts of the Diameter of the bodies of those two Planets, for the Astronomers doe diuide the Diameter aswell of the Sunne, as of the Moone into 12. and some into 24. parts, which they call points, and therefore are woont to say, that the Sunne or Moone are darkned or eclipsed 7. points, 8. points, 9. points, &c.

Of the two Colures, and why they are so named, and wherto they serue: also of the foure Cardinall points, that is, of the two Equinoctiall, and the two Solstitiall points, and of the entrance of the Sunne into any of those points or into any other signe.

Chap. 16.

What be Colures?

They be great mouable circles passing through both the Poles of the world, which the Astronomers do otherwise call Circles of declination, wherof they make 180. which are halfe so manie as there be degrées in the Equinoctiall, applying them to diuers vses not néedfull here to be rehearsed, for sith that there are but two Colures accustomablie set downe in the Spheare, without the which a materiall Spheare cannot bée made, I minde heere therefore onelie to treate of them.

Shew first what this name Colure signifieth.

This word Colure béeing compounded of Colos and Oura

is as much to say as vnperfect or maymed, the taile being cut off, because none of those Circles are euere séene whole aboue our Horizon, but parte thereof, for some part is alwayes séene, and some part is alwaies hidden, as that part which is aboue the Horizon, and nigh vnto the Pole, is alwayes séene, because it neuer goeth downe vnder the Horizon: likewise that which is nigh vnto the South pole is alwayes hidden from vs, because it neuer riseth aboue our Horizon, as by turning the Spheare a-bout, you may easilie perceiue the same.

Which be those Colures that are commonly set downe in the Spheare, and how are they named?

They are two great moouable circles, passing through the Poles of the world, crossing one another in the said Poles with right Sphearicall angles, by meanes whereof they deuide the whole Spheare into foure equall partes, of which two Colures the one is called the Colure of the Equinoxes, and the other the Colure of Solstices.

Describe these two Circles, & shew why they are so named.

The Colure of the Equinoxes is so called because it cutteth the Zodiaque in the beginning of Aries, which is called the ver-nal Equinoxe: and also in the beginning of Libra, which is cal-led the Autumnall Equinoxe, at which two times the dayes and nightes be equall, as hath bene said before when we did speake of the Equinoctiall circle, and this circle deuideth the Ecliptique in-to two halfes, the one Septentrionall, and the other Meridionall and thereby sheweth the signes wherein the Sunne maketh the dayes longer and shorter than the nightes, for whilest hee is in the 6. Northern signes, he maketh the dayes with vs longer than the nights, and when he is in the 6. Sowthern signes he maketh the nights longer than the dayes: now you haue to vnderstand that the Colure of the Solstice, is so called because it cutteth the Zo-diaque in the two Solstitiall points, that is to say, in the begin-ning of Cancer, and the beginning of Capricorne, as you may sée in beholding and turning the Spheare about with your hand.

Why are these two points called *Solstitiall*?

They take their name of these two Latine words Sol and sta-tio, that is to say, the Sun and standing, for when the Sunne is in any of the two points, hee séemeth to stand still, or at the least

The first Booke

mooueth so little, as his proper moouing from West to East cannot be easilie perceiued, during the space of twelue dayes. And you haue to note, that when the Sunne entreth into the first degrée of Cancer, which is about the 12. of June, then heé is at the highest, and the dayes be at the longest, and therefore, it is called the Summer Solstice. Againe, when hee entreth into the first degrée of Capricorne, which is about the 12. of December, then the Sunne is at the lowest, and the nightes are at the longest, and therefore it is called the Winter Solstice. And in this Colure there are set downe the two Poles of the Ecliptique line being distant from the Poles of the worlde 23. degrées and 28′. Moreouer on this colure is measured the greatest declination of the Sunne, which is alwayes equall to the distance of the Pole of the Ecliptique, from the Pole of the world, as hath bene said before. And you haue to note that the 4. former points, that is to say, the 2. Equinoxes, and the 2. Solstices, are commonlie called the foure cardinall or principall pointes, and of some they are called, the foure points of Change, signifying the 4. beginnings of the four diuers seasons of the yeare: for betwixt the beginning of Aries and beginning of Cancer, is contained the Spring time, and betwixt the beginning of Cancer and the beginning of Libra, is the Summer time: and from the beginning of Libra to the beginning of Capricorne is the time called Autumne, or fal of the leaf: and from the beginning of Capricorne to the beginning of Aries is contained the winter season, albeit the Sun entreth not into any of these signes alwaies at one self day or time of the yeare, for at Christ his incarnation, the Sun entred into Aries the 25. of March, and into Cancer the 24. of June, and into Libra the 27. of September, and into Capricorn the 25. of December, which was then the shortest day in the yéer, and the beginning of Winter, and therefore is called of the Latines, dies brumalis, on which day Christ our Sauiour was born, so as from the time of his birth vnto this present yeare, there are run almost 13. daies, wherefore, vnlesse the kalenders bee reformed as well here in England as els where (for the Romane reformation is not so exactlie true as it might be) wee shall haue in processe of time, the Spring in Winter, and the Winter in Autumne

How

of the Spheare. 150

How shall I know this present yeare, or any yeare to come hereafter, at what day and houre the Sunne entreth into any of the 12. Signes?

First you must learne by some good Ephemerides, or other Table, the true entrance of the Sunne into euerie signe in any yeare passed before, then from the time of the entrance of the Sun into the signe which you desire to know, consider how manie years are betwixt, & how many leap years are in the same contained, and substract for so many times as there be leape yeares, 44′. of an houre, and ad to the hours remaining, so many times fiue houres, and 49′. as there be yeares remaining ouer and besides the leape yeares, and that summe shall shewe you the day, houre, and minutes of the true entrance of the Sun into that signe in the same yeare that you desire to knowe.

Of the Horizon both right and oblique, making thereby three kindes of Spheares, that is, the right, the Paralell, and the oblique Spheare.

Chap. 17.

What is the Horizon?

It is a great immoouable circle which deuideth the upper Hemisphear, which is as much to say, as the upper half of the world which we see, from the nether Hemispheare which we see not, for standing in a plaine field, or rather vpon some high mountaine voyd of bushes & trees, and looking round about, you shall see your selfe inuironed as it were with a circle, and to be in the verie midst or centre thereof, beneath or beyond which circle, your sight cannot passe, and therefore this circle in Greeke is called Horizon, and in Latine Finitor, that is to say, that which determineth, limiteth or boundeth the sight, the Poles of which circle are imagined to be two points in the firmament; whereof the one standeth right ouer your head, called in Arabick Zenith: and the other directlie vnder your feete, called in the same tongue Nadir, that is to say the point opposite, and from point to point you must imagine that there goeth a right line passing through the centre of the world, and also through

The first Booke

through your bodie both head and féet, which is called the Axletrée of the Horizon, and you haue to vnderstand that of Horizons ther be 2. kinds, that is, right, & oblique, making 3. kinds of Sphears, that is to say, the right Spheare, the paralel Spheare, and the oblique Spheare,

When is the Horizon said to be right, and thereby to make a right Spheare?

It may be said to be right two manner of waies, first, when the Horizon passeth through both the Poles of the world, cutting the Equinoctiall with right angles, in which Spheare they that dwel haue their Zenith in the Equinoctiall, which passeth right ouer their heads, to whom the daies and nights are alwaies equal. Secondly, they are said to haue a right Horizon, & to dwell in a right Spheare, to whom one of the Poles of the world is their Zenith, and their Horizon is all one with the Equinoctial, cutting the Axletrée of the world in the very midst with right angles, and because the Horizon & the Equinoctial are Paralels, this kind of Spheare is called a paralel Spheare, in which Sphear they that dwel haue 6. moneths day, and 6. moneths night, as you may easily perceiue by placing the Sphear, so as one of the Poles may stand right vp in the midst of the Horizon, by means wherof you shall sée 6. signes of the Zodiaque to be alwayes aboue the Horizon, and 6. signes to be alwayes vnder the Horizon: Againe by placing the Sphear so as both the Poles may lie vpon the Horizon, you shall sée the shape of the first right Spheare, wherein the Horizon passeth through both the Poles of the worlde, and the Equinoctiall passeth through the Poles of the Horizon, which are the two points called before the Zenith and Nadir.

When is it said to be an oblique Horizon, and thereby to make an oblique Spheare?

When the Pole of the world is eleuated aboue the Horizon, be it neuer so little, so as the Horizon do cut the Equinoctiall with oblique angles, and look how much the Pole of the world is eleuated aboue your Horizon, so much is your Zenith distant from the Equinoctial, and the nigher that your Horizon approcheth to the Pole, the nigher your Zenith approcheth to the Equinoctial. Againe, look how much the Equinoctiall is eleuated aboue your Horizon, so much is your Zenith distant from the Pole, all which things

of the Spheare.

things this figure here following doth plainly shew, whereby you may easily perceiue that the latitude, which is the distance of your Zenith from the Equinoctial, is alwaies equal to the altitude of the Pole, which is the distance betwixt your Horizon & the Pole, as for example, knowing the latitude of Norwich to be 52. degr. lay the zenith of this figure vpō the 52. degrees, reckoning from the Equinoctial towards the pole Arctique on your left hand, and look what distance is betwixt the said zenith and the Equinoctial, the selfsame distance you shall find to be betwixt the Horizon and the foresaid Pole on your right hand, and you may doe the like vpon the Sphere it selfe by raising the moouable Meridian aboue the Horizon at that altitude, so as the 52. degr. may be euen with the Horizon.

A Figure shewing the latitude of any place to bee equall to the eleuation of the Pole.

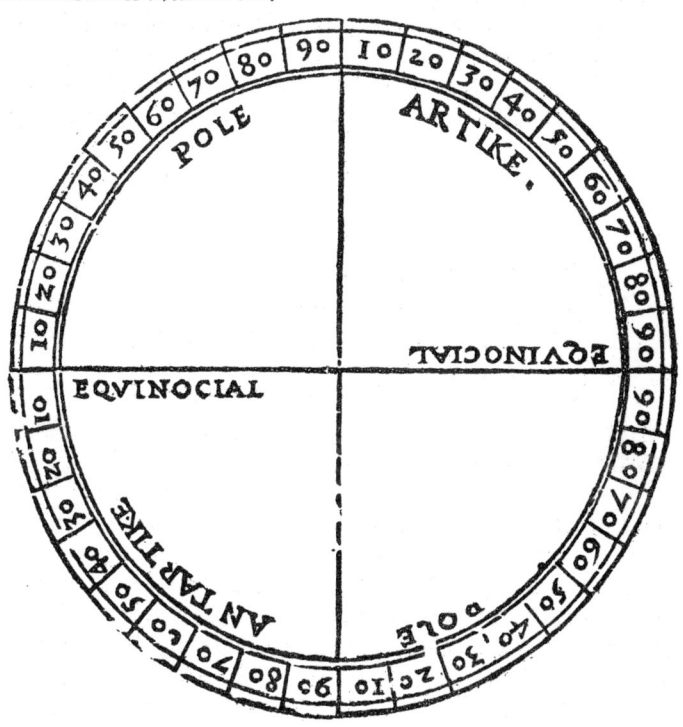

What

The first Booke

What other vses hath this circle?

In this circle are set down the foure quarters of the world, as East, West, North and South, and the rest of the winds: Againe, this circle deuideth the artificial day from the artificial night, for al the while that the Sun is aboue the Horizon it is day, and whilest it is vnder the same it is night. And by this circle we knowe what starres doe continually appeare, and which are continually hidden, also what starres doe rise and goe downe. Againe, in taking the eleuation of the Pole, this circle is chieflie to be considered, for when we know how many degrées the Pole is raised aboue the Horizon, then we haue the eleuation therof for that place. For to euery seueral place, yea to euery little moment of the earth in an oblique Spheare, belongeth his proper Horizon and seueral altitude of the Pole, whereby it appeareth that the Horizons are infinite and without number.

How shall I knowe in any place, hauing an oblique Horizon, how much the Pole is eleuated aboue the Horizon?

That is declared in the second booke of this Treatise, whereas I speake of the latitude and longitude of the earth, in the eight chapter.

Of the Meridian, and of the vses thereof.

Chap. 18.

What is the Meridian?

It is a great immoouable circle passing through the Poles of the world, and through the Poles of the Horizon.

Why is it called the Meridian?

Because that when the sun rising aboue the Horizon in the East, cometh to touch this line with the center of his body, thē it is midday or noontide to those, through whose Zenith that circle passeth And whē the sun after his going down in the west cometh to touch the self line again in ye point opposit, it is to them midnight, & note that diuers cities, hauing diuers latitudes, that is to say, being distant one from another North and South bee it neuer so farre, may haue one selfe Meridian: but if they be distant one from another East and West, bee it neuer so little, then they must néedes haue diuerse Meridians, and such distance betwixt the two seuerall

uerall Meridians, is called the difference of longitude, whereof we shall speake hereafter more at large when we come to treat of the longitude and latitude of the earth, which something differeth from the longitude and latitude of the Starres or Planets, whereof we haue alreadie spoken in the 11. chapter.

How many Meridians be there?

The Astronomers doe appoint for euery two degrees of the Equinoctiall a Meridian, so as they make in al 180. Albeit most commonly in the Spheare they set downe but one, which serueth for all by turning the bodie of the Spheare to it, which for that cause is called the moouable Meridian. And in such Spheares as haue not a foote and a standing Horizon, there is no Meridian at all, but the two Colures are faine to supply their want, but all terrestriall Globes are commonly described with twelue Meridians, cutting the Equinoctiall in 24. pointes, and deuiding the same into 24. spaces, euerie space contayning 15. degrees, which is an houre, by meanes whereof we know how much sooner or latter it is noonetide in any place, for it is noontide sooner to those whose Meridian is more Eastward then to them whose Meridian is more Westward. And contrariwise the Eclipse of the Sun or Moone appeareth sooner to those whose Meridian is more Westward.

What other vses hath this circle?

This circle deuideth the East part of the world from the West and also it sheweth both the North and South, for by turning your face towardes the East, you shall find the Sunne being in that line at noone tide to be on your right hand right South, the opposit part of which circle sheweth on your left hand the North. Also this circle by reason that it passeth through both the Poles of the world, deuideth both the Equinoctiall and all his Paralels into two equall parts aswell aboue the Horizon as vnder the Horizon, and by that means it deuideth the artificial day and artificiall night each of them into two parts, that is to say, into two semidiurnal & into two seminocturnal partes. For betwixt that part of the Horizon where the sun riseth, mounting stil vntil he come to this circle, which is at noontide, is contayned the first halfe of the day, & the other half is from the same circle to the going down of the Sun vnder the Horizon. And the first part of the night is

the

the space betwixt the Suns going down and his comming again to the Meridian, which is at midnight, and from thence to the time of his rising is the other half of the night, and also the Astronomers take the beginning of their naturall day from this circle, counting either from noontide to noontide, or els from midnight to midnight. Againe, this circle sheweth the right ascentions and declinations of the starres, and the highest altitude, otherwise called the Meridian altitude of the Sun or of any starre, or degree of the Ecliptique, or of any other point in the firmament, all which vses and many others more you shal better vnderstand hereafter, when we come to shew the vses of the globe aswell terrestriall as celestiall.

Of the vertical circle, and vses thereof.
Chap. 19.

But here you haue to note that though the most part of Geographers do set down in their Spheares but 6. great circles, yet there is another great circle called the circle Verticall, which passeth right ouer our heads through our zenith, wheresoeuer we be vpon the land or sea, crossing our Horizon in 2. points opposite, and deuiding the same into two equal parts, and such kind of circles are called in Arabicke Azimuthes, whereof you may imagine that there be so many as there be rombes or winds in the Marriners Compasse which are in number 32. yea, and if you will, you may make halfe so many as there be degrees in the Horizon, which are in number 360. the halfe whereof is 180. If you be right vnder the Equinoctiall, and doe goe or saile right East or West, then the Equinoctiall is your Verticall circle, and if you goe or saile right North or South, then the Meridian is your verticall circle, which two circles notwithstanding do alwayes keepe their names. But in sayling by any other rombe, that circle which is imagined to passe from the true East point right ouer your head vnto the true West point, or which crosseth your Meridian in the zenith point with right Sphericall angles, is most properly called the verticall circle, and the learned seamen haue great respect to two speciall kinds of Verticall circles, that is, the Magnetical Meridian, and the Azimuth of the Sun.

What

of the Spheare.

What manner of verticall Circles be those, and whereto serue they?

M. Borrough in his discourse of the variation of the Compasse, defineth the Magneticall Meridian to be a great Circle, which passeth through the Zenith and the Pole of the load stone called in Latine Magnes, and deuideth the Horizon into two equall parts, by crossing the same in two points opposite. Againe the Azimuth of the sunne is a great Circle, passing through the Zenith & the Centre of the sunne in what part of the heauen so euer he be, so as he be aboue the Horizon, which Circle deuideth the Horizon into two equall parts, by crossing the same in two points opposite, And by helpe of these two Circles and a certaine instrument made of purpose to giue a true shadowe, he teacheth to finde out the true Meridiã of any place: And also to know how much any Mariners Compasse doth varie from the true North and South, in North-easting or Northwesting, whereof I shall speake more at large hereafter in my treatise of Nauigation.

What vse is there of the verticall Circles or Azimuthes?

The verticall Circle sheweth what time the Sunne or any other starre rising beyond the true East point, is passed before the Sunne or saide starre, commeth to the true East or any other rombe. Also in what Coast or part of heauen, the Sunne, Moone, or any other starre is at any time being mounted aboue the Horizon, as whether it be Southeast or Northeast, or in any other rombe: Also by helpe of the verticall Circle most properly so called, are the twelue houses of heauen set, according to Campanus and Gazula. And by helpe of these Circles you may also knowe how any place vpon the earth beareth one from another eyther Eastward or Westward, and so forth, for euery place hath his seuerall Azimuth answerable to the Horizon and Zenith of the saide place.

Of certaine Circles called Almicanterathes.

Since I haue spoken here somewhat of the verticall Circles called Azimuthes, it shal not be amisse to shew you also that there be other Circles to be considered of in the Spheare aswell as in the Astrolabe called Almicanterathes, that is to say, Circles of Altitude, which though they be not all great Circles, for euery one

X   is

is lesser then other proceeding from the oblique Horizon of anie place to the Zenith of the saide place, yet the first Almicanterath which is the verie oblique Horizon it selfe, is a great Circle deuiding the Spheare into two equall partes, and all the rest are lesser and lesser, vntill you come to the very Zenith, and are paralels to the Horizon, euen as the Tropiques and the other lesser Circles are paralels vnto the Equinoctiall and the Zenith in Sphericall bodies is the Centre of them all, though it be not so in Astrolabes, for there euery Almicanterath is faine to haue his seuerall Centre, of which Circles there be in all 90. according to the number of 90. degrees contayned betwixt the oblique Horizon and the Zenith, and these Circles doe serue to shewe the Altitude of the Sunne or Moone, or of any other starre fixed or wandring, being mounted at any time aboue the oblique Horizon, which is easie to be founde by any Quadrant, Crosse-staffe, or Astrolabe. But leauing to speake any further of these Circles, because they are not vsed to bee described in Spheares but onely in Astrolabes, I will nowe treate of the foure lesser Circles before mentioned, which are commonly set downe in euery Spheare or Globe.

Of the foure lesser circles, that is to say, the circle Arctique, the circle Antarctique, the Tropique of Cancer, and the Tropique of Capricorne, and also of the fiue Zones, that is to say, two cold, two temperate, and one extremely hoat.

Chap. 20.

Hich call you the lesser Circles?

They are those that doe not deuide the Spheare into two equall parts, as the great Circles doe, and of such there be foure, that is the two Polar circles, and the two Tropiques, that is to say, the Tropique of Cancer, and the Tropique of Capricorne, of which Polar circles the one is called Arctique, and the other Antarctique, and are made by the turning about of the two Poles of the Zodiaque, which Poles being situated in the Colure of the

Solstices are so farre distant from the Poles of the world, as is the greatest declination of the Sunne from the Equinoctial, which is 23. degrées, 2'8. as hath béene said before.

Which is the Arctique Circle, and why is it so called?

The Arctique Circle is that which is next to the North pole, and hath his name of this worde Arctos which is the great beare or Charles wayne, which are seuen starres placed next to this Circle on the outside thereof, and it is otherwise called the Septentrionall Circle of this worde Septentrio, which is as much to say as seuen Oxen, signified by the seuen starres of the little Beare, which doe mooue slowely like Oxen, and are placed all within the sayde Circle, and the bright starre that is in the tippe of the tayle of the sayde little Beare, is called of the Mariners the loade starre or North starre, whereby they sayle on the Sea, and the Centre of this Circle is the North Pole of the world which is not to be séene with mans eye.

What is the Antarctique Circle?

It is that which is next vnto the South Pole, and it is so called, because it is opposite or contrarie to the Circle Arctique.

Now describe the two Tropiques.

The Tropique of Cancer is a Circle imagined to bée betwixt the Equinoctiall and the Circle Arctique, which Circle the Sunne maketh when he entreth into the first degrée of Cancer, which is about the twelfth or thirtéenth day of June being then in his greatest declination from the Equinoctiall Northwarde, and nighest to our Zenith, béeing ascended to the highest point that hee can goe, at which time the dayes with vs be at the longest, and the nightes at the shortest. And so from thence hée declineth to the other Tropique called the Tropique of Capricorne, which is a Circle imagined to bée betwixt the Equinoctiall and the Circle Antarctique, which the Sunne maketh when he entreth into the first degrée of Capricorne, which is about the twelfth or thirtéenth day of December, at which time he is againe in his greatest declination from the Equinoctiall Southwarde, and furthest from our Zenith: whereby the dayes with vs bée then at the shortest, and the nights at the longest: And note that these two Circles are called Tropiques of this Greeke worde Tropos, which is as much to say as a conuersion or turning, for when the Sunne arriueth to any of these two Circles, hee turneth backe againe eyther ascending or descending, by reason of which foure Circles as well the firmament as the earth is deuided into fiue Zones, that is to say, two colde, two temperate, and one extremely hoat, otherwise called the burnt Zone, of which fiue Zones, the foresaide foure Circles are the true boundes. For of the two colde Zones, the one lyeth betwixt the North pole and the Circle Arctique, and the other lyeth betwixt the South Pole and the Circle Antarctique, and of the two temperate Zones, the one lyeth betwixt the Circle Arctique, and the Tropique of Cancer, and the other lyeth betwixt the Circle Antarctique, and the Tropique of Capricorne, and the extreme hoat Zone lyeth betwixt the two Tropiques, in the middest of which two Tropiques, is the Equinoctiall line, as you may see in this figure, and also in the Spheare or Globe it selfe.

A figure shewing the fiue foresaid Zones.

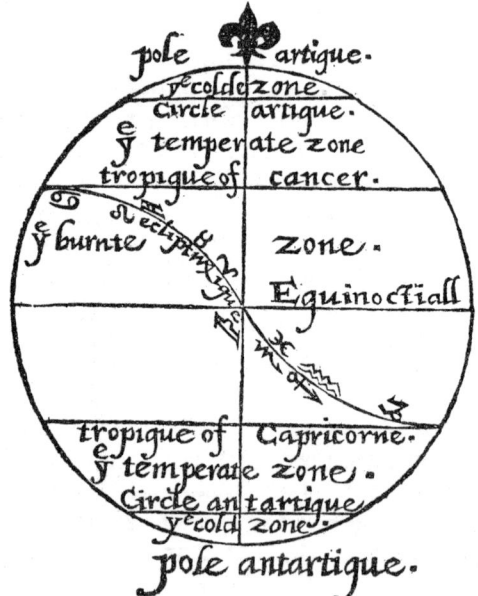

Of which Zones the aunciēt men were wont to say that thrée were vnhabitable, that is, the two colde, and the extreame hoat, which experience sheweth in these latter dayes to bee vntrue, as we shall declare more at large when we come to treate of the diuision of the earth: Againe you haue to vnderstande that euerie one of these lesser Circles doth containe in length, 360. degrées as well as euery one of the greater Circles, but the degrées are not of like bignesse, no more then the Circles themselues are like in compasse or circuit, for the lesser the Circles are in circuit, the lesser their degrées must néedes be.

Sith euery of the lesser Circles differ one from another in circuit, and thereby the degrees of euery Circle be lesser then other, how shall I know the true quantitie of euery degree in ech Circle, and how many minutes are required in euery lesser degree proportionally to answere one degree of the Equinoctiall.

For the better knowledge hereof, you must first imagine that there may be as many Circles made from the Equinoctiall towards any of the Poles, as there be degrées of Latitude, which are in number 90. as hath béene saide before: And the nigher that any Circle is to the Equinoctiall, the greater it is in circuit, and the further from the Equinoctiall towards any of the Poles, the lesser in circuit, and therefore more or lesse minutes are requisite to answere to one degrée of the Equinoctiall, as you may easily perceiue by this Table following, consisting of 6. collums, euery front or head whereof is noted with thrée great letters, D. M. S. signifying degrées, minutes and seconds, sixe times repeated, and in the beginning of the first colum on the left hand is set downe one degrée, which is the first degrée of 90. & nighest vnto the Equinoctiall, right against which one degrée is placed towards the right hand, 59. minutes, and 59. seconds : and so procéeding from degrée to degrée successiuely, vntill you come to 90. you shall finde how many minutes and seconds doe answere to one degrée of the Equinoctiall, and this Table will also serue to shew the difference of miles in euery sundry clyme or paralell, whereof we shall speake hereafter when we come to treat of the earth.

The first Booke

A Table shewing how many minutes are requisite to make one degree, in euery lesser circle answerable to one degree of the Equinoctiall.

D	M	S	D	M	S	D	M	S	D	M	S	D	M	S	D	M	S
1	59	59	16	57	41	31	51	26	46	41	41	61	29	5	76	14	31
2	59	58	17	57	23	32	50	53	47	40	55	62	28	10	77	13	30
3	59	53	18	57	4	33	50	19	48	40	9	63	27	14	78	12	28
4	59	51	19	56	44	34	49	45	49	39	22	64	26	18	79	11	27
5	59	46	20	56	23	35	49	9	50	38	34	65	25	21	80	10	25
6	59	40	21	56	1	36	48	32	51	37	46	66	24	24	81	9	23
7	59	33	22	55	38	37	47	55	52	36	56	67	23	27	82	8	21
8	59	25	23	55	14	38	47	17	53	36	7	68	22	29	83	7	19
9	59	16	24	54	49	39	46	38	54	35	16	69	21	30	84	6	16
10	59	5	25	54	23	40	45	58	55	34	25	70	20	31	85	5	14
11	58	54	26	53	56	41	45	17	56	33	33	71	19	32	86	4	11
12	58	41	27	53	28	42	44	35	57	32	41	72	18	32	87	3	8
13	58	28	28	52	59	43	43	53	58	31	48	73	17	33	88	2	5
14	58	13	29	52	29	44	43	10	59	30	54	74	16	32	89	1	3
15	57	57	30	51	58	45	42	26	60	30	0	75	15	32	90	0	0

Of the starres and celestiall bodies contained in the firmament, and first of their substance.

Chap. 21.

Auing briefely described all the Circles as well greater as lesser that are imagined to be in the 8. heauen, I thinke it good now to speake somewhat of the starres and celestiall bodies placed in the saide heauen, And first of their substance, & then of their mouing, figure, shape, number, magnitude or greatnes, also of their Longitude, Latitude

of the Spheare.

Latitude, declination, ascention, descention both right and oblique, and of the ascentionall difference, and finally of the threefold Poeticall rising and going downe of the starres, but first of their substance.

Of what substance are the starres?

The starres bee of the same substance that the heauens are wherein they are placed, differing onely from the same in thicknes, and therefore some defining a starre doe say, that it is a bright and shining bodie, and the thickest part of his heauen, apt both to receiue and to reteine the light of the Sunne, and thereby is visible and obiect to the sight: for the heauen it selfe being most pure, thinne, transparent, and without colour is not visible, and for this cause the milke-white impression in heauen like vnto a white way called of the Astronomers Galaxia, and of the common people our Ladies way is visible to the eye, by reason that it is thicker then any other part of the heauen.

Why are not the starres seene as well in the day, as in the night.

Because they are darkened by the excellent brightnesse of the Sunne from whome they borrowe their chiefest light.

Of the moouing and shape of the starres.

Chap. 22.

What moouing haue the starres?

The selfe same moouing that the heauen hath wherein they are placed.

Whereby are the heauens mooued?

Some saye that the first mooueable is turned by God himselfe, and all the rest of the heauens euery one by his proper intelligence, which though it turneth his heauen about, yet it giueth neither life, sense, nor vnderstanding thereunto, as some haue vntruely holden, affirming the heauens to be liuing and intelligible bodies.

The first Booke

If the starres haue no moouing of themselues, wherof commeth it then, that some seeme to our sight sometime nigher and sometime further off.

All the fixed starres of the firmament are alwayes of like distance, notwithstanding by reason of the manifold moouing of the firmament, wherein they are placed, they seeme to change their places, and sometime to bee more towardes the East or West, North or South. And whereas the vii. Planets called the wandring starres, do change their places now here now there, that chanceth not by their owne moouing, but by the moouing of the heauens wherein they are placed: for a starre being round of shape hath no members meete to walke from one place to another, but onely changeth his place through the motion of his Spheare or heauen wherein such Planet is fixed.

Of the number of the starres, and of their magnitude or greatnesse, and into how many Images they are deuided, and how many starres euery image containeth.

Chap. 23.

May the starres be numbred by man? No, for as Dauid saith, that belongeth onely to God, who as he created them, so he can number them and call them all by their names, notwithstanding the Astronomers by their industry and diligent obseruation, haue attained to the knowledge of many: as first they know the seuen planetes, otherwise called the wandringe starres, and haue made manifest demonstrations of their motions, and by continuall obseruation haue found out the manifold vertues, powers and influences of the same, but of the fixed starres they could neuer finde more then 1022. and because the starres are not equall in greatnes or bignes, they make sixe differences of greatnesse, appointing to the first difference 15. starres, which are bigger then all the rest, whereof euery one containeth the earth 107. times, to the second difference 45. starres, whereof

euery

euery one containeth the earth 90. times. To the third they appoint 208. starres, whereof euerie one containeth the earth 72. times. To the fourth difference they appoint 474. starres, whereof euery one containeth the earth 54. times. To the fift they assigne 217. starres, whereof euery one containeth the earth 57. times. To the sixt or last greatnesse they appoint 49. small stars, whereof euery one containeth the earth 18. times, and some say 20. times. Besides these there be 14. others, whereof 5. be called clowdy and the other darke, because they are not to be seene but of a very quicke and sharpe sight. And you haue to note that the antient Astronomers do deuide all the fixed starres to them knowne into 48. images, whereof they liken some to liuing things as to men, women, beastes, monsters, foules, fishes, and créeping wormes, and some to things without life, hauing some artificiall shape, of which 48. images, they appoint 12. to the Zodiaque, commonly called the 12. signes, as Aries, Taurus, Gemini, Cancer, Leo, Virgo, Libra, Scorpio, Sagittarius, Capricornus, Aquarius and Pisces. Againe they place in the North part of the firmament 21. images, and in the South part thereof 15. images, which make in all 48. The description of all which images, together with their names hereafter followeth: and first I will describe vnto you the twelue Images contained in the Zodiaque.

Of the xij. Images or signes of the Zodiaque.

Chap. 24.

The twelue signes (as some affirme) doe containe of the foresaide number of fixed stars 273. For the first signe called Aries, that is to say, the Ramme, contayneth 13. starres, which Image or signe being placed in the coniunction of the Zodiaque with the Equinoctiall, hath his backe turned towards the North, and his head towards the East, and riseth with his head, and goeth downe with his féete. The second signe called Taurus, that is to say, the Bull, containeth 33. starres, whereof there is one bright starre of the

first

The first Booke

first bignesse called Oculus Tauri, that is to say, the Bulles eye, who hath his head enclyning towardes the West as though hee looked towardes the earth, and riseth and goeth downe with his heeles vpwarde. The thirde Signe called Gemini, that is to say the twinnes, doe contayne 18. starres, their heades looking towardes the North, and their backes being ioyned together doe embrace one another, and doe rise lying, and doe goe downe with their feete. The fourth Signe called Cancer, that is to say, the Crabbe contayneth nine starres, extending his feete towardes both the Poles, and looking towards Leo, hath his bellie turned towardes the earth, and hee riseth and falleth with his hinder part or backe part of his bodie. The fifth Signe called Leo, that is to say the Lyon, contayneth ten starres, whereof there be two bright starres of the first bignesse, the one in his breast called Cor Leonis, and Regulus, that is to say the Lyons heart, and the other in his tayle called Cauda Leonis, that is to say, the Lyons tayle, who looketh towardes Cancer, and hauing his backe turned towardes the North, he riseth and goeth downe with his head. The sixth Signe called Virgo, that is to say, the Virgine, whose head is behinde the Lyon and toucheth the Equinoctiall line with her left hande, holding in the same hande an eare of corne, shee both ryseth and goeth downe with her head: this image contayneth sixe and twentie starres, whereof there is one bright starre of the first bignesse called Spica Virginis, that is to say, an eare of corne. The seuenth Signe called Libra, that is to saye, the Ballance, contayning eight starres hath two skales, whereof the one hangeth towardes the North, and the other towardes the South. The eight signe called Scorpius, that is to say the Scorpion, contayneth one and twentie starres, who looketh towardes Virgo, and extendeth his feete towardes both the Poles, he boweth his tayle towardes the North, hauing his belly turned towardes the earth, and he riseth and goeth downe bowing. The ninth Signe called Sagittarius, that is to say the Archer, contayning one and thirtie starres, hath his head towardes the North, and looketh towards the Scorpion, hauing a bowe and a shaft, whereof the bowe toucheth his left hande and left foote, he riseth right vppe, and goeth downe headlong. The tenth Signe called Capricornus,

that

that is to say, the Goate contayning eight and twentie starres, hath his backe turned towardes the North, and his heade towardes the Archer, and turning himselfe towardes Aquarius, he riseth right vp, and goeth downe headlong. The eleuenth Signe called Aquarius, that is to saye, the water-bearer contayning two and fourtie starres, hath his heade towardes the North, extending his left hande vppon the backe of Capricorne, and with his right hande powreth out water out of his potte, which bendeth towardes the East, runneth euen to Pisces, hee riseth and goeth downe with his heade before anie other of his members. The twelueth Signe called Pisces, that is to say, two Fishes, doe containe foure and thirtie starres, whereof the backe of the first is towardes the North, and the backe of the seconde towardes the West arme of Andromeda, and one of the Fishes looketh towardes Aquarius, and the other towardes the North, and betwixt these two Fishes is a certaine little line wherewith their tayles are bounde together as it were with a bond, the lower part of which Fishes, doth alwayes both ryse and goe downe, and not the vpper part: And though the 12. Signes of the Zodiaque are saide to bee equall both in length and breadth, that is to say, hauing thirtie degrees in length, and twelue degrees in breadth, as hath béene sayde before, yet these Images are not equall, for some doe extende further then the Zodiaque in breadth, and some are more then thirtie degrees in length, As the Tables of Alphonsus doe manifestlie shewe, who sayeth there that the twelue Signes doe contayne three hundred and fiftie starres, for he appointeth to Aries eighteene, to Taurus fourtie foure, to Gemini twentie fiue, to Cancer thirteene, to Leo thirtie fiue, to Virgo thirtie two, to Libra seuenteene, to Scorpio twentie foure, to Sagittarius thirtie and one, to Capricornus twentie eight, to Aquarius fourtie fiue, to Pisces thirtie eight, which make in all three hundred and fiftie, in which Tables are also set downe the Longitude, Latitude, and Magnitude of the saide starres, but the Longitude of the sayde starres, is farre altered from that Longitude which they had in his tune, whereof wée shall speake héereafter more at large.

Of

The first Booke

Of the xxj. Northerne Images.

Chap. 25.

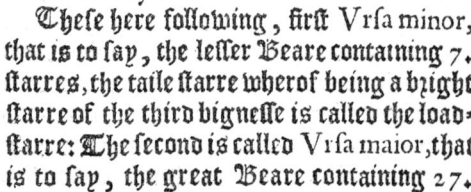Hich be they?

These here following, first Vrsa minor, that is to say, the lesser Beare containing 7. starres, the taile starre wherof being a bright starre of the third bignesse is called the loadstarre: The second is called Vrsa maior, that is to say, the great Beare containing 27. starres, whereof there be 7. principall, making a shape like vnto a Cart with foure wheeles, and therefore it is commonly called Charles waine. The third is called Draco, that is to say the Dragon that kept Iunos Orchard from robbing, containing 31. stars. The fourth is called Cepheus, the proper name of a King of Ethiope, containing 11. starres. The fifth is called Bootes, that is to say the roaring keeper of the Beare, containing 22. starres, whereof there is one bright starre betwixt his legges, of the first bignesse called Arcturus. The sixth is called Corona Ariadnæ, that is to say the Crowne of Ariadna the daughter of king Minos, containing 8. starres. The seuenth is Hercules, who lyeth groueling with his heeles towardes the North pole, holding a clubbe in his right hand, and a Lions skinne hanging on his left arme containing 28. starres. The eight is called Lira, that is to say an instrument of musicke like a Harpe placed in heauen in the memorie of Orpheus, containing 10. starres. The ninth is called Cignus, that is to say the Swanne, into which Iupiter transformed himselfe to deceaue the Nymph Lɤda, containing 17. starres. The tenth is called Casiopeia sitting in her chaire, wife to Cepheus, and mother to Andromeda, containing 13. starres. The eleuenth is called Perseus, holding his sworde in the one hand, and the head of Gorgon or Medusa in the other hand, whose haires were all Serpents, containing 26. starres. The twelfth is Auriga & Ericthonius, who was the inuentor of the first Chariot that euer was made, containing 13. starres. The 13. is called Serpentarius and Anguitenens, that is to say, he that holdeth the Serpent, who as some thinke was Æsculapius the famous Phisitian,

Phisitian, containing 24. starres. The 14. is called Serpens or Anguis, that is to say, the Serpent of Esculapius containing 18 starres. The 15. is called Sagitta or Telum, that is to say the shaft or dart wherewith Hercules slew the Eagle tormenting Prometheus in the mount Caucasus, containing 5. stars. The 16. is called Aquila, that is to say, the Eagle which caried Ganimedes into heauen containing 9. starres. The 17. is called Delphinus that is to say the Dolphyn, which saued Arion that excellent Musitian, being cast by Pyrats into the sea, containing 10. stars. The 18. is called minor Equus, that is to say, the lesser Horse, containing onely 4. little darke starres in his head. The 19. is the great Horse called Pægasus, that is to say the winged Horse, wherewith Bellerophon conquered the monstrous beast called Chimera, which was halfe a Lyon and halfe a Dragon, and is adorned with 20. starres. The 20. is called Andromeda the daughter of Cepheus by Cassiopeia, and wife to Perseus, who for her constant loue towards her husband, was placed in heauen nigh vnto him, and was adorned with 23. starres. The 21. is called Triangulum, that is to say a Triangle or three cornerd figure, which being like in shape to the Ile Cicilia the Goddesse Ceres obtained to be placed in heauen, and was adorned with 4. starres, that is to say, euery corner one, and the fourth in the midst of the shortest side: To these (in mine opinion) ought to be added Bernices haire, called crinis Bernices, which in al celestial globes is placed not farre from the right hinder foote of the great Beare, and this Image containeth 4. little starres.

Of the 15. Southerne Images.

Chap. 26.

WHich be they?

These hereafter following, whereof the first is called Cætus, that is to say the Whale, that monstrous fish which by the appointment of Neptune, would haue destroyed Andromeda, whom Perseus deliuered by killing the Fishe, and afterwardes tooke Andromeda to wife, which Fish Neptune placed in heauen, adorning the same with two and twentie starres.
The

The first Booke

The second is called Orion with his sword by his side, who afterward was slaine by Diana by mis-hap against her will, for the which she placed him in heauen, adorning him with 38. starres, whereof there be two bright starres of the first bignesse, the one in his right shoulder called Bed Alguze, and the other in his left foote called Rigell Alguze. The third is the flood called Eridanus, into the which Phaeton the sonne of Apollo was stroken with a thunder bolt by Iupiter, for burning the earth by rashly driuing his fathers Chariot, which he was not able to guide, in memorie whereof the flood was placed in heauen, and adorned with 34. starres, whereof one is a bright starre of the first bignesse called Acarnar. The fourth is called Lepus, that is to say the Hare, placed nigh vnto Orion, because he was a hunter, adorned with 12. starres. The fift is called Canis maior, that is to say the great Dogge, passing all others in swiftnesse, which was giuen by Aurora to Cepheus the sonne of Eolus, and is placed next to the Hare, being adorned with 18. starres, whereof there is in his mouth a very faire starre of the first bignesse called Syrius. The sixt is called Canis minor, that is to say, the lesser Dogge, without the which Orion his maister would not be placed in heauen, which hath but two starres, whereof the one is in his flanke, and is a bright starre of the first bignesse called Canicula and Procion. The seuenth is the ship called Argos, in the which Iason and his companions sayled to Cholcos to winne the Golden Fléese, which is adorned with 45. starres, whereof there is one bright starre of the first bignesse in the left oare called Canopus. The eight is called Hydra, that is to say, the Water-serpent which Hercules slewe, or as some say which kept the water-bowle, and would not suffer the thirstie Crowe to drinke, which Crowe Apollo sent for water to doe sacrifice, and is adorned with 25. starres. The ninth is called Crater, that is to say the Cuppe or Bowle which the Crowe brought too late vnto Apollo, and therefore his feathers were made all blacke whereas before they were white, contayning 7. starres. The tenth is called Coruus that is to say the Crowe before mentioned, adorned with 7. starres. The 11. is called Chiron siue Centaurus, the son of Saturne passing all others in iustice and religion, and therefore is figured in heauen as though he were offring sacrifice vppon an altar adorned

with

of the Spheare   160

with 37. starres whereof there is one bright starre of the first bignesse in his right foote called Chiron and Centaurus. The twelfth is called Lupus, that is to say the Wolfe, into which Licaon the cruell Tirant was turned by Iupiter, or as some say that Wolfe which Centaurus killed to doe sacrifice vpon the altar containing 19. starres. The thirtéenth is called Ara, that is to say the Altar made by the Smithes of Vulcane, whereupon all the Goddes sware to reuenge the insolencie and pride of the Giants, which altar is placed next to Centaurus being adorned with 7. starres. The fourtéenth is called Corona Australis that is to say, the Southerne Crowne which Bacchus did weare when hee fetched his mother Semele from hel, which is placed in heauen, & is adorned with thirtéene starres. The fiftéenth and last of the Southerne signes is called Piscis Australis, that is to say, the Southerne Fish which was placed in heauen in memoriall of the fishes which the people of Syria did worship as their Goddes, and is adorned with 12. starres, whereof there is one bright starre of the first bignesse in his mouth called Fomahant: All which Images and Signes before mentioned, aswell Northerne as Southerne, you may sée plainely described in euery celestiall Globe, and also set foorth in Plano in the nether ende of Vopellius his vniuersall Map, that is to say the Northerne signes on the left side, and the Southerne signes on the right side of the Mappe: Or in the front of Planctius his great vniuersall Mappe, who in the ronole representing the Southerne halfe of the celestiall Globe, setteth down also certaine Southerne starres lately found out by the trauellers into the Indies, as the Crosse, the Southerne Triangle, Noe his Doue or Pigeon, and another in shape of a man called Polophilax, so as there be now in all 19. Southerne Images.

Of the longitude of the fixed starres, and of the procession of the vernall Equinoctiall point, and what it is.
Chap. 27.

What is the Longitude of a starre?

The Longitude of any starre, is that Arke or portion of the Ecliptique line which is contayned betwixt the first point of Aries, and that Circle which passeth through the Poles of the Zodiaque, and

and also through the bodie of the starre, as for example the starre called Cor Leonis is distant in these dayes from the vernall Equinoctiall point of the Ecliptique line 143. degrées and 3'2. and thereby is found to be in the 23. degrée and 3'2. of Leo, againe the starre called Spica Virginis in these dayes is distant from the first point of Aries 198. degrées, and so is found to be in the eightéenth of Libra.

Why do you say in these dayes?

Because the fixed starres in processe of time, doe change their Longitude by reason of their proper moouing vpon the Poles of the Zodiaque, which is from West to East, for whereas Spica Virginis in Ptolomies time was in the 26. degrée of Virgo, it is found now to be in the 18. of Libra, the cause whereof is the precession of the Equinoctiall point or section.

Define what that Precession is?

It is an Arke or portion of the Ecliptique line, contayned betwixt two great Circles, both passing through the Poles of the Zodiaque, in such sort as the one passeth through the first minute of the vernall Equinoctiall point of the saide Ecliptique, and the other Circle passeth through the first or former starre of the Rams horne, from which starre the Astronomers doe make all the celestiall motions and reuolutions to take their first beginning, and this starre in olde time past was knowne to be before the vernall Equinoctiall point, which is the first moment of Aries, but now it is found to haue passed that point so farre towardes the Solsticiall point, as in these dayes it is knowne to be in the 27. degrée and 4'2. of Aries, and in processe of time it will be cleane out of Aries, and enter into Taurus.

Of the Latitude of the fixed starres.

Chap. 28.

What is the Latitude of a starre?

The Latitude is none other thing, but the distance of any starre from the Ecliptique line either towards the North or South pole of the Zodiaque, and such Latitude neuer changeth or altereth, for as the starre Spica Virginis is at this present

present two degrées distant from the Ecliptique line towards the South, so it euer hath béene and euer shall be, and the like is to be said of all the rest of the fixed starres which do alwaies kéepe their Latitude, be it Northward or Southward, néere to the Ecliptique or farre from the same.

Of the Declination of the fixed starres.

Chap. 29.

What is the declination of a starre?

The declination is none other thing, but the distance of any fixed starre from the Equinoctiall, eyther Northward or Southward, which is mutable as well as the Longitude: for as the fixed starres do change their Longitudes, so also by little and little they decline either more or lesse from the Equinoctiall: As for example, the declination of the starre called Canicula, that is to say the lesser Dogge in the yeare of our Lord, 138. when Ptolomie liued, was 15. degrées 4′4. and 3″8. towardes the South. But in these dayes the declination of the said starre is but sixe degrées and 7′. towardes the South, and by reason that the fixed starres in processe of time doe change their Longitude and declination, they are not alwayes vnder one selfe signe, but doe flit out of one signe into another.

How is the Longitude, Latitude, and declination of anie starre to be knowne, and how are the starres themselues to be knowne in the firmament.

The Longitude, Latitude, and declination of any starre is to be knowne most truely by the Astronomicall Tables calculated of purpose, and you may know the same also without hauing regard to euery small minute, by helpe of the celestiall Globe, all the necessarie vses whereof I haue set downe in a little Treatise to be added hereafter to this Booke, and there also I shew you how to find out any starre in the firmament that is described in the globe, in which Globe are set downe as many stars as euer were known, a few excepted towards the South pole, which were founde out but of late dayes, of which starres I shall haue occasion to speake hereafter

The first Booke

hereafter in my Treatise of Nauigation. In the meane time I will proceede to the ascention and descention of the starres both right, meane, and oblique.

Of the ascention and descention, that is the rising and setting of the starres, aswell according to the Astronomers, as according to the Poets.

Chap. 30.

Oe the Astronomers and Poets differ touching this matter?

Yea they differ greatly, aswell in name as in matter: for whereas of the Poets it is called ortus & occasus Signorum, that is to say, the rising and falling of the Signes, so of the Astronomers it is called Ascentio & descentio Signorum, that is to say, the ascention and descention of the Signes, againe they differ in matter, or rather in manner, for that the Astronomers do consider the rising and falling of the starres more exactly then the Poets, for the Astronomers do consider the degrees and minutes of the same, and also do ground their ascention and descention vpon more certain demonstrations then the Poets. Moreouer whereas the Poets by their manner of rising and falling, doe simply set downe the time of thinges done or to be done, the Astronomers doe the same a great deale more exactly, and by their manner of ascention and descention doe consider the increase and decrease of the dayes, of which Astronomicall ascention and descention, I mind here to treate first in generall and then in particular.

Of the Astronomicall ascention and descention in generall both right, meane, and oblique, and what a giuen Arke is.

Chap. 31.

Efine what the Astronomicall ascention and descention is.

Astronomicall ascention is that portion or Arke of the Equinoctiall line which riseth together with some giuen

of the Spheare. 162

giuen Arke of the Ecliptique line aboue the Horizon, and the descention is that portion or Arke of the Equinoctiall that goeth downe or setteth together with some giuen Arke of the Ecliptique line vnder the Horizon, according to the moouing of the worlde which is from East to West.

What meane you by a giuen Arke?

A giuen Arke is as much to say as some supposed portion of the Ecliptique or of any other Circle, as if you woulde knowe the ascention of some supposed portion of the Ecliptique, contayning for example 25. or 30. degrees, here this portion of the Ecliptique contayning that number of degrees, is called the giuen Arke, of which Arkes some are called continuall, and some descrete or deuided, which I minde to English here whole and broken, for so I doe English quantitas, continua, & discreta in my Logicke, That Arke is sayde to bee continuall or whole which taketh his beginning from the first point of Aries, and so proceeding orderly, endeth at some other degree of the saide Ecliptique. And that Arke is called discrete or broken, which doth not take his beginning from the first point of Aries, but beginneth at some other degree of the Ecliptique, as for example, suppose that it beginneth at the fourteenth of Taurus, and endeth at the fifteenth of Gemini, this Arke is called a deuided or broken Arke, because it doth not beginne at the first point of Aries, and so procéede successiuely: moreouer you haue to vnderstande that the auncient Astronomers doe commonly make but two kindes of ascention and descention, that is, right and oblique, but there be in déede thrée kindes of ascention, that is to say, right, oblique, and meane ascention.

When is any ascention saide to bee right, oblique or meane?

It is saide to be right, when that portion of the Equinoctiall which riseth or goeth downe together with the Ecliptique, is greater or more in circuit then that of the Ecliptique. And it is sayde to be oblique, when that portion of the Equinoctiall which riseth or falleth together with the Ecliptique, is lesser then that of the Ecliptique, Againe that is sayde to be meane ascention, when that portion of the Ecliptique which ascendeth, is neyther greater nor lesser then that of the Equinoctiall, for as in

The first Booke

the right Spheare euery quarter of the Ecliptique hath a meane ascention, and equall to euery quarter of the Equinoctiall, beginning the quarters at any of the foure principall pointes, so if you take thrée signes in any other part of the Zodiaque, their ascentions will not agrée with a quarter of the Equinoctiall, sith there is no one signe that doth equally agrée with the like portion of the Equinoctiall, and all this matter dependeth vppon the knowledge of the vse of certaine Circles before defined.

Which be they?

These thrée, the Zodiaque, the Equinoctiall, and the Horizon: for first the Zodiaque doth shewe the place of the Sunne, that is to say in what degrée it is of any signe together with the minutes of the same, and turning about euery day by the diurnall motion, doth both appeare aboue the Horizon, and also is hidden vnder the Horizon. Secondly the Equinoctiall with his equall rising and going downe, doth measure the time of the sunne whilest he maketh his abode vnequally and diuersely aboue the Horizon. Thirdly the Horizon deuideth the one Hemispheare from the other, on which Horizon is to be considered what Angle any signe or starre maketh therewith, in his ascention or descention, and according as any portion of the Ecliptique riseth or setteth rightly or obliquely, so in respect of the Angle which it maketh with the Horizon, it is called a right or oblique ascention or descention.

Why should the ascentions and descentions be measured rather by the equinoctiall line then by the Ecliptique, sith the course of the sunne measureth all times?

The cause thereof is the obliquitie of the Zodiaque, hauing diuerse and variable situations, whereby the sunne abideth sometimes a great while aboue the Horizon, and sometimes but a little while, all which inequalitie is onely to be measured by the Equinoctiall, which is alwayes equally moued vppon his Poles. Hitherto of the Astronomicall ascention and descention in generall, nowe of all thrée ascentions and descentions in particular.

Of

of the Spheare. 163

Of the right, oblique, and meane ascention in particular, and of the chiefest causes of such diuersitie of ascentions.

Chap. 32.

ND for the better vnderstanding of the Astronomicall ascention and descention, wee will make this diuision, for either it is of some point or starre, or else of some portion of a Circle chiefely of the Ecliptique line. In the ascention of any point or starre, wee consider two thinges. First what Angle it maketh with the Horizon either right or oblique. Secondly the time from the rising of the first minute of Aries, which is the first beginning of the Longitude of any starre or Circle in heauen, and in respect of the Angle euery ascention is sayd to be right in a right Spheare, and oblique in an oblique Spheare: Againe the time of the ascention is to be measured by the degrees of the Equinoctiall from the first minute of Aries, vnto that degree and minute of the Equinoctiall which ascendeth together with the starres, And note by the way that 15. degrees of the Equinoctiall do make an houre, and foure doe make one degree of the same Equinoctiall, for foure times 15. doe make 60. minutes, which is an houre, Againe euery ascention considered, according to the time of his gate, is eyther right, oblique, or meane: if it be right, it is slowe: if it be oblique, it is quicke: if it be meane, it is equall. Now the ascention of any Arke or portion of a Circle is also eyther right, oblique, or meane: if it be right, it ascendeth slowely: if oblique, it ascendeth quickely: if meane, it ascendeth equally. And the better to vnderstand all these three kindes of ascentions, I will set downe these twelue rules heere following, whereof fiue doe belong to the right Spheare, and seuen to the oblique.

In the right Spheare all the foure quarters rising from the foure principall points, haue a meane ascention, and so hath all the foure points themselues.

In the right Spheare all those signes that be equally distant from the foure principall points haue equall ascentions.

P 3   In

3 In the right Spheare all starres or pointes that be in the Solsticiall Colure haue meane ascention.

4 In the right Spheare those signes that do ascend rightly, doe descend rightly, and those that doe ascend obliquely, doe descend obliquely.

5 In the right Spheare, Gemini, Cancer, Sagittarius, and Capricornus, doe ascende rightly, and all the rest obliquely.

6 In the oblique Spheare, the two Equinoctiall points haue meane ascention.

7 In the oblique Spheare ech halfe of the Spheare, beginning at eyther of the Equinoxes haue meane ascention: but this rule holdeth not, if that you beginne any other where.

8 In the oblique Sphere, those signes that do ascend rightly, do descend obliquely, & those which ascend obliquely do descend rightly.

9 In the oblique Spheare, the ascention of any supposed signe is equall to the descention of his opposite signe, and the descention of any supposed signe, is equall to the ascention of his opposite signe.

10 In the oblique Spheare, the ascention of any signe being added to his descention, is equall to the ascention and descention of the same signe being in the right Spheare.

11 In the oblique Sphere euery two signes equally distant from the two points of the Equinoctiall, haue equall ascentions and descentions.

12 In the oblique Sphere vnder the pole Arctique, al signes from Cancer to Capricorne, do ascend rightly, and all the rest obliquely, but contrariwise vnder the Pole Antarctique.

What is the chiefe cause of the diuersity of ascentions and descentions, aswell in the right as in the oblique Spheare?

The chiefe cause is the diuersity of the Angles which the Zodiaque maketh with the Horizon, for the sharper that the Angles be the lesser portion of the Equinoctiall riseth together with the Ecliptique, and the righter that the Angles be, the greater portion of the Equinoctial riseth, but the Equinoctiall by reason of his vniformitie, maketh his Angles alwayes equall one to another, that is to say, in the right Spheare, it maketh right Angles, and in the oblique Spheare, though not right, yet in euery signe it maketh like Angles.

How to know the diuersities of the ascentions and descentions, as well in the right as oblique Spheare.
Chap. 33.

THat is to be knowne most exactly by the Tables of ascentions calculated of purpose by Iohannes de Monte Regio, and by Reinholdus called in Latine Tabulæ directionum, and you may knowe it also without hauing respect to euery minute by marking and obseruing the same in a materiall Spheare or Globe, that hath a standing foote with a firme Horizon, for if you will know the diuersities of ascentions in a right Spheare, then you must lay the Spheare or Globe so as the Horizon may passe through both the Poles, and in turning about with your hand the Equinoctiall, together with the Ecliptique from East to West, marke with what degrée of the Equinoctiall any signe beginneth to ascend, & marke that degrée of the Equinoctiall with a little péece of waxe, then turne the Globe or Spheare towardes the West, vntill the last degrée of the sayd signe doe appeare iust with the vpper edge of the Horizon, and then marke what degrée of the Equinoctiall is aunswerable to the said last degrée of the foresaid signe, and there set another péece of waxe, then count the degrées of the Equinoctiall contayned betwixt those two markes, and if it be more then 30. that signe is said to ascend rightly, if it be lesse then 30. then that signe ascendeth obliquely, if it be iust 30. then it hath a meane ascention, & by allowing 15. degrées of the Equinoctial to an houre, and 4′. to a degrée, you shall know in what time that signe riseth. As for example if you would know what ascention the whol signe Taurus hath in a right Spheare, and also in what time it riseth, doe thus, First lay both the Poles of the Spheare iust vpon the Horizon, so as the same Horizon may passe through both ỹ poles, then bring the first point of *Taurus* to the East part of the Horizon, so as it may touch the vpper brimme or edge of the Horizon, and staying it there with your hande, looke what degrée of the Equinoctiall doth also touch the Horizon at that instant, which you shall finde to be 27. degrées 54. and marke that degrée of the Equinoctiall with a little péece of waxe, or some other thing

that

The first Booke

that may be easily put out or taken away, that done, put forwarde the foresaid signe Taurus still towardes the West, vntill the last degree of the saide signe be ascended vp euen to the vpper edge of the Horizon, and there staying it with your hande, looke againe what degree of the Equinoctiall doth rise withall, which you shall finde to be 57. degrees 4'8. and there set another marke vppon the Equinoctiall, then by telling the degrees conteyned in the Equinoctiall betwixt the two markes, you shall finde the number of degrees to be 29. degrees 5'4. and by allowing 15. degrees to one houre, and 4'. to a degree, you shall finde that the whole signe Taurus spendeth in his rising one houre, 5'9. 3 6. But now sith the Meridian in any place (as hath bæne said before) doth alwaies shew the right ascention of any starre, signe, arke, or point, because that cutteth both the Equinoctiall and the Horizon with right Angles: you may therefore find the right ascention of the sayd signe, or of any other signe or starre without remoouing the Spheare from your owne eleuation or Latitude in this manner following, bring the first degree of Taurus close to the mooueable Meridian, and there staying it marke what degree of the Equinoctiall the Meridian cutteth at that present, which you shall find to be 27. degrees 5'4. which is the right ascention of the first point of Taurus, then hauing brought the last point of Taurus to the foresaide Meridian, marke what degree of the Equinoctiall the sayd Meridian cutteth at that present, and you shall finde it to bæ the 57. degrees 4'8. now by counting vpon the Equinoctiall, the degrees conteyned betwixt those two markes, you shall finde the number to be 29. degrees 5'4. and you may finde the selfe same number by subtracting the right ascention of the first point of Taurus, out of the right ascention of the last point of Taurus, & thereby you shal know the time of his rising to be the same that you found in the right Spheare. Now if you would know the ascention of any signe in an oblique Spheare, then hauing placed your Spheare according to your Latitude, which for example sake suppose to be 52. degrees, and that in such Latitude you would know what ascention the whole signe Taurus hath, and in what time he riseth, you must first bring the first degree of Taurus to the East part of the Horizon, so as it may mæte euen with the vpper edge of the Horizon, and there staying it, marke what

degree

degrée of the Equinoctiall riseth therewith, which you shall finde to be 12. degrées 4′ 8. and hauing marked that degrée, put forward the foresaid signe Taurus towards the West vntill the last degrée thereof be ascended vp to the vpper edge of the Horizon, and then make another marke vpon the point of the Equinoctiall, which riseth at that instant with the last degrée of Taurus, which you shal finde to be 29. degrées 4′ 2. and by counting the degrées contained in the Equinoctiall betwixt the two markes, or by taking the lesser ascention out of the greater, you shall finde the number of degrées to be 16. and 5′ 4. whereby you may conclude that the ascention of Taurus in that Latitude is oblique, and that he spendeth in his rising one whole houre 7′. 3″ 6. And looke what order is hére taught to finde out the ascention of any signe, the same order is to be obserued for the finding out of the descention of any signe, sauing that you must séeke for the descention of any signe in the West part of the Horizon of the Spheare or Globe, and not in the East part. As for example, if you would knowe what descention Taurus hath, and in what time he descendeth in the foresaide Latitude: here hauing brought the first degrée of Taurus to the West part of the Horizon, so as it may touch the vpper edge thereof, and hauing also marked what point or degrée of the Equinoctiall toucheth the same Horizon at that instant, which you shal find to be 42. degrées 3′ 0. cease not to turne the Spheare or Globe, vntill all the whole signe of Taurus be descended vnder the Horizon, and that the last degrée thereof doe méete iust with the vpper edge of the Horizon, and there stay it vntill you haue againe marked that point of the Equinoctiall which toucheth the Horizon at that instant, which you shall finde to be 84. degrées 5′ 4. and by counting the degrées contained betwixt the two markes on the Equinoctiall, you shall find the number of degrées to be 42. degrées 2′ 4. so as you may conclude that the descention of Taurus in that Latitude is right, and that he spendeth in his going downe two houres 4′ 8.

How shall I knowe the right or oblique ascention of any of the fixed starres, and also at what houre of the day or night they rise & set, and how long they abide aboue the Horizon: finally when they are at the highest, and when they are at the lowest, called the depression or lowest Meridian Altitude of the

The first Booke

the starres?

All these things are most truely knowne by Tables calculated of purpose, and also they are to be knowne by helpe of the celestiall Globe in such manner as shall be declared hereafter when we come to treate of the said Globe.

Of the ascentionall difference and vses thereof.

Chap. 34.

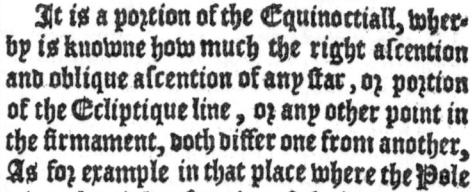

Hat is the ascentionall difference?

It is a portion of the Equinoctiall, whereby is knowne how much the right ascention and oblique ascention of any star, or portion of the Ecliptique line, or any other point in the firmament, doth differ one from another, As for example in that place where the Pole is eleuated 52. degrees, the right ascention of the first point of Taurus is 27. degrees and 5′4. and the oblique ascention of the same point is 12. degrees 4′8. here by taking the lesser out of the greater, that is 12. degrees 4′8. out of 27. degrees and 5′4. there will remaine 15. degrees and 6′. which is the ascentionall difference.

What vses hath the ascentionall difference?

The ascentionall difference being knowne, all the oblique ascentions and descentions of the starres are easily knowne by the Tables of directions, againe by this difference is knowne the increase and decrease of the artificiall day in euery Latitude, and therefore it is called of some incrementum diei. Moreouer it sheweth the semi-diurnall Arke of the artificiall day, for in euery oblique Spheare the artificiall day is alwayes either longer or shorter then the Equinoctiall day throughout the yeare, vnlesse the sunne be in either of the Equinoctiall points.

How is the increase or decrease of the day to be knowne by the ascentionall difference?

That shall be declared hereafter in the 50. Chap. of this first booke, whereas wee treat of the artificiall day and night, in the meane time we will speake somewhat of the poeticall rising and setting of the starres.

Of

Of the Poeticall rising and setting of the starres.

Chap. 35.

Define what the Poeticall rising and setting is.

The Poeticall rising is the appearing of some starre aboue the Horizon, determined by the sunne, and the Poeticall setting, is either the going downe of some starre vnder the Horizon, or else the hiding thereof vnder the beames of the sunne.

How manifold is the Poeticall rising and setting?

Threefold, that is, Cosmicus, Acronicus, and Heliacus, the signification of which wordes shall appeare vnto you by the definitions of the foresaid three kindes here following: For ortus Cosmicus, called in Latine mundanus, which is as much to say here as the worldly or morning rising, is when any starre riseth in the morning aboue the Horizon, together with the sunne, or rather with that point of the Ecliptique line wherein the sunne is at that time. And the Cosmicall setting, called in Latine occasus Cosmicus, is when a starre goeth downe vnder the Horizon at such time as the sunne riseth, so as this kind of rising and setting is wholly to be referred to the rising of the sunne.

What is *Ortus* and *occasus Acronicus*?

Ortus Acronicus which is as much to say as the Euening or temporall rising, is when any starre riseth aboue the Horizon in the Euening at the going down of the sunne: And occasus Acronicus, that is to say, the Euening setting is when any starre goeth downe vnder the Horizon, together with the sunne, and therefore this kind is alwaies to be referred to the going downe of the sunne, and not to his rising: And whatsoeuer signe or starre doth rise Acronicè, the same goeth downe Cosmicè, and whatsoeuer starre doth rise Cosmicè, the same goeth downe Acronicè. And generally all starres that rise in the day time, are saide to rise Cosmicè, and all those that rise in the Euening after the sunne set, are said to rise Acronicè.

The first Booke

What is *Ortus & occasus Heliacus*.

Ortus Heliacus, that is to say, the Solar rising, is when any star by departing from the beames of the Sun appeareth, & may be séen, which before being darkned by the sunne could not be séen. And occasus Heliacus, is when any starre by the nigh approching of the sunne ceaseth to bee séene, for by reason that the sunne by his yearely course & oblique motion of the Ecliptique, doth sometimes approch to diuerse starres, and sometime by little and little retireth backe againe from the same, it falleth out that those stars to whom he approcheth, are by nighnesse of his great light, darkned and not séene, and by his departing from them, and especially when the sunne is in the East or West part of the firmament, they beginne againe to be séene. And therefore as in the other 2 kinds, the Horizon together with the rising and setting of the sunne, are to be considered as chiefe causes thereof, so in this last kinde the chiefe cause is to be referred to the highnesse or farrenesse of the sunne from the starre.

Whereto serueth the knowledge of this threefold Poeticall rising and setting of the starres?

It serueth chiefely to vnderstand thereby those Poets and Histroriographers, which in shewing the time of any act done or to be done, doe not set downe the day of the moneth, but are wont to describe the time by the rising or setting of some notable starre, which they thinke most méete for their purpose, and thereby doe greatly adorne their stile, and specially being poeticall: And because that the times wherein such starres did rise or set, do greatly differ in these dayes from the auncient times. Many therefore of our moderne writers, as Garceus and others, haue made diuerse Tables of purpose to finde out the difference, and thereby to come to the true knowledge of the times by the auncient men described, of which matter I leaue to speake, thinking it not méete to trouble young Saplers therewith, for whom I chiefely wrote this Treatise of the Spheare.

Yet some affirme that the auncient men did vse the foresaid poeticall rising and setting of certaine starres, and specially of the *Pleiades, Hiades, Orion, Arcturus, Capella,* and *Lira,* (which starres were to them best knowne) as a Kalender not onely to know thereby the difference of times, and seasons of the

the yeare: but also by their manner of rising, setting, hiding, and appearing to prognosticate and to fore-see tempests and stormes, yea and that in these daies wee also (as some write) might doe the like, though there were neither Kalender nor Ephemerices, and in that respect the knowledge hereof seemeth most necessarie for Mariners.

All such things are to be knowne more exactly by the Astronomicall ascention and descention, then by the Poeticall rising or setting of the starres: And you haue to vnderstande that the starres since those dayes haue changed their places, their longitudes, and declinations, and thereby in diuerse respects haue altered their Natures and qualities, yea and the very signes themselues: As for example, neither Taurus, Gemini, nor Cancer, is so hoat and dry now, as in times past, neither doth Scorpio cause so much thunder now, as in times past, some againe are more or lesse cold and moyst then they haue béene heretofore, the causes whereof I leaue to the discussing of the Astrologers, and so once againe ende with this matter.

Of time, what it is, and into what parts it is deuided.

Chap. 36.

MOst men that write of the Spheare, after they haue spoken of the ascentions, doe immediatly treate of the diuersitie and inequalitie of dayes and nightes, but sith dayes, nights, and houres, are but parts of time, like as be wéekes, moneths, and yeares, I mind here therefore first briefly to treat of time, and then of all his chiefest parts in order, for if you will be instructed at large of these matters, then read the booke of Iohannes de sacro Busto de anni ratione, and also Iohannes Garceus his booke de tempore.

How define you time?

Leauing to speake of time, without time, that is to say euerlasting and infinite, called of the Latines Eternitas, ascribed

chiefely

The first Booke

chiefely to God, & therefore not contained within the mooueable Spheares or heauens: I mind to speake here onely of that time which is a number measuring the mouing of the first mooueable, and of all other mutable thinges, which time had his beginning with the world, and shall ende with the same, and this time consisteth of two parts, that is first, and last, or rather before or after, successiuely following one another, and these two partes are knit together with a common bound called of the Latines Nunc, that is to say now, or at this present, which is the end of that which went before, and the beginning of that which followeth after, and therefore some doe deuide time into thre parts, that is, time past, time present, and time to come, but the time present is a moment indiuisible, and is the beginning of time, euen as a point or pricke is the beginning of all Magnitudes, & yet least part therof it selfe: Againe time is deuided of some into greater and lesser parts, the greater are such as these: Kalendes, Nones, Ides, a wéke, a month, a yeare, the space of fiue yeares, called of the Romaines Lustrum, and of the Gréeks Olympias, the Romaines did call it Lustrum a lustrando, that is to say, of going about, because that they vsed in the end of euery fiue yeares, with lights and torches of waxe to goe in precession round about the Citie, and did purge the same by sacrifising a Dogge, a Sowe, and an Oxe, and at that time also they did chuse their Dictator in a place called the fielde of Mars, but the space of fiue yeares called Olimpias, tooke his name of the high mount Olympus in Greece, whereas in the end of euery fiue yeares were celebrated all kind of martiall playes, as Fencing, Wrestling, Running, and such like in the honour of Iupiter Olympicus, also the space of 15. yeares called indictio, in which space those forraine Nations that dwelt farre off, and were tributary to the Romaine Empire, payd their tributes, that is to say, in the first fiue yeares they payde onely gold, in token of their obedience to the Empire: In the second fiue yéeres they paid siluer for Souldiers wages, and in the last fiue yeares they paide brasse towardes the reparation of armour and munition. Item the space of an hundred yeares, called in Latine seculum, and in English an age wherof the playes that were celebrated in Rome euery hundred yeare, were called Ludi seculares, and last of all the space of a thousand yeares, called æuum, contayning tenne ages,

of the Spheare. 168

ages, againe the lesser partes (as Iohannes de sacro Busto saith) are these fiue, the first is called in Latine quadrans, which is the fourth part of a day, that is sixe houres: The seconde punctus, which is the fourth part of an houre in the sunnes account, but in the Moones account the fift part of an houre: The third is called momentum, which is the tenth part of punctus: The fourth is called vncia, which is the twelfth part of momentum: The fift is called Atomus, which is the 48. part of vncia. But because in all the greater partes of time, there is no greater variation or difference, then in that which in Latine is called annus, and in English a yeare. I minde here therefore first to treat of a yeare, and then of monethes, weekes, dayes, nights and houres.

Of the yeare, and of his diuerse kindes, and of the diuerse computations had thereof in diuerse ages, and amongst diuers Nations.

Chap. 37.

BE there diuerse kindes of differences of yeares?

Yea indeede, but I will speake here onely of three kindes or differences, that is of the great yeare, the Solar yeare, and the Lunar yeare, whereof the two last are most necessarie for our purpose.

What is the great yeare?

The great yeare is a space of time in the which not onely all the Planets, but also all the fixed starres that are in the firmament, hauing ended all their reuolutions do returne againe to the selfe same places in the heauens, which they had at the first beginning of the world: And therefore it is called of some the yeare of the world, and of some the great yeare of Plato, which contayneth according to Alphonsus, 49 {000}. yeares, whereof we haue spoken before, yet some affirme that the perfect yeare of the worlde containeth but 36000. yeares, whose reuolution is after one degrée in 100. yeares, but leauing this matter as not greatly profitable, we will speake now of the yeare Solar.

Of

The first Booke

Of the sunnes yeare called in Latine *annus solaris*, and of the diuerse kinds thereof, and first of the Tropicall yeare, both equall and vnequall.

Chap. 38.

Hat is the yeare Solar?

It is that space of time in which the sunne departing from any point of the Ecliptique line, or from some fixed starre of the Zodiaque, goeth round about the Zodiaque by his owne proper moouing, which is from West to East, and so returneth againe to the selfe same point or starre from which he first departed, and the Astronomers doe make diuerse diuisions of the Solar yeare, for first they say that it is eyther Astronomical or Politicall: Secondly that the Astronomicall yeare is eyther Tropicall or Syderall. Thirdly that the Tropicall is either equall or vnequall, which vnequall yeare they otherwise call the apparant yeare, and true yeare, all which kindes haue in a manner one selfe definition, sauing that the Tropicall yeare taketh his beginning from the vernall Equinoctiall point, and the Syderall yeare from the former starre of the Rams horne, and doe differ chiefely in quantitie.

Shew then what quantitie, that is to say, how many daies, houres, and minutes euery such yeare conteyneth.

The equall Tropicall yeare being counted alwayes from the middle point of the vernall Equinoxe, contayneth 365. dayes, fiue houres, 4′9. 1″5. and 4‴6. But the vnequall or apparant Tropicall yeare conteyneth sometime more and sometime lesse then the equall yeare, for sometime besides the 365. dayes and fiue houres, it amounteth to 5′6. 5″3. and ‴. so as it is more then the equall Tropicall yeare by 7′. 3″7. and 1‴5. and sometime ouer and besides the foresaid 365. dayes and fiue houres, it onely contayneth 4′2. 3″8. and 2‴7. which is lesse then the equall Tropicall yeare by 6′. and ‴. which inequalitie chiefely chanceth by reason of the vnequall precession of the two Equinoctiall points before defined in the 26. Chapter.

of the Spheare.

Of the *Syderall* yeare, and how much it containeth.

Chap. 39.

WHat is the Syderall yeare?
The Syderal or starry yeare is that space of time wherein the sunne walking vnder the firmament, departeth from the first or foremost starre of the Rams horne, and returning to the same starre againe, which space of time alwayes and equally contayneth 365. dayes, sixe houres, 9′. and 3″9. so as this yeare is alwayes greater then the Tropicall yeare, and by his equalitie doth alwayes rule and rectifie the inequalitie of the Tropicall yeare.

Of the Politicall yeare, and diuerse kinds thereof.

Chap. 40.

WHat is the Politicall yeare?
It is a yearely space of time which any people or nation attributeth to the course of the Sunne or of the Moone or of eyther of them which is diuerse and manifold, according to the diuerse customes of the Nations, of all which I meane not to speake at this present particularly, but of certaine speciall and necessary to be knowne, as of the Julian yeare, of the Egyptian yeare, of the Jewes yeare, and of the Athenian yeare.

Of the Iulian yeare, and why it is so called.

Chap. 41.

WHat is the Iulian yeare?
The Julian yeare is that which we vse at this present day which of all other yeares draweth nighest vnto the Tropicall yeare, for this yeare consisteth of

Z   365.

The first Booke

365. dayes and sixe houres, which sixe houres, if it should be reckoned euery yeare, it would make a great confusion, and therefore it is reckoned at the end of euery fourth yeare, which yeare consisteth of 366. dayes, for foure times sixe doth make 24. houres, which is one whole naturall day, whereof that yeare is called the leape yeare, And thereby the Iulian yeare is sayd to be two fold, that is common, contayning 365. dayes, and the other bissextile or leape yeare, contayning 366. dayes. This worde bissextile is deriued of bis and sextus, because the sixth day next before the Kalends of March is twice repeated or reckned, which indeede is the 25. of Februarie, vpon which day the feast of Saint Mathias commonly falleth.

Why was it called the Iulian yeare?

Because Iulius Cæsar the first Monarch of the Romaine Empire caused the yeare (accor ding to the course of the Sunne) to be reduced to the number of dayes and houres before expressed, who brought an excellent Astronomer with him at his comming from Egypt, aswell for that purpose as to teach the Mathematicall disciplines vnto the Romaines, yet you haue to consider that the Iulian yeare being greater then the Tropicall yeare, doth cause great diuersitie in that it maketh aswell the Equinoctiall & Solsticiall points, as also the entrance of the Sun into the other signes by little & little to anticipate or to runne before, for whereas in Iulius Cæsars time the vernall Equinox was the 23. day of March, the same Equinox is now about the 11. day of March, which is sooner by 12. dayes.

Of the Egyptian yeare, and how many daies it contayneth.
Chap. 42.

The Egyptian yeare contayneth the iust number of 365. dayes, by reason of which equalitie this yeare is very fit to serue the Astronomers turne in making their Astronomicall computations, but the Egyptian yeare hath no certaine place of beginning: For by omitting the sixe houres which is in the Iulian yeare, it doth anticipate in the space of 4. yeares

yeares one whole day in such sort, as 1460. Julian yeares doe make of the Egyptian yeares. 1461.

How many Moones the Iewes yeare, and the Athenians yeare doth containe.
Chap. 43.

He Iewes yeare contayneth for the most part twelue Moones, and sometimes thirtéene Moones, which kinde of yeares did agrée with the yeares of the Greekes, and of the Athenians, and also of the auncient Romaines before Iulius Cæsars time: and the auncient Romaines did beginne their yeare from March, but the latter Romaines from the winter Solstice. Againe the Iewes did beginne their yeare at the first new Moone that followed next after the vernall Equinox: But the Athenians began their yeare at the new Moone that followed next after the Sommer Solstice: The most people of Asia began their yeare at the Autumnall Equinox: But the most part of those that dwel in these partes of the world, following the custome of the Romaine Church, doe beginne their yeare at the Kalends of Ianuary, which in old time was not much distant from the Winter Solstice, which Solstice at Christs birth was the 25. of December, but nowe the same Solstice is about the twelfth day of December, so as the Winter Solstice falleth sooner by thirtéene dayes then it did at that time. But we here in England do begin the yeare at the 25. of March.

Of the yeare Lunar, and of the kinds thereof.
Chap. 44.

Ow many kindes of Lunar yeares be there, and which be they?

Of Lunar yeares there be two kindes, whereof the one is ordinary called in Latine Annus communis, and the other extraordinarie or excessiue, called by a Gréeke name Embolismalis, The ordinary

dinary or common yeare, is the space of twelue Moones or changes, passing by course within the yeare Solar, and is called common because it hath twelue Moones Lunar, euen as the Solar yeare hath twelue monthes Solar, and consisteth of 354. dayes and a little more, so as the Solar yeare excædeth the Lunar yeare by 11. dayes, for the yeare Solar conteyneth (as hath bæne saide before) 365. dayes, in which account the Fractions in both yeares are omitted: And therefore if these two yeares should begin together at one selfe time, the Lunar yeare would end his course sooner by 11. dayes then the yeare Solar.

What is the extraordinarie Lunar yeare called Embolismalis?

It is the space of thirtéene Moones or changes contayning 384 dayes, so as this yeare excædeth the common Lunar yeare by 30. dayes, and is more then the yeare Solar by 19. dayes.

Of the diuers kinds of monethes, and into what parts euery Solar moneth is deuided according to the Romanes, that is, into Kalends, Nones, and Ides.

Chap. 45.

How many kindes of monethes be there, and which be they?

There be thrée kinds, that is, the month Solar, the moneth Lunar, and the monet Usuall.

The moneth Solar is that space of time which the Sunne spendeth in passing through any one of the 12. Signes.

The Lunar moneth is that space of time which the Moone spendeth whilest she departing from the Sun returneth to him againe.

The Usuall moneth is that number of dayes which are set downe in our common Kalenders, whereof some containe thirtie dayes, some thirtie and one, and the moneth of February hath but 28. dayes. But if you will readily know which containe more dayes and which lesse, kéepe alwaies in memorie these old English verses here following.

Thirtie

of the Spheare.

Thirtie dayes hath Nouember,
Aprill, Iune, and September.
Februarie hath 28. alone,
And all the rest haue thirtie and one.

But when it is leape yeare, February hath 29. dayes, Againe the Romaines deuided the Solar moneth into Kalends, Nones, and Ides.

What be Kalends?

The Kalends are the first day of euery moneth from which the Romaines counted the daies of the moneth procéding backward, As for example the first day of Aprill they named the Kalends of Aprill, and the last day of March next before they called in Latin pridie Kalendas Aprilis, that is the day before the Kalends of Aprill, and the next day before that, the third Kalends of Aprill, and the next day before that the fourth Kalends, and so foorth vntill they come to the Ides.

Whereof sprange this name Kalends?

Of the Grǽke verbe Calo, which is as much to say as call, for the first day of euery moneth the cryer standing in a high place made foure cals or more to signifie thereby to the people howe many dayes in that moneth the faires or markets called Nundinæ should endure, and of Nundinæ sprange this word Nonæ, that is to say the dayes of the faires: For looke howe many Nones there were in euery moneth, so many faires there were, during which time the Romaines neuer worshipped any God because there was no holy day during that time, and therefore Ouid saith that Nonarum tutela Deo caret, that is to say, no God had tuition of the Nones.

What are Ides?

They are those dayes by which the Nones are deuided from the rest, and these Ides doe deuide in a manner the whole moneth into two equall parts, for the first Ides most commonly falleth eyther on the 13.14. or 15. day of the moneth.

How many Ides, Nones, and Kalendes doe belong to euery moneth?

Of Ides euery moneth hath eight, but of Nones March, May, July, and October, haue sixe, and all the rest of the moneths haue

Z 3 but

The first Booke

but foure Nones, but they differ most in the number of Kalends, as you may perceiue by this Table following, which sheweth how many Kalendes, Ides, and Nones, doe belong to euery moneth. Thus farre of the moneth Solar, now I will speake of the moneth Lunar.

The Table.

Moneths.	Kal.	Ide.	None.	Moneths.	Kal.	Ide.	None.
Ianuarie.	19	8	4	Iuly.	17	8	6
Februarie.	16	8	4	August.	19	8	4
March.	17	8	6	September	18	8	4
Aprill.	18	8	4	October.	17	8	6
May.	17	8	6	Nouember.	18	8	4
Iune.	18	8	4	December.	19	8	4

Of the diuerse kindes of moneths Lunar.

Chap. 46.

How many kindes be there, and which be they?

Iohannes de Sacro Busto sayth that there be foure kindes, that is the moneth of paragration, the moneth of apparation, the moneth medicinall, and the moneth of consecution.

The moneth of Paragration, is that space of time in which the Moone departing from any one point of the Zodiaque, goeth by her proper moouing about the Zodiaque, and returneth againe to the said point from which she first departed, which her reuolution is accomplished in 27. dayes, & 8. houres. And this reuolution of some is called a yeare, and by this account the Moone tarrieth in euery signe two dayes sixe houres and 2′9.

The moneth of Apparation consisteth of 28. dayes, deuided commonly by foure weekes, euery weeke contayning seuen daies,

for

for foure times seuen maketh 28. of which foure wéekes the first is counted from her first apparance vnto the ende of the seuenth day, and so foorth from wéeke to wéeke, so as the fourth wéeke endeth at the 28. day, in which account the odde houres during the Moones aboue vnder the beames of the sunne, when as she is said to be combuste, are not reckned.

The moneth Medicinall contayneth but 26. dayes, and a halfe (as Galen sayth) and is deuided also into foure wéekes, the diuision being made by minutes.

The moneth of Consecution is that space of time wherein the Moone being in coniunction with the sunne, goeth about her Circle and returneth againe to the same point, and not finding the sunne there because he hath in that while passed through one whole signe, she hasteth after and in two dayes and foure houres 44 and a little more she ouertaketh the sunne, and is againe with him in coniunction, of which her following and ouertaking, the sunne this moneth is called the moneth of Consecution, which moneth consisteth of 29. dayes and a halfe: during which time as the sunne by his owne proper course passeth through one signe or there abouts, so the Moone by her course in the selfe same time passeth through the whole Zodiaque and one signe more: And note by the way that the sunne in making his owne proper course doth not enter into any signe in the very beginning of any moneth, but rather about the midst of euery moneth, or at the least not much ouer or vnder that day

Of a weeke.

Chap. 47

Now because that euery moneth aswell Solar as Lunar is deuided into foure wéekes, I will speake somewhat of a wéeke.

What is a weeke?

A wéeke called most commonly in Latine Septimana, which is as much to say as seuen mornings, is the space of seuen dayes, whereof the first is called Sunday, the second Munday and so foorth to Saturday,

turday, which names the Gentiles gaue to these seuen dayes in honour of the seuen Planets whom they worshipped as Gods, for they called the first day the day of the sunne, the second the day of the Moone, the third the day of Mars, the fourth the day of Mercury, the fifth the day of Iupiter, the sixth the day of Venus, the seuenth the day of Saturne.

Are there any more names belonging to this word weeke?

Yea, it is called also by a Græke name Hebdoma, that is to say, contayning seuen dayes, and in the scripture it is called sometime Sabbatum, as when the Pharisie sayd that he fasted Bis in Sabbato that is to say, twice a weeke, for the Iewes called Sunday the first of the Sabaoth, and Munday the second of the Sabaoth, and so foorth in order vntill Saturday, which in deede was their true Sabaoth or day of rest.

Why is it not still so counted amongst vs Christians, but changed into Sunday?

For two causes, first to auoide the superstition of the Iewes, and partly it was done in the honour of Christ, whose day of birth, resurrection, and sending of the holy Ghost, was on the Sunday.

Of dayes and nights both naturall and artificiall.

Chap. 48.

The Astronomers doe deuide the dayes into two kindes, whereof the one is called naturall, and the other artificiall.

Which call you a naturall day?

A naturall day is one entire reuolution of the Equinoctiall about the earth, whereunto must be added such portion of the Zodiaque, as the sunne in the meane while maketh by his proper motion, which is from West to East.

In what time doth the Equinoctiall euery day make his reuolution?

In 24. houres, which space contayneth both day and night, according to which reuolution and number of houres, the most part of Horologies or clockes in the East countryes doe goe, and are set to shew the houres of the day, but yet diuersly, for some begin
their

their naturall day at the rising of the sunne, as the Bohemians, and the Persians, and some at the going downe of the sunne, as the Italians, the Athenians, the Iewes, the Egyptians, and the Arabians, but the Astronomers reckon their naturall day from noonetide to noonetide.

If the Equinoctiall doth make his reuolution iust in 24. houres, then all naturall dayes are equall?

That is true if ye onely consider the motion of the Equinoctiall, but if you adde thereunto (as I saide before) that portion which the sunne in the meane while maketh by his owne proper motion, you shall finde them to be vnequall, because that portion is sometime more, sometime lesse, according to the swift or slowe ascention of the signe wherein the sunne is.

What is the artificiall day?

It is the distance or space that is betwixt the rising of the Sun, or going downe of the same.

The causes why the time betwixt the rising & going downe of the sunne is vnequall.

Chap. 49.

It must nædes be vnequall, because the abode of the sunne aboue the Horizon is variable, aswell for the vnequall ascention and descention of the signes, as also for the obliquitie of the Horizon and Zodiaque: and therefore the spaces of the artificiall dayes must nædes be vnequall: for the sun aswell in ascending from the beginning of Capricorne, to the beginning of Cancer, as also in descending from the beginning of Cancer to the beginning of Capricorne, describeth on ech part 182. Circles or Paralels, the middlemost whereof is the Equinoctiall. All which Paralels are deuided into two parts by the Horizon, and the Arkes which are aboue the Horizon are called the artificiall daies, and the Arkes beneath the Horizon are called the artificiall nights, which Arkes to those that dwell vnder the Equinoctiall in a right Spheare are alwaies equall, that is to say, the diurnall Arke is equal to the nocturnal,

because

The first Booke

because their Horizon passeth through the Poles of the world, but to those that dwell in an oblique Spheare, aboue whose Horizon the Pole is any thing eleuated, be it neuer so little, the Arkes or Paralels are vnequall one to another, that is to say, eyther making short daies and long nights, or else long dayes and short nights, the Equinoctiall onely excepted, which aswell in the oblique Spheare as in the right, is alwayes deuided by the Horizon into two equall partes, and so maketh the dayes and nights equall in all places of the world. All which thinges you shall easily comprehend by these two figures following, whereof that on the left hande representeth the right Spheare, and the other on the right hande representeth the oblique Spheare.

A figure of the right Spheare.   A figure of the oblique Spheare.

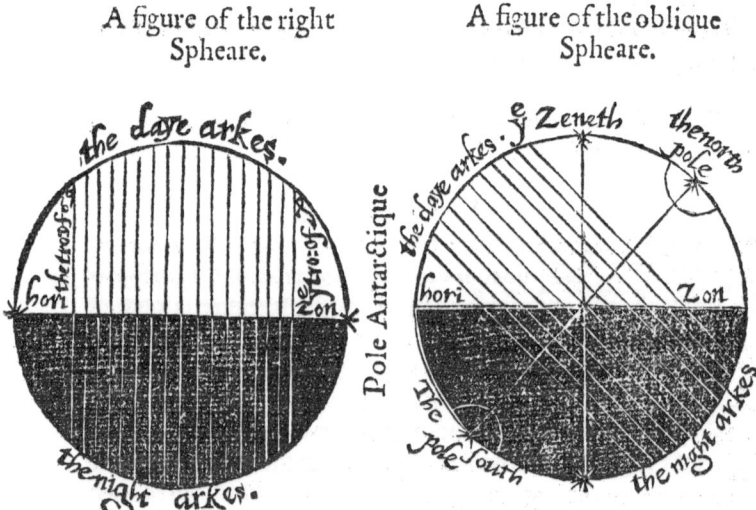

And here I thinke it not amisse to shewe you howe to finde out the length of euery artificiall day and night throughout the yeare in euery Latitude by the materiall Spheare, and likewise howe to know at what part of the Horizon the sunne riseth and setteth euery day, and also how to finde out his Meridian Altitude, wherby you shall know how nigh or how farre from your Zenith hée is euery day.

How

of the Spheare. 174

How to find out by the materiall Spheare or Globe, and by helpe of the afcentionall difference before defined, the increafe and decreafe of euery day throughout the yeare in euery feuerall Latitude, and at what houre the funne rifeth and fetteth.

Chap. 50.

Irſt hauing ſet the Spheare at your Latitude, learne to knowe by ſome Table or inſtrument in what ſigne and degrée thereof the ſunne is at ſuch day of the moneth and yeare as you ſéeke, and bring that degrée cloſe to the moueable meridian, & there marke in what point or degrée the ſaid Meridian cutteth the Equinoctiall, & what number it hath, for ſo ſhall you haue the right aſcention of the degrée of the ſunne for that day. That done, bring the ſaid degrée of the ſunne to the Eaſt part of the Horizon, ſo as it may méete euen with the vpper edge thereof, and ſtaying it there, marke what degrée of the Equinoctiall at that inſtant doth alſo touch the Horizon, and what number it hath, and that is the oblique aſcention of the foreſaid degrée of the ſunne, then ſubtract the leſſer number out of the greater, and that which remaineth ſhall be the aſcentionall difference, as for example in this preſent yeare 1590. the xi. day of April the ſunne is in the firſt degrée of Taurus, whoſe right aſcention in the Latitude 52. by doing as before is taught, you ſhall finde to be 27. degrées and 5′ 4. and the oblique aſcention thereof to be 12. degrées 4′ 8. and the aſcentionall difference to be 15. degrées 6′. which difference you muſt firſt double, and then conuert the ſame into houres and minutes by allowing 15. degrées to an houre, and 4′. to a degrée, that done, adde thoſe houres & minutes to the Equinoctiall day, which is alwaies 12. houres, & the ſumme of that addition will ſhew you the length of the day, when the ſunne is in the firſt point of Taurus, which is 14. houres & 1′ 2. and that is the true length of that day. But you haue to note by ỹ way that as you haue to adde the ſum of ỹ houres to ỹ Equinoctial day, the ſunne being in any of the 6. Northerne ſignes, ſo muſt you ſubtract the ſaid houres from the Equinoctiall day, the ſunne being in any of the ſixe Northerne ſignes, As for example, ſuppoſe the ſunne to be in the firſt point of Scorpio,

the

The first Booke

the right afcention whereof is 207. degrees 5′4. and the oblique afcention is 223.degrees, here by taking the leſſer out of the greater you ſhall finde the afcentionall difference to be 15. degrees 6′. which being doubled, maketh 30. degrees and 1′2. that is two houres and 1′2. which if you ſubtract from 12. there will remain 9. houres 4′8. which is the length of the artificiall day, when the ſunne is in the firſt degree of Scorpio, the one halfe whereof is called the ſemi-diurnall Arke of that artificiall day, which is 4. houres 54. minutes, whereby you may gather that the ſunne at that time riſeth foure minutes after ſeuen of the clocke, and ſetteth againe foure minutes before fiue: ſo likewiſe in the former example where the afcentionall difference was 15. degrees ſixe minutes, which being doubled made 30.degrees and 1′2. that was two houres and 1′2. which being added to 12. houres made 14. houres and 1′2. the halfe whereof is ſeuen houres and ſixe minutes, whereby you may gather that the ſunne did then riſe 6′. before 5. of the clocke in the morning, and did ſet 6′. after 7.of the clocke in the Euening, and ſo you haue both the forenoone & afternoone of the day, which two times are beſt to be reckoned alwayes from 12. a clocke at noone, that is to ſay, the forenoone houres and minutes, from twelue backward, and the after moone houres and minutes from twelue forward, & by ſubtracting the whole length of the artificiall day from 24. houres, you ſhall haue the length of the artificiall night, as in the former example take 14.houres and 1′2. from 24. houres, and there will remaine for the length of the artificiall night 9.houres and 4′8.

How to know by the materiall Spheare or Globe, in what part of the Horizon the ſunne riſeth and ſetteth euery day, and thereby the length of the day. Alſo how to know the Meridian altitude of the ſunne euery day throughout the yeare, & being at his Meridian altitude, to know how farre diſtant he is from the Zenith euery day.

Chap. 51.

Firſt then hauing ſet the Spheare or Globe at your Latitude, which ſuppoſe to be 52. degr. bring the degree of the ſun for that day, to the Eaſt part of the Horizon, ſo as it may meete iuſt with the upper edge therof, and there

of the Spheare.

there set a little péece of waxe vpon the Horizon, that done turne the said degrée of the sunne to the West part of the Horizon vntill it méete againe with the vpper edge of the Horizon, and there set another péece of waxe vpon the Horizon, and those two markes will shewe you in what part of the Horizon the sunne riseth and setteth: As for example, I would know this present yeare 1594. in what part of the Horizon the sunne doth rise and set the twelfth of June, here finding by the Ephemerides that the sun is entred 16. into Cancer at that day, I bring that point of the Ecliptique to the East part of the Horizon, so as it may méete iust with the vpper edge thereof, and there I set a little péece of waxe vpon the Horizon: that done I turne the said point of the Ecliptique to the West part of the Horizon, and whereas that point toucheth the vpper edge of the said Horizon, I set there another péece of waxe vpon the Horizon, then counting the degrées vppon the Horizon from the true East point thereof to the first péece of waxe Northward, I finde the number of the degrées to be 40. degrées or therabouts, which containeth thrée points and a halfe and somewhat more, of the Mariners Compasse, whereby I gather that the sunne riseth at that time very néere to the Northeast, and setteth néere to the Northwest, because the number of the degrées in both parts of the Horizon are like.

How shall I know how many points of the Mariners compasse are contained in any number of degrees, exceding the number of degrees and minutes contained in one point?

The Mariners Compasse containeth 32. points, and euery point containeth 11. degrées and $\frac{1}{4}$. of a degrée which is 15. minutes, wherefore whensoeuer you would know how many points of the Compasse are contained in any number of degrées be it great or small, multiply that number by foure, and deuide the product thereof by 45. and the quotient will shewe the number of the points, and if there be any remainder, then because the Mariner doth make euery point to haue foure quarters, multiply that remainder by foure, and deuide the product by 45. which is the common diuisor, and the quotient will shew the quarters, or if the remainder be but small, then multiply that remainder by eight, which are halfe quarters, and deuide the product thereof by 45. as before, and the quotient will shew the halfe quarters of a point:

As

The first Booke

As in the former example in multiplying 40. degr. by 4. the product is 160. which if you deuide by 45. you shall finde in the quotient 3. whole points, & the remainder to be 25. which being multiplyed by 4. the product will be 100. which if you deuide by 45. you shall find in the quotient 2. quarters of a point, & the remainder to be $\frac{10}{45}$. of a quarter, that is to say, if you can deuide one quarter into 45. parts, then you must take 10. of those parts, and adde them to the former summe, which being of small importance is not to be regarded. But now to returne to my first matter, I say that by counting the degrees vppon the Horizon, from the first peece of waxe to the South point of the Horizon, I find the number of degrees to be 130. degr. & by allowing 15. degr. to an houre, I finde the halfe day to containe 8. houres 1′0. which being doubled maketh 16. houres 2′0. Then to know the Meridian altitude of the sunne, you must bring the degree of the sunne right vnder the Meridian, and the number of the degrees contained in the Meridian betwixt the South point of the Horizon, and the said degree of the sunne will shewe the Meridian Altitude of the sunne, so shall you find the Meridian Altitude of the sunne, being in the first point of Cancer to be 61. degrees, which once had, the distance from the Zenith is soone knowne: for if you subtract 61. degrees from 90. you shall find the distance of the sunne from the Zenith to be 29. degrees. Againe contrariwise the distance from the Zenith being subtracted from 90. the remainder will shew the Meridian Altitude, for those two numbers being added together, doe alwaies make a iust quarter of the great Circle, which is 90. degrees.

Of houres as well equall as vnequall, and into what partes they are deuided.

Chap. 52.

S of daies and nights there be 2. kinds, that is naturall and artificiall before defined, so likewise are there 2. sorts of houres, that is equal, and vnequall: An equall houre is the 24. part of a naturall day, and euery such houre containeth 15. degrees of the Equinoctiall, for 15. times 24. maketh 360. degrees, which is the whole

whole circuit or Longitude of the Equinoctiall, which according to the diurnall moouing of the first mooueable maketh his reuolution in 24. houres as hath béene said before, and therefore this equall houre is also called an Equinoctiall or naturall houre. The vnequall houre is the twelfth part of an artificiall day or artificiall night, which daies and nights as they be sometimes long and sometimes short, according to the time of the yeare, so are the houres of the same. For both day and night, be it neuer so short, is deuided by the Astronomers into twelue houres, so as when here in some part of England towards the North, the artificiall day is 17. houres long, and the night but seuen houres, if you deuide either of these into twelue parts, you shal find that euery such twelfth part of the day shall containe more then the naturall or Equinoctiall houre by 2′5. and the twelfth part of the said artificiall night to containe but the one halfe of an equall houre and fiue minutes, which notwithstanding being both added together, will make in all 24. equall houres, which is a naturall day, so as by this meanes you may easily perceiue that a naturall day comprehendeth both kindes of houres, aswell vnequall as equall. Note also that the vnequall houres are called sometime artificiall and sometime temporall houres, artificiall, because they are dayly changed by the varietie of the artificiall dayes and nights, and are neuer equall but twice in the yeare, when the sunne is in eyther of the Equinoxes: and they are called temporall, because the auncient obseruers of time were woont to make diuerse clockes and Horologies to shewe these vnequall and temporall houres, of which clockes there are yet some to be séene at this day. Moreouer the houres both equall and vnequall are deuided not onely into quarters of houres, but also into minutes, for euery houre be it long or short, is deuided into 60. minutes, and euery minute into 60. secondes, and euery seconde into 60. thirdes, and so forth to fourths, fifts, sixthes, and into as many as you will, so as you make your diuision alwayes by 60. But you haue to note, that as the Astronomers doe deuide the artificiall day and artificiall night into houres both equall and vnequal, so the Iewes doe deuide ech of them into foure quarters, in manner and forme following.

How

The first Booke

How and in what manner the Iewes doe deuide the artificiall day and night, ech of them into foure quarters.

Chap. 53:

Hey deuide the artificiall day into foure quarters by allowing to euery quarter 3.houres, accounting the first houre of the first quarter at the rising of the sunne, and the third houre of the said quarter they called the third houre, and the third houre of the second quarter they called the sixth houre, which was midday or noonetide, againe they called the third houre of the third quarter, the ninth houre, and they called the second houre of the fourth quarter the eleuenth houre, and they called the last houre which was the twelfth houre of the day, euentide because then the sunne went downe.

Whereto serueth the knowledge hereof?

By knowing this account you may the better vnderstande certaine places of the Scripture making mention of thinges done at certaine houres not like vnto our common houres, For whereas it is said in the Gospell of Saint Iohn, that Christ healed the Rulers sonne that was sicke of an Ague in Capernaum, and that the Ague left him the seuenth houre, is as much to say, as the Ague left him at one of the clocke in the afternoone: Againe where as mention is made in the Gospell of S. Matthew, of the labourers that came to worke in the Vineyard at the eleuenth houre is to be vnderstood at fiue of the clocke in the afternoone, or rather one houre before the sunne set, for you must thinke that the sunne riseth not nor yet goeth downe in Iewrie alwaies at sixe of the clocke, for then they should haue no artificiall day nor night, but all dayes and nights should be a like. Also they deuided the artificiall night into foure quarters, otherwise called by them the foure watches of the night, for the first three houres was the first watch, during which time all the Souldiers both young and old of anie fortified Towne were wont to watch, the second three houres they called the second watch, which was about the dead of the night, at which time the young Souldiers onely watched, and the third

quarter

of the Spheare.

quarter of the night containing also three houres, and was called the third watch the Souldiers of middle age did watch, and the last three houres called the 4. watch which was about the breake of day, the old Souldiers onely watched. But nowe because the auncient Astronomers doe appoint the gouernment of the vnequal houres to the seuen Planets, it shall not be amisse to shewe you here what Planet raigneth euery houre both day and night.

How to knowe what Planet raigneth in euery houre of the day or night artificiall, as well by helpe of a table, as by a rule contained in one verse.

Chap. 54.

But first I will describe vnto you the Table, and then briefly set downe the vse thereof. In the first collum of this Table on the left hand are set downe the seuen dayes of the wéeke, whereof the first is Sunday, and the second Monday, and so foorth downeward to Saturday, against euery which day the Planets are placed towardes the right hand, euery one in course one after another, and in the first rowe of this Table which is the head or front thereof, are placed the houres of the day written in Arithmeticall figures, and in the next rowe of the saide front, are set downe the houres of the night written as you sée in common numerall letters.

The Table.

Houres of the day.	1	2	3	4	5	6	7	8	9	10	11	12		
Houres of the night.	iij	iiij	v	vj	vij	viij	ix	x	xj	xij			i	ij
Sunday.	☉	♀	☿	☽	♄	♃	♂	☉	♀	☿	☽	♄	♃	♂
Munday	☽	♄	♃	♂	☉	♀	☿	☽	♄	♃	♂	☉	♀	☿
Twesday.	♂	☉	♀	☿	☽	♄	♃	♂	☉	♀	☿	☽	♄	♃
Wednesday.	☿	☽	♄	♃	♂	☉	♀	☿	☽	♄	♃	♂	☉	♀
Thursday.	♃	♂	☉	♀	☿	☽	♄	♃	♂	☉	♀	☿	☽	♄
Friday	♀	☿	☽	♄	♃	♂	☉	♀	☿	☽	♄	♃	♂	☉
Saturday.	♄	♃	♂	☉	♀	☿	☽	♄	♃	♂	☉	♀	☿	☽

The first Booke

The vse of the Table.

Now the vse of the saide Table is thus: whensoeuer you would knowe in what houre of the day or night any Planet raigneth, you must first seeke out the houre of the day or night, and if it be of the day, then you shall finde it in the first rowe of the front, if of the night, in the second rowe of the front, as hath bæne saide before: and from that houre descende with your finger to the common Angle standing right against the day which you seeke, and that will shewe you what Planet then raigneth. As for example, if you would know on Wednesday at 8. of the clocke of the day what Planet raigneth, then hauing found the number of 8. in the front, written in Arithmeticall figure, come straight downe from thence with your finger to the common Angle standing right against Wednesday, and you shall finde that Mercurie raigneth. And if you would knowe what Planet raigneth the same day at the eight houre of the night, then descende from the houre of the night downe to the common Angle, and you shall finde that the sunne raigneth, and so forth of all the rest.

The rule contained in one verse, and the vse thereof.

The rule in verse is thus:
Sol, Ve, Mer, Luna, Saturnus, Iupiter, & Mars.

These vii. wordes (the coniunction & being left out) doe signifie the seuen Planets: For Sol is the Sunne, Ve standeth for Venus, Mer for Mercurius, Luna is the Moone, and the other thræ Planets following, as Saturnus, Iupiter, and Mars, doe make vp the number of seuen, which must alwayes followe one another, in such order as they are here set downe in the foresaide verse, and to haue the true vse of this rule, you must first apply euery Planet to his owne proper day, as Sol to Sunday, Luna to Munday, Mars to Tuesday, Mercurius to Wednesday, Iupiter to Thursday, Venus to Friday, and Saturnus to Saturday: for euery one of these Planets gouerneth the first houre of his owne proper day, and the Planet placed next to him in the verse, gouerneth the second houre of the same day, and so forth orderly, as for example, if you would knowe what Planet shall raigne on Sunday at the third houre of the day, you must first say that Sol doth raigne

raigne the first houre because that is his day, and Venus raigneth the second houre, and Mercurie the third houre according to your rule, & so by keeping the order of the verse, you shall easily appoint to euery houre both of the day and of the night artificiall his owne gouernour: For though both day and night be deuided ech of them into 12. houres, making in all 24. houres, and that there be but seuen Planets, yet by appointing euery Planet to his owne proper day as gouernour of the first houre of the same day, and by obseruing the order of the verse in repeating the said Planets, you shall not faile to giue to euery houre his proper Planet. Thus hauing sufficiently spoken of the celestiall part, I will now proceede to the Elementall part of the worlde, contained in the second Booke of this Treatise.

The second part of the Spheare.

Of the Elementall part of the world.
Chap. 1.

Hat doth the Elementall part containe? I told you before that as the celestiall part doth containe the eleuen heauens before described, so the Elementall part containeth y̌ 4. Elements, that is to say, fire, Aire, Water, & Earth, which are of themselues pure substances, and the first & next beginnings whereof all mixt bodies are compound, and therefore not to be seene with our outward eyes: for as we our selues are bodies compound, so with our outwarde senses we can discerne nothing but that which is compound: and therefore the fire, aire, water or earth which we daily feele or see, are not the Elements themselues, but things compounded of them. The natures and properties of which Elements I mind here but briefely to touch, sith the exact handling thereof belongeth rather to naturall Philosophers & to Physitians, then to Geographers, who haue to deale only with y̌ situations of the earth with Zones, Paralels, Climes, Longitudes, Latitudes, distances & such like thinges belonging to the measure and description of this earth here which we inhabite.

Define these Elements.

Of the Fire and of his nature and motion.
Chap. 2.

He Fire is an Element most hoat and dry, pure, subtill, and so cleare as it doth not hinder our sight looking through the same towards the stars, and is placed next to the Spheare of the Moone, vnder the which it is turned about like a celestiall Spheare.

Aa 3 Of

The second Booke

Of the Aire and into how many Regions it is deuided.
Chap. 3.

Next to the Fire is the Aire which is an Element hotte and moyst, & also most fluxible, pure & cleare, notwithstāding it is farre thicker & grosser as some say, towards the Poles thē elswhere, by reason that those parts are farthest from the sun: And this Element is deuided of the naturall Philosophers into thrée Regions, that is to say, the highest Region, the Middle Region, and the lowest Region, which highest Region being turned about by the fire, is thereby made the hotter, wherein all fierie impressions are bredde, as lightnings, fire drakes, blazing starres and such like.

The middle Region is extreame cold by contra opposition by reason that it is placed in the midst betwixt two hotte Regions, and therefore in this Region are bred all cold watry impressions, as frost, snow, ice, haile, and such like.

The lowest Region is hotte by the reflexe of the sunne, whose beames first striking the earth, doe rebound backe againe to that Region, wherein are bred cloudes, dewes, raynes, and such like moderate watry impressions, which thrée Regions of the aire with the rest of the Elements this figure doth plainely shew.

of the Spheare. 180

Of the Water, and whether it be round or not.

Chap. 4.

Ext to the Aire, is the Water which is cold, moist, and fluxible, and being lighter then the Earth would of his owne nature surmount and couer the whole earth, had not God in the creation of the world deuided waters from waters (as the Booke of Genel. saith) and gathered together those waters that are vnder the firmament into certaine concauities of the earth, leauing other partes of the earth dry, and discouered, that man and beast might inhabite the same, & haue foode necessarie for their behoofe, so as now both water and earth doth make one entire and Sphericall bodie, which is enuironed with the Aire.

Is not the water a round bodie of it selfe without the earth?

Many late writers doe deny the whole bodie to be round, affirming onely the Conuexe superficies or vpper face of the water to be round, for (say they) the earth being not altogether round the Concaue superficies of the water cannot be round, notwithstanding the most part of the auncient writers do affirme the whol bodie of the water to be round, saying that the water hath the like shape in his whole, that it hath in his parts: For the parts which are drops, are round, ergo the whole is round. Againe they proue the water of the sea to be round by demonstration thus, suppose a ship to depart from the shore whereon some marke is set, which you may see with a right leauelled line standing at the stearne of the said shippe, but sayling further from the shore, you cannot see the marke any more standing vpon the sterne, but shall be faine to goe vp to the toppe of the mast to see it, by reason that the water being a round bodie riseth and swelleth in the midst, and so letteth your sight as this figure plainely sheweth.

Aa 4   Of

The second Booke

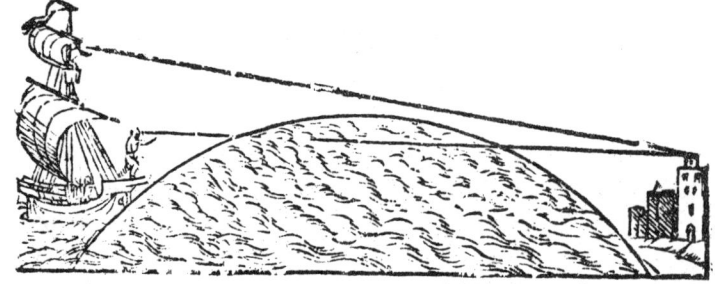

Of the Earth and whether it be all round or not.

Chap. 5.

NExt to the Water is the Element of the Earth, which of his nature is thicke, heauie, cold, dry, and not fluxible as is the water and aire, but is firme and apt to kéepe his place, and though some deny the earth to bee round because of the high mountaines, and déepe vales and vallyes therin, which are nothing in comparison of the whole earth to alter that roundnes which it hath by nature, yet Aristotle affirmeth in his second booke de coelo & mundo, the fourtéenth Chapter, that the earth of her own nature is round, prouing the same as well for that the Moone when she is eclipsed in part, could not haue such horned shape as this figure representeth.

Unlesse the earth were also round by the interposition whereof shée is eclipsed, either totally or in part as hath bén said before, Againe he prooueth the roundnes of the earth by the altering of the Horizon, for in going from North to South, our Horizon altereth in such sort as wee discouer those starres which we could not sée before, but were cleane hidden from our

our sight, some also deny that the earth is in the middest of the world, and some affirme that it is mooueable, as also Copernicus by way of supposition, and not for that he thought so in déede: who affirmeth that the earth turneth about, and that the sunne standeth still in the midst of the heauens, by helpe of which false supposition he hath made truer demonstrations of the motions & reuolutions of the celestiall Spheares, then euer were made before, as plainely appeareth by his booke de Reuolutionibus dedicated to Paulus Tertius the Pope, in the yeare of our Lord 1536. But Ptolomie, Aristotle, and all other olde writers affirme the earth to be in the middest, and to remaine vnmooueable and to be in the very Center of the world, proouing the same with many most strong reasons not néedefull here to be rehearsed, because I thinke fewe or none do doubt thereof, and specially the holy Scripture affirming the foundations of the earth to be layd Psal. 104. so sure, that it neuer should mooue at any time: Againe you shall finde in the selfe same Psalme these words, Hee appointed the Moone for certaine seasons, and the Sunne knoweth his going downe, whereby it appeareth that the Sunne mooueth and not the earth. But leauing this matter, we will now speake of the compasse of the earth, and of the Longitude and Latitude thereof.

Of the compasse of the earth, and of the diuersitie of measures according to diuerse countryes.

Chap. 6.

Can the whole earth be measured?

Yea very well, for sith the earth and the water (as hath béene said before) doe make together, one whole Sphericall or round bodie, and that euery great Circle as well thereof as of the heauens, containeth 360. degrées, there is no more then to be done, but to allow for euery such degrée 60. Italian miles, which differ not much from our English miles, so as in multiplying 360. degrées by 60. you shall finde the whole compasse of the earth to be 2160. miles, of which compasse if you would knowe the true Diameter, then hauing multiplyed
the

The second Booke

the said compasse or circuit of 21600. miles by seuen, deuide the product thereof by 22. and the quotient together with the remainder, will shew the true Diameter which is 6872. miles, fiue furlongs, and $\frac{9}{11}$ of a furlong, and the halfe of that is the semi-Diameter of the earth, which is 3436. miles, and $\frac{4}{11}$ of a mile: and as the Italians and we English men doe measure great distances on the earth by miles, so the French, the Spanish, and the high Almaines, doe measure such distances by leagues both by lande and sea, and euery one differeth from other: for the French league containeth two of our miles, the Spanish league three, and the common league of Germany foure, and the great league of Germanie containeth fiue of our miles, yea in some places of Germany, as in Sueuia, the leagues are so long as a man shall skant ride three of them in a whole day. Againe the Grecians did measure the distances of the earth by furlongs, the Egyptians by signes, and the Persians by parasanges, all which measures doe greatly differ euen in the smallest partes, from whence all measures doe take their first originall: for as well amongest the auntient men, as amongest them of latter dayes, foure barley kernels couched close together side by side, and not end long, are saide to make a finger breadth, and three finger breadthes an inch, and foure finger breadthes a palme or hand breadth, and three palmes or nine inches a span, and foure palmes or hand breadthes a foote, and two foote and a halfe to make a common pace, and fiue foote to make a Geometrical pace, of which kinde of paces, 125. do make a furlong, and eight furlongs doe make an Italian mile, and foure such miles doe make a common Germaine league, as hath bæne saide before, but by reason that the barley kernels be not in all Countryes of like bignesse, neither finger breadthes, inches, hand-breadthes, fœte, nor any of the other measures are founde any where to be equall: for the French foote of Paris is longer then ours by an inch, and the Italian foote is longer by two inches and more, and yet their miles are somewhat shorter then ours: and the Germaine foote (according to Stophlerus) is lesse then ours by two inches and a halfe. But to shew the diuersitie of measures would require a long discourse more intricable then profitable, and therefore I leaue to talke any further thereof, wishing you when we speake of

miles,

of the Spheare.   **182**

miles, furlongs, paces, or féete, to confider the meafure thereof
according to the inch or foote of our English ftandard.

Of the Longitude and Latitude of the earth.

Chap. 7.

WHat meane you by the Longitude and Latitude of the earth?

Longitude is as much to fay as length,
and Latitude fignifieth breadth, for fith the
earth is a bodie, it muft néedes haue both
length, breadth, and depth.

How define you fuch Longitude and Latitude, and how is it to be counted?

The Longitude of the earth in generall is that fpace or vpper
face of the earth which extendeth from Weſt to Eaſt, and againe
frō Eaſt to Weſt: And the Latitude in generall is that fpace which
extendeth North & South euen from the one pole to thother. Now
to know how fuch Longitude and Latitude is to be accounted, you
muſt firſt vnderſtand that ẏ Equinoctiall Circle girding the earth
in the very midſt, is deuided into 360. degrées by reafon of certaine Meridians which paffing through the Poles of the world,
doe cut ech halfe of the Equinoctiall in eightéene pointes, which
being doubled doe make 36. fpaces, euery fpace containing tenne
degrées: and fome doe deuide the Equinoctiall with 36. Meridians, cutting ech halfe thereof in 36. points, which being doubled
doe make 72. fpaces, euery fpace containing 5. degrées, which
commeth all to one reckoning, for fiue times 72. doe make 360.
as well as tenne times 36. of which Meridians, be there neuer fo
fewe or many (for you may if you will make halfe as many Meridians as there be degrées in the Equinoctiall which amounteth
to a 180.) yet according to Ptolomie, that Meridian is faide
to be firſt and furtheſt Weſtwarde which paffeth through the I-
lands called Infulæ fortunatæ, for the Weſt Indies were not
knowne nor difcouered in his dayes, nor yet long time after, ſince the difcouerie whereof, the late Cofmographers of
theſe dayes doe make the firſt Meridian to paffe through the
Ilandes called Azores, which Ilandes, as appeareth by their

Cartes

The second Booke

Cardes are situated more Westwarde from the foresaid Insulæ fortunatæ, by fiue degrees, the reason that moueth them so to do, is because the Mariners Compasse as they say, wil neuer encline to the true North pole, but when they sayle either by the Ile S. Mary or S. Michaell, affirming that in euery other place the Compasse doth varie from the true North, eyther by Northeasting or Northwesting. And by thus altering the auncient placing of the first Meridian, they must likewise alter all the Longitudes set downe heretofore by Ptolomie or any other auncient writer, notwithstanding the matter is easilie holpen, for by adding to euery Longitude Eastward fiue degrees, or by subtracting fiue degrees from euery Longitude Westward, you shall not greatly vary from those auncient Longitudes that be truely set downe. But to returne againe to my first purpose, I say that wheresoeuer this first Meridian cutteth the Equinoctiall, there beginneth the first degree of Longitude, which proceedeth Eastward vntill you come to 180. degrees, which being the one halfe of the earth is as farre as you can goe Eastwarde, for then the earth being round, you must nædes turne againe Westward vntill you come to the 360. degree, which is the last degree of Longitude, and endeth where the first degree beginneth: and therefore the Cosmographers measuring alwaies the Longitude by the degrees of the Equinoctiall, do define Longitude to be that portion of the Equinoctiall Circle, which is contained betwixt the first Meridian and the Meridian of any place supposed, but the distance betwixt any two supposed Meridians (neither of them being the first Meridian) is not called of them Longitude, but the difference of Longitude: For suppose the distance of the one Meridian to be twentie degrees distant from the first Meridian, and the other but tenne, these Longitudes you see are not like but doe differ, and therfore the distance betwixt any such two places may be very well called the difference of Longitude, and not Longitude it selfe, which hath alwaies regard to the first Meridian and to none other.

Define once againe what Latitude is?

Latitude is none other thing but the distance of any place from the Equinoctiall either towards the North pole or towards the South pole, so as there be two kinds of Latitudes, the one Northerne

of the Spheare.   183

therne, and the other Southerne: And such Latitude is measured vpon the Meridian which passeth through any place supposed. For euery Meridian is also deuided into 360. degrées, and by reason that the Equinoctiall girdeth all such Meridians in the very midst, it deuideth them all into foure equall quarters, euery quarter contayning 90. degrées, which is the greatest Latitude that any place can haue, as you may sée in this figure following, whereof the first Meridian on the left hand is put to signifie according to Ptolomie, that which passeth through the Fortunate Ilandes, or by the Azores according to the moderne Cosmographers (if you will haue it so) contayning the degrées of the Latitude both Northward and Southward, and through the midst of all the Meridians passeth the Equinoctiall contayning the degrées of the Longitude.

The figure of the longitude and latitude
of the world.

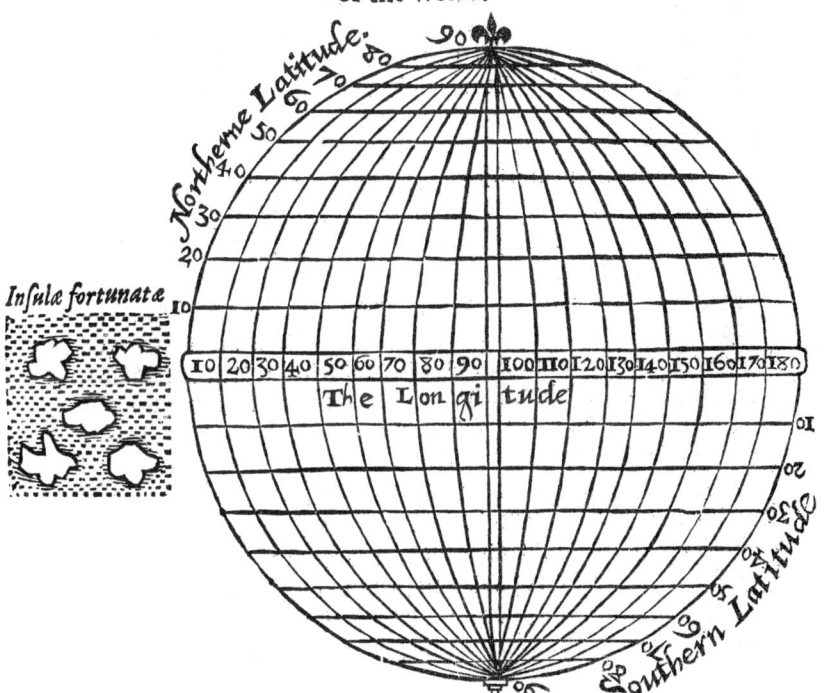

The second Booke

What be the other manifold Circles in this figure deuiding the Meridians on ech side of the Equinoctiall, as well towards the North pole, as towards the South pole.

They be called Parallels, whereof we shall speake in the next Chapter, in the meane time marke well this figure that thereby you may the better conceiue what is the Longitude and Latitude of the earth.

Vnderstanding now the Longitude & Latitude of the whole earth, I am desirous to know how the Longitude and Latitude of euery seuerall place of the earth or sea is to be found out, & how farre any place is distant one from another.

How to know the Latitude of any place, aswell in the day as in the night.

Chap. 8.

Ecause the Latitude of any place is more easie to be found (as most men thinke) then the Longitude, I will first treat of Latitude. The Latitude then is to be knowne by the Astrolabe, Quadrant, Crossestaffe, and by such like Mathematicall instruments, & that diuerse wayes whereof the most easie is thus: first with your Astrolabe or Quadrant, or any such like instrument, take the height of the sunne right at noone, when the sunne is in the first point of Aries or of Libra, which height if you subtract from 90. that which remaineth is the true Latitude of that place. But if you would knowe the Latitude at any other day or time of the yeare, then after that you haue taken the height of the sunne at noone, otherwise called the Meridian altitude, you must first learne to knowe the true degreé of the sunnes declination by the Table of the declinations before set downe, together with the vse thereof in the 13. Chap. of the first booke, or by some other Table more lately calculated, and if such declination be Northernly, then you must subtract that from the foresaid Altitude or height: but if the declination be Southernly, then you must adde the same vnto the foresaid height, and by such subtraction or addition, you shall haue the height of the Equinoctial aboue your Horizon, which being subtracted from 90. that which remaineth is the true Latitude of that place: and

to

to be sure in taking the Meridian altitude, it shall be needefull to take it diuerse times one after another, with some little pawse betwixt, to see whether it increaseth or decreaseth, for if it doth increase, then it is not yet full noone, but if it decreaseth then it is past noone. This last way of finding out the Latitude, is, and hath béene most commonly taught, as well by the auncient, as moderne writers, as a most sure and readie way of finding the Latitude of any place.

What if the sunne doe not shine at noone, nor perhaps all that day?

Then you must tarrie untill night that some starre appeare, which you perfectly knowe, and such a one as both riseth and setteth. And hauing taken the Meridian altitude of that starre with your Astrolabe or Quadrant, you must learne what declination he hath, and whether it be Northerne or Southerne. For if the starre hath North declination, then you must subtract his declination from his Meridian altitude, and the remainder shall be the Altitude of the Equinoctiall, which being taken out of nintie, shall be the Latitude of the place, or eleuation of the pole: but if the declination of the starre be Southernly, then you must adde his declination to his Meridian altitude, and that summe shall be the Altitude of the Equinoctiall, which being taken out of 90. the remainder shall be the eleuation of the Pole. As for example, supposing that you knowe the starre called Arcturus, or Bubulcus, and that you finde his Meridian altitude by your Astrolabe or Quadrant to be 59. degrées 30. minutes and also that you haue learned by some Table, that his declination to the Northwarde is 21. degrées 30. minutes: here by taking his saide declination because it is Northernly, out of his Meridian altitude, you finde the remainder to be 38. degrées which is the Altitude of the Equinoctiall, which béeing taken from 90. the remainder wilbe 52. degrées, which is the Latitude of the place whereas you made your obseruation, and this is a farre more readie waye then to waite all night to take the Meridian altitude, and also the depression of such a starre as neuer setteth, which is séeledome done in one selfe night. And therefore I would wishe all Mariners to acquaint themselues with manie starres that doe both rise and set, and so shall they

be

be sure to finde one such starre or other, to be at his Meridian altitude at any houre of the night that they desire, if the starres doe shewe.

How to know the true Longitude of any place.

Chap. 9.

Though the Longitude may be found out by diuers waies not easie for euery mans capacitie, yet because Gemma Frisius thinketh none so sure as to knowe the same by the eclipse of the Moone, (which also as he saith) may sometime faile by reason of the diuersitie of aspects and Latitude of the Moone, and for that cause hath inuented a more readie way to finde out at all times the Longitude of any place, I minde here therefore briefely to shewe you first the order of finding out the Longitude by the Eclipse of the Moone, and then how to finde out the same by that readie way which he hath inuented: the order then to knowe it by the Eclipse of the Moone is thus: First you must learne by some Ephemerides at what houre the Eclipse shall be in some place, where you knowe alreadie by some Table the Longitude, that done, you your selfe or some other for you, must the same day of the Eclipse obserue by the Astrolabe at what houre the Eclipse beginneth in that place, whereof you knowe not the Longitude: For if the Eclipse doe beginne in both places at one selfe houre, then assure your selfe that both places haue one selfe Longitude, but if it beginne sooner or later, then there is difference betwixt them, according to the varietie of the time, which difference is thus to be knowne: Take the lesser summe of houres out of the greater, and there shall remaine either houres, or minutes, or both, if houres, then multiply the same by fifteene, if minutes, deuide the same by foure, (for in this account fifteene degrees doe make one houre, and foure minutes doe make one degree) and adde the difference so found to the Longitude, if the Eclipse doe appeare there sooner: if later then subtract the sayd difference from the knowne Longitude, and that which remaineth will shew the

vnknowne

vnknowne longitude. But note by the way, that if there remaine any minutes after the diuision, you must multiply those minutes by 15. and so shall you haue the minutes of degrees.

Shew the vse of this rule by some example.

For example I finde by the tables of Ptolomey that the longitude of Paris in France is 23. degrées, and by some Almanacke or Ephemerides I finde that the Eclipse doth begin there at thrée houres after midnight, now by this I would know the longitude of Tubing a famous citie in Sueuia, which is a region of Germany, at which towne in the verie day of the Eclipse I cause to bee obserued by Astrolabe at what houre the Eclipse beginneth there, and I finde that it beginneth at 3. of the clocke and 24'. after midnight, then by subtracting the lesser number of time out of the greater, I find the remainder to bée 24'. which béing deuided by 4'. which doe make one degrée, the quotient shall be 6. degrées, and that is the difference, which being added to the knowne longitude of Paris (because the Eclipse is sooner there than at Tubing) it maketh in all 9. degrées, whereby I gather that the longitude of Tubing is 29. degrées, by this meanes all the Tables of Cosmographers are most commonlie made, and yet manie times they greatly differ in their longitudes for lacke perhaps of vsing diligence in taking the right houre and moment of the Eclipse, and for not dulie considering the diuers aspects, and what latitude the Moone hath at that instant which may cause great error.

A readie way to finde out the longitude of any place, inuented by *Gemma Frisius*.

Chap. 10.

That way is done by the helpe of some true Horologie or watch apt to bee carried in iournying, which by an Astrolabe is to bée rectified and set iust at such houre as you depart from the place where you are, to goe to any other place whereof you are desirous to know the longitude, in which your going you must bee diligent to sée that

The second Booke

that your watch neuer cease going, and being arriued at that place whereof you seeke to know the longitude, you must tary vntill the Index do iustlie touch the pricke of some perfect houre, and also at that instant, to see what houre it is by your Astrolabe, for if your Astrolabe and watch do both agree in one, then assure your selfe that there is no difference of longitude, but that you haue trauelled still vnder one selfe Meridian eyther towards the North or South. But if they differ one houre or certaine minutes, then reduce them to degrees, or to the minutes of degrees in such order as is before taught, and thereby you shall finde the longitude which you desire to know. But to take the longitude of anie place vpon the sea by this manner of way, most men thinke it were a great deale better to doe it by the helpe of a great houre glasse, made to run 24. houres, which must be watched when it is ready to run out, that it may be immediately turned: for watches made of Iron or Steele will soone rust vpon the sea. This way of taking the longitude and many others I doubt not but that M. Borrough hath tryed, & therby is able to iudge which is best, wherfore in mine opinion hee may do his country great good by setting downe in writing that way which hee by experience knoweth to be best and readiest, for want whereof the Mariners cardes do make the sea men oft to erre in their voiages.

Another way taught by Appian to finde out the longitude of any place with the crosse staffe by knowing the distance betwixt the Moone and some knowne starre that is situated nigh vnto the Ecliptique line.

Chap. 11.

Irst seeke to knowe by the Astronomicall Tables the true moouing of the Moone according to the longitude at the time of your obseruation at some certaine place, for whose Meridian the rootes of those Tables are calculated and verefied. Also you must know the degree of longitude of some fixed Starre nigh vnto the Ecliptique, going eyther next before or els next after the moouing of the Moone, then you

you must séeke out the distance of the moouing of the Moone and of the said Starre, which distance once had, apply the crosse staffe to your eye, and mooue the crosse vp and downe vntill you may sée the center of the bodie of the Moone with the one end of the crosse and the foresaid fixed Star with the other end of the crosse, so shall the crosse shew you by the degrées and minutes marked vpon the staffe, the distance of the Moone and of the foresaide Starre aunswerable to the place of your obseruation, which being set downe, set downe also the distance betwixt the Moone and the foresaid Starre that was first calculated, and then take the lesser out of the greater, so shall remaine the last difference, which may be rightly called the diuersitie of aspectes, which difference if you deuide by the moouing which the Moone maketh in one houre, you shall knowe thereby the time in which the Moone is or was ioyned with the first distance of the foresaid Starre, then hauing conuerted that time into degrées and minutes, adde or subtract the product thereof to or from that Meridian, vnto which the Tables (whereby you first calculated the moouing of the Moone) were verefied, that is to say, if the distance betwixt the Moone and the fixed Starre of your obseruation be lesser, then adde the degrées and minutes to the knowne longitude, so shall you find the place of your obseruation to be more Eastward, but if it bée greater, then subtract the degrées and minutes from the known longitude, and the place of your obseruation shall bee more westward. All which rules Gemma Frisius affirmeth to be true, so as the Moone be more westward than the fixed Starre: for if at the time of your obseruation the Moone bee more Eastwarde, then you must worke cleane contrarie, that is to say, if the distance betwixt the Moone and the fixed Starre bee lesser, you must subtract the degrées and minutes from the knowne longitude, so shall the place of your obseruation bee more Westward: but if it be greater, then you must adde the degrées and minutes vnto the knowne longitude, and you shall finde the place of your obseruation to be more Eastward.

The second Booke

How to know the distance of places, that is to say, how many miles one place is distant from another, and howe many wayes places are said to differ in distance one from another.

Chap. 12.

THe distance may be knowne diuerse wayes, that is, either Arithmetically, Geometrically, or by the Tables of Sinus. But before I shew you the order of any of these wayes, you haue to vnderstand that any two places do differ in distance one from another one of these 3. maner of wayes, that is, either in latitude onely, or in longitude one'y, or els in both: if two places hauing one selfe longitude do differ onely in latitude, then according to the Arithmetical way you must subtract the lesser latitude out of the greater, and the remainder shall be the difference, which being multiplied by 60. will shewe the number of miles: as for example, London and Roan hauing in a manner one selfe longitude, doe differ onely in latitude, for the latitude of London is 51. degrées 32'. & the latitude of Roan is 49. degrées and 10'. which being the lesser latitude and therefore to bee taken out of 51. degrées 32'. there remaineth 2. degrées 22', which 2. degrées being multiplied by 60. maketh 120. whereunto if for the 22'. annexed to the degrées you adde 22. miles (for euery minute is a mile) it shal make in all 142. miles, which by a right line is the true distance betwixt London and Roan. But you haue to note that the difference of 2. sundry latitudes is not to be knowne by subtracting the lesser out of the greater, vnles both the places be so situated, as both may haue either North latitude or South latitude, for if the one place haue North latitude and the other South latitude, the difference is to be known by addition, and not by subtraction: as for example, Naples in Italy hath 41. degrées of North latitude, and la Madalena in Aphricke not farre from Manicongo, hath 8 degrées of south latitude, both places hauing one self Meridian, here the difference of these two latitudes is to be knowne by addition, and not by subtracting the lesser out of the greater, for 8. and 41. being added together do make 49. and that is ƥ true difference, which being multiplied by 60. maketh 2940. miles. So likewise the difference of the two longitudes is not

alwaies

of the Spheare.

alwayes knowne by subtracting the lesser out of the greater, vnlesse the two places haue both East longitude or els both West
longitude: As for example Lisbona in Spaine hath in East longitude 13. degrees and Cap de los flauos in the West Indies
hath in west longitude 334. degrees: here the difference of these
two longitudes is not to be known by taking the lesser out of the
greater, but thus, first take 334. out of 360. and there will remaine 26. degrees, wherunto if you ad the East longitude for Lisbona, which is 13. degrees, it wil make in all 39. degrees, which is
the true difference of the 2. longitudes, for if you should take 13.
degrees out of 334. there would remaine 321. which is not the
true difference. But to know the distance of two places differing
in longitude, this Table here following is need full.

The Table of miles aunswerable to one degree of euery
seuerall latitude.

	1			2			3			4			5			6		
	D	M	S	D	M	S	D	M	S	D	M	S	D	M	S	D	M	S
1	59	59		16	57	41	31	51	26	46	41	41	61	29	5	76	14	31
2	59	58		17	57	23	32	50	53	47	40	55	62	28	01	77	13	30
3	59	55		18	57	4	33	50	19	48	40	9	63	27	14	78	12	28
4	59	51		19	56	44	34	49	45	49	39	22	64	26	18	79	11	27
5	59	46		20	56	23	35	49	9	50	38	34	65	25	21	80	10	25
6	59	40		21	56	1	36	48	32	51	37	46	66	24	24	81	9	23
7	59	33		22	55	38	37	47	55	52	36	56	97	23	27	82	8	21
8	59	25		23	55	14	38	47	17	53	36	7	68	22	29	83	7	19
9	59	16		24	54	49	39	46	38	54	35	16	69	21	30	84	6	16
10	59	5		25	54	23	40	45	58	55	34	25	70	20	31	85	5	14
11	58	54		26	53	56	41	45	17	56	33	33	71	19	32	86	4	11
12	58	41		27	53	28	42	44	35	57	32	41	72	18	32	87	3	8
13	58	28		28	52	59	43	43	53	58	31	48	73	17	33	88	2	5
14	58	13		29	52	29	44	43	10	59	30	54	74	16	32	89	1	3
15	57	57		30	51	58	45	42	26	60	30	0	75	15	32	90	0	0

The second Booke

Describe this Table.

This Table is diuided into 6. collums, euerie collum containing, first the degrees of latitude, and then the miles and seconds of miles answerable to euerie degree, for euerie degree of the verie Equinoctiall it selfe is in value 60. miles, but the further you goe from the Equinoctiall either Northward or Southward, euery degree of latitude is lesser in value than other, and containeth fewer miles, as you may easilie see by the said Table, proceeding from one degree to 90. which is the greatest latitude that any place can haue.

How to know by the helpe of the foresaide Table the distance of two places differing onely in longitude.

Chap. 13.

You must multiplie the difference of the longitudes by the number of miles answerable to the latitudes of the said places, omitting alwayes the seconds of miles set down in the said Table, because in this account they are of small importance: as for example, London and Antwerp, hauing both in a maner one selfe latitude, do differ onlie in longitude 6. degrees 42'. which difference being multiplyed by 37. miles answerable to 51. degrees of latitude, as you see in the Table, doe make in all 247. miles and 54". of a mile. But in making this multiplication, you must first multiply the 6. whole degrees by 37. and the product thereof will amount to 222. then by the rule of proportion you may find out the value of the minutes annexed to the degrees of the difference of longitude in saying thus, if 60'. which is one degree, do require 37. miles, what shall 42'. require. And by working according to the rule of proportion, you shall find the fourth number which you seeke to bee 25. miles and 54". which being added to the first product 222. maketh in all 247. miles and 54". of a mile, which wanteth but 6". to make vp another mile.

How

How to finde out the true distance of two places differing both in longitude and latitude by the Arithmeticall way.

Chap. 14.

HOw is that done?

First, take the difference of the longitudes and latitudes of both places by subtracting the lesser out of the greater, then conuert the same into miles by multiplying the difference of the two longitudes into the miles that be answerable to the latitude of each place, which miles you shall finde in the Table aforesaid, and if there be any minutes annexed to the degrées of the difference of longitude, then reduce the same also to miles by the rule of proportion, as before is taught, and hauing added the two products together, halfe the summe, and set it by it selfe. Then multiply the difference of the latitudes into 60. miles, and adde thereunto the fraction of minutes annexed to the said difference if it hath anie fraction, allowing for euerie minute one mile, and set that number also by it selfe: that done, square the summes reserued, that is to say, multiplie ech one part by it selfe into it selfe, and hauing added the two products together into one summe, séeke out the square roote thereof, & that shal be the true distance of the two places. As for example, if you would knowe the true distance betwixt London and Venice, first you must knowe by some Table the longitude and latitude of both townes, wherefore finding the longitude of London to be 19. degrées, 54′. and the latitude thereof to be 51. degrées, 32′. And the longitude of Venice to be 35. degrées, 30′. and the latitude thereof to be 44. degrées, 45′. Now by subtracting the lesser longitude out of the greater, I finde the difference of longitude to bee 15. degrees, 36′. and by subtracting the lesser latitude out of the greater, I find the difference of latitude to be 6. degrées, 47′. Then knowing the latitude of London to be 51. degrées, I resort to the table of miles appointed for euerie degrée of latitude before set downe, and theer I find that to 51. degrées of latitude do answere 37. miles and cer-

Bb 4   tiane

The second Booke

taine seconds, which being of small moment are not woont to be reckoned. Then in multiplying the difference of the longitudes which is 15. degrees 36'. by 37. miles I find the product of the 15. degrees so multiplied to be 555. and because there bee 36'. annexed to the foresaid 15. degrees, I seeke by the rule of proportion to knowe how many miles that fraction containeth, in saying thus, if 60. require 37. what shall 36. require? and I find 22. miles, which being added to 555. maketh 577. then by seeking in the foresaid table how manie miles be answerable to the latitude of Venice, which is 44. degrees, I find the number of miles to be 43. by which number I multiply once againe the difference of longitude, which is 15. degrees. 36'. the product whereof, together with the fraction annexed thereunto being conuerted into miles by the rule of proportion, as before, doth amount to 670. which sum being added to the former conuerted longitude, which is 577. maketh in all 1247. the halfe whereof is 623. which halfe number I reserue by it selfe, that done I multiply the difference of the latitude, which is 6. degrees 47'. by 60. miles, in saying 6. times 60. maketh 360. whereunto I adde for the 47'. annexed 47. miles, & it maketh in al 407. which sum I reserue also by it self. Then I multiplie the first reserued number into it selfe, the product whereof is 388129. That done, I multiplie the seconde reserued number also into it selfe, the product whereof is 165649. which two last productes being added together, do make in all 553778. whereof the square roote being taken, is 744. miles, which is the true distance of Venice by a right line from London. And to the intent that the order and working herein may more plainly appeare vnto you, I haue set downe all the particular numbers of the same here by them selues, as it were in a Table.

	longitude.	latitude
London.	19. degr. 54'.	51. degr. 32'.
Venice.	35. degr. 30'.	44. degr. 45'.
The difference of their longitudes and latitudes.	15. degr. 36'.	6. degr. 47'.

of the Spheare.

The difference of the longitudes conuerted into miles: for London is	577.
For Venice.	670.
The summe of the two conuerted longitudes added together, is	1247
The halfe whereof, which is the first reserued number, is,	623
The second reserued number, which is the difference of the latitudes, conuerted into miles, is	407
The summe of the first reserued number multiplied into it selfe, is	388129
The summe of the second reserued number multiplied into it selfe, is	165649
The summe of both added together, is	553778
The square roote whereof, which is the summe of the miles, is	744

How to finde out the distance betwixt two places, differing both in longitude and latitude by help only of a demicircle deuided into 180. degrees without any Arithmetike.

Chap. 14.

But now because the way before taught to finde out such distance by the Arithmeticall way may seeme perhaps to some folkes very busie and tedious, I haue thought good therefore to set downe this other way which was sent me not long since from my louing friend M. Wright of Cayes colledge in Cambridge, who is well learned in the Mathematicals, & is so apt therunto by nature, as he is like inogh to attain to such perfit

know-

The second Booke

knowledge therein as he may be able thereby hereafter greatly to perfect his country, if for want of sufficient exhibition he bee not forced to leaue so noble a studie, wherefore I wish with al my heart that all Gentlemen of abilitie were minded to shew their liberalitie towards him in that behalfe. But to returne to my matter, I say that the way to find out the foresaid distance is this here following: first, hauing drawne a demi-circle vpon a right Diameter (the larger that the demi-circle is the better,) and diuided the same into 180. degrees, like vnto this hereafter described, and marked with the letters a b c d. whereof d. is the center and a. c. the Diameter. Then learne first by some Table to know the longitude and latitude of both places, and the difference of their longitudes, as you did before in seeking to know by the Arithmetical way the distance betwixt London and Venice, the difference of whose two longitudes is 15. degrees, and 36'. as you may see in the former Table: for in working by this way, you haue chiefly to seeke out in the circumference of the demicircle but three things, that is, first, the difference of the two longitudes, secondlie, the lesser latitude, and last of all the greatest latitude. Knowing therefore the difference of the said two places in longitude to be 15. degrees, 36'. seeke out the same in the demicircle, beginning to count at A. and so proceed towards B. And at the end of those degrees and minutes set downe a pricke marked with the letter e. vnto which pricke drawe a right line by your ruler from D. the center of the demicircle. That done, seeke out the lesser latitude, which is 44. degrees, and 45'. in the foresaid demicircle, beginning to account the same from the pricke e. and so procede towards the letter B. and at the end of the said lesser latitude, set downe another pricke marked with the letter g. from which prick or point draw a perpendicular line which by help of your squire or compasses may fal with right angles vpon the former right line drawne from D. to e. and where it falleth, there set downe a pricke marked with the letter h. That done, seeke out the greater latitude, which is 51. degrees, and 32'. in the foresaid demicircle, beginning to account the same from A. towards B. and at the end of that latitude set downe another prick marked with the letter I. from whence draw another perpendicular line that may fal by help of your squire or copasses with right angles

vpon

of the Spheare. 190

vpon the Diameter A. C. and there make a prick marked with the letter K. That done, take with your compasse the distance that is betwixt k. and h. which distance you must set downe vpon your said Diameter A. C. setting the one foote of your compasse vpon k. and the other towardes the center D. and there make a pricke marked with the letter L. Then take with your Compasse the length of the shorter perpendicular line g. h. and apply that widenesse vpon the longer perpendicular line I. K. setting the one foot of your Compasse at I. which is the end of the greater latitude, and extend the other foote towards K. and there make a pricke marked with the letter M. That done, take the distance betwixt L. and M. with your compasse, and apply the same to the demicircle, setting the one foote of your Compasse in A. and the other towards B. and there make a prick marked with the letter N. And the number of degrees contained betwixt A. & N. will shew the true distance of the two places, which you shall find to be 12. degrees and almost 24'. Now by allowing for euery degree 60. miles, and for euery minute a mile, the summe of miles will agree with the former distance found out by the Arithmeticall way which was 744. miles. And thus you haue to deale to know the distance of any other two places whatsoeuer, differing both in longitude and latitude. But you haue to note by the way, that if the difference of the longitudes doth excede the number of 180. then you must subtract that exceding difference out of 360. and the remainder shall be the difference of the longitudes, and then work in all points as is before taught. And this way is as Geometricall, as that which Appian setteth downe in his booke of Geographie, to be done by the helpe of the terrestriall Globe, the order whereof here followeth.

How

The second Booke

	Longitude.	latitude.
London.	19. deg. 54′.	51. deg. 32′.
Venice.	35. deg. 30′.	44. deg. 45′.
The difference of their longitudes and latitudes is	15. deg. 36′.	6. degr. 47′.

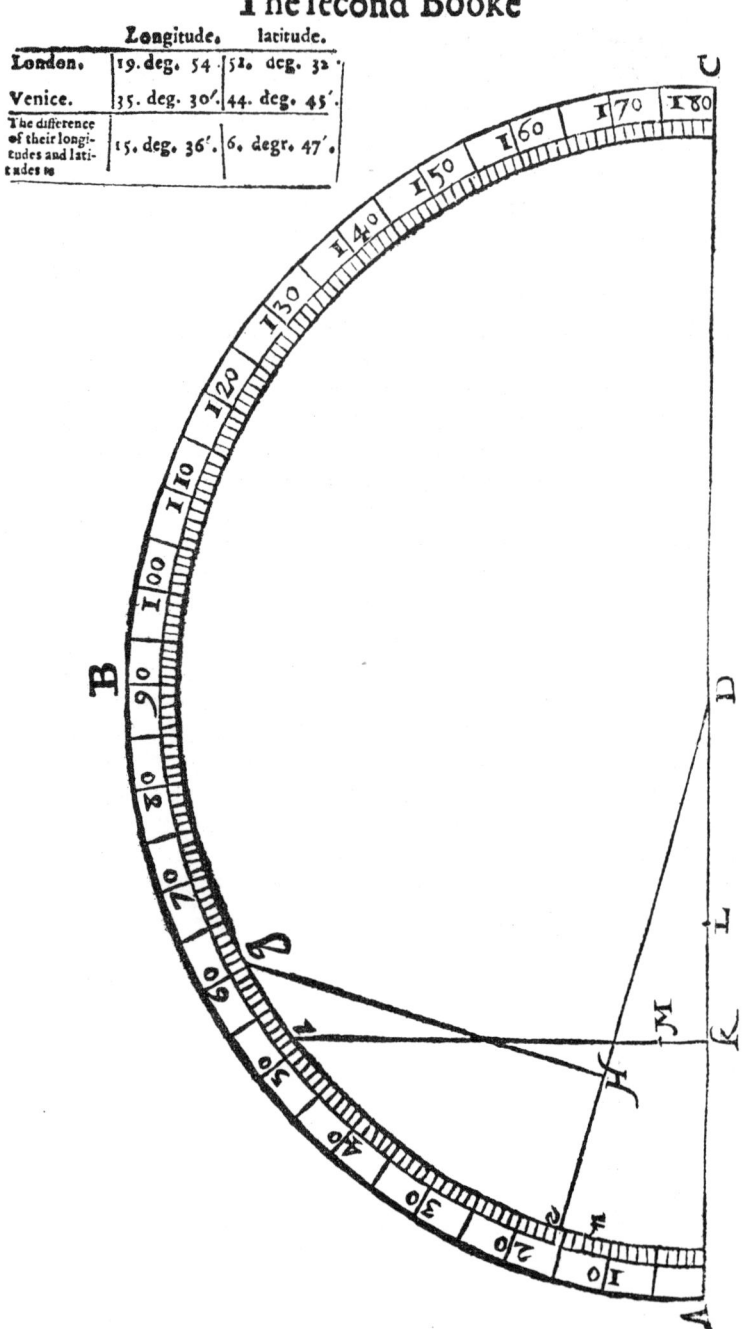

How to find out the distance of the places by the Geometricall way.

Chap. 15.

How is that done?

Most readily and easily by helpe of a terrestrial globe in this maner following. First take the distance of the 2. places by extending your compasse vpon the globe, from the one place to the other, which if you would know how many miles it comprehendeth, apply the same distance so taken vnto the Equinoctiall line, setting the first foot of your compasse vpon the first Meridian in that point, whereas it cutteth the Equinoctiall, then sée how many degrées of the Equinoctial are comprehended betwixt the two féet of your Compasse, and multiply those degrées by 60. & the product thereof shall shewe you howe many Italian miles such distance is in length. But if either of the places or both, be wanting and not expressed in the Globe, then you must learne by the tables of Ptolomey or of some others, as of Appian, Gemma Frisius, Orontius or such like, the longitude and latitude of the said places, that done, hauing sought out the longitude of the first place in the Equinoctiall, turne the globe about with your hand vntill you haue brought the longitude right vnder the brazen Meridian, which being stayed there, séeke out in the said Meridian the latitude of the said place, and there set a marke vpon the globe, for there the place should stand, and doe in like maner to finde out the seconde place: Then by extending your Compasse from the one marke to the other, you shall haue the true distance, which distance if you apply to the Equinoctiall like as before is taught, the degrées thereof being multiplyed by 60. will shewe you how many miles those two places are distant one from another.

May not the distance of places be found out aswell by an vniuersall Map as by the globe terrestriall?

Yes indéed and more readily by reason that for the most parte euery Map hath his proper skale, so as you néed to do no more but to take the distance of the 2. places with your Compasse, and to apply

The second Booke

apply the same to the skale shewing the miles or leagues.

What if the Map haue no skale?

Then you must seeke out the distance by such meanes as I do shew in my Treatise of the vse of vniuersall Mappes, and also in my description of Planctius his Map.

I pray you in the meane time proceed in shewing me the third way of finding out of the distance of places which you said was *per tabulas Sinuum*.

The order of finding out the distance of 2. places differing both in longitude and latitude per tabulas Sinuum, is plainly set downe before in the ende of my Arithmeticke, whereas I doe make a plaine description of the said Tables, and do shew the vse thereof aswel by this as by diuers other examples, wherefore I wish you to resort to that treatise and you shal haue your desire. For hauing for this time sufficientlie spoken of the longitude, latitude and distance of places, and how the same is to bee found out, I thinke it meet nowe to treat of the 5. Zones, of Climes and Parallels, whereinto the ancient Cosmographers thought good to deuide the earth, to the intent that euery part thereof might bee the better known how it is situated either Northward or Southward, whether it be hot or cold, or betwixt both, and of what length the day and night is in euery place, and what manner of shaddow the Sun yeeldeth euery where, and such like accidents, and first of the 5. Zones.

Of the 5. Zones.

Chap. 16.

He most ancient Cosmographers considering how the Sun by his oblique, and variable course did warme with his beames one part of the earth more than another, gathered thereby that the earth had three temperatures that is to say, extreame hot, extreame cold, and a meane temperature, that is, neither too hotte, nor too colde: And therefore to shew vs vnto which of these temperatures, any part of the earth was subiect, they deuided the earth into 5. Zones, answe-

of the Spheare.   192

answerable to the 5. Zones of the firmament, by helpe of the Equinoctiall and the 4. lesser circles before described in the first part of this treatise, the 20. Chapter, and are there set foorth in figure, which figure I thought good to set downe againe in this place, to the intent you might the better remember what was said there touching the Zones.

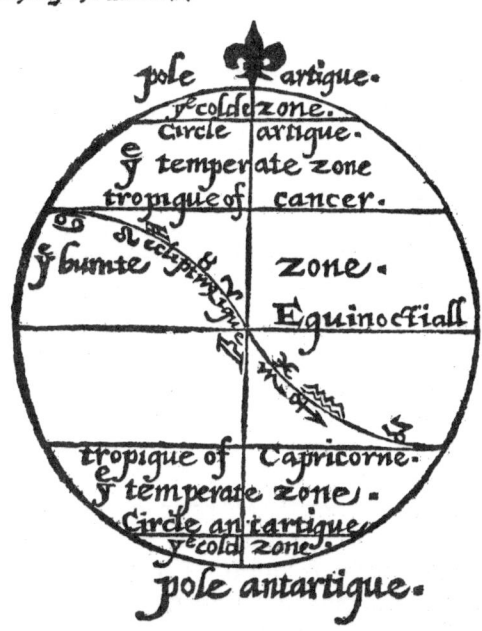

But other Cosmographers comming afterward, not satisfied with the 5. Zones, because they shewe nothing but the situation and the three temperatures of the earth, did deuide the earth into certaine Climes and Parallels, to find out thereby the length of the day and night in euery place, and the true latitude thereof as Ptolomey and many others after him haue done, making such diuisions as we shal speake of hereafter. But trulie I think with Mercator, that the best and most exact way of deuiding the earth to serue all purposes is to bee made by degrees and minutes, wherein is lesse error than in Climes and Parallels, neither can Climes or Parallels be so well described when you drawe nigh to any of the Poles, for that the spaces as well betwixt

the

The second Booke

the Parallels as betwixt the Meridians doe growe continually straighter & straighter, as you may see in the figure of Parallels hereafter following.

I remember that in the 20. chapter wheras you described the 5. Zones, you said that the ancient men did greatly erre in affirming 3. of the Zones to be vnhabitable, that is, the two colde Zones and the hot Zone, I pray you therefore shew me here the cause why they erred.

They erred for lacke of experience, because they had neuer traueled into those regions, but in these latter dayes men of diuers nations, specially the Spanish, English, French, & Flemmish haue trauelled very farre, some towards the North pole, and some towardes the South pole, and also through the burnt Zone, for those that sayle from the North partes towardes the South pole, or from the South partes towards the North pole, must needs passe in their voiage through the burnt Zone: and these men doe affirme that they haue found all the 3. Zones, that is to say, the two cold and hotte Zones to be well inhabited. And our late Cosmographers doe not let to render cause why they should be habitable, for (say they) though the cold be very extream in those regions that lie next vnto the Poles, yet the Sun appearing and giuing shine vnto them both day and night, doth greatlie qualifie and moderate the extreame cold of those regions. But truly, in mine opinion, they haue small comfort of the Sun, sith it striketh almost round about their feet, without yeelding any warm reflexe from aboue, and especially to those that doe inhabite nigh to either of the Poles. Againe, they say that the burnt Zone is habitable, by reason that the night to them is continually as long as the day, the coolenesse whereof doth greatly refresh the extream heat of the day. But now let vs returne to our purpose, and speak somewhat of the Climes and Parallels, & because euery Clime consisteth of 2. Parallels, I think it best to speak first of Parallels.

Of Parallels.
Chap. 17.

What be Parallels.

Parallels be lines either right or circular alwaies equallie distant one from another so as they can neuer meete. And of Parallels that are to be considered in the spheare,

of the Spheare.

Spheare, some make 3. kinds, according to their threefold signification, for some are called the Parallels of the sunne, who in departing from the Equinoctiall towards any of the Poles, maketh every day throughout the yeare one Parallel, so as in going from the Equinoctiall to the Tropique of Cancer, he maketh 182 Parallels. And as many againe in going from the Equinoctiall to the Tropique of Capricorne. The second kinde of Parallels are called the Parallels of Latitude, And the third the Parallels of the longest day, which two last are in effect both one, for the further that any Parallel is situated from the Equinoctiall towards either of the Poles, the more Latitude it hath, & so by consequent maketh the day longer to those that dwell vnder that Parallel, of which Parallels the auncient Cosmographers do make in all but 21. procéeding proportionally either towards the North pole or South pole, as you may sée by this figure here following the middle line or Circle whereof is the Equinoctiall.

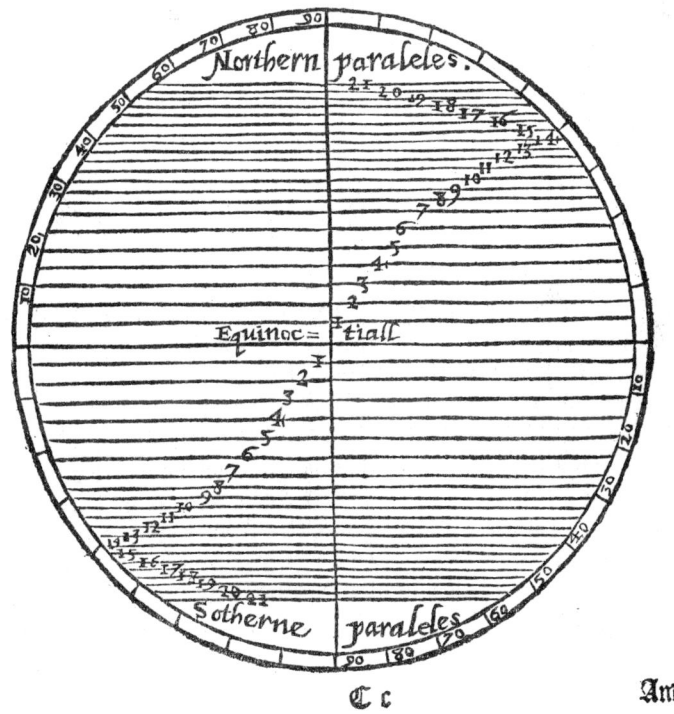

The second Booke

And every Parallel proceeding from the said Equinoctiall, eyther Northward or Southward, doth lengthen ye day by one quarter of an houre in such proportion as this Table here following sheweth, appointing to the first Parallel, and next vnto the Equinoctiall 4.degr. & 15. and to the second parallel 8.degr. 30. and so foorth vntill you come to the 21.Parallel passing through ye Iland, which Parallel is distant from the Equinoctiall 63.degr. 16. further then which Parallel Northward, the Tables of Ptolomie doe not extend, and Southward they extend no further then to that Parallel which hath lesse then 20.degrees of Latitude.

The Table of Parallels shewing how many degrees and minutes every one is distant from the Equinoctiall, made according to the rule of Ptolomy.

Parallels	D	M	Parallels	D	M	Parallels	D	M
the first	4	15	the eight	30	45	fifteenth	48	40
the second	8	30	the ninth	33	40	sixteenth	51	50
the third	12	45	the tenth	36	24	seuenteenth	54	30
the fourth	16	35	eleuenth	39	0	eighteenth	56	30
the fifth	20	30	twelfth	41	20	nineteenth	58	20
the sixth	24	15	thirtenth	43	15	twentieth	61	10
seuenth	27	30	fourtenth	45	24	the xxj.	63	16

Of Climes both old and new.

Chap. 18.

What is a Clime?

A Clime is a space of the earth comprehended betwixt two Parallels, in which space the longest day doth vary by halfe an houre.

How many Climes be there?

The auncient Cosmographers deuided aswell that part of the earth which lyeth betwixt the Equinoctiall and the North pole, so much I say as they thought to be habitable, as also the habitable part which lyeth betwixt the Equinoctiall and the South pole, ech of them into 7. Climes, to euery of which Northerne Climes they gaue a seuerall proper name, according to the name of the place through which the midst of the said Clime did passe, for they called the first Clime Diameroes. Dia is a Greeke preposition signifying

ing in English By or Through, and Meroes is a Citie of Egypt situated in a certaine Ile enclosed with the flood Nilus, which Ile hath also the like name. The second Clime is called Dia Syenes, which is also a Citie in Egypt situated right vnder the Tropique of Cancer. The third Clime is called Dia Alexandrias, which is another Citie of Egypt situated vpon the West mouth of Nilus, falling into the sea of Egypt. The fourth Clime is called Dia Rhodou, Rhodos is the chiefest Citie of an Ile called Rhodos, standing in the sea called Mare Carpathium, washing the South-west end of Natolia, sometime called Asia minor, which Ilande, together with the Citie Rhodos, Suliman the great Turke wan not many yeares since from the Christians. The people of this Ile in S. Paules time were called Colossians, to whom he wrote one Epistle, and they were so called of a great brazen Image called in Latine Colossus, containing in Altitude 105. feete, which was dedicated to the Sunne, or as some say to Iupiter. The sift Clime is called Dia Rhomes, that is to say by Rome that famous Citie of Italy, and sometime head of all the world. The sirth is called of the old men Dia Boristhenes, notwithstanding the moderne writers thinke that it may be more rightly called Dia Pontou, Pontus is both a Sea and Country lying Eastwarde right against Constantinople. The seuenth Clime is called of the olde men Dia Ripheos, but of the moderne writers Dia Boristhenes which is a great flood of Scithia in the South part of Sarmatia, which falleth into the sea called Mare Euxinum. To these seuen Climes Appian addeth two others, so as in all he setteth downe nine Climes, making the eight Clime to passe through Ripheos which are mountaines enuironing Sarmatia on the North side, and the ninth through Denmarke.

What names did they giue to the Southerne Climes?

The selfe same that the Northerne Climes haue, sauing that they put before euery such name this Græke word Anti, which is as much to say as Contrarie or Right against, as Anti Meroes, Anti Syenes, &c. as you may easily perceiue by this figure of Climes here following, notwithstanding none of those Southern Climes were knowne to Ptolomie more then Anti-meroes, and hardly all that, the furthest part whereof hath not twentie degrees of Latitude.

The second Booke

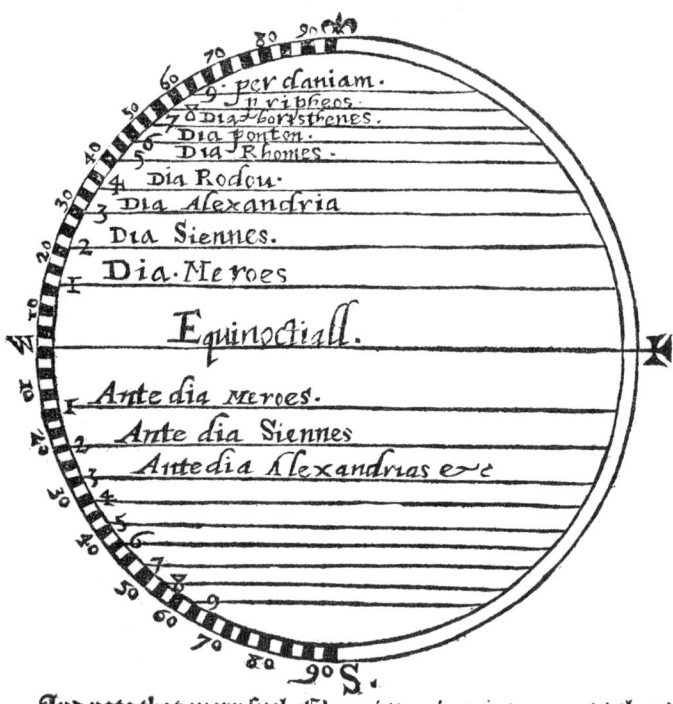

And note that euery such Clime is deuided into 3. parts, that is the beginning, the midst, & the end. And if you would know what degrees and minutes of Latitude euery such part hath, that is to say, how many degrees euery such part is distant from the Equinoctiall, then consider wel this Table following, which both briefly and plainely sheweth the same.

	The beginning.		The midst.		The end.	
Degrees and Minutes of houres.	D	M	D	M	D	M
The first Clime.	12	45	16	35	20	30
The second Clime.	20	30	24	15	27	30
The third Clime.	27	30	30	45	33	40
The fourth Clime.	33	40	36	24	39	0
The fift Clime.	39	0	41	20	43	30
The sixt Clime.	43	30	45	24	47	15
The seuenth Clime.	47	15	48	40	50	30
The eight Clime.	50	30	51	50	53	10
The ninth Clime.	53	10	55	30	56	30

of the Spheare.

May not the North and South part of the world ech of them be deuided into more then 9. Climes?

Yes indéede as our latter writers affirme, for betwixt the Equinoctiall and the 66. degrée and 30. of Latitude in which the longest day containeth full 24. houres without hauing any night, they make 48. Parallels, and thereby 24. Climes, for euery Clime containeth two Parallels, and euery Parallell maketh the day to increase by one quarter of an houre as hath béene saide before. But from thence forth, though you may continue the Parallels almoste to the very pole, yet you can make no more but 24. Climes by reason that the spaces of the Parallels towardes either of the Poles, do grow more narrow euery one then other, so as from 66. degrées and 30. minutes of Latitude vnder which the Circle Arctique passeth, the longest day is not to be counted any more by houres, but by whole dayes, wéekes, and monethes, in so much as they which dwell right vnder the North pole, whose Zenith is the pole it selfe, haue sixe monethes day whilst the Sunne abideth in the sixe Northerne signes, and 6. moneths night, the Sunne being in the sixe Southerne signes. And contrariwise they that dwell right vnder the South pole haue 6. months day, the Sunne being in the 6. Southerne signes, and 6. months night whilest the Sunne remaineth in the 6. Northerne signes. But because I thinke it a more readie way to account the length of the day by the degrées of Latitude, then by Climes or Parallels, I thought good here to set downe Orontius his Table made for that purpose, which from the 67. degrée of Latitude to the North pole agréeth in all points with the Table of Iohannes de Sacro Busto, in which Table from the saide 67. degrée is not onely set downe the longest day, but also what portion of the Zodiaque alwaies appeareth aboue the Horizon, which portion when it containeth one whole signe, then the day is one moneth long, and the night as much, if two signes, then the day is two months long and the night as much, and so foorth successiuely vntill you come to 6 signes, which is the one halfe of the Zodiaque, making both day and night ech of them sixe monethes long, as I haue saide before, and as you may plainely sée with your eye by placing the Spheare on his Horizon at euery such Latitude.

The second Booke

A Table shewing the longest day in euery degree of La-
North pole, by whole degrees without minutes from one de-

Degrees of latitude.		The longest day.			Degrees of latitude.		The longest day.		
D	M	H	M	S	D	M	H	M	S
1	0	12	3	28	34	0	14	16	24
2	0	12	6	56	35	0	14	21	52
3	0	12	10	24	36	0	14	27	20
4	0	12	14	0	37	0	14	33	4
5		12	17	28	38		14	37	36
6		12	20	56	39		14	44	56
7		12	24	48	40		14	51	12
8		12	28	0	41		14	57	44
9		12	31	36	42		15	4	24
10		12	35	12	43		15	11	20
11		12	38	48	44		15	18	40
12		12	42	44	45		15	26	8
13		12	46	8	46		15	34	8
14		12	49	44	47		15	42	24
15		12	53	28	48		15	51	4
16		12	57	20	49		16	0	8
17		13	1	4	50		16	9	44
18		13	4	46	51		16	19	52
19		13	8	56	52		16	30	32
20		13	12	48	53		16	41	52
21		13	16	48	54		16	54	8
22		13	21	4	55		17	7	4
23		13	25	4	56		17	21	4
24		13	29	20	57		17	36	16
25		13	33	35	58		17	52	48
26		13	38	0	59		18	10	48
27		13	42	24	60		18	30	56
28		13	46	16	61		18	53	20
29		13	51	30	62		19	18	24
30		13	56	16	63		19	48	40
31		14	1	12	64		20	24	24
32		14	6	8	65		21	10	32
33		14	11	12	66		22	20	40

of the Spheare.

titude, proceeding orderly from the Equinoctiall to the gree to 90.

Degrees of latitude		The longest day			The arch of the Zodiaque alway appearing aboue the Horizon	
D	M	Daies	H	M	De.	M
67	0	24	1	40	22	52
68	0	42	1	16	40	0
69	0	54	16	25	52	0
70	0	64	13	46	61	26
71		74	0	0	70	26
72		82	6	39	78	22
73		89	4	58	84	56
74		96	17	0	92	12
75		104	1	4	96	20
76		110	7	27	105	16
77		116	14	22	111	20
78		122	17	6	117	6
79		127	9	55	122	46
80		134	4	58	128	22
81		139	31	36	133	50
82		145	6	43	139	6
83		151	2	6	144	22
84		156	3	3	149	36
85		161	5	23	154	42
86		166	11	23	159	50
87		171	21	47	164	52
88		176	5	29	169	58
89		181	21	58	174	58
90		187	6	39	180	0

What things elſe do the Coſmographers teach to be conſidered in Climes and Parallels?

Diuerſe things, as firſt how many Italian miles euery clime hath in breadth and length, alſo what ſeaſons of the yeare, and what ſhadowes the ſunne yeeldeth to thoſe that dwel vnder diuers Climes and Parallels.

The second Booke

What breadth and length do the auncient writers appoint to euery one of the seuen climes.

They appoint such breadth and length as this Table folowing sheweth, in which also is set downe the degrées and minutes of Latitude through which the middle Parallell for euery Clime passeth, and also the number of miles answerable to one degrée of euery such Latitude.

Climes.	miles in bredth.	miles in length.	The degrees and minutes of Latitude through which the middle parallell of euerie clime passeth		The number of miles answerable to one degree of euery such Latitude.
			De.	M.	
1	465	20555	16	37	57
2	420	19453	24	15	54
3	370	18398	30	45	51
4	350	17299	36	24	48
5	270	16215	41	20	45
6	225	15136	45	24	42
7	195	14426	48	40	40

How the bredth & length of euery Clime is to be known.

The breadth is to be knowne by multiplying the degrées of difference contained betwixt the beginning & end of euery Clime by 60. miles: And if the degrées of difference haue any minutes annexed thereunto, then you must adde so many miles as there be minutes, to the product of the former multiplication. Nowe to haue the length of euery Clime, you must séeke to knowe by the Table of miles, how many miles be answerable to one degrée of that Parallell of Latitude which passeth through the midst of the Clime, and by that number of miles multiply 360. and the product thereof shall be the length. But if the degrées of that Latitude haue any minutes annexed thereunto, then you must finde out the miles of those minutes by the rule of proportion in saying thus: if 60. doe require so many miles as you founde in the Table of miles, what shall so many minutes require, and adde the quotient thereof to the product of the former multiplication, so shall you haue the true length of the Clime, all which thinges are obserued in the foresaid Table, which Table by obseruing like

order

order, you may extend (if you will) to the number of 24. Climes, set downe by Orontius and by diuerse other moderne writers.

Of the diuerse seasons and shadowes incident to diuerse Climes and Parallels, and first what seasons and shadowes they haue that dwell right vnder the Equinoctiall.

Chap. 19.

Those that dwel right vnder the first Clime, and specially right vnder the Equinoctial, which people at one instant may see both the Poles, haue two sommers and two winters, for the sun hauing to passe right ouer their heads twice in the yeare, which is when he is in Aries, and againe in Libra, then they must needes haue two sommers, because the sunne at both those times is nighest vnto them, but when he is in Cancer or in Capricorne, then he is furthest frō them, and thereby maketh vnto them two Winters. But yet neither of them so cold as our winter is, whereby it appeareth that our two times of Spring and Autumne are to them 2. sommers, & our two times of sommer and winter are to them two winters.

What shadowes haue those inhabitants?

They haue fiue sundry shadowes, for when the Sunne is in either of the Equinoxes, they cast their shadow in the morning whē the sunne riseth towards the West. And at night when he goeth downe towards the East, and at noone day they haue no shadowe at all, but a perpendicular shadowe, which stretcheth right downe from head to foote, because the sunne being then in the Equinoctiall, must needes at that time of the day be right ouer their heads, as you may plainely see in euery material Spheare, hauing a foote with a firme Horizon, and dewly placed to shew a right Spheare, but when the sunne is in any of the Southerne signes, then the foresaid inhabitants doe cast their shadowe towardes the North, And when he is in the Northerne signes, they cast their shadowe towards the South, and because they dwell in a right Spheare, their daies and nights be alwaies equall.

Of

The second Booke

Of the seasons and shadowes which they haue that dwell betwixt the equinoctiall and the Tropique of Cancer.

Chap. 20.

Hese haue also two sommers and two winters, and 5. shadowes like vnto the others because the sunne passeth twice in the yeare right ouer their heads, first in his declining from the Equinoctiall towards Cancer, and againe in his returning from Cancer towards the Equinoctiall, vnder which Clime Arabia fœlix is said to be situated. For Lucan writeth that the Arabians comming to Rome to aide Pompey, meruailed much to see that the trees did neuer cast their shadowes on the left hand, because in their countrey their shadowe is sometime on the right hande, and sometime on the left, sometime perpendicular, sometime orientall, and sometime occidentall, but in Rome and in all other places beyond the Tropique of Cancer the shadowe alwayes at noonetide tendeth Northward, the words of Lucan be thus, Ignotum vobis Arabes venistis in orbem vmbras mirati nemorum non ire sinistras.

How is Lucan to be vnderstood here in vsing this speech on the left hand?

As all other Poets are, for they hauing regard to the West, do alwaies make the North part their right hande, and the South part the left. But the Astronomers hauing regard to the South, do make the West their right hand, and the East their left hand: againe the Geographers contrariwise hauing regarde to the North, doe make the East their right and the West their left hand, whereby you may see that the right hand and left hand may be taken three manner of waies, that is according to the manner of the Poets, of the Astronomers, and of the Geographers.

Sith these inhabitants haue seasons and shadowes like to those that dwel right vnder the Equinoctiall, wherein then do they differ?

They differ in that their dayes & nights be not alwaies equall, by reason that their Horizon declineth from the poles of the world, and by cutting the Parallels of the sunne with oblique Angles it deuideth those Parallels into vnequall parts, neither can their

two

two sommers be so extremely hotte as the others, because the sunne for the most part is further from them.

Of the seasons and shadowes which they haue that dwell right vnder the Tropique of Cancer.

Chap. 21.

They haue but one sommer and one winter, by reason that the sun neuer passeth right ouer their heads but once in the yeare, and that is when he entreth into the first degree of Cancer: In which time when the sunne riseth, the shadowe tendeth towardes the West, and at noone the shadow is perpendicular enclining on neither side, but falling right downe. And at night when the sunne goeth downe, the shadow tendeth towards the East: and at all other times of the yeare the shadow at noonetide falleth alwaies Northward, and as touching their daies and nights, they shorten and lengthen according as the sunne either approcheth towards Cancer, or retireth towards Capricorne.

Of the seasons and shadowes which they haue that dwell betwixt the Tropique of Cancer and the circle Arctique.

Chap. 22.

They haue also one sommer and one winter as we haue here in England, for the sunne neuer passeth right ouer their heads, by reason that so soone as he hath made his last and highest Parallell in Cancer, he returneth againe Southward, and therefore their shadow is neuer at noonetide perpendicular but shooteth Northward, and the dayes and nights do lengthen and shorten according as the sunne maketh his course either through the Northerne or Southerne signes. For whilest he passeth through the Northerne signes the dayes are longer then the nights, and in passing through the Southerne signes he maketh the nights longer then the dayes.

Of

The second Booke

Of the seasons and shadowes which they haue that dwell right vnder the Circle Arctique, and how long their day is.

Chap. 23.

They haue but one Winter and one Sommer, and their shadow alwaies tendeth sidelong & Northward. And because their Zenith being in the circle Arctique, is at all times of the yeare all one with the Pole of the Zodiaque: the Ecliptique line therefore must nædes be all one with their Horizon, whereby the one halfe of the Zodiaque in a very moment doth rise aboue the Horizon, and the other halfe in the same instant goeth downe, and as the whole Tropique of Cancer appeareth alwayes aboue the Horizon: So the whole Tropique of Capricorne is alwaies hidden vnder the Horizon, so as when the sunne entreth into the first degrẽ of Cancer, their day is 24. houres long and their night but a moment, so contrariwise when the sunne entreth into the first degrẽ of Capricorne, their night is 24. houres long, and their day but a moment, as you may plainely sẽ by placing the Spheare at the 66. degrẽ and 3′0 of Latitude.

What seasons, shadowes and length of day they haue that dwell betwixt the circle Arctique and the pole Arctique.

Chap. 24.

These haue like seasõs and shadowes as those that dwell vnder the circle Arctique, sauing that their Winter is colder and longer, because the nigher they approch to the Pole, the further are they frõ the sunne, and also haue both longer dayes and nights, for the more that the Pole is eleuated aboue their Horizon, the greater portion of the Zodiaque doth alwaies appeare aboue the same, which portion if it containe one whole signe, then their day is a moneth long, and their night as much, if two whole signes, then their day is two moneths long, and their night as much, and so foorth as hath bẽne said before.

Of those that dwell right vnder the Pole.
Chap. 25.

WHat seasons, shadowes, and length of day haue they that dwell right vnder the Pole, if there be anie such people?

Truely you doe well to doubt thereof, for in mine opinion humane nature is not able to suffer the extreme cold that by reason must needes be in those parts, neither do I thinke that euer any man either Christian or Heathen did euer sayle so farre as to discouer any land there. Notwithstanding if there be any such people their season is alwayes so extreme cold as no part thereof is worthy to be called a sommer, but rather a continuall winter, and as for their shadow sith the sun when he is at the highest doth neuer mount aboue their Horizon more then 23. degrees 30 at the most, their shadow must needes go round about them nigher, for the most part to their feete then to their heads, for the pole is their Zenith, and the Equinoctiall their Horizon, whereby 6. signes which is the one halfe of the Zodiaque, doth alwaies appeare aboue their Horizon, and the other halfe is alwaies hidden vnder their Horizon, and thereby they haue 6.moneths day & 6.moneths night, day I say whilest the sunne is in the 6.Northerne signes, & night whilest the sunne is in the 6.Southerne signes, and yet the night can not be so darke there as elswhere, by reason that the sunne is neuer distant from their Horizon aboue 23. degrees 30. which is only when he entreth into the first degree of Capricorne. So likewise the day with them can neuer be so cleare as elswhere, by reason that the sunne mounting no higher aboue their Horizon then 23. degr. 30. which onely is when he entreth into the first degree of Cancer, hath no power to dissolue their grosse, thicke, cloudie and mistie aire, yet haue they some preheminence in that they may (if their cloudie aire be not the let) alwaies see all the fixed starres that are placed in the sky betwixt the Equinoctiall and the Pole, because they neuer go down but are alwaies remaining aboue their Horizon: whereas in all other parts of the worlde, the said starres can not be seene all at once, for that they both rise and set, more or lesse in number according as the Zenith of the inhabitants of euery place is more or lesse distant from the Equinoctiall.

By

The second Booke

By what names certaine inhabitants of the earth are called, as well according to the diuersitie or likenesse of shadowes as of situation.

Chap. 26.

Ou shall vnderstand that according to the diuersitie or likenesse of shadowes, the auncient Cosmographers haue giuen to the inhabitants certaine Græke names, whereof some are called Aphiscij, some Heteroscij, & some Periscij.

Shew what these names do dignifie.

Amphiscij be those that cast their shadowes both wayes, that is sometime towardes the North, and sometime towardes the South, as those that inhabite the burnt Zone.

Heteroscij be those that cast their shadowe onely one way, as those that dwell in either of the temperate Zones, for if they dwell in the North temperate Zone, they do cast their shadow alwaies at noonetide towards the North. And if they dwell in the South temperate Zone they cast their shadowe at noonetide towardes the South.

Periscij are those that cast their shadow round about them as those that dwell in either of the cold Zones, to whom the Pole is their Zenith: againe they giue them certaine names according to the diuersitie or likenes of their situation and of the seasons incident to those places, whereof some are called Antœci, some Periœci, and some Antipodes siue Antichtones.

What doe these names signifie?

Antœci be two sundry Nations, the one dwelling towardes the North pole, and the other towards the South pole, hauing one selfe Meridian and one selfe Latitude, that is to say be of like distance from the Equinoctiall, the one Southward and the other Northward, as the letters a. b. in the figure folowing do shew.

Periœci be those that dwell in one selfe Parallell, how distant soeuer they be East and West as the letters b. c. in the said figure doe shew.

Antipodes be those that dwell féete to féete, so as a right line being drawne from the one to the other, passeth through the Centre of the world as you may sée by the Letters a. c. in the figure following.

And

of the Spheare.

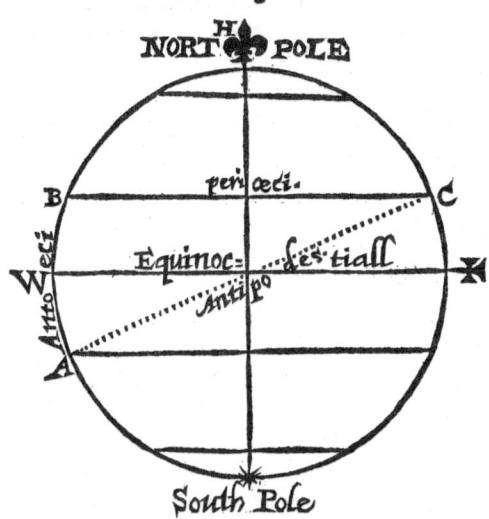

And the first of these called Antœci haue contrarie seasons of the yeare, for whē it is sommer to the one, it is winter to the other: Againe Pericœci, though they haue like seasons of the yeare, yet because they are so farre distant in length one from an other, it is therfore mid-day to the one when it is mid-night to the other. But Antipodes be contrary in seasons and in all other things, hauing nothing common, more then that they haue one selfe Horizon.

By what names certaine parts of the earth are called by reason of their diuerse shapes.
Chap. 27.

Now besides the foresaid names attributed to the inhabitants of the earth for such respect as is aboue said, the Cosmographers doe giue also to diuerse parts of the earth according to ye diuers shapes therof diuers names also: for if any part of the earth be enuironed round with water either salt or fresh, it is called Insula, that is an Iland, as England, Ireland, & such like, but if the water goe round about it sauing in one part, thē it is called peninsula, that is to say, almost an Iland, as Denmarke, Italy, Morea, and such like, and if it be a narrow straight enclosed with ye sea on both sides, then it is called Isthmus

The second Booke

as the narrow straight of Corinth lying betwixt Boetia and Achaia in Gréece which diuers Emperours of Rome haue in vaine attempted to cut, to thintent to make there through a nauigable passage: finally, when it is neither insula, peninsula, nor Isthmus, then it is called Continens, that is to say firme land, as Saxonie, Boemia, Sueuia, and such like, but these be speciall continents, for the Cosmographers of these daies do make but thrée general continents, that is first so much as was knowne to Ptolomie, and to the rest of the auncient writers, secondly the West Indies, lately found out, and thirdly the South part of the world not yet wholly discouered: Againe the auncient men deuided that portion of the world, which was knowne in their daies into 3. parts, that is, Europe, Asia, and Afrique, whereunto the moderne writers haue added a fourth part called America, containing the West Indies. Now if you would knowe what Kingdomes, Regions, Cities, Townes, Seas with their Hauens, Ports, Bayes, and Capes, Iles, Floods, Marishes, and Mountaines, are contained in euery one of these foure parts, then peruse often the vniuersall Maps, and Terrestriall Globes, as well of the moderne as of the auncient writers, and also the Tables of Ptolomie and of Ortelius, which I wish that they had béene made in such forme as the Tables of Ptolomie are: for hauing the North alwaies set in the front, it should be the readier to compare the shape or situation of any place or Region to the vniuersall Mappe, and by knowing the Longitude and Latitude of any place, it should be the easier to finde the same, as well in the speciall Table as in the vniuersall Mappe or Globe. The vse of which vniuersall Maps, I haue alreadie written in a seuerall Treatise by it selfe printed not long since. But now for so much as the knowledge of the windes hath béene alwaies thought a thing méete to be treated of by him that writeth of Cosmography, and specially of the Spheare, I will héere speake somewhat of them, neither doe I mind to make any great discourse thereof, as naturall Philosophers are woont to doe, but onely to define what the winde is, and to shew into how many parts the same is deuided, as well by the auncient as moderne writers, and by what names they are called, and somewhat of the qualities thereof.

Of

Of the wind, what it is, what motion it hath, and of the diuers names and diuisions thereof.

Chap. 29.

Irst then you haue to vnderstand that Aristotle and the rest of his sect, doe define the wind to be an exhalation hotte and dry, engendred in the bowels of the earth, and being gotten out, is caried sidelong vpon the face of the earth.

Why is not his motion right vp & down, aswel as sidelong?

Because that whilest by his heat he striueth to mount vp, and to passe through all the three Regions of the aire, the middle Region by his extreme cold, doth alwaies beate him backe, so as by such strife & by the meeting of other exhalations rising out of the earth, his motion is forced to be rather round then right.

What is the cause why he bloweth more sharpely at one time then another, and in one place more then another, and sometime not at all.

As the fumes that rise of new exhalations, & out of flouds and waters, may encrease his force, so lacke of heate and fumes doth diminish the same. Againe the roundnes of the earth is cause that he bloweth somtime more in one place then in another: also mountaines, hils, and great woods may hinder his force in some place of the earth wheras vpon the plain or vpō the broad sea he bloweth most sharpely, and as for his not blowing at all, it may chance diuerse waies, as either for lacke of sufficient heate to open the pores of the earth to let himselfe out, or for that some extreme frost and cold doth close the pores of the earth so straite as he cannot get foorth, or for that the sunne with his extreme heate consumeth the fumes and vapours that should maintaine him. But leauing to answere any more to these questions, I will shew you how many windes were obserued by saylers in old time, and how many are obserued at this present day, and how they were named. True it is that all Nations doe agree in placing the foure principall winds, according to the foure quarters of the Horizon or Angles of the world, that is to say, East, West, North and South: but in the subdiuisions of the saide foure quarters they differ, for some deuide euery quarter of the Horizon into two, making onely eight winds, and some into three, and thereby do make 12. winds which

The second Booke

the auncient Grækes and the Romaines did chiefely obserue, the names of the 8. windes are commonly expressed in the Italian tongue thus: Tramontano, North, Mezzodi, South, Leuante, East, Ponente, West, Griego, Northeast, Garbino, Southwest, Maistro, Northwest, Syrocco, Southeast, which names are often vsed by Christopher Columbus, Albertus Vesputius, and others that sayled first into the East and West Indies, and if you will know the names of the 12. windes vsed by the auncient Greekes and Romaines, then behold this figure here following, wherein you shal find them set downe in English, Græke, and Latine, that is to say, the English names without the Circle, and the Greeke and Latine names within the Circle, the Greeke vppon the right side of euery line pointing the winde, and the Latine name vpon the left side of every such line.

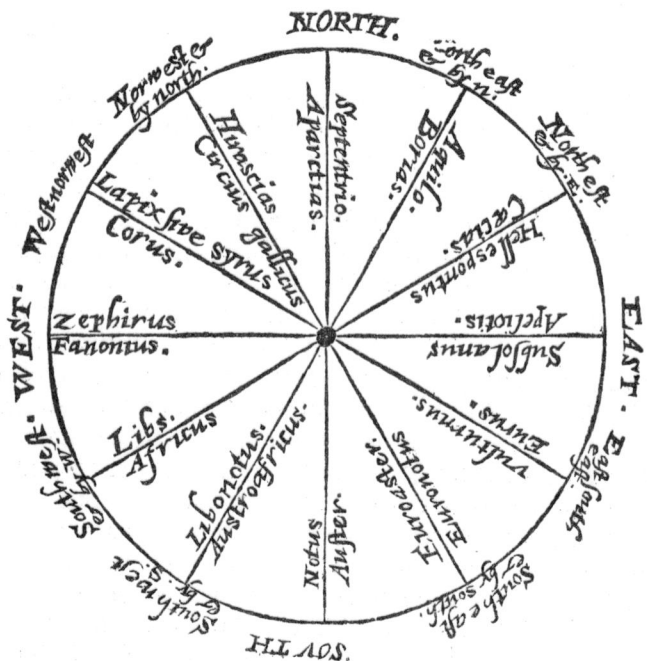

Of the nature and qualities of the foresaid 12. windes.

Chap. 29.

The North wind called Septentrio or Aparctias is extremely colde and dry, prohibiting raine, it preserueth health by clensing the Aire of all pestiferous infections, but it causeth drie coldes, and hurteth the fruites and floures of the earth.

2 The Northeast and by North, called Aquilo or Boreas, is also colde and dry without raine, it hurteth the flowres and fruites of the earth, and specially the Uines when they bud.

3 The Northeast and by East, called Helispontus or Cæcias, is hotte drying vp all things.

4 The East winde called Subsolanus, is hotte and dry, temperate, sweete, pure, subtill, and healthfull, and specially in the morning when the sunne riseth, by whom he is made more pure and subtill, causing no infection to mans bodie.

5 The East Southeast, called Eurus or Vulturnus, is also hotte and dry, he bloweth lowde, and therefore is called of Lucretius, Altitonans vulturnus.

6 The Southeast and by South called Euroauster or Euronotus, is hotte and moyst, and bredeth clouds and sicknes.

7 The South wind called Auster or Notus, is hotte and moist breeding thicke cloudes, great raines, and pestiferous aire.

8 The Southwest and by South, called Austro aphricus, is temperately hotte, & yet breedeth sicknes & raine as some write.

9 The Southwest and by West named Aphricus or Libs, is cold and moyst causing raine.

10 The West winde called Fauonius or Zephirus is temperately hotte and moyst, and wholsome in the Euening, it dissolueth frost, yce, and snow, and maketh flowres and grasse to spring, and some write that it causeth Thunder.

11 The West Northwest, called Corus or Syrus, is cold and moyst without any great rigour.

12 The Northwest and by North called Syrus or Trachias is cold and dry of earthly nature, breeding snow and windes.

Of

The second Booke

Of the moderne diuision of the windes.

Chap. 30.

But the Mariners of these our latter daies, to be the better assured of their routes and courses on the sea, do deuide euery quarter of the Horizon into 8. seuerall windes, so as they make in all 32. windes, which of the Spaniards are called Rombes, which windes together with their names, you may see plainely set foorth in this figure following representing the Mariners Compasse.

A figure of the 32. windes representing the
Mariners Compasse.

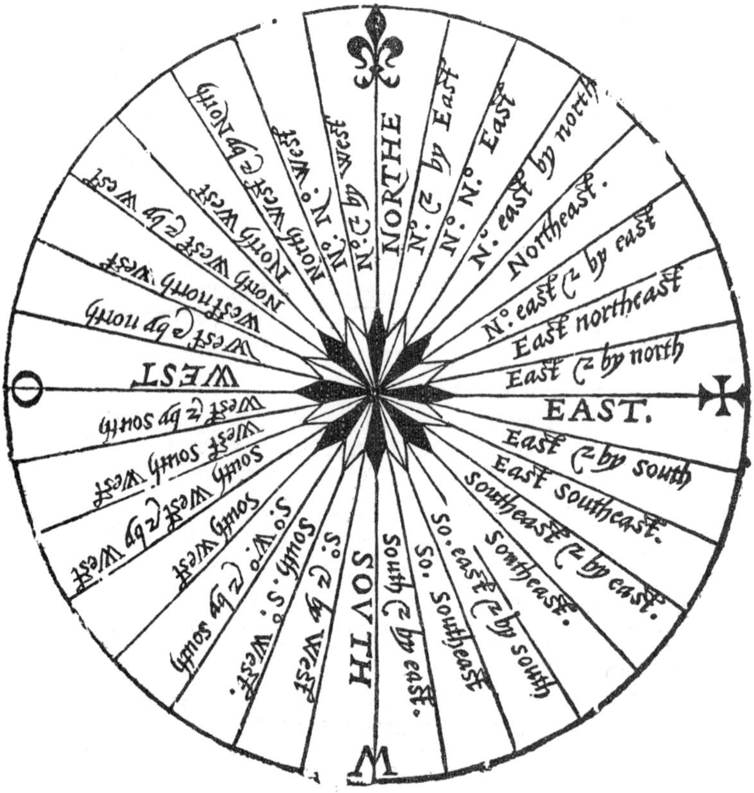

And note that eight of these windes are called principall windes, that is North, South, East, West, Northeast, Southwest, Northwest, and Southeast, and all the rest are called Collaterall windes, but the first foure are chiefest, from whence the names of all the rest are deriued, neither doe the learned Pilots in their Tables, call the first foure windes, Rombes, but will say the first, second or third Rombe from North, from East, South, or West, vnto the number of seuen, for so many Rombes they make in euery quarter betwixt the foure chiefe and principall windes, of which matter I shall speake hereafter in my treatise of Nauigation. In the meane time I heartily pray all those that shall vouchsafe to reade this my treatise of the Spheare, to take my labour therein bestowed in good part, and where any fault is, friendly to correct the same without any skorne or disdaine.

A plaine description of Mercator his two

Globes, that is to say, of the Terrestriall Globe and of the Celestiall Globe and of eyther of them: Together with the most necessary vses therof written by M. *Blundeuill*.

Whereunto is added a briefe description of the two great Globes lately set foorth by M. *Molinaxe*: and of Sir *Frances Drake* his first voyage into the Indies.

He telleth the number of the starres and calleth them by their names. Psal. 147.

Imprinted at London by *Iohn Windet*.

A plaine description of the two globes

of Mercator, that is to say, of the Terrestriall Globe, and of the Celestiall Globe, and of either of them, together with the most necessary vses thereof, and first of the Terrestriall Globe, written by M. Blundeuill.

He Terrestriall Globe is a round body couered with an vniuersall Map containing both sea and land, which is deuided by the later Cosmographers into foure principall parts, that is, Europe, Affrique, Asia, and America, the Longitude and Latitude of euery which part is alreadie set downe in my description of vniuersal Maps and Cards. And in this Globe are also set downe certaine stars, some towards either of the poles, & some nigh to the Ecliptique line, whereof we shall speake hereafter when we come to treate of the Celestiall Globe. But you haue to vnderstand that the one end of this Globe is called the pole Arctique, that is to say, the North pole, and the other ende the pole Antarctique, that is to say, the South pole: vpon which two poles it is turned about. And this Globe is traced with certaine circular lines, whereof some be greater and some lesser, those greater which passe through both the poles, are called Meridians of this word Meridies, which is as much to say as noonetide, for when the Sunne toucheth anie of those Circles, it is noonetide to all those that dwell right vnder that Circle, of which Meridians though you may imagine to bée halfe so many as there be degrées in the Equinoctiall, which amounteth to the number of 180. yet there are set downe in this Globe no more but twelue which doe cut the Equinoctiall in 24. points, making thereby 24. spaces, euery space containing 15.

degrées

The vse of the Globe.

degrées of the Equinoctiall, which fiftéene degrées doe make one houre, for the Equinoctiall is a great Circle which guirding the Globe in the very middest betwixt the two Poles, and representing the motion of the first mooueable, maketh his daily reuolution from East to West in foure and twentie houres. And this great Circle by deuiding the Globe into two equall parts, causeth the same to haue two Latitudes or breadthes, the one Northerne and the other Southerne, for that space of the halfe Globe which lyeth betwixt the Equinoctiall and the North pole, is called the North Latitude, and the other halfe of the Globe which lyeth betwixt the Equinoctiall and the South pole, is called the South Latitude, ech Latitude contayning euerie where from the Equinoctiall to either of the Poles nintie degrées. Againe this Circle is deuided into 360. degrées containing the whole Longitude of the earth, and euery degrée is 60. Italian miles, which being multiplied into 360. maketh in all 21600. miles, and such Longitude is to be counted from West to East, beginning the same at the first Meridian which passeth through the Iles of Canaria, otherwise called Insulæ fortunatæ, crossing the Equinoctiall and the Ecliptique line, in the first degrée of Aries, and also in the first degrée of Libra, which are the two Equinoctiall points, so as the one halfe of the Equinoctiall which goeth from West to East, containeth 180. degrées, and the other halfe returning againe from East to West contayneth also 180. degrées, which maketh in all 360.degrées, ending where the first degrée of Longitude did beginne: For though Mercator and others in their seuerall Mappes doe in these dayes make the first Meridian to passe through the Iles Azores which are fiue degrées more to the West. Yet Mercator in his great Terrestriall Globe dedicated first to the Lorde Granuella, and afterwardes to the Emperour Charles the fift, Anno Dom. 1541. placeth the first Meridian as before is said, both whose Globes my Worshipfull friend Sir Thomas Knyuet of Ashwelthorpe Knight, did courteously lend me I thanke him, who as he is very well learned in all the liberall Sciences himselfe, so is hee a great fauourer and furtherer of all such as delight in any learned or vertuous exercise: And by helpe of those Globes I wrote this Treatise. But to returne

to

The vſe of the Globe. 206

to my matter, there is another great ſloping and ouerthwart Circle, called the Ecliptique line, vnder which the Sunne continually walketh, and this line is marked with the Characters of the twelue ſignes, euery ſigne containing in length thirtie ſuch degrees as the Equinoctiall hath, ſo as in all, the Ecliptique line contayneth 360. degrees, which is the Longitude of heauen, and the firſt degree of the Longitude of anie ſtarre beginneth at the firſt point of Aries, and endeth at the ſame point. Nowe of the leſſer Circles there be foure principall, that is, the two Tropiques, that is to ſay, the Tropique of Cancer towardes the North pole, and the Tropique of Capricorne towardes the South pole, betwixt which two Tropiques, the Sunne continually goeth right vnder the foreſaide Ecliptique line, neuer mounting higher then the Tropique of Cancer, nor deſcending lower then the Tropique of Capricorne. Of the other two leſſer Circles, the one is called the Circle Arctique enuironing the North Pole, and the other the Circle Antarctique enuironing the South Pole: all which foure Circles are called Parallels, that is to ſaye, equally diſtant one from another, and by reaſon of theſe foure Circles, the Globe is deuided into fiue Zones, that is to ſay, two colde Zones, two temperate, and one extremely hotte: of the cold Zones, the one lyeth betwixt the Circle Antarctique, and the South pole, and the other betwixt the Circle Arctique and the North pole.

And of the temperate Zones, the one lyeth betwixt the Tropique of Cancer and the Circle Arctique, and the other lyeth betwixt the Tropique of Capricorne and the Circle Antarctique. The hotte Zone is that which lyeth betwixt the two Tropiques, through the middeſt of which Zone paſſeth the foreſayde Equinoctiall line, right vnder which, they that dwell, haue no Latitude at all, and therefore their dayes and nightes are alwayes equall, (that is to ſaye) twelue howres day, and twelue houres night. Alſo beſides theſe foure foreſayde leſſer Circles or Parallels, the Globe is traced with eight other Parallel Circles on each ſide of the Equinoctiall towards eyther of the Poles, making on each ſide nine equall ſpaces, euery ſpace contayning tenne degrees of the

Equinoc-

The vse of the Globe.

Equinoctiall, so as they doe make in all 90. degrées, which is a quarter of the great Circle. Moreouer there be certaine Mariners Compasses deuided into 32. circular lines, signifying the 32. winds, whereby Mariners doe saile, and do direct their ships from port to port, which lines doe diuersely crosse the other Circles before rehearsed. Besides these circular lines before described, there is fastned vnto the two poles of the Globe, a Meridian of brasse commonly called the moueable Meridian within the which the Globe turneth about, & this Meridian is deuided into foure quarters euery quarter containing 90. degrées, so as the whole circuit thereof is 360. degrées, one of which quarters towardes the North pole is deuided with circular lines into thrée seuerall spaces, in the first and lowest whereof being next to the bodie of the Globe, are set downe the numbers of degrées of Latitude. In the next space aboue that, are grauen the Parallels and numbers of houres of the longest day in euery Latitude, and the spaces of euery houre are deuided with little stræks into foure parts, signifying the foure quarters of an houre: and these houres doe encrease till you come to 24. houres, and from thence the day incereaseth by monethes, from one to sixe, ouer which is grauen this Latine word Menses, (that is to say) monethes, and the third and vppermost space contayneth the Climes, beginning at one, and so procædeth to 8. The like Climes, Parallels, and houres of the longest day, are to be accounted, also in the said Brazen meridian procéeding from the Equinoctial towards the South pole, though they be not there set downe. And vpon this Brazen meridian is placed at the North pole another little brazen Circle, together with his Index called the houre whéele, euery halfe wherof contayneth twelue houres, the Index of which whéele being set vppon the North pole turneth about as the Globe turneth, and yet you must imagine the Pole it selfe to be immoueable. Also there is a quarter Circle of thinne brasse plate, deuided into 90. degrées fastened in such sort vppon the brazen Meridian, as you may remoue it too and fro, yea and also take it cleane off if you will, which quarter Circle hath a square head of brasse, signifying the Zenith or verticall point of any place, and this Circle is called the quarter of altitude, which quarter is deuided into 90. degrées, procéeding from the Horizon vpward to the Zenith, wher-

by

by may be described vpon the Globe the 90. Almicanterathes or Circles of Altitude. And this quarter serueth for diuerse purposes, as to finde out thereby the Altitude of any starre or point of the Ecliptique line, or of any other point in heauen, and the very edge thereof on the right hand, sheweth the Azimuth or verticall Circle of any place, and in what coast and part of heauen any star is. Also how any place of the earth beareth one from another as you shall more plainely perceiue hereafter by the examples of certaine propositions thereto belonging. Then there is another Circle of brasse plate somewhat thicker called the semi-Circle of position, which serueth chiefely for matters of Astronomie, as to finde out the twelue houses of heauen, and is to be fastned on the Horizon, so as it may be remooued and set vpon which side of the Globe you will. Also there belongeth to the Globe a little round squire of brasse, made with right Sphericall Angles, the head or stile whereof is to shewe the shadowe of the Sunne being set vpon the Globe, in stead whereof a needle being set right vp, will sometimes serue the turne. Finally to the Globe belongeth another Circle called the Horizon, which is a broad Circle of wood, hauing a foote of wood, within which Horizon, the Globe together with the brazen Meridian is to be set at what Altitude you list. And this Horizon is deuided with diuers circular lines into seuen vnequall spaces, whereof the first and narrowest space next vnto the bodie of the Globe, containeth the degrees of the 12. signes of the Zodiaque, euery signe containing 30. degrees. The seconde space containeth the Characters and names of the said signs, and also the number of the degrees, as 10.20.30. The third the daies of euery moneth. The fourth the names of euery moneth, and of certaine Festiuall daies. The fift, the names of the 12. windes, which the ancient Græekes and Romaines were woont to obserue, whereof I haue alreadie spoken in my Spheare. The sixth and seuenth do shew the 12. houses, together with their significations necessarie for the setting of figures and calculating of natiuities, and in the foote of the said Horizon is a little Compasse with a needle to shew the North and South, according to which the Globe is to be placed. But yet not before the said needle be truely rectified, according to the variation which it hath in that place where you are to vse the Globe: for otherwise the needle may cause great

error

The vſe of the Globe.

error, whereof we ſhall treate in the next Chapter. Thus hauing briefely deſcribed the Terreſtriall Globe, and all the parts thereof, I thinke it good now to ſhew you how to place the Globe according to the foure quarters of the world.

This Treatiſe of the two Globes, containeth 50. Propoſitions as followeth.

HOw to place the Globe truely, according to the foure quarters of the world, and according to the Latitude of any Region. Propoſition. 1.

To know vnder what Clime any place or Region is, and of how many houres the longeſt day is there, and alſo what Latitude any place deſcribed in the Mappe hath. Propoſition. 2.

How to know what Longitude any place deſcribed in the Mappe hath. Prop. 3.

How to know the diſtance betwixt any two places deſcribed in the Mappe. Prop. 4.

To know how one place beareth from an other. prop. 5.

How to find out by the Globe the place of the Sunne, that is to ſay, the degree and minute of that ſigne wherein the ſun is euery day throughout the yeare. prop. 6.

How to rectifie the Index of the houre-wheele for euery ſeuerall day throughout the yeare. prop. 7.

How to knowe euery day at what houre the ſunne riſeth or ſetteth. Propoſition. 8.

How to know in what part of the Horizon the ſunne riſeth and ſetteth euery day. Prop. 9.

How to know the length of euery day and night throughout the yeare, aſwell by helpe of the houre-wheele, as by counting the degrees vpon the Horizon. Prop. 10.

How to know by the Globe how much the ſunne declineth euery day throughout the yeare from the Equinoctiall. Propoſition. 11.

How to know by the Globe the Meridian altitude of the ſunne,

The vſe of the Globe.

ſunne, that is to ſay, his height at noone tide euery day thorough out the yeare, and how far he is then diſtant from your Zenith. Prop. 12.

How to know the altitude of the ſun at euery other houre of the day. Prop. 13.

How to know the houre of the day by the Globe. Propoſition. 14.

To know how much the vnequall houres otherwiſe called the Planetary houres, doe differ from the artificiall houres throughout the yeare, and how many minutes euery vnequall houre containeth. Prop. 15.

How to know euery day when the dawning of the day, and the twilight of the night beginneth and endeth, and the time of their continuance. Prop. 16.

How to knowe the aſcention of the ſunne, both right and oblique. prop. 17.

The 18. propoſition, containing the deſcription of the Celeſtiall Globe, and ſhewing wherin it is like or differing from the terreſtriall Globe.

The 19. propoſition containing a particular deſcription of the 48. Images of the fixed ſtarres that are in the Celeſtiall Globe, together with their ſundry names, and alſo the names of ſo many ſtarres as are named in the Globe. prop. 19.

How to finde out in heauen any vnknowne ſtarre deſcribed in the Globe two manner of wayes, that is, either by the helpe of ſome knowne ſtarre, or elſe by knowing the true houre of the night. prop. 20.

How to knowe by the Globe the Meridian altitude which is the higheſt altitude of any ſtarre, and alſo how high or low he is at any other time. prop. 21.

How to know by the Globe what ſtarres are aboue the Horizon at any time of the day or night. prop. 22.

How to know by the Globe at what time any ſtarre riſeth aboue the Horizon, mounteth to the higheſt, and ſetteth, and with what degree of the Ecliptique he riſeth, mounteth, and ſetteth: and alſo in what part of the Horizon he riſeth and ſetteth. prop. 23.

How to know in what part of the firmament any ſtarre is,
and

The vse of the Globe.

and how many degrees it is diftant from the Meridian at any houre, and being right vnder the meridian to know how farre it is diftant from your Zenith. Propofition. 24.

How to finde the houre of the night by the Globe. pro.25.

How to know the verticall ftars in euery latitude. Prop. 26.

How to know the true place of any ftar, that is to fay, in what figne and in what degree thereof any ftarre is. Prop. 27.

How to finde the place and Longitude of any ftarre by the Globe. Propofition. 28.

How to find out the Latitude of any ftarre. Prop.29.

How to find out by the Globe, the declination of any ftar: Propofition. 30.

How to know the magnitude or greatnes of any ftarre, and his nature and qualitie by the Globe, and alfo the right afcention of the Arque of the Ecliptique, which accompanieth the right afcention of any ftarre. Prop. 31.

A Table of the fixed ftarres.

How to find out the right and oblique afcention of any ftar and alfo of the afcentionall difference. Prop. 32.

To know in what quantitie of time any whole figne or any other Arque of the Ecliptique doth rife or fet. Prop. 33.

How to know by the Globe what ftars doe rife or fet euerie day Cofmically, Acronically, or Heliacally. Prop. 34.

To know in what time of the yeare any ftarre rifeth or fetteth, either Cofmically or Acronically. Prop. 35.

Of the Horofcope and the reft of the 12.houfes. Prop. 36.

How to find out the Horofcope or afcédent at any time of the day or night by the Globe, and thereby to know the foure principall Angles of heauen. Prop. 37.

How to erect a figure by the Globe according to *Regio Montanus* his way, which is called the reafonable way, and is counted the beft of all others. Prop. 38.

How to know the Latitude of any place or region by any of the fixed ftars defcribed in the Globe. Prop. 39.

Another way to find the eleuation of the pole. Prop. 40.

A third way how to find out by 2. ftars the eleuation of the pole, not knowing their Meridian altitude. Prop. 41.

A fourth way to find out the Latitude of any Region by any knowne known fixed starre or Planet that may be seene. Prop. 42.

A briefe description of the diurnal Table set down in *Stadius* his Ephemerides, together with the vse therof. prop. 43.

Howe to finde out the place of any Planet by the Ephemerides. prop. 44.

A briefe description of the Table of *Stadius*, set downe in the 112. page of his Ephemerides, to find out thereby the dayly latitude of the Moone be it North or South, together with the Cannon or rule thereof, plainely declared by example. prop. 45.

How to know the true place of the Sunne or Moone or of any other Planet euerie houre of the day throughout the yeare. prop. 46.

How to find out the place of the Moone by the Globe, when she is aboue the Horizon, without the helpe of any Ephemerides or other table whatsoeuer. prop. 47.

Another way to find out the place of the Moone without taking the altitude of any starre. prop. 48.

How to find out the longitude of any region. prop. 49.

Another way to find out the vnknowne longitude of any place by the Globe. prop. 50.

How to place the Globe trulie according to the 4. quarters of the world, and according to the latitude of any region.

The first. Proposition.

First in bedding your Globe together with the brazen Meridian into the 2. nickes of the Horizon and also into the slit of the pinne which standeth in the middest of the foot, so that the North pole of your Globe be answerable to the North quarter or north wind of the world, described vpon the Horizon, and see that that part of the brazen Meridian, wherein are described the Climes, Parallels & houres of the longest day, may stand aboue the Horizon.

The vse of the Globe.

zon, and also that the one halfe of the brazen Meridian may iustlie and euenlie appeare aboue the Horizon, and the other halfe vnder the Horizon: Againe, you must sée that the Equinoctiall line of the Globe doe méete iust with the middle point or stréeke of the brazen Meridian whereas the first degrée of latitude doth beginne, and also that the body of the globe doe not leane to the one side of the Horizon more than to the other, but to bee equally distant from the same in all places, and in any wise to sée that the Horizon stande alwaies leuell, to which ende some Globes haue a plummet of leade hanging by a little chaine or thréde, which because it will mooue with euery winde, I for my part do think it better for you to haue such a little leuell made of purpose, as you may set the same vpon anie place of the Horizon wher you list, and therby make the Horizon to stand leuel on euerie side as you will your self, and it may be made of a little péece of thicke boorde, like a Triangle, thus.

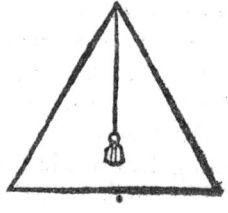

Then with your 2. handes laying holde of the 2. next pillers, turne the foot of the Globe vntil it stand right North and South, which is to be done thus. First find out the true Meridian of the place whereas you are to vse the Globe by such meanes as M. Burrough teacheth in his discourse of the variation of the Compasse, the 7. chapter. which vndoubtedlie is a most certaine way, by which you shall finde thrée thinges at one instant, that is, the true Meridian, the variation of the néedle, and the true latitude of any place. But if you haue not the instrument of variation, by helpe whereof this matter is to bée accomplished, then in some open place betwixt 8. and 9. of the clocke in the forenoon or sooner, vpon a smooth table or plancke, standing leuell, draw a good large circle with your Compasses, in the Center whereof must be fixed a round and straight pinne of Iron or Latton wyer, in length a good deal shorter than the semidiameter of that circle, and to the intent that the pinne or stile may stande right vp, without inclining on either side, it would be rectified by a true Squire: That done, waite diligently vntill the shadow of the pinne head doe iustlie touch the circumference of the circle, so as it neyther passe beyond the circle, nor come short of the same, and there make a pricke, and so let it stay vntill about thrée of the clocke in the afternoone, about which time the shaddow of the pin will beginne to approch nigh vnto the sayd circle, and so soone as it toucheth the same, make there another prick: that done, diuide the Ark or portion of the circle, contained betwixt those 2. pricks into two equall parts, and in the middest therof set another prick, then laying your ruler to the middle pricke, and to the Center of the circle, draw a right line through the Center & also through the middle pricke from the one side of the circle to the other, and beyond if you list: for you may make the said line of such length as you shall thinke most méete to serue your turne, and that line shall bee the true Meridian for that place, shewing the right North and South part of your Horizon, and by crossing the said line with another right line in the very middest with right angles, you shall haue the true East and West, and to auoyd long wayting, you may draw diuerse circles one within another. Thus hauing found the true North and South, East, and West, place your Globe accordingly, so as the brazen Meridian of the Globe may aunswere the Meridian line alreadie drawne vpon the boorde or plancke, and to the intent that afterwarde you may knowe at all times both day and night, howe to place your Globe right North and South, it shall bée necessarie to drawe a right line vpon the foote of the Horizon, answerable to the foresaid Meridian first drawne vpon the boord or planck, and vpon that line to fasten a pretie handsome Compasse, hauing a needle of an inch long at the least, which is much more certaine, than such a little néedle as is woont to bee set in the foote of euerie Globe: And when the néedle standeth still, marke how much the North point thereof declineth eyther East or West from the true Meridian before fastned, and whereas you see the Needle to decline, bee it East or West, there set fast a little pinne of Latton to serue as a marke, whereto you may alwayes direct the North point of the needle when you would haue the Globe to stand right North and South, for there is no needle touched with the load stone, bee it neuer so good a stone, but it wil vary from the true Meridian line, either more or lesse, & therefore no trust is to be giuen to the néedle vntil you know the true

vari-

The vse of the Globe.

variation therof, the finding of which variation, as I said before, is most trulie taught by M. Borough in his book before mentioned, wherunto I once againe refer you, and the rather for that it is written in our mother tongue. But Gemma Frisius teacheth to set the Globe right North and South thus, first goe into an open place whereas the Sunne shineth, and vpon some table standing leuel, and also the Globe standing leuell, the Pole being eleuated aboue the Horizon according to your latitude, fixe a right nædle in the degrée of the signe, wherein the Sunne is that day, so as the nædle may stande right vp without inclining anie maner of way, and if it be in the forenoone, turne the East side of the Globe towardes the Sunne, moouing both the Globe and his seat to and fro, vntill you sée the nædle to cast no shaddow at all, for so shall the Globe stand right North and South. But if it bee in the afternoone, you must turne the West side of the Globe towards the Sun, and then worke as before, and by anie of these wayes before taught you shall not one y place the Globe answerable to the foure quarters of the world, but also you shall find that the Circles, Poles, and Axletrées of the Globe, are answerable to the Circles, Poles and Axletrees of the heauens. Now hauing placed your Globe answerable to the 4. quarters of the world, learne to know by some Table or moderne Map, or els by such wayes as are set downe both in my Spheare, and also in the latter ende of this Treatise, the latitude of the region wherein you dwell: As for example, the latitude of these partes here about Norwich is 52. degrées, then taking holde of the brazen Meridian aswell with your left hand aboue the Horizon, as with your right hand beneath the Horizon, turne the same vp and downe in the nickes of the Horizon, vntill the North pole bee eleuated aboue the Horizon 52. degrées, the last of which degrées must méete euen with the vpper brimme of the Horizon, that done, séeke for England in the Globe, not leauing to turne the bodie of the Globe to and fro vntil you haue brought that Meridian which passeth through England right vnder the brazen Meridian and holding it stil at that stay, drawe the head of the brazen quarter of altitude right ouer the place, and ouer that Meridian vnder which you dwell, so shall the head of your brazen quarter stands for your Zenith in the verie middest of the Horizon

and

and kéeping al things thus at a stay, you may sée how euery region or countrey is situated, and how it beareth from you, and which be vnder your Climate and which be not, and to be sure that the brazen Zenith may stand in his right place, it shall be néedfull to set it so as the left square side therof made with a long notch may touch the selfe same degrée of latitude in the brazen Meridian, which it hath of altitude, for looke how many degrées the Pole is eleuated aboue the Horizon, which is called the altitude, so many degrées must the brazen Zenith be distant from the Equinoctial of the globe, which is called the latitude, and is euer all one with the altitude of the Pole. Thus hauing shewed you howe to place the Globe according to the foure quarters of the world, I thinke good now to declare vnto you, first the vses of the terrestriall globe, whereof some are common also to the celestiall globe, as you shall perceiue hereafter by the propositions following, which propositions may be very well diuided into 3. sorts, whereof the first do properly belong to the description of the vniuersall Map, wherewith the terrestriall globe is couered, which are but few in number. The second kind of propositions do chiefly belong to the Sunne, and to his apparances, which are also common to the celestial globe, and may be found out by either of the globes. But the third kind do belong most properly to the fixed starres and to their apparances, and therefore I mind not to set downe them vntill I come to describe the celestiall globe, and to shewe the vses thereof.

The chiefest propositions belonging to the vniuersall Map, wherewith the terrestriall Globe is couered, are these following.

To know vnder what Clime any place or region is, and of how many houres the longest day is there, and also what latitude any place described in the map hath.

The 2. Proposition.

Hauing set the Globe at your latitude, bring the place or region which you séek right vnder the brazen Meridian, and the vpper space of the said Meridian will shew the Clime. And the second or middle space will shewe the houres of the longest day. And the thirde space the latitude: As for example, if you would knowe vnder

The vse of the Globe.

what Clime London is, then hauing found London, fixe a long needle in the red spot next to the name of London (for all townes for the most parte in the Globe are marked with red spottes) that done, turne the Globe with your hand vntill the needle do touch the brazen Meridian, and by staying the Globe there with your hand, you shall find London to be vnder the 8. Clime, and that the longest day in the yeare is there 16 houres and 20. minutes, and the third and nethermost space doth shew the latitude of the place, which is 51. degrees, 32′. Againe, by obseruing this order without changing your latitude, you shall finde Venice to be vnder the 6. Clime, and their longest day to be 15. houres and somewhat more, and the latitude of that Cittie to bee 44. degrees, 30. And you shall finde Ierusalem to be vnder the third Clime, and the longest day there to be 14. houres, and the latitude thereof to be 31. degrees. 30.

How to know what longitude anie place described in the Map hath.

Proposition. 3.

The Globe standing still at your owne latitude pricke the needle in the place whereof you seeke the longitude, and bring it as before, to the brazen Meridian, and staying it there, looke at what number of degrees the brazen Meridian cutteth the Equinoctiall, and that is the longitude of the place, counting the degrees from the first degree of the Equinoctiall, which beginneth at the first point of Aries vnto the place of the section: by doing thus, you shall finde the longitude of London to bee 19. degrees, and the longitude of Venice to be 36. degrees, and the longitude of Ierusalem to be 67. degrees and 30. minutes.

How to know the distance betwixt any two places described in the Map.

The vse of the Globe. 212

Proposition. 4.

Pen your Compasses so wide as you may set the one foote thereof iust in the one place, and the other foote in the other place, and applie that wydenesse to the Equinoctiall line, counting how many degrees of the Equinoctiall are contained betwixt the two fæte of your compasses, & by allowing to euery degræ 60. miles, you shall haue the true distance of the places. Thus you shall finde the distance betwixt London and Venice to be 13. degrees, 30′. which being multiplied by 60. maketh 810. miles, and the distance betwixt London and Ierusalem to be 40. degrees. 30′. which being multiplied by 60. maketh 2430. miles. And it maketh no matter to what part soeuer of the Equinoctial you do apply the widenes of your cōpasses, so as you set the first foot at the very beginning of some one degræ, and let the other foot fall out as it will, either at a whole degræ at a half, or at a quarter of a degræ, which smal partes are to bee counted by minutes by coniecturall discretion. And note here that no 2. places can bee distant one from another East and West more than 180. degrees which is iust the one half the circuit of the Earth, beyond which half or on this side thereof the places must nædes be nærer together by means of the roundnesse of the Earth, and if either of the 2. places wherof you would know the distance be not expressed in the Globe, then learne to find out the longitude and latitude thereof by some table, whereof the first meridian is supposed to passe through the Ilands of Canariæ and mark vpon the Globe where the said longitude and latitude do crosse, for there ought that place to stand which is missing, to which place direct the one foot of your compasses & then worke as before is taught. And by this means you may also know the distance betwixt any two starres contained in the Globe.

To know how one place beareth from another.

Proposition. 5.

Though you may partly finde out this by marking the direction of the lines, proceding from the marriners Compasse set downe in the Globe: yet in mine opinion it is readier to do it by applying the Flie described in my little treatise of vniuersal Maps, vnto the Globe by setting

Ee 4   the

The vse of the Globe.

the center thereof vpon the first place from whence you goe, and by drawing a thred through that place whereto you would goe, in such order as is there taught, so you shall finde Venice to beare from London South East and by East, and Ierusalem to beare from London East, South East, and two quarters more towards the South. But Gemma Frisius teacheth to knowe by the Globe how one place beareth from another thus, hauing set the Globe and also the Zenith of the quarter of altitude at such latitude as the first place or region hath, from whence you would goe, bring that first place vnder the brazen Meridian, and there stay the Globe vntil you haue brought the quarter of altitude to the second place, and the nether end of the quarter of altitude will shew you vpon the Horizon amongst the windes how the second place beareth from the first, so shall you find Hispaniola to beare from Spaine right West, these be the chiefest propositions belonging to the vniuersall Map, wherewith the terrestrial Globe is couered, and therefore I will now set downe those that belong to the Sun, which may be done aswell by the one Globe as by the other, and first how to find out the place of the Sunne.

How to find out by the Globe the place of the Sunne, that is to say, the degree and minute of that signe wherein the Sun is euery day throughout the yeare.

Proposition. 6.

Though the surest way be to find it out by the Ephemerides, which sheweth the verie minute, yet without hauing respect to the minutes, you may find it out by the Globe thus, séeke out the day of the moneth vpon the Horizon, and that will point you to y̓ degrée of the signe wherin the sun is that that day: As for example, the 6. day of May pointeth right to the 25. degrée of Taurus the Globe standing leuel, and your rule being rightly laid vpon the Horizon, but during the leape yeare you must adde one degrée more than the Horizon sheweth euery day from the beginning of the leape yeare throughout al that yeare.

How

The vſe of the Globe.   213

How to rectifie the Index of the houre wheele for euerie ſeuerall day throughout the yeare.

Propoſition. 7.

Auing placed the Globe at your latitude and alſo found out the degrǽ of the Sun, as is before taught, bring that degrǽ of the Sunne to the brazen Meridian, and there ſtaying it with the one hand, turne with your other hande the index of the houre whǽle to the higheſt part of the ſaid whǽle, marked with the number of 12. ſetting the point of the index iuſt with the ſtrǽke of the whǽle made to ſhew the houre of 12. or nœne-tyde, and that will ſerue your turne for al that day, and thus muſt you doe euery day in which you haue to vſe the helpe of the ſaid houre whǽle for any purpoſe.

How to knowe euery day at what houre the Sunne riſeth or ſetteth.

Propoſition. 8.

Auing ſet the Globe at your latitude, and rectifi-ed the index of the houre whǽle by the 7. propoſiti-on, turne the Globe to the Eaſt, ſo as the degrǽ of the Sun may touch the Horizon, and then the in-dex of the houre whǽle will ſhew you at what houre the Sunne riſeth. Againe, if you bring the ſaide degrǽ of the Sunne vnto the Weſt part of the Horizon, the index of the houre whǽle will ſhew you at what houre he goeth downe. As for exam-ple, Anno 1590. the third of June, the Sun being in the 21. de-grǽ 33′. of Gemini. I bring that point to the verie edge of the Horizon on the Eaſt part thereof, and there ſtaying it, the index of the houre whǽle ſheweth that the Sunne riſeth 8′. before 4. of the clocke in the morning: and the ſaid point of the Ecliptique being turned to the Weſt part of the Horizon, the index ſhew-eth that he ſetteth 8′. after 8. of the clocke at night.

How

The vse of the Globe.

How to know in what part of the Horizon the Sunne riseth and setteth euerie day.

Proposition. 9.

Eeke the degrée of the Sun in the Ecliptique line, & turn it to the East part of the Horizon, then you shal sée whether it riseth iust East or not, and whether it inclineth towards the South, or towards the north, and likewise by bringing the saide degrée to the West part of the Horizon, you shall sée in what part of the said Horizon he goeth downe: As for example, in the last proposition the Sunne being in the 21. degrée, 33′. of Gemini, and brought to the East part of the Horizon, I finde that the Sunne did rise distant from the East towards the North 38. degrées, 30′. of the Horizon, which is thrée points and somewhat more of the Marriners Compasse from East, towardes the North, so as the Sunne riseth Northeast and by East, and a little more Northward, and the said place of the Sunne being brought to the West part of the Horizon, I find that hee setteth North West and by West, and somewhat more Northward.

How to know the length of euery day and night throughout the yeare, aswell by help of the houre wheele as by counting the degrees vpon the Horizon.

Proposition. 10.

Irst you must know when the Sunne riseth and setteth by the 8. proposition, then looke how many houres the index doth go from the Sun rising to the Sun setting, and that is the length of the day, which number if you take from 24. the night wil appeare: as for example, knowing by the 8. proposition, that the Sun being in the 21. degrée, 33′. of Gemini, riseth 8′. before 4. in the morning, and goeth downe 8′. after 8. of the clocke at night, I find by counting

the

The vſe of the Globe.   214

the houres which the index of the houre wheele hath run, that the length of the day is 16. houres and 16′. which you may know alſo by counting the degrees vpon the Horizon from the place of the Suns riſing, vnto the South point of the ſaid Horizon, which you ſhall find to be 128. degrees, which being doubled maketh 256. degrees, which if you diuide by 15. it will make 16. houres and 16′. as before.

How to know by the Globe how much the Sun declineth euery day throughout the yeare from the Equinoctiall.

Propoſition. 11.

Hauing found the place of the Sunne, bring the ſame to the brazen Meridian, and by counting how manie degrees are betwixt that place and the Equinoctiall, you ſhall know what declination the Sun hath that day: As for example, ſuppoſing the Sunne that day you ſeek to be in the firſt degree of Gemini, bring the ſaid degree of the Ecliptique to the brazen Meridian, and you ſhall find vpon the ſame Meridian the Sun to be declined from the Equinoctiall Northward almoſt 20. degrees. Againe, ſuppoſing the Sun to be in the firſt degree of Aquarius, if you bring the ſame degree of the Ecliptique vnto the Meridian, you ſhal finde the declination of the Sun to be almoſt 20. degrees Southward.

How to know by the Globe the Meridian altitude of the Sun, that is to ſay, his height at noontyde euery day throughout the year, and how far he is then diſtant from your Zenith.

Propoſition. 12.

Bring the place of the Sun that day you ſeeke to the brazen Meridian, and ſtaying it there, count vpon the ſaid brazen Meridian how many degrees are contained betwixt the place of the Sun and the South point of the Horizon, and that is the Meridian altitude of the Sun for that day, which if you ſubtract from 90. the remainder wil ſhew how many degrees he is diſtant that day at noontide from your Zenith: as for example, ſuppoſing the Sunne to be that day you ſeeke in the firſt degree of Taurus, bring that degree of the Ecliptique to the brazen Meridian, and ſtay it there vntil you haue

coun-

The vſe of the Globe.

counted how many degrees of the said Meridian are contained betwixt the place of the Sunne, and the South point of the Horizon, and you shall find the number of degrees to be 50. which is the Meridian altitude of the Sunne for that day, which 50. degrees being taken out of 90. there remaineth 40. and so many degrees the Sun is that day at noontide distant from your Zenith. The like order is to be obserued in seeking to know the Meridian altitude of any Starre, or any other point in heauen.

How to know the altitude of the Sunne at any other houre of the day.

Propoſition. 13.

Hauing rectified the index of the houre wheele by the 7. propoſition, if the houre which you ſeek be in the forenoone, turne the Globe ſo as the index of the houre wheele may touch that houre of the forenoone, at which you deſire to knowe the altitude of the Sun, and there ſtay the Globe, vntill you haue brought the quarter of altitude on the Eaſt ſide of the globe vnto the place of the Sunne, ſo ſhall you find vpon the ſaid quarter the altitude of the Sun at that houre. And if you deſire to know the altitude of the Sunne at any houre in the afternoone, then turne the Globe ſo as the index of the houre wheele may touch the houre of the afternoone, and there ſtay the Globe vntill you haue brought the quarter of altitude on the Weſt ſide of the Globe vnto the place of the Sun, and the ſaid quarter will ſhew you the altitude of the Sun at that houre. As for example, I would know how high the Sun is at 8. of the clocke in the morning, the Sunne being in the firſt degree of Taurus: here hauing rectified the index of the houre wheele, I turne the Globe ſo as the index of the houre wheele may lie vpon the 8. houre of the forenoone, and there I ſtay the globe vntill I haue brought the quarter of altitude on the Eaſt ſide of the Globe vnto the place of the Sunne, whereby I finde the altitude of the Sunne at that houre to be almoſt 28. degrees. Again, if you wil know how high the Sunne is being in the ſame degree of the Ecliptique at 5. of the clocke in the afternoone, then turne the index of the houre wheele, ſo as it may touch that houre, and ſtay it there vntill you haue brought the quarter of altitude on the

the West side of the Globe vnto the place of the Sunne, so shall you find the altitude of the Sun to be at that houre 21. degrées.

How to know the houre of the day by the Globe.
Proposition. 14.

This is to be done two maner of wayes, the first is thus, set the Globe in some open place whilest the Sunne shineth, and you must sée that it stand both leuell and also right North and South, as is taught in the first proposition, and that the index of the houre whéele be rectified according to the degrée of the signe wherein the Sunne is that day by the 7. proposition, that done, fixe a néedle in the place of the Sunne, and turne the bodie of the Globe to and fro vntill the néedle cast no shaddow at all, and there staying the Globe, the index of the houre whéele will shew you the houre of the day. But if you séeke the houre in the forenoone, remember to turne the East side of the Globe towards the Sunne, if in the afternoone, then turne the West side of the Globe towards the Sunne. The second way is thus, hauing rectified the index of the houre whéele, take the altitude of the Sun with some Quadrant or Astrolabe, and hauing marked the same altitude vpon the quarter of altitude, applie it to the degrée of the Sun on the East side of the Globe if it bée in the forenoone: but if it be in the afternoone applie the quarter of altitude to the degrée of the Sun on the West side of the globe, and the index of the houre whéele will shewe you the houre which you séeke. As for example, the 6. of June 1590. the Sunne being in the 24. degrée, 24'. of Gemini, I found by my Astrolabe the altitude of the Sunne to be 48. degrées, which I marked vpon the quarter of altitude, and because I tooke the altitude of the Sunne in the forenoone, I brought the quarter of altitude marked with that degrée to the place of the Sunne on the East side of the Globe, and there staying the Globe I found that the index of the houre whéele did point to the 9. houre of the forenoone and somewhat past. The like is to be done to know any houre of the afternoone, so as you forget not to apply the quarter of altitude vnto the place of the Sunne on the West side of the Globe.

And

The vse of the Globe.

And by taking the altitude of any knowne Starre, and working in like manner as before, you shall know the houre of the night, as shall bee taught hereafter when wee come to treate of the Starres.

To knowe how much the vnequall houres otherwise called the planetarie houres do differ from the Artificiall houres throughout the yeare, and how many minutes euery vnequal houre containeth.

Proposition. 15.

Irst you must knowe by the 10. proposition the length of the day, that is to say, how many houres it is long, and reduce those houres into minutes, and diuide the product by 12. and the quotient, together with the remainder (if there bee any left after the diuision) will shewe you the quantitie of the vnequall houre of the day, that is to say, howe many minutes it contayneth. The like is to bee done to knowe the vnequall houre of the night, for hauing the length of the artificiall night, worke as before, and you shall haue your desire: As for example, knowing by the 10. proposition the length of the day when the Sunne is in the 21. degree. 33'. of Gemini to bee 16. houres 16'. hereby reducing those houres into minutes, and by deuiding the product thereof by 12. you shall finde the vnequall houre of the daye to containe 81. minutes and ⅓ of a minute or 20. seconds, which is more than one whole artificial houre of the day by 21. minutes and 20. secondes. Againe, knowing the length of the artificiall night by the said proposition to beé 7. houres, 44. minutes, if you do reduce the same into minutes, and diuide by 12. you shall finde thereby the vnequall houre of the night to containe no more but 38. minutes, and 40. seconds, which is lesse than the artificiall houre of the day by 21. minutes, and 40. seconds.

How

The vſe of the Globe.

How to know euery day when the dawning of the day and the twilight of the night beginneth and endeth, and the time of their continuance.

Propoſition. 11.

Hauing rectified the index of the houre whæle by the 7. propoſition, firſt finde out the oppoſite point to the degree of the Sunne, and turne the Globe together with that oppoſite point, and alſo together with the quarter of altitude towards the Weſt, ſo as the oppoſite point may mæte euen with the 18. degræ of the quarter of altitude, and ſtaying the Globe there, the index of the houre whæle will ſhew at what houre the dawning beginneth. As for example, I would know at what houre the dawning of the day beginneth the 19. of Aprill 1590. when as the Sunne is in the 8. degree of Taurus, the oppoſite point whereof is the 8. degree of Scorpio, wherefore I turne the Globe together with the ſaide point oppoſite and alſo together with the quarter of altitude towards the Weſt, ſo as the ſaid point oppoſite may mæte euen with the 18. degree of the quarter of altitude, and there ſtaying the Globe, I finde by the index of the houre whæle that the dawning of the day beginneth at 2. of the clocke in the morning and 20. minutes after, which dawning alwayes endeth when the Sunne riſeth, as in the former example, the Sunne being in the 8. degræ of Taurus doth riſe 45. minutes after 4. of the clocke in the morning, ſo as the continuance of the dawning is two houres and 25 minutes, for by taking two houres and 20. minutes out of 4. houres and 45. minutes, there remaineth 2. houres and 25 minutes. Againe, the twilight beginneth when the Sun goeth downe, which in the former example is at 7. of the clocke and 15′. Nowe to know when the twilight endeth, you muſt doe thus, turne the Globe and the quarter of altitude towardes the Eaſt, ſo as the oppoſite point, which is the 8. degree of Scorpio may mæte euen with the 18. degræ of the ſaid quarter, and the index of the houre whæle will ſhewe you that the twilight endeth at 9. of the clocke and 45′. after, ſo as the continuance of the twilight is two houres, and 30′ for by taking 7. houres & 15′, out of 9. houres & 45′. there remaineth

The vse of the Globe.

neth 2. houres and 30'. But you haue to vnderstand, that the dawning twilight is not alwaies to bee knowne throughout the yeare by the Globe, for from the 11. day of May to the 10. of Julie, you shall find that the opposite point of the Sunne will not agræ iust with the 18. degreé of the quarter of altitude: because that no opposite point during that time will amount to aboue 16 or 17 degrees of the quarter of altitude at the most, because the Meridian altitude it selfe of any such opposite point is not aboue 17. degrees, for during all that time both dawning and twilight had næd in this our latitude to be accounted as night, vnlesse you will make no night at all.

Howe to knowe the ascention of the Sunne both right and oblique.

Proposition. 17.

Auing set the Globe at your latitude, bring the degree of the Sunne to the brazen Meridian, and there staying it, marke at what number of degrés the said Meridian cutteth the Equinoctiall, counting that number from the first point of the Vernall Equinoctiall point to that section, and that is the right ascention, As for example, the Sunne being in the first degreé of Gemini and brought to the Meridian, you shall finde that the Meridian cutteth the Equinoctiall in the 58. degreé thereof, and that is his right ascention. Now if you would knowe his oblique ascention, being in the first degree of Gemini, bring that degree to the East part of the Horizon, so as it may touch the vpper edge thereof, and staying the Globe there, looke what degree of the Equinoctiall toucheth the Horizon at that instant, which you shall finde to bee the 30. degree of the Equinoctiall, and that is his oblique ascention. These are the chiefest propositions that belong to the Sunne, and are to be found by eyther of the Globes, wherfore I will now proceed to those propositions that are to bee knowne most properlie by the celestiall Globe, But first I will make a description of the saide celestiall Globe, whereby it shall plainelie appeare wherein the one Globe is like the other, and wherein the one differeth from the other.

The

The 18. Propofition, contayning the defcription of the celeftiall Globe, and fhewing wherein it is like or differing from the terreftriall Globe.

Propofition. 18.

The Celeftial globe is like to the terreftrial globe, in that it is round, hauing both like Axletrees, Poles, hourewheele with his index, brazen Meridian, quarter of altitude of braffe or Latton with his fquare head or Zenith, and a halfe circle of braffe or Latton, called the femicircle of pofition, alfo a ftanding foot with an Horizon of wood diuided into feuen feueral fpaces, containing in a manner the felf fame things that are before defcribed in the Horizon of the terreftriall Globe, alfo in the bodie of the celeftiall globe are fet downe certaine circles like vnto the terreftriall Globe, that is to fay, the Equinoctiall and the Ecliptique line. Moreouer the foure leffer circles, that is to fay, the two tropiques, the circle Arctique, and the circle antarctique. But the celeftiall Globe differeth from the terreftriall Globe in thefe foure things following. Firft the celeftiall Globe hath one thinne Demicircle of braffe or Latton more than the terreftriall Globe hath, which Demicircle is diuided into two quarters, each quarter containing 90. degrees, made fo at ech end as it may be faftened when need is, vpon the 2. Poles of the Zodiaque, to find thereby the longitude and latitude of euery Star defcribed in the Globe, and therefore may very well bee called the Semicircle of longitude and latitude. The fecond difference is that whereas the terreftrial Globe is traced with 12. Meridians, diuiding the Equinoctial into 24. fpaces, euery fpace containing 15. degrees. The celeftial Gobe is onely traced with 6. Meridians, diuiding the Equinoctiall into 12. fpaces, euery fpace contayning 30. degrees. The thirde difference is that the celeftiall Globe hath not thofe 8. Parallels of latitude wherewith the celeftiall Globe is traced. The fourth difference is that whereas the terreftriall Globe is couered with an vniuerfall Map contayning the foure principal diuifions of the earth, that is , Europa, Africa, Afia and America: the celeftiall Globe is couered with a Map, wherein are painted al the fixed Starres that were

Ff   knowne

The vse of the Globe.

knowne to the ancient Astronomers diuided into 48. Images, with which Images, to the intent you might bee the better acquainted, and that you might the more readily find out any Star described in the Globe, I thought good to set downe a particular description of the said 48. images as followeth.

The 19. Proposition, containing a particular description of the 48. Images of the fixed Starres that are in the celestiall Globe together with their sundrie names and also the names of so many starres as are named in the Globe, of which 48. Images 21. are ascribed to the North part of the firmament, 12. to the Zodiaque, and 15. to the South part of the firmament.

Proposition. 19.

His description is diuided into two parts according to the twofold declination of the fixed stars, that is to say, Northern and Southern, for those Stars are said to haue North Declination, which are situated betwixt the Equinoctiall and the North pole, and those to haue South declinattion which are situated betwixt the Equinoctiall, and the South pole, and because that sixe great Circles or Meridians, passing through the poles of the world, doe deuide the Equinoctiall into twelue equall spaces, euerie space containing 30. degrees, I will begin my description at the first point of *Aries*, which is the Vernal Equinoctiall point, and so proceed towards the right hand round about the Globe, setting downe all such Images, or partes of Images as are situated towards the north pole, and are contayned in euery seuerall space betwixt two Meridians, & hauing described al the north part, I wil vse like order in describing the South part. And you may behold all the Northern Images by turning the Globe about with your hand without taking the same out of his bed or seat, the pole being eleuated aboue the Horizon 50. or 60. degrees, but to view the Southerne Images, it shall be needfull to take the Globe cleane out of his seat, and to hold it so as the north pole may stand right vp, so shal

you

you ſée euery Southern Image and Starre at your pleaſure. And yet to know how the Starres are ſituated in heauen, you had néed to imagine your ſelfe to bee within the Globe, in the verie center thereof and not without the Globe foꝛ otherwiſe thoſe ſtars that are ſituated in heauen on your right hand, if you haue regard to the outſide of the Globe, will ſéeme to be on your left hand.

The Northern Images contained in the firſt ſpace, intercepted betwixt the firſt Meridian and the ſecond Meridian.

IN this ſpace you ſhall firſt ſée next vnto the Equinoctiall the following Fiſh of the ſigne *Piſces* together with the bond, both Southerne and Noꝛtherne, called in Latine Linum auſtrale & *Septentrionale*, alſo the knot of the bonde, which is called Nodus, Syndeſmon, and hipouraion, which is a faire Star of the thirᴅ bignes.

Item the firſt part of *Aries* with the two Starres in his right hoꝛne from the foꝛmer Starre, whereof the Aſtronomers do alwayes make their computation.

Item the whole Image of *Andromeda*, her head and right arme excepted, in whoſe girdle is a Starre of the third bignes called Mirach, and in her left foote a Starre of the third bignes called Alamac.

Item the Triangle called *Triangulus* and Deltoton with his foure ſtarres.

Item the whole Image of *Caſsiopeia* ſauing her right arme, and the vpper parte of the back of her chaire, in whoſe bꝛeaſt is a ſtarre of the thirᴅ bignes called Shedar.

The Northern Images contained in the ſecond ſpace.

FIrſt the head of the Whale called Cetus in whoſe ſnout is a ſtarre of the third bigneſſe called Menkar.

Item all the hinder parts of *Aries* called in Gréeke Chrios, in Engliſh the Ram.

Item the right legge, necke, bꝛeaſt, right eare, and moʒell of the Bull, in whoſe right thigh towardes the ſhoulder point is a ſtar of the fourth bigneſſe called Alfon and in his bꝛeaſt a ſtarre

The vse of the Globe.

of the third bignesse called Alfo, and in his Mozel lying vpon his right legge, is another starre of the 3. bignesse called Alfon, and in his necke toward the Withers are 7. Starres of diuers bignes called by these diuers names, that is, Vigilie, Atlantides, Pleiades, and Athoratæ, commonlie called the 7. Starres.

Item the whole head of Medusa, called Caput Medusæ vel Gorgonis, and Ras Algol.

Item the whole image of Perseus otherwise called Chelube, his right hand, sword, and right foot excepted.

The Northern Images contayned in the 3. space.

First the left leg of the Bull hauing 2. starres thereon: moreouer his head, hornes and most part of his right eare, on whose lefte eie is a Star of the first bignes, called Oculus Tauri, Palilicium, and Aldebaran, also in his face are certaine lesser starres called succulæ and Hyades.

Item the vpper part of Orion otherwise called Alguze, holding a club in his right hand, & a Lios skin in his left hād on whose right shoulder is a Star of the first bignesse called Bed Alguze, and on his left shoulder a star of the second bignes called Bellatrix

Item the left foot of the former Gemini, contayning two stars of the fourth bignesse, whereof the one is called Propous.

Item the whole Image of Auriga, otherwise called Ericthonius and Heniochos, holding a raine and a whippe in his right hand and hauing a goat hanging on his backe, which hath two little Goates sucking her behind, which be two starres of the second bignes, called by diuerse names, as Hedi, heriphoi, and Sadaeni, and in the flanke of the Goat is a starre of the first bignes called hircus, aix, holenie and Alhaior.

The Northern Images contained in the 4. space.

First the whole image of the little Dog, in whose left flanke is a Starre of the first bignes called Canis minor, Procion, Algomeisa, and Alsahere.

Item the whole image of the two twinnes called Gemini or Didimoi, the left foot of y̌ former twin only excepted, which former twin is called Apollo, Castor, Anhelar, and the other is called Pollux, and Abrachaleus, in whose left eare is a star of the second bignes called Ras Alguze.

Item

The vse of the Globe.

Item the tayle and halfe bodie of Cancer.
Item the Mozell of the great Beare, whereon is a Starre of the fourth bignesse.

The Northern Images contayned in the 5. space.

First the head and necke of Hydrus.
Item the fore part of Cancer, called in Grǣke Carchinos, vpon whose right Clea is a Starre of the fourth bignes called Acubene, and betwixt his head and his right Clea is a starre called Presepe, Phatue, and Meelleph, and on his backe are two starres called Aselli and Onoi.

Item the fore part of Leo against whose heart is a star of the first bignesse, called with these names, Cor Leonis, Regulus, Basileus, and Calb alezet.

Item the fore part of the great Beare called Vrsa maior, Arctos Eliche, and Calisto.

The Northren Images contayned in the sixt space.

First the hinder parts of Leo otherwise called Alezet, in whose taile is a starre of the first bignesse called Cauda Leonis, and Deneb Alezet.

Item the head and shoulders of Virgo.
Item the darke starre of Bernices haire.
Item the hinder partes of the great Bear, his tayle excepted.
Item the hinder parts of the Dragons taile, containing two starres of the fourth bignes.

The northern Images contained in the seuenth space.

First, the most part of Virgo, who is otherwise called Parthenos, Erigone, Preuindemiator, Protigiter, Almucedie, and Alaraph.

Item the left leg and left arme of Bubulcus, otherwise called Bootes vociferator, Arctophilax and Lanceator, betwixt whose legges is a starre of the first bignes called Arcturus, Asimech, and Alramech.

Item Bernices haire, called Cincinnus, Cesaries, plochamos, Berenices crinis, and Trica.

Item the tayle of the great Beare contayning 3. starres of the third bignes, wherein that next his rumpe is called Aliot, & that

Ff 3   which

The vſe of the Globe.

which is in the tippe of his tayle, is called Benenacz.

Item a part of the Dragons tayle contayning 2. Starres stan-ding nigh together nigh vnto the Circle Arctique.

The northern images contayned in the eight ſpace.

First the head and neck of the ſerpent called Anguis, ſerpens, Engchelis and Ophis.

Item the crowne of Ariadna, called Corona gnoſia, Stephanos Ariadnis, and is commonly called Corona ſeptentrionalis, that is, the Northern Crowne, in which is a Starre of the ſecond bignes called Malfelcare, alpheta, and Muniir.

Item the moſt part of the Image of Bubulcus, hauing a club in his right hand, in whoſe left ſholder is a Star of the fourth bigneſſe called Ceginus, and there is another of the fourth bigneſſe in his club, right againſt his face called Incalurus.

Item the fore part of the little Beare called Vrſa minor, arctos and Cinoſura.

The northern images contayned in the 9. ſpace.

First the vpper part of Serpentarius, otherwiſe called Ophioucos, and a langue in the Crowne of whoſe head is a Starre of the third bigneſſe called iras alangue.

Item the whole Image of Hercules with the Lyons ſkin hanging on his left arme, otherwiſe called Engonaſi, algethi, Neſſus, and ignotum idolum (his right hand holding the club, and his right leg excepted) in whoſe head is a ſtarre of the third bigneſſe, called Ras Algethi, & this Image lieth groueling with his heels towards the North pole, and his head towards the Equinocttal, which mæteth almoſt with the head of Serpentarius.

Item the head of the Dragon called Draco & Aben, in whoſe head is a Starre of the third bignes called Ras Aben.

Item a part of his tayle containing 6. ſtarres.

Item the hinder part of the little Beare, contayning 2. ſtars.

The northern images contained in the tenth ſpace.

First, the vpper part of Antinous, hauing at each elboe a ſtarre of the third bignes.

Item.

The vſe of the Globe.

Item the laſt end of the Serpents taile, in the tip whereof is a ſtarre of the fourth bignes.

Item the whole image of the Eagle called Aquila, Vultur volans, Aetos and Alcair.

Item the whole image of the Shaft called Sagitta, Telum, and Hoiſtos.

Item the whole image of the Harp called Lira and Alohore, that is to ſay, Vultur cadens, and Chelis, in the vpper part whereof towards the left hand is a faire Starre of the ſecond bignes called Fidicula, Lira, Alangue, Vega, and Brineck.

Item the head, necke, and left wing of the Swan, called Auis, Cignus, Olor, Hornis, Adigege, and of ſome Gallina.

Item the neck, bodie and fore part of the Dragon contayning 11. Starres.

Item part of the little Beares taile, contayning one Starre next to his rump of the fourth bignes.

The Northerne Images contayned in the 11. ſpace.

Firſt, the Crowne of Aquarius his head, contayning one Star of the fift bignes.

Item the little Horſe called Equus and Hippos, whoſe neck is incloſed with a cloud, and in his head are foure little ſtarres.

Item the head and two fore féet of the winged horſe called Pegaſus, on whoſe right noſtrell is a Starre of the third bignes called Eniph alpharaz.

Item the Dolphin called Delphinus, containing ten Stars, whereof one is of the third bignes.

Item the bodie, legs, and the right wing of the Swan, which lyeth on her back with her bellie vpward, in whoſe bodie towards the tayle, is a faire Starre of the ſecond bignes called Deneb Adigege, and Arided.

Item the right arme and right leg of Cepheus, on whoſe right ſhoulder is a ſtar called Alderaimim.

Item part of the little Beares taile contayning the middle ſtarre of his tayle.

The Northern images contained in the 12. ſpace.

Firſt, the moſt part of the former fiſh of the ſigne Piſcis together with part of her band.

Item the necke, body, and wing of Pegaſus, otherwiſe called

Ff 4   Equus

The vse of the Globe.

Equus Gorgoneus, and Alpharaz, rising out of a cloud, in which cloud is the head of Andromeda, hauing on the right side thereof a faire starre of the second bignes, and in the right wing of *Pegasus* is a starre of the second bignes called Marcab Alpharaz and on his right shoulder another Starre of the second bignesse called Scheat Alpharaz.

Item the right arme and hand of Andromeda holding part of her chaine, in the ring whereof is a Star of the fourth bignes.

Item the left arme of Cepheus.

Item the tip of the little Beares taile, in which is the North starre called Alrucuba of the third bignesse.

The names of the Images contayned in the celestiall globe betwixt the Equinoctiall and the South pole together with so many starres as are named in that part of the globe, beginning as I did before in describing the northern Images at the Vernall Equinoctiall point, and so proceede from space to space contayned betwixt euery two Meridians towardes the right hand.

The Southern Images contained in the first space, beginning at the Vernall Equinoctiall point.

First, the most part of the Whale called Cetus, *Pistrix*, and Balena (his head and fore part of his bellie excepted) in the mid bodie whereof towards the back, is a starre of the fourth bignes called Baten kaetos, and in the lower part of his taile is another starre of the third bignes, called Deneb Kaitos.

The Southern Images contayned in the second space.

The fore part of the Whales belly and his ghilles, contayning fiue Starres.

Item the most part of the Floud Eridamus, called of some Nilus, and in Græke *Potamos Eridanos*, contayning 22. Starres, whereof one is called Angetenar, which is about the middest of the Floud nigh vnto the Whales bellie, and there is in the verie end of the Floud another starre of the first bignes, called Acarnar.

The Southerne Images contained in the third space.

First,

The vse of the Globe.

First, the nether part of Orion or Alguze from the middle of his backe downward, in whose girdle are three faire starres, whereof the middle starre is of the second bignesse called Orion or Alguze.

Item another part of the floud Eridanus which seemeth to come from the left foot of Orion, which starre in his left foote is called Algebar, Rigel, Alguze.

Item the whole Image of the Hare, called Lepus and Lagos containing 12. little starres.

Item the rest of Eridanus contayning foure Starres, whereof there is one called Beemum of the fift bignesse.

The Southerne Images contained in the fourth space.

First, the whole image of the great Dog called Canis maior, and Syrios in whose mouth is a Starre of the first bignesse called Asceher and Alhabor.

Item the fore part of the great shippe Argos, with her two Oares hauing a scutchen with 4. Starres, the greatest whereof being of the third bignes is called Markeb, and vnder the vpper hatches in the fore part of the ship is a Starre of the fift bignesse called Alphard, and in the least oare towards the South pole, is a faire Starre of the first bignes, called Canopus and Suhel.

The Southern Images contained in the fift space.

First, the mid part of the Serpent called Hydrus and Asuia, in the which is a faire Star of the 2. bignes called Alphard.

Item the hinder part, mast and toppe of the ship Argos, which seemeth to come out of a cloud, containing diuers Starres of diuers bignes without name.

The southern Images contained in the sixt space.

Item another part of Hydrus whereupon standeth the image of the cuppe or boule called Crater, vas and patera, and also the Crowes head.

Item the hinder part of Centaurus, in euery part whereof are diuers Starres without name.

The southern Images contayned in the 7. space.

First,

The vſe of the Globe.

First, the left wing of the ſigne Virgo and her left hand, holding an eare of wheat, whereon is a Starre of the firſt bignes, called Spica Virginis, ſtachis, Acimon, Alacel, and Azimech.

Item the Crowe called Coruus and Corax, his head & necke excepted, in whoſe left wing is a ſtar of the fourth bigneſſe called Algorab.

Item the reſt of Hydrus, whereon the Crow ſtandeth, containing thræ ſtarres without name.

Item the reſt of Centaurus or Chiron with his boreſtaffe trimmed with boughs, his right hand and right foot excepted.

The ſoutherne Images contained in the eight ſpace.

The whole image of Libra, the ring only excepted.

Item the fore part of Scorpio, whoſe fore cleas do lie vpon the two ballances, that is to ſay, his right clea vpon the North ballance, and his left clea vpon the South ballance, hauing vpon ech clea a ſtar of the ſecond bignes.

Item part of the Serpent called Anguis or Ophis, hauing one Star of the fourth bignes.

Item the left hand of Serpentarius, holding part of the Serpent, vpon which hand are two ſtarres of the third bignes called Yedd.

Item vpon the head of Scorpio are 3. ſtars of the third bignes, ſtanding all in a row and diuers others as well vpon his back as vpon his left little clea without name.

Item the whole Wolfe called Fera, Lupus and Therion.

Item the right hand of Centaurus holding the ſaid Wolf by the belly in both which are diuers ſtarres without name.

Item the right foot of Centaurus, in which is a faire ſtarre of the firſt bignes, and is called by the name of Centaurus.

The ſoutherne images contayned in the 9. ſpace.

First, the nether part of Serpentarius, that is to ſay, from his mid backe downward, hauing the ſerpent winding betwixt his legs and aboue his right arme, in both which are diuers Starres without name.

Item the hinder part of Scorpio from his mid bodie to the outermoſt end of his tayle, who hath diuers names, as Scorpius Nepa, Alatrab, in the midſt of whoſe bodie is a faire Starre of the

firſt

The vse of the Globe. 222

first bignesse called cor Scorpionis, Antares, and chalb Alatrab, and in the tip of his tayle are two Starres of the third bignesse, called Alascha and scomlec Alatrab.

Item the most part of the Bowe with the heade of the shaft of Sagittarius and his right foot, in all which parts are diuers stars without names

Item the Altar with the flame and smoake called Ara, thuribulum, lar, sacrarium, thimiaterion.

The southern images contained in the tenth space.

The lower part of Antinous from the breast downward kneeling vpon an Altar contayning foure Starres without name. Item the forepart of Capricornus his head.

Item the image of Sagittarius, otherwise called Crotus, and toxeuter (his bow, the end of his arrow, and his right foot excepted) hauing diuers Starres without name.

Item the Southerne Crowne, called Corona Australis, and Notios stephanos, and of Aratus, it is called Dinotos Cyclos, that is to say, the Southern circle which Crowne is placed betwixt the two fore legges of Sagittarius, and in the said Crowne are diuers Starres, amongst which there is one of the second bignes touching the left knee of Sagittarius called Corona Australis, who also hath on his left foote another Starre of the second bignes without name.

The southern Images contayned in the 11. space.

The fore part of Aquarius otherwise called Ganimedes and Hydrochos, holding a handkercher in his left hand, wherein are three Starres, and he hath diuers Starres in his bodie without name.

Item the whole Image of Capricornus, otherwise called Pan Aigoceros, and Algedi, the fore part of his head onely excepted in whose taile is a Starre of the fourth bignes called Deneb Aldegi.

Item the hinder part of the Southern Fish hauing diuers starres without name.

The Southerne Images contained in the twelfth space.

First Aquarius his right hande holding the water pot called Vrna and Chalpi, out of the which he powreth the water down

The vse of the Globe.

into the mouth of the Southern fish, which water is called Aqua and hyder in which are diuers Starres without name.

Item the lower bellie parte of the former fish of the signe Piscis wherein are two Starres without name.

Item both the thighes and legges of Aquarius, vpon the calfe of whose right leg is a Starre of the third bignesse called Scheat Aquarii, and Crus Aquarii.

Item the head of the southern fish called Piscis Meridionalis, and Icthis notios, in whose mouth is a faire Starre of the first bignes, called Fomahant.

But you haue to vnderstand that besides the 15. Southern Images before mentioned, ther ar lately found out by the Portugales and others that haue sailed into the East and West Indies 4. other images towards the south Pole, as the crosse or Crosier, the south triangle, Noahs doue or Pigion, & another image made like a Philosopher called Polophilax, all which are set downe in the celestial Globe, lately set forth, first at the great charges of M. Sanders, and now at the like charges of M. Molinax of Lambeth of whome I lately bought both the Globes, that is, the terrestriall and celestiall, and I wish that the longitude, latitude and declination of euery Star contained in the said 4. images were trulie set downe, for Plancius maketh some doubt thereof. Notwithstanding if you be desirous to know the longitude, latitude, and declination of the said Starres by help of the foresayd great Globe, then you must worke as I doe shewe you hereafter in seeking for any Starre contayned in Mercator his globe, so shall you haue your desire.

Moreouer, to most of the Stars described in the Globe are annexed the Characters of some of the 7. Planets, to shew the nature & qualitie of the Stars & some stars ar also marked with some one letter or other, the more readily to find out therby the foresaid characters, As for example to Cor Leonis are annexed the Characters of Iupiter and Venus, & vnder the sayd Star is set the letter m. to shew the Characters which are not alwayes set hard by the Starre, but sometime a good distance off, for where the characters are set nigh vnto the star, there nédeth no letter, as in the Star called spica Virginis, whereunto are annexed the characters of Mars and Venus, without any letter to signifie the same,

and

The vſe of the Globe. 223

and where diuers Starres be of one ſelfe qualitie, they are ſeuerally marked with letters of one ſelfe ſame ſhape, as about the Starre Spica Virginis you ſhall finde diuers little Starres, each one marked with the letter h, ſignifying their nature to bée all one, that is, to participate of Mars and Venus, to whoſe characters is alſo ioyned the letter h, ſignifying that they be of that nature and quality.

The Characters are theſe here following ſet ouer euery Planets head.

♄　　♃　　♂　　☉　　♀　　☿　　☽
Saturne. Iupiter. Mars. Sol. Venus. Mercurie. Luna.

The nature of euery one of the Planets here followeth.

SAturne is colde and drie. Iupiter is temperately hot and moiſt. Mars is extreamly hot and drie. Sol hot and ſomewhat drie. Venus is temperately colde and moiſt. Mercurie is of changeable nature, and plyant to the nature, bee it good or bad of euery other Planet or fixed Starre whereto it is ioyned. Luna is cold and moiſt.

Beſides the images and ſtarres both Northern and Southern aboue mentioned, there is alſo ſet downe in the celeſtiall Globe a certaine impreſſion called in Græke Galaxia, that is to ſay, the milke white way, the deſcription whereof here followeth.

A briefe deſcription of the milke white way, called in Greeke *Galaxia*, and in Latine *Via lactea*.

THis way as Garceus writeth, procéedeth from the ſigne Gemini, and ſo paſſeth through the legs and loynes of Ericathonius, and from thence through the right ſide of Perſeus, and then through the whole image almoſt of Caſsiopeia, and from thence through the left wing of the Swan called *Auis*, Gallina and Cignus, and from thence through the Image called in Latine *T*elum in Engliſh a Dart, ſhaft, or quarell, and from thence through the flying Eagle called in Latine Vultur volans, and from thence through the greateſt part of Sagittarius his bowe, & from thence
through

The vſe of the Globe.

through the Alter called *Ara* and *Thuribulum*, and from thence through the legges of Centaurus, and ſo to the ſhip called Argos, from whence riſing againe, and paſſing through part of the great Dog called Canis maior, it returneth againe to Gemini.

Thus hauing deſcribed vnto you all the 48. Images, and ſhewed the names of as many Stars as are named in the Globe, and alſo the milke white way, I mind now to procéede to the propoſitions belonging to the fixed Starres deſcribed in the Globe, as followeth.

How to find out in heauen anie vnknowne Starre deſcribed in the Globe two manner of wayes, that is, either by the helpe of ſome knowne ſtarre, or els by knowing the true houre of the night.

Propoſition. 20.

The firſt and ſureſt way is thus, take with your Quadrant or Aſtrolabe the altitude of the knowne Star, marking therewith in what part of the heauen the ſame is ſituated Eaſt or Weſt, North or South, and then hauing ſet your Globe right North and South, and at the true latitude of the place wher you are, bring the quarter of altitude to the ſaid Starre, and therewith turne the Globe vntill you ſée that the ſaid Starre hath the like place and the like altitude in the Globe that it had in heauen, then kéeping the globe ſtil at that ſtay, ſéek in the globe the ſtar that you would find out in the firmament, & marke wel in what part of the Globe it is ſituated, and how it beareth from the knowne Star, either Eaſt, Weſt, North or South, and bring the quarter of altitude to that Starre, that you may know the altitude therof by help of the ſayd quarter, which altitude once taken, turne your eyes towardes that part of the firmament, and hauing placed the Diopter of your Aſtrolabe, at that altitude, look what Star in that part of the firmament doth anſwere to ſuch altitude, and that is the Starre which you ſéeke, whoſe name for the moſt part is ſet downe in the Globe. The ſecond way is thus, hauing learned the true houre of the night by ſome clocke or watch, bring the degrée of the Sun vnto the brazen Meridian, & holding it there, ſet the index of the houre whéele

iuſt

The vse of the Globe.   224

iuſt at the 12. houre, that done, turne the body of the Globe to or fro vntil the index of the houre wheél fal iuſt vpon the houre which you ſéeke, and kéeping the Globe at that ſtay, ſéek out the vnknown Starre in the Globe, and conſider how it beareth from you in the Globe either Eaſt, Weſt, North or South, then bring the quarter of altitude to the Star, that you may know thereby what altitude it hath in the Globe, which once found and hauing ſet the Diopter of your Aſtrolabe at that altitude, turne your eie towards that part of the firmament whereunto the place of the Starre found in the Globe directeth you, and that Star which anſwereth to that altitude is the Star which you ſéek. But this way is not ſo ſure as the other way firſt taught, vnles you know the true houre indéed.

How to know by the Globe the Meridian altitude which is the higheſt altitude of any ſtarre, and alſo how high or lowe he is at any other time.

Propoſition. 21.

Hauing ſet the Globe at your latitude, bring the ſtar to the brazen Meridian, and there ſtaying it, number vpon the brazen Meridian the degrées contayned betwixt the ſaide Star and the South or north point of the Horizon, according as the Starre is ſituated either Northward or Southward, for ſo in the latitude 52 you ſhall find the Meridian or higheſt altitude of the Star Arcturus or Bubulcus to be 60. degrées. But if you would know his altitude at any other time, then you muſt rectifie the index of the houre whéele by the 7. propoſition, and hauing ſet the index at the houre wherein you ſéeke, ſtay the Globe there, vntill you haue brought the quarter of Altitude vnto the Star, and that wil ſhew the altitude of the Star at that houre if it be aboue the Horizon.

How to know by the Globe what Starres are aboue the Horizon at any time of the day or night.

Propoſition. 22.

OF Starres according to diuers latitudes ſome are alwaies aboue the Horizon & ſome are alwaies vnder the Horizon, & ſome do both riſe and ſet, if you would know what Starres be aboue the Horizon in the day

time

The vse of the Globe.

time, then hauing rectified the Index of the houre whéele by the 7. proposition, take the altitude of the Sun with your Astrolabe or quadrant, and therewith consider whether the Sun be in the East part, or in the West part of the firmament. Then bring the quarter of altitude on the East or West side of the Globe according as you saw the Sun at that present to be in the firmament, and make the degree of altitude, marked in the quarter of altitude to méete euen with the degrée of the Ecliptique line wherein the Sunne is that day, and there staying the Globe, you shall sée all the starres that be aboue the Horizon at that present, aswell on the East side as on the west side of the Globe, and the Index will shew you at what houre you took the aforesaid altitude. But if it be in the night season, and that the starres doe appeare, take with your Astrolabe the altitude of some known star, and by doing as is before taught, you shal haue your desire. But you must not forget first of all to rectifie the Index of the houre whéele by the 7. proposition.

How to know by the Globe at what time any star riseth aboue the Horizon, mounteth to the highest, and setteth, and with what degree of the Ecliptique he riseth, mounteth, and setteth, and also in what part of the Horizon he riseth and setteth.

Propofition. 23.

Hauing rectified the Index of the houre whéele by the 7. proposition, bring the Star to the East part of the Horizon, so as it may touch the edge thereof, and the Index of the houre whéele will shew at what houre hée riseth and by looking at that instant to the Ecliptique line you shall sée what degrée of the Ecliptique riseth then with him. That done, bring the said starre to the brazen Meridian, and the Index of the houre whéele will shew at what houre hee is at the highest, and there staying the Globe, marke what degrée of the Ecliptique line doth fall right vnder the brazen Meridian at that instant, for that degrée is sayde to accompany him when hée is at his highest. Then bring the sayd Starre to the West part of the Horizon, and you shall finde by the Index of the houre whéele at what houre hee setteth, and what degree of the

Eclip-

Eclyptique doth accompany him at his setting. As for example, I would knowe the sixteenth day of June 1590. the sunne being in the fourth degrée of Cancer, when Arcturus otherwise called Bubulcus and Lanceator doth rise, mounteth to the highest, and setteth, here hauing rectified the Index of the houre-whéele by the seuenth Proposition, I bring the said starre to the East part and very edge of the Horizon, and I finde that he riseth a little before twelue of the clocke at noone, and that the 28. degrée and 3'0. of Virgo, riseth with him, and by looking amongst the windes vpon the Horizon right against the place of his rising, I find that he riseth Northeast and by East. Secondly by bringing the said star to the brazen Meridian, the Index of the houre-whéele sheweth that he is at his highest halfe an houre after seuen of the clocke at night, and is then plaine South, and that the 29. of Libra, doth then accompany him. Thirdly by bringing the saide starre to the West part of the Horizon, the Index of the houre-whéele sheweth that he setteth or goeth downe a quarter of an houre before foure in the morning, and that the fourth degrée of Capricornus doth accompany him at his setting: and by looking there vpon the Horizon, I find amongest the winds that the said star setteth Northwest and by West.

How to know in what part of the firmament any starre is, and how many degrees it is distant from the Meridian at any houre, and being right vnder the Meridian to knowe how far it is distant from your Zenith.

The 24. Proposition.

Hauing rectified the Index of the houre-whéele by the seuenth Proposition, turne the Globe vntill the Index touch the houre wherein you séeke. And staying the Globe there, bring the quarter of Altitude to the starre be it on the East or West side of the Globe, and the nether end of the said quarter will shew vpon the Horizon among the winds, in what part of the firmament the star is: now if you would know how farre that starre is distant from the Meridian doe thus: looke what degrée of the Equator is at that instant vnder the Meridian, and there make a marke, & then turne

The vſe of the Globe.

turne the Globe vntill you haue brought the ſaid ſtarre vnder the brazen Meridian, and marke what degrée of the Equinoctiall the ſaid Meridian cutteth at that inſtant, that done, count the degrées contained betwixt the two markes vpon the Equinoctiall, for ſo many degrées is the ſtarre diſtant at that time from the Meridian eyther towards the Eaſt or Weſt, and by allowing 15. degrées to an houre, and 4′. to a degrée, you ſhall know in what time the ſaid ſtarre hath to approch to the Meridian, or howe much he is paſt the Meridian, and hauing brought the ſaid ſtarre right vnder the Meridian, you ſhall knowe how farre it is diſtant from your Zenith, by counting the degrées that are contained in the Meridian betwixt the ſaid ſtar and your Zenith: As for example, the ſixth of October 1591. (the ſunne being then in the 22. of Libra) I finde the ſtarre Cor Leonis at thrée of the clocke in the after noone to be Weſt, Northweſt, & there ſtaying the Globe, I ſée that the Meridian cutteth the Equinoctiall at the 246. degrée, whereas I make a marke, that done, I bring the ſtar Cor Leonis to the ſaid Meridian at which inſtant the Meridian cutteth the Equinoctial in the 146. degrée, which being taken out of 246. there remaineth 100. degrées, which is his diſtance from the Meridian, which being deuided by 15. I find in the quotient 6. houres and 4′0. for ſo much as he is paſt the Meridian towards the Weſt, and by bringing the foreſaid ſtarre to the Meridian, I finde him to be diſtant from our Zenith 39. degrées, and 30. minutes.

How to find the houre of the night by the Globe.
The 25. Propoſition.

Auing ſet the Globe at your Latitude, rectifie the Index of the houre-whéele by the 7. Propoſition, then hauing taken the Altitude of ſome ſtarre that you knowe, and is in the Globe with your Aſtrolabe or Quadrant, bring the quarter of Altitude vnto the ſtarre, be it in the Eaſt or Weſt, according as you found the ſtarre to be in the firmament, not leauing to turne the Globe vntill you haue made the ſtarre to haue the like Altitude in the Globe vpon the quarter of Altitude, and alſo the like ſituation that you found it to haue in the firmament

The vſe of the Globe. 226

ment by your Aſtrolabe or Quadrant, & ſtaying the Globe there, the Index of the houre-whéele wil ſhew the houre: As for example, in the yeare 1590. the firſt of Januarie the ſunne being in the 22. of Aquarius, I hauing with my Aſtrolabe found that the ſtarre called Canis maior, that is to ſay the greater dogge, was eleua= ted aboue the Horizon in the Eaſt part of the firmament 20. de= grées, I brought the quarter of Altitude to the Eaſt ſide of the Globe, not leauing to turne the Globe vntill I had made the ſtar to méete euen with the 20. degrée of the quarter of Altitude, and there ſtaying the Globe, I found by the Index of the houre-whéele that it was 8. of the clocke at night, and a quarter paſt.

How to know the verticall ſtarres in euery Latitude.

The 26. Propoſition.

The Globe and the brazen Zenith being ſet accor= ding to the Latitude of the place where you are, turne the Globe from Eaſt to Weſt, and as ma= ny ſtarres as paſſe right vnder your Zenith are ſaide to be verticall as in the Latitude 52. you ſhall ſée the taile of the great Beare, the head of Perſeus, and diuerſe others to paſſe through your Zenith in tur= ning the Globe from Eaſt to Weſt.

How to know the true place of any ſtarre, that is to ſay, in what ſigne and in what degree thereof any ſtarre is.

The 27. Propoſition.

Auing faſtned the ſemi-Circle of Longitude and Latitude vppon the two Poles of the Zodiaque, the North pole whereof is in the Circle Arctique not farre from the right clawe of the Dragon, and the South pole therof is in the circle Antarctique right oppoſite to the other, then bring the ſaide ſemi-Circle to the ſtarre, whoſe place you ſéeke, and marke there= with what point of the Eclitique the ſaide ſemi-Circle cutteth, and that is the place of the ſtarre. As for example, by bringing

G g 2   the

The vſe of the Globe.

the ſaid ſemi-Circle to the ſtarre called Hircus, that is to ſay, the Goate, I finde it to be in the 15. degrè 30. minutes of Gemini, and that is his place.

How to find the Longitude of any ſtarre by the Globe.

The 28. Propoſition

I Haue told you in my Spheare, that the Longitude of any ſtarre is that arch or portion of the Ecliptique line, which is contained betwixt the firſt point of Aries, and that Circle which paſſeth through the Poles of the Zodiaque, and alſo through the bodie of the ſtarre, which Circle, the ſemi-Circle of Longitude and Latitude here repreſenteth: and by making the ſaide ſemi-Circle to paſſe through the ſtarre called Hircus before mentioned, I finde that the number of degrées of the Ecliptique, contained betwixt the ſaid Circle and the firſt point of Aries to be 75. degrées & 30. which is the Longitude of the ſaid ſtarre, and thereby maketh his place to be in the 15. degrè 30. of Gemini, as before is ſet downe in the laſt Propoſition.

How to find out the Latitude of any ſtarre.

The 29. Propoſition.

I Told you alſo in my Spheare, that the Latitude is none other thing, but the diſtance of any ſtarre from the Ecliptique line, eyther towardes the North or South pole of the Zodiaque, which diſtance the ſemi-Circle of Longitude and Latitude made to paſſe through the bodie of the ſtarre, and cutting the Ecliptique line doth alwaies ſhew, as in the former example, in making the foreſaide ſemi-Circle to paſſe through the foreſaide ſtarre called hircus, I find by counting the degrées of the ſaid ſemi-Circle contained betwixt the Ecliptique line and the bodie of the ſtarre, that the Latitude of that ſtarre is 22. degrées and 30. minutes towardes the North, likewiſe by bringing the ſaid ſemi-Circle to the ſtarre which is in the right ſhoulder of Orion called Bed Alguze, I find his Latitude to be 17. degrées towards the South.

How

How to find out by the Globe the declination of any starre.

The 30. Proposition.

The declination is none other thing but the distance of any starre from the Equinoctiall, either towards the North pole or South pole of the world, which is to be found thus: First hauing brought the starre right vnder the brazen Meridian, and there staying the Globe, count the degrees of the said Meridian contained betwixt the saide starre and the Equinoctiall point or stroke of the said Meridian, and that shall be the declination of the starre: As for example, bring the starre Hircus vnto the Meridian, and you shall find the declination thereof to be 45. degrees towardes the North pole of the world, and that the starre which is in the foote of the Hare called Lepus, hath of South declination 23. degrees.

How to know the magnitude or greatnes of any starre, and his nature and qualitie by the Globe, and also the right ascention of the Arch of the Ecliptique which accompaneth the right ascention of any starre.

The 31. Proposition.

Marcator describeth the sire magnitudes of the starres by making sire figures or shapes of stars placed not farre from the head of the great beare, whereof the first and greatest hath 16. pointes or beames, the second eight, the third sire, the fourth fiue, the fift sire, which indeede by that account would be but foure: and the sixth hath fiue, which would haue but three: but in mine opinion it had bene better to haue made the first magnitude with ten points, the second with nine, the third with eight, the fourth with seuen, the fift with sire, and the sirt with fiue: so should the magnitude of euery starre described in the Globe haue bene the more easily knowne. But to the intent that you might exercise your selfe in finding out by the

Gg 3   Globe

The vse of the Globe.

Globe, the place, Longitude, Latitude, and declination of any starre that is described in the Globe: I haue thought good to adde hereunto the Table of Garceus, shewing not only the Longitude, Latitude and declination of the most notable starres that are both Northward and Southward, but also the right ascention, magnitude or bignesse, the qualitie or nature of euery such starre, and also the Arch of the Ecliptique line, which accompanieth the right ascention of euery star, which Table though by the saide Garceus was calculated out of the Astronomicall Tables for the yeare of our Lord 1564. and not by the Globe, yet for your better exercise in matters of the Globe, I thought good to set downe such Longitude, Latitude, declination, magnitude, and right ascention, and all other things contained in the said Table, according as they are to be found out by the Celestiall Globe of Mercator, and not calculated by any of the Astronomicall Tables.

Though the Longitudes and declinations of the fixed starres set downe in this Table to serue Mercator his Globe, doe not altogether agrée with the great Celestiall Globe, lately set foorth by M. Sanderson and by M. Molinaxe, by reason that the Longitudes and declinations of the saide starres are more lately calculated, that is to say, for the yeare of our Lord, 1592. yet it will serue to shewe you the way how to exercise your selfe in the said Globe, and thereby you may correct this Table, and make it to agrée in all points with the new great Globe, whereby you shall reape more pleasure then griefe or paine. For I had this Table (as I haue saide before) out of Garceus, and did here set it downe more for your exercise, and to acquaint you with the fixed starres that are described in the Celestiall Globe, then for any other purpose.

The vse of the Table next ensuing.

This Table is deuided into eight collums, The first whereof on the left hande, containeth the names of the starres, the seconde the degrées and minutes of Longitude, together with the Characters of the 12. signes: the third the degrées and minutes of Latitude both Northerne and Southerne, the North Latitude being marked with the

the letter N. and the South Latitude with the letter S. the fourth containeth the degrées and minutes of declination both Northern and Southerne, marked with the letters N. and S. as before. The fift sheweth the magnitude or greatnesse of the starre, whether it be of the first, second or third bignesse, &c. The sixth containeth the degrées and minutes of the right ascention of the foresaide fixed starres. The seuenth containeth the Characters of the Planets, signifying the nature and qualitie of the starres. The eigth containeth the degrées and minutes of the right ascention of the Ecliptique line and the signes of the Zodiaque.

Here followeth the Table.

The vse of the Globe.

The names of the starres.	The Longitude & place.		The signes.	The Latitude		North & South.
The first starre of the Rammes horne.	27	40	♈	7	40	Nor.
The right shoulder of Cepheus.	8	0	♈	69	0	Nor.
The last starre of Eridanus.	22	30	♈	53	30	Sou.
The shoulder of Andromeda.	16	20	♈	24	30	Nor.
The girdle of Andromeda.	24	50	♈	26	30	Nor.
The right side of Perseus	25	0	♉	30	0	Nor.
The head Algol or Meduza.	20	40	♉	23	0	Nor.
The South starre of Pleiades.	23	29	♉	4	20	Nor.
The North starre of Pleiades.	23	10	♉	5	0	Nor.
The Buls eye.	2	50	♊	5	0	Sou.
The right shoulder of Orion.	23	0	♊	17	0	Sou.
The left shoulder of Orion.	12	0	♊	17	0	Sou.
The right shoulder of Autiga	24	0	♊	20	0	Nor.
The Goate, or Hircus.	16	0	♊	22	30	Nor.
The former of the Kiddes	13	0	♊	18	0	Nor.
The latter of the Kiddes	13	30	♊	18	0	Nor.
The left foote of Orion.	10	10	♊	31	30	Sou.
The tayle of the little Beare	21	10	♊	66	0	Nor.
The middle starre of Orions girdle.	18	0	♊	24	30	Sou.
Canopus in the shippe Argo.	8	30	♋	75	0	Sou.
The great dogge.	9	0	♋	39	10	Sou.
The little dogge.	20	30	♋	16	0	Sou.
The former head of Gemini.	14	20	♋	9	40	Nor.
The following head of Gemini	17	40	♋	6	30	Nor.
The North Asellus.	1	30	♌	3	0	Nor.
The South Asellus in Cancer.	2	20	♌	0	10	Sou.
Praesepe, the Manger. in Cancer.	1	20	♌	0	40	Nor.
The shoulder of the great Beare.	11	40	♌	49	0	Nor.
The bright starre of Iydrus.	21	0	♌	20	30	Sou.
The Lyons heart.	23	30	♌	0	10	Nor.
The Lyons necke.	23	10	♌	8	30	Nor.
The Lyons tayle.	15	30	♍	12	0	Nor.
The middle starre of the great Beares tayle.	9	50	♍	56	0	Nor.
The right wing of Virgo.	3	10	♎	15	10	Nor.

The vse of the Globe. 229

Declination		Nor. & South	Magnitude or greatnes	The right Ascention		The nature of the stars shewed by the charecters of the Planets	The right ascention of the arch Ecliptique accompaning the starre		The Charecters of the signes
				Degr.	Min.				
17	20	Nor.	3	23	0	♂ ♄	25	0	♈
61	0	Nor.	3	317	15	♄ ♃	15	0	♒
41	0	Sou.	1	43	30	♃ ♀	15	50	♉
28	26	Nor.	3	5	0	♀	5	0	♈
33	0	Nor.	3	12	10	♀	13	0	♈
48	0	Nor.	2	43	0	♄ ♃	15	29	♉
39	30	Nor.	2	40	20	♄ ♃	13	0	♉
22	0	Nor.	5	50	10	♂ ☽	23	10	♉
23	10	Nor.	5	50	0	♂ ☽	22	20	♉
15	30	Nor.	1	63	0	♂	4	20	♊
6	20	Nor.	1	83	20	♂ ☿	23	50	♊
4	40	Nor.	2	72	0	♂ ☿	13	0	♊
43	0	Nor.	2	82	0	♂ ☿	22	50	♊
45	0	Nor.	1	71	30	♂ ☿	13	0	♊
40	0	Nor.	4	68	50	♂ ☿	10	28	♊
39	40	Nor.	4	69	0	♂ ☿	11	0	♊
9	0	Sou.	1	73	0	♃ ♄	14	20	♊
86	0	Nor.	3	4	30	♄ ♀	4	30	♈
1	50	Sou.	2	79	25	♃ ♄	20	20	♊
52	0	Sou.	1	94	24	♄ ♃	3	30	♋
16	0	Sou.	1	97	0	♂ ♃	6	30	♋
6	10	Nor.	1	109	30	♂	18	0	♋
32	0	Nor.	2	106	47	☿	15	28	♋
28	30	Nor.	2	110	5	♂	18	20	♋
22	30	Nor.	4	124	0	♂ ☉	1	40	♌
19	20	Nor.	4	124	50	♂ ☽	2	0	♌
20	0	Nor.	neb.	124	0	♂ ☽	1	30	♌
62	30	Nor.	2	161	0	♂	9	30	♍
5	0	Sou.	2	137	0	♄ ♂	14	30	♌
13	30	Nor.	1	145	93	♃ ♂	23	30	♌
21	40	Nor.	2	148	40	♄	26	0	♌
16	20	Nor.	1	171	10	♄ ♀	20	30	♍
56	55	Nor.	2	196	0	♂	17	0	♎
12	30	Nor.	3	189	0	♄ ☿	9	40	♎

The vse of the Globe.

The names of the starres.	The Longitude & place.		The signes.	The Latitude		North & South.
The left shoulder of Bubulcus.	10	0	♎	49	30	Nor.
The Rauens bill.	6	30	♎	21	40	Sou.
Arcturus the great starre betwixt Bubulcus his legs	18	0	♎	32	0	Nor.
Spica virginis.	17	40	♎	2	0	Sou.
The middle starre in the front of Scorpio	26	40	♏	1	30	Sou.
The South ballance of Libra.	9	0	♏	0	40	Nor.
The North starre of the ballance Libra.	13	10	♏	9	0	Nor.
The left hand of Serpentarius.	27	0	♏	17	0	Nor.
The bright starre of Ariadnes crowne.	5	30	♏	44	30	Nor.
The heart of the Scorpion.	3	30	♐	3	30	Sou.
The head of Hercules.	8	40	♐	37	30	Nor.
The head of Serpentarius.	15	30	♐	36	0	Nor.
The head of the Dragon	20	20	♐	76	0	Nor.
The bright starre of Lyra	9	0	♑	62	0	Nor.
The Eagle, alias vultur volans.	24	50	♑	29	20	Nor.
The first starre of the taile of Capricorne.	16	0	♒	2	0	Sou.
The following starre thereof.	17	10	♒	2	0	Sou.
The taile of the Dolphin.	8	10	♒	29	30	Nor.
The Fomahant.	28	0	♒	23	0	Sou.
The point of the dart.	1	10	♒	39	30	Nor.
The tayle of the Swanne.	0	20	♓	60	0	Nor.
The right thygh of Pegasus.	22	50	♓	31	0	Nor.
The taile of the Whale.	27	0	♓	20	20	Sou.

How to finde out the right and oblique ascention of any starre, and also of the ascentionall difference.

The 32. Proposition.

He right ascention of any star is an arch or portion of the Equator, to be counted from the first point of Aries according to the succession of the signes, with which portion in a right Sphere any starre both riseth mounteth to the Meridian, and setteth: and in an oblique Sphere it is a portion of the Equator, wherewith the star is mounted to the Meridian: as for example, in a great Sphere the star called Cor Leonis, that is to say, the hart of the Lion,

The vse of the Globe. 230

Declination	Nor. & South	Magnitude or greatnes	The right Ascention Degr. Min	The nature of the stars shewed by the charrects of the Planets	The right ascention of the arch Ecliptique accōpaning the starre	The Charrects of the signes.
40 0	Nor.	3	213 0	♄ ☿	5 0	♍
22 30	Sou.	3	176 43	♄	26 20	♍
22 0	Nor.	1	209 0	♃ ♂	1 0	♍
10 0	Sou.	1	195 10	♂ ♀	16 30	♎
21 0	Sou.	3	234 0	♂ ♄	26 30	♍
14 0	Sou.	2	217 0	♄ ♂	9 20	♍
8 0	Sou.	2	223 10	♃ ♀	15 46	♍
3 30	Sou.	3	238 30	♄ ♀	30 0	♍
28 0	Nor.	2	229 0	♀ ☿	21 20	♍
25 0	Sou.	2	241 0	♃ ♂	3 0	♐
15 0	Nor.	3	252 35	☿	14 0	♐
12 50	Nor.	3	258 30	♂ ☿	19 10	♐
52 0	Nor.	3	266 10	♄ ♂	26 30	♐
38 20	Nor.	1	275 20	♀ ☿	5 0	♑
7 30	Nor.	2	210 40	♄ ♂	20 4	♑
18 0	Sou.	3	319 0	♃ ♄	17 0	♒
17 30	Sou.	3	320 10	♃ ♄	18 0	♒
10 0	Nor.	3	304 0	♄ ♂	1 10	♒
33 10	Sou.	1	339 0	♀ ☿	7 0	♓
18 10	Nor.	4	295 0	♂ ♀	23 8	♑
43 30	Nor.	2	307 10	♂ ☿	5 0	♒
25 0	Nor.	2	341 0	♃ ♂	9 30	♓
19 50	Sou.	3	5 35	♄	6 0	♈

Lion, both riseth, mounteth and setteth with the 145. degr. 3′0. of the Equinoctiall. But the right ascention of the said starre in an oblique Spheare, is to be found onely by bringing the said starre to the Meridian, and you shall find it to be all one with the right ascention in a right Spheare, for by bringing the starre called Cor Leonis to the Meridian in an oblique Spheare, you shall finde the right ascention thereof to be all one with that which it had in a right Spheare, that is 145. degrēs 3′0. of the Equinoctiall. But if you would know the oblique ascention of any star, then hauing set the Globe at your Latitude, bring the starre to the East part of the Horizon, & marke what degrē of the Equinoctial
riseth

The vſe of the Globe.

riſeth therewith and that is the oblique aſcention: As for example, the Globe ſtanding at the Latitude 52. bring the ſtarre, Cor Leonis to the Eaſt part of the Horizon, & by ſtaying the Globe there, you ſhall finde the 127. degré of the Equinoctiall to riſe with that ſtarre. Now if you would know the aſcentionall difference, that is to ſay, the difference betwixt the right and oblique aſcention, you haue no more to doe but to ſubtract the leſſer out of the greater, and the remainder ſhall be the aſcentionall difference, as in the former example, take 127 degrées out of 145. degrées and 3′0 and there ſhall remaine 18. degrées 3′0. and that is the aſcentionall difference, by helpe wherof you may know the encreaſe and decreaſe of the artificiall day and night throughout the yeare in any Latitude, if you obſerue that order which I haue already ſet downe in the firſt part of my Spheare, the 50. Chapter.

To know in what quantitie of time any whole ſigne or any other Arch of the Ecliptique doth riſe or ſet.

The 33. Propoſition.

Auing ſet the Globe at your Latitude, and rectified the Index of the houre whéele by the ſeuenth Propoſition, bring the beginning of any Arch or ſigne to the Eaſt part of the Horizon, and marke what degré of the Equinoctiall riſeth therewith, that done, bring the end of the ſaid ſigne or arch to the Eaſt part of the Horizon, and marke there alſo what degré of the Equinoctiall toucheth the Horizon, and ſtaying the Globe there, looke how many houres or parts of houres the Index of the houre-whéele hath runne betwixt the beginning and ending of the ſaid ſigne or arch, and ſo you ſhall know the quantitie of that time, you may know it alſo by the number of degrées of the Equinoctiall contained betwixt the beginning and ending of the ſaid ſigne or arch by allowing 15. degrées to an houre, and 4′. to a degré. As for example, ſuppoſing the ſunne to be in the firſt degré of Taurus, you ſhall finde by working according to the rule before ſet downe, that the whole ſigne of Taurus doth ſpend in riſing one houre 8′. Now if you will knowe howe much time he ſpendeth in deſcending, bring the firſt degré of Taurus to the Weſt part

part of the Horizon, marking what degrée of the Equinoctiall toucheth the Horizon at that instant, and also to what houre the Index pointeth, then turne the Globe still Westwarde vntill the last degrée of Taurus méeteth euen with the edge of the Horizon, and then marke againe as well the degrees of the Equinoctiall that toucheth the Horizon, as also to what houre the Index pointeth, and you shall find the number of degrées of the Equinoctiall to be 42. degrees, which maketh two houres 48. minutes, which is answerable to the houre of the houre-whéele, and so much time the whole signe of *Taurus* spendeth in his descention or going downe.

*H*ow to know by the Globe what starres doe rise or set euery day Cosmically, Acronically, or Helically.

The 34. Proposition.

AS for the thréefold Poeticall rising and setting of the stars, you shall find them plainly defined in the first part of my Spheare, Chap. 35. but to finde out the same by the Globe, you must doe thus. First hauing set your Globe at your Latitude, and sought out the place of the sunne for that day, bring the degrée of the sunne to the East part of the Horizon, and stay the Globe there, that you may sée what stars doe rise a little before the sunne, and which rise together with the sunne. For those that rise a little before the sunne, are saide to rise Helically, and those that rise together with the sunne, are sayde to rise Cosmically, and those starres that are in the very West part of the Horizon at the rising of the sunne, are saide to set Cosmically. Againe those starres that rise immediatly after the sunne, doe set Helically, that done, turne the degree of the sunne vnto the West part of the Horizon, and staying the Globe there marke what starres are readie to goe downe with him, for those are said to set Acronically, and staying the Globe still there in the West, marke what starres at that present doe rise in the East part of the Horizon, for those are said to rise Acronically.

To

The vſe of the Globe.

To know in what time of the yeare any ſtarre riſeth or ſetteth, either Coſmically, or Acronically.

The 35. Propoſition.

Ere hauing ſet the Globe at your Latitude, and knowing the degrée of the ſunne, bring the ſtarre to the Eaſt part of the Horizon, and therewith conſider what degrée of the Ecliptique the Horizon cutteth at that preſent, that done, finde out the ſelfe ſame degrée vpon the Horizon in the narrow ſpace of degrées next vnto the bodie of the Globe, and right againſt that degrée, you ſhall finde in what day and moneth that ſtarre doth riſe Coſmically, As for example, I would knowe at what time of the yeare Cor Leonis riſeth Coſmically in the Latitude 52. I bring the ſtarre Cor Leonis to the Eaſt part of the Horizon, and I finde that the Horizon cutteth the Ecliptique in the 23. degrée 30. of Leo, which degrée being found againe vppon the Horizon, pointeth to the ſixth day of Auguſt, ſo as I conclude that Cor Leonis doth riſe that preſent day Coſmically, for then both he and the ſunne are in a manner in one ſelfe degrée of the Ecliptique: nowe to know the Coſmicall ſetting of the ſaide ſtarre, turne the ſame ſtarre to the Weſt part of the Horizon, and marke what degrée of the Ecliptique doth then riſe in the Eaſt, and you ſhall finde the ſame to be the 23. degrée of Aquarius, which degrée béeing found againe vpon the Horizon in the narrow ſpace of degrées next to the body of the Globe, containing the degrées of the Zodiaque, will point to the 31. day of Ianuarie, at which time the ſunne is oppoſite to the ſaid ſtarre, and therefore it is ſaide to goe downe Coſmically, becauſe it goeth downe when the ſunne riſeth. Now to know the Acronicall riſing of any ſtarre at any time, bring the ſtarre to the Eaſt part of the Horizon, and marke therewith what degrée of the Ecliptique goeth downe in the Weſt at that inſtant, for the ſunne being in that degrée is oppoſite to the ſtarre: As for example, by bringing the ſtarre Cor Leonis to the Eaſt part of the Horizon, you ſhall find that the 23. degrée of Aquarius goeth downe at that inſtant, which degree being found againe vpon the Horizon, will ſhewe the day and moneth when the ſtarre riſeth

Acroni-

Acronically, and so you shall find the starre Cor Leonis to rise Acronically the 31. day of Ianuary. Contrartwise if you woulde know when the said starre setteth Acronically, bring the said star Cor Leonis, to the West part of the Horizon, & therewith marke what degrée of the Ecliptique then setteth in the West, which degree being found againe vpon the Horizon, will shew that the said starre setteth Acronically the 6. of August.

Of the Horoscop and the rest of the twelue houses.

The 36. Proposition.

This word Horoscop doth not onely signifie the degree of the Ecliptique, otherwise called the ascendent which riseth aboue the Horizon in the beginning of any thing that is to be sought or knowne, but also sometimes the whole figure of heauen containing the 12. houses, and doth shew the very secrets of nature, so that there is nothing that chanceth to the inferiour bodies, but some cause thereof doth appeare by meane of the Horoscope in heauen, and therefore the Astrologians haue deuided the whole heauen into 12. houses, which are numbred from the Horoscope, which is the East Angle, and so forth according to the succession of the signes, of which 12. houses the foure principall are foure points of the Zodiaque whereof two doe fall vpon the Horizon, and the other two vpon the Meridian, and are called principall points, poles, or Angles, that is the beginning of the first house, of the fourth house, of the seuenth house, and of the tenth house, and those that doe follow next any of these principall Angles, are called succéding houses, in Latine Succedentes, as the second, the fifth, the eight, and the eleuenth house. And those that goe next before any of the foure principall Angles are called falling houses, in Latine Cadentes, as the 12. the third, the sixth, and the ninth: and such houses as haue no familiaritie with the Horoscop or ascendent, as the second, the sixth, the eight, and the eleuenth houses are said to be slow and deiect, all which things this Table here following doth shew, containing the number and names of the houses, and also their significations.

The

The vse of the Globe.

The 12. houses.	The names of the houses.		The significations of the houses.
1	Angle	East	life
2	succeeding	the lower gate	gaine
3	falling	the Goddesse	brethren
4	Angle	the bottom of heauen	parents
5	succeeding	good fortune	children
6	falling	euill fortune	health
7	Angle	the West	wife
8	succeeding	the higher gate	death
9	falling	God	religion
10	Angle	the middle of heauen	kingdome
11	succeeding	the good spirit	benefactor
12	falling	the euill spirit	prison

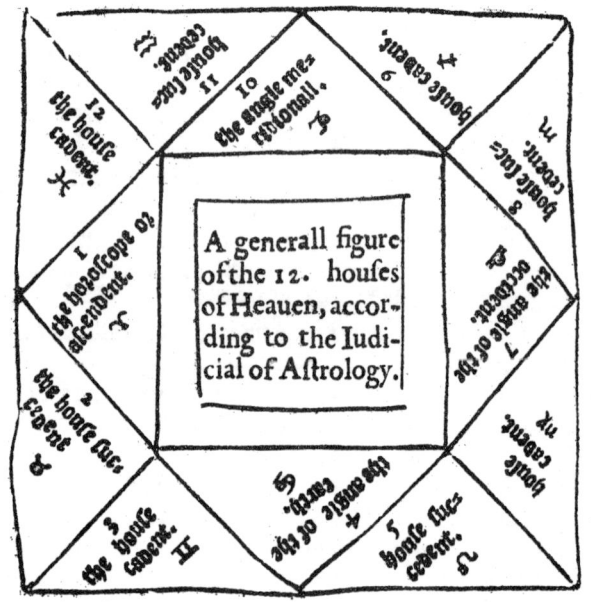

How

The vſe of the Globe.

How to finde out the Horoſcop or aſcendent at anie time of the day or night by the globe, and thereby to knowe the 4. principall angles of heauen.

Propoſition. 37.

Irſt hauing ſet the Globe at your latitude, and rectified the index of the houre wheele according to the degree of the ſigne wherein the Sun is that day you ſeeke by the 7. propoſition, if it be in the day time, take with your Aſtrolabe or Quadrant, the altitude of the Sun. But if it be in the night, take the altitude of ſome knowne Starre, that thereby you may knowe the houre of the day or night in which you ſeeke the aſcendent. But if it happen that neither Sunne nor ſtarre is to be ſeene that day or night, then learne by ſome true clocke or watch what houre it is, and hauing ſet the Index of the houre wheele at that houre, ſtay the Globe there, and therewith marke what degree of the Ecliptique riſeth in the Eaſt parte of the Globe aboue the horizon at that inſtant, and that degree is the Horoſcop or aſcendent for that houre. As for example, you woulde finde out the aſcendent the 16. of June 1590. at 8. of the clock in the forenoone, at which day the Sunne is in the 4. degree of Cancer, here hauing firſt ſet the Index of the houre wheele at that houre by ſtaying the Globe there, you ſhal finde that the 21. degree 30′. of Leo is the aſcendent, which is the Eaſt angle or firſt houſe, whereby you may alſo at that inſtant find out the other three angles, that is, the weſt, South, and North angles, for the oppoſit point of the Zodiaque to the aſcendent is the Weſt angle or ſeuenth houſe. And that degree of the Zodiaque, which is at that inſtant right vnder the Meridian aboue head in the South angle or 10. houſe, and the oppoſite point to that beneath, is the north angle or fourth houſe, for hauing found the aſcendent, which is the Eaſt angle or firſt houſe to be the 21. degree, 30. minutes of Leo, the Weſt angle muſt needs be the 21. degree. 30. minutes of Aquarius. Againe the ſouth angle or 10. houſe is the 8. degree of Taurus, and the oppoſite point to that is the North Angle or fourth houſe which is the 8. of Scorpio.

H h

The vse of the Globe.

How to erect a figure by the Globe according to *Regio Montanus* his way which is called the reasonable way, and is counted the best of all others.

Proposition. 38.

Irst you must find out the degree of the ascendent as is taught in the last chapter, which is alwayes the first house, then staying the Globe with some pretie wedges of wood being thrust betwixt the Horizon and the bodie of the Globe, marke there what degree of the equinoctial doth touch the Horizon at that instant, and number from thence vpward vpon the said Equinoctiall 30. degrees, to the ende of which 30. degrees, bring the semicircle of position being first fastened in his due place vpon the East side of the Horizon, and looke what degree and signe of the Zodiaque the circle of position cutteth at that present, and that degree shall be the twelfth house: then number againe other 30. degrees vpon the Equinoctiall vpwardes towards the brazen Meridian, and to that bring the semicircle of position, marking what degree of the Zodiaque the said semicircle cutteth, and that shall be the 11. house, that done, looke what degree of the Zodiaque is right vnder the brazen meridian aboue heade, and that shall be the 10. house. Then hauing set the semicircle of position vpon the West side of the Globe, number from the brazen Meridian westward vpon the Equinoctiall other 30. degrees, to the end whereof bring the semicircle of position, and marke what degree of the Zodiaque the circle of position cutteth, and that shall be the 9. house, then from thence downward number vpon the Equinoctiall other 30. degrees, to the end whereof bring the semicircle of position, marking there what signe and degree of the Zodiaque the saide semicircle cutteth, and that shall be the 8. house. Now hauing these sixe houses, the opposite points of the sayd sixe houses will shewe you the other sixe houses, and if you will know which houses, and also which signes are opposite one to another, marke well this Table following.

The

The vse of the Globe.   234

The houses op- $\left\{\begin{array}{ll} 1 & \text{to } 7 \\ 12 & 6 \\ 11 & 5 \\ 10 & 4 \\ 9 & 3 \\ 8 & 2 \end{array}\right\}$ The Signes op- $\left\{\begin{array}{ll} \Upsilon & \text{to } \libra \\ \taurus & \scorpio \\ \gemini & \sagittarius \\ \cancer & \capricorn \\ \leo & \aquarius \\ \virgo & \pisces \end{array}\right\}$

And to make al these things the more plain vnto you, I thought good to set downe this example following. Suppose that the 16. of Iune, Anno 1590. and at 8. of the clocke in the morning you would erect a figure to knowe how the 12. houses of heauen are situated at that present, first hauing drawne such a square figure as this here following, representing the 12. houses, learne by the last proposition, who is the ascendent at that instant, and you shall finde it to be the 21. degrée 30′. of Leo, which must bée set in the first house, and the 28. degrée of Cancer in the 12. house and the 22. degrée of Gemini in the 11. house, and the 9. degrée of *Taurus* in the 10. house, and the 5. degrée of Aries to be in the 9. house, and the 12. degrée of *Pisces* in the 8. house. Nowe the opposite house to the first house or ascendent is the 7. house which is the 21. degrée. 30′. of Aquarius, for the opposite signe must alwayes haue like number of degrées, then the opposit to the 12. house is the 6. house, which the 28. deg of Capricorne, and the opposit to the 11. is the 5. house, which is the 22. degrée of Sagittarius, and the opposite to the 10. house is the 4. house, which is the 9. degrée of Scorpio, and the opposite to the 9. house is the 3. house, which is the 5. degrée of Libra, and the opposite to the 8. house is the second house which is the 12. degrée of Virgo, all which things this figure here following doth plainely shewe. And if you would know what Planets should be placed in euerie house, you must learne that out of the Ephemerides, or out of some of the Astronomicall Tables.

H h 2   Now

The vse of the Globe.

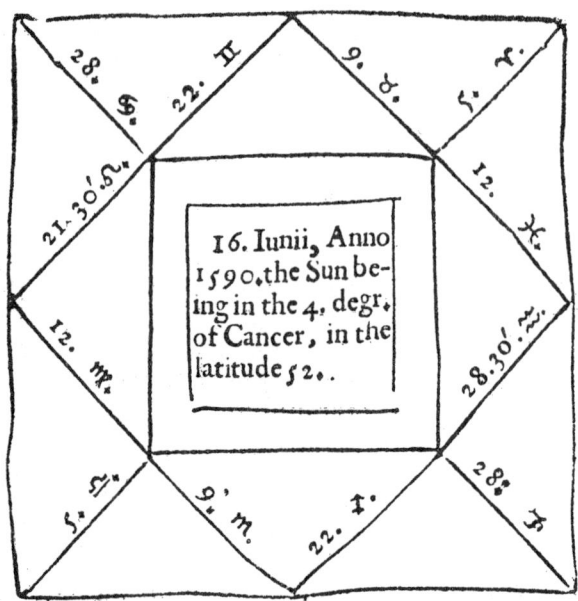

Now because most men that doe shewe the vse of the Globes, doe also teach therein how to finde out the latitude and longitude of anie region, I thought good therefore to set downe here some of their manifolde waies touching the finding out of the same, notwithstanding that I haue alreadie written something thereof in the second part of my Spheare, in the 8. 9. 10. & 11. chapters.

How to know the latitude of any place or region by any of the fixed starres described in the Globe.

Proposition. 39.

Take in the night season with your Astrolabe the Meridian altitude of some knowne starre that is to be found in the Globe, then hauing brought the starre vnder the brazen Meridian, turne the Meridian vp and downe in the nickes of the Horizon, vntill the same Starre haue the

saide

said altitude in the brazen Meridian, which you found it to haue in the firmament by your Astrolabe, that done, number the degrées of the Meridian contayned betwixt the Pole and the Horizon, and that is the latitude of that place.

Another way to find the eleuation of the Pole.

Proposition. 40.

Hauing brought the Globe into an open place where the Sunne shineth at noonetide, and placed the same right North and south as is taught in the first chapter, set a nédle in the degrée of the Sunne, and bring the same to the brazen Meridian, not leauing to turn the same Meridian vp and downe in the nicks of the Horizon, vntill the nédle cast no shadow at all, and there staying the Globe, looke how many degrées the pole is eleuated aboue the Horizon, for that is the latitude of that place.

A third way to finde out the latitude of any place without taking the Meridian altitude of any Starre.

Proposition. 41.

Take at any houre the altitude of 2 knowne stars, and such as are to be found in the Globe at one selfe instant whereof the one must be situated towards the East, and the other towards the West, then turne the Globe together with the Meridian vp and downe in the nicks of the Horizon, vntill you finde by helpe of the quarter of altitude each Starre to haue the selfe same altitude in the Globe, that it had in heauen, which being done, looke how many degrées the North pole is eleuated aboue the Horizon, and that is the latitude of that place. But if you would knowe the latitude of any place that is towardes the South Pole, then you must first place the South pole aboue the Horizon, and then worke as before.

The vse of the Globe.

A fourth way to find out the latitude of any region by any knowne fixed starre or Planet that may be seene.

Proposition. 42.

Irst take his Meridian altitude, and then learne to know eyther by the Globe or by some table his declination, which if it bee Northerne you must subtract the same from the Meridian altitude, if Southern adde it to the Meridian altitude so shal you haue the eleuation of the Equinoctiall, which being subtracted from 90. the remainder shal be the latitude of that place as for example I find the Meridian altitude of Oculus *Tauri* to be 53.degrees, 50'.32". and his declination Northward to be 15. degrees, 50'. 32". which declination being subtracted from the Meridian altitude, there remaineth 38.degrees, and that is the eleuation of the Equinoctiall, which being subtracted from 90. there remaineth 52. for the latitude of that place.

Hitherto I haue set down the chiefest propositions that are to be done by the Globe, touching the Sun and the fixed stars: & lastlie, shewed how to find out the latitude of any place, wherefore now I thinke good to shew you how to find out the place of the Moone, and of euery one of the rest of the Planets in the Globe by help of the Ephemerides, and thereby to know when euery planet riseth and setteth, and first of the Moone.

A briefe description of the diurnall table set downe in *Stadius* his Ephemerides, together with the vse thereof.

Proposition. 43.

Ut for as much as the diurnall Table of the Ephemerides shewing the dayly motion of the Planets is very nedful to serue diuers turnes I think it not amisse here briefly to describe the same, and specially that of Iohannes Stadius whose diurnall table or Almanacke beginning at the 202. page of his booke, and at the yere of our Lord 1583. continueth to the yeare 1606.

Of which table euery page on the left hand is diuided into 9.
Col-

Collums. In the first collum wherof on the left hand are set down the dayes of the moneth, first the Gregorian dayes according to the Romane account, and next to that the dayes of the moneth according to our English account, then in the front of euery other collum are set downe the characters first of the Sunne, and then of other the sixe Planets that is to say, of the Moone, Saturne, Iupiter, Mars, Venus, Mercurie, and last of all, the head of the Dragon figured thus ☊. And right vnder these seuen Planets, and also vnder the head of the Dragon are set downe the signes and degrees wherein euery of these is euery day of the moneth throughout the yeare at noomtyde, and in the foot of the sayd Table is set downe the latitude of euerie one of the 5. Planets, proceeding by the dayes of the moneth diuided into three partes. And in the margent of euery left page are set downe the chiefest feastes and Saints dayes that fall in euery moneth throughout the yeare. Moreouer, there is a Table on the right hand right against the left Table, in which are set downe first the dayes of the moneth, and then what coniunction or any other aspect the Moone hath with any of the other sixe Planets, that is, with the Sunne, with Saturne, with Iupiter, with Mars, with Venus and with Mercerie, which Planets are set downe in the front of the saide Table, and vnder them the characters of such aspertes as the Moone hath that day with any of the other Planets. The characters of which aspects are these here following.

☌ ☍ △ ☐ ✶

Whereof the first signifieth a coniunction: the seconde an opposition, the third a trine aspect, the fourth a quadrat aspect, and the fift a sextile aspect.

Two Planets are said to be in a coniunction when they are both in one selfe signe. And to be in an opposition when they are in two seuerall signes opposite one to another. For then they be distant one from another 6. signes. And they are saide to bée in a trine aspect when they be distant one from another by four signes. And to be in a quadrat aspect when they are distant one from another by thrée signes. And to be in a sextile aspect when they are distant but two signes one from another.

The vse of the Globe.

How to finde out the place of any *Planet* by the Ephemerides.

Proposition. 44.

Ow to find out the place of any Planet, or of the head of the Dragon by this diurnall Table, you must first seeke out the day of the moneth in the fist collum of the left table, right against that on your right hand in the sayd left Table, you shall finde in the common angle right vnder the Planet or Dragons head (which so euer of them you seeke) the signe and degree wherein the said Planet or Dragons head is the said day at noontyde.

And to find out the aspects which the Moone hath with any of the Planets the same day, you must resort to the other table on the right hand, obseruing like order as before.

The example.

As for example the 21. of Aprill 1592. which is the first of May, according to the Romain account, I find by the table on the left hand, the Sunne to be in the 10. degree 50'. of *Taurus*, the Moone to be in the 3. degree, 47'. of Capricorne, Saturne, to be in the 8. degree, 10'. of Cancer, Iupiter to be in the 18. degree. 28'. of Sagittarius, Mars to be in the 12. degree, 6'. of Gemini, Venus to be in the 2. degree, 0'. of Aries, and Mercurie to be in the 6. degree, 20'. of *Taurus*, and the head of the Dragon to be in the 29. degree, 45'. of Gemini.

And right against this in the Table on the right hande you shall find the Moone to bee in a trine aspect with the Sunne, to bee in an opposition with Saturne, to bee in a trine aspect with Mercurie.

Thus hauing brieflie shewed you the vse of the diurnall Table, I will shew you now how to find out the latitude of the Moone as well North as South, by help of *Stadius* his Table, set down in the 112. page of his Ephemerides, and first I will briefly describe the said Table.

A

The vse of the Globe.   237

A briefe description of the Table of *Stadius* set downe in the 112. page of his Ephemerides to finde out thereby the dayly latitude of the Moone be it North or South together with the Canon or rule thereof plainly declared by example.

Proposition. 45.

His Table is diuided into 8. collums, wherof the first on the left hande contayneth the degrées of euery signe set downe in the front of the Table, which degrées are to be counted descending from one to 30. for so manie degrées there be in euery signe: and the last collum on the right hand contayneth the like number of degrées belonging to the signes set downe in the base or foote of the sayd Table, and this number ascendeth vpward from 1. to 30. and for that purpose it woulde not haue bene amisse to haue set ouer each head of those 2. collums this word gradus, next vnder the word Signa. And of the other sixe collums the first thrée on the left hand doe containe the degrées, minutes and seconds of the North latitude, and the other thrée towards the right hand do containe the degrées, minutes, and seconds of South latitude. Moreouer, the 12. Signes are to bée numbred in the front from the third collum on the left hand from 1. to 5. forward toward the right hand, and at the foot from 6. to 11. backward towards the left hand, set downe in arithmeticall figures.

The rule or Canon together with a plaine example shewing the vse of the Table.

First knowing the day of the moneth, resort vnto the diurnall Table of motion of the Planets in the Ephemerides, and hauing there found out the motion or place of the Moone, and also of the Dragons heade answerable to the day wherein you séeke, subtract the place of the Dragons head from the place of the Moone, which is easily done so often as the arke of the Moone is greater, that is to say, contayneth more signes and degrées than the arke of the Dragons heade, beginning your account in both arkes from the first point of Aries. But if the arke of the Moone bée

The vſe of the Globe.

be leſſer than the ark of the Dragons head, ſo as you cannot make your ſubtraction, then you muſt add to the place of the Moone 12. ſignes, which is 360. degrees, and you muſt adde alſo thereunto the number of ſo many ſignes as are contayned betwixt the firſt point of Aries, and the firſt point of that ſigne wherin the Moone is at that preſent, which ſigne it ſelfe is not to bee numbred, and when you come to take out of that whole ſumme the place of the Dragons head, you muſt firſt adde to the ſaid place of the Dragons head the number of ſo many ſignes as are contained betwixt the firſt point of Aries, and the firſt point of that ſigne wherein the Dragons head is at that inſtant, but not the ſigne wherein it is, and then hauing made your ſubtraction, remember alwayes to take out of that remainder 90. degrees, which is three ſignes, ſo oft as you haue neede to adde 12. ſignes to the place of the Moone, and not otherwiſe, and with that remainder you muſt reſort to the foreſaid table of the Moones latitude, as for example.

The Example.

Suppoſe that you would knowe what latitude the Moone had the firſt of Nouember 1590 here reſorting to the diurnall Table of the Ephemerides, you find according to the day propoſided, the place of the Moone tobe in the 16. degree 49′. of Taurus, and the place of the Dragons head to be in the 28. degree 14′. of Cancer. Now according to the rule before giuen, you muſt take the place of the dragon here, which is 28. degrees, 14′. of Cancer out of the 16. deg. 49′. of Taurus, which is the place of the Moone, and becauſe you cannot take the greater ſum out of the leſſer, you muſt adde to the leſſer ſum 12. ſignes, which make 360. degrees, and alſo one ſigne, for Aries going next before Taurus, in which ſigne the Moone is, ſo ſhall you make the whole ſum to be 13. ſignes 16 degrees 49′. out of which ſumme you muſt ſubtract the Dragons head, which with the ſignes that goe next before Cancer, counting from the firſt point of Aries, do make 3. ſignes 28. degrees, 14′. which being ſubtracted out of 13. ſignes, 16. degrees, 49′ there remaineth 9. ſignes, 18 deg 35′. out of which you muſt alſo ſubtract 90. deg. which is 3. whole ſignes, & ſo you find the remainder to be 6. ſignes, 18. degrees 35′. with which laſt remainder you haue to enter into the table of the Moones latitude, in the foot where-

The vſe of the Globe.   238

wherof you ſhal find 6. ſignes, and in the laſt collum on the right hand 18. & in the next collum towards the left hand, & in the common angle anſwerable aſwel to the ſaid 18. degrée, as alſo to the 6 ſignes, you ſhal find the latitude of the Moone to be 4. deg. 45′. & 17″. (which ſeconds may be very wel omitted) and her latitude to be ſouth. But now becauſe there are 35′. more annexed to the 18. degrées of the foreſaid remainder, you muſt find out a proportionall part anſwerable to thoſe minutes which is to be done thus.

Take out of the table the whole latitude anſwerable to 6 ſignes & 19. deg. which is one degrée more, ſo as now the latitude of the Moone is 4. deg. 43′. omitting the ſeconds. Then ſubtract 4. deg. 43′. out of 4. degrées 45′. & there remaineth 2′. Now to find out a proportionall part anſwerable to the former 35′. you muſt ſay thus. If 60′. require 2′. what ſhall 35′. require? and the quotient yéeldeth 1′. 10″. which being ſubtracted out of 4. degrées 45′. there wil remaine 4. degrées. 44′. and ſo much was the ſouth latitude of the Moone at that preſent day.

How to know the true place of the Sun or Moone, or of any other planet euery houre of the day throughout the yeare.

Propoſition. 46.

BVt now becauſe that to finde out the true place of the Moone or of any other Planet in the Globe to work thereby certainly and truly, it is not inough to knowe their places in the Zodiaque at noonetyde, vnleſſe you knowe the ſame at the verie houre in which you ſecke. Stadius teacheth a briefe rule to find out the true place of any Planet at what houre ſo euer you deſire by helpe of a ſhort Table ſet downe in the 109. page of his Ephemerides, which rule and Table he borrowed of Reinholdus. And this Table conſiſteth of 3. collums, in euery front whereof are ſet firſt degrées minutes ſeconds, and thirds, and vnder them minutes, ſecondes, thirds, and fourths. And note that the firſt row of numbers on the left hand ſignifieth ſometime degrées, and ſometime minutes, eyther of which do extend in this Table to 60. & to 61. right againſt which rowe in euery collum is ſet down on the right hand the proportionall part for one houre, the rule is thus.

The

The vſe of the Globe.
The Rule.

First finde out by the Ephemerides the place of the Planet wherein it is at noontyde the same day that you ſeeke, and also the place of the same Planet, wherein it is the next day following at noone, and subtract the leſſer out of the greater, if the two places at noonetyde be ſtill in one ſelfe ſigne, but if the two places beé in two ſeuerall ſignes, then you muſt count how many degrées of the Ecliptique the one place is diſtant from the other, and that ſhall be the difference, with which difference you muſt reſort to the Table, and right againſt that on the right hande in the same collum wherein you found the difference, you ſhall also find the proportionall part for one houre, as by this example you ſhall more plainly vnderſtand.

The Example.

Suppoſe then that you deſire to knowe the true place of the Sunne or of the Moone, or of any other Planet at fiue of the clocke in the after noone the 30. of December 1591. at which day you find the Sunne at noontyde to be in the 18. degrée, and 7. minutes of Capricorne, and the next day at noone to be in the 19. degrée and 8'. of the same ſigne, the difference whereof you finde by subtraction to be one degrée and one minute, with which difference you muſt enter into the table, and ſéeking for one degrée in the firſt collum on the left hande, you finde next vnto that on the right hand 2'. 30". then for one minute which before was named one degrée, you find next vnto it 2". 30'". which being added to the former ſumme laſt found, maketh 2'. 32". 30'". which is the proportionall part for one houre. Then hauing multiplyed that by 5. houres to ſerue 5. of the clocke in the afternoone, you ſhall find the product to be 12'. 42". 30'". which being added to the firſt Meridian place, which is 18. degrées and 7. minutes of Capricorne, maketh in all 18. degrées 19'. 42". 30'". And remember that if your difference be only minutes, then the bodie of the Table doth ſhew the proportionall part. But if the difference doth containe degrées, then the fourths muſt bee made thirdes, and the thirds ſecondes, the ſecondes minutes, and the minutes degrées. And this table ſerueth for the 7. Planets for euer. Notwithſtanding Stadius ſetteth downe another Table to finde out thereby the proportional moouing of the Moone ſeruing as wel for houres

as

The vſe of the Globe.   229

as for minutes of houres, which table beginneth at the 144. page, and endeth at the 184. page of his booke.

In the front of which table in euery page are ſet downe the differences of the two places of the Planets which you haue found by ſubtracting the leſſer out of the greater, if the planet be at both noonetydes in one ſelfe ſigne, but if ſhe be at the two noontides in two ſeuerall ſignes, then ſuch difference is to bee accounted vpon the Ecliptique line, to know howe many degrees the one place is diſtant from the other, and as for the minutes, if there be any, you may know the difference thereof by ſubtracting the leſſer number out of the greater. And in the outermoſt collum on the left hande are ſet downe the houres and minutes marked in the foote with the letter H. ſignifying houres, and vnder that with the letter M ſignifying minutes, whereof the houres proceede but to 24. but the minutes extend to 60. which make one houre proceeding by euen numbers in this ſorte. 1. 2. 4. 6. 8. &c. and therefore not finding that houre or minute which you ſeeke in the ſayd collum, you muſt take the next number which is leſſer by one, and ſo make it vp by adding the firſt and onlie odde one vnto it, and in each common angle you ſhall find the proportionall part that is to bee added to the firſt place of the Planet anſwering to the day of the moneth wherein you ſought. All which thinges you ſhall more plainelie vnderſtand by this one example.

The example.

Suppoſe then that you would knowe the true place of the Moone at 5. of the clocke. 17′. in the afternoone the 28. of December 1591. and looking in the Ephemerides you find the place of the Moone to be the ſame day at noonetide in the 26. degree 12′. of Libra. and the next day at noontide to be in the 8. deg. 17′. of Scorpio. here by numbring how many degrees are contayned in the Ecliptique line betwixt thoſe two places, and by ſubtracting the leſſer number of minutes out of the greater, you ſhall finde the difference to be 12. degrees 5′. with which difference you muſt reſort to the table of proportionall parts, and there hauing found the ſaid difference in the front of the 13. page of the ſaid table, look in the firſt collum on the left hand, for the houres and minutes according to the rule before ſet downe. As here in this example, becauſe you cannot find 5. houres, you take foure, right againſt

which

The vſe of the Globe.

which on the right hand in the common angle you ſhal find 2. degrees 0'. 50". then for one houre to make the 5. houres, you shall find in the common angle 30'. 12", that done, ſeeke out the 17. in the firſt collum, which number not being there you muſt take 16. againſt which in the common angle you shall find 8'. 3". 20'''. and for the one minute which is to be added to 16. you ſhall find in the common angle againſt that one minute 30". 12''', and by adding all theſe ſummes together, you ſhall find the proportionall part to be 2. degrees, 39'. 35". 32'''. which being added to the place of the Moone at noontyde firſt found in the Ephemerides, maketh in all 28. degrees, 51'. 35". 32'''. of Libra, which is the true place of the Moone for that houre. But you haue to note that the denominations of the numbers contained in the common angles of proportional parts be not alwaies like, for when they are to anſwer houres then the firſt number on the left hand in euery angle ſignifieth degrees, and the reſt minutes & ſeconds, but if they haue to anſwere minutes of houres, then the firſt numbers doe ſignifie minutes, and the reſt ſeconds and thirds, as you may eaſilie perceiue by examining the former example. But nowe to returne to my firſt intention, which was to ſhewe you how to find out the true place of euerie Planet in the Globe, you haue to vnderſtand that hauing found out the true place of any Planet at the day and houre wherein you ſeeke, by ſuch meanes as is before taught, then reſort to the celeſtiall Globe, and hauing ſet the same at your latitude, and alſo rectified the houre wheele according to the day wherein you ſeeke. Suppoſe that the 26. of May 1592. you would knowe in what part of the Globe the Moone is to be found at 5. of the clocke in the afternoone. Here firſt you muſt knowe her place at noone the ſame day, which is the 14. degree, 28'. of Aries, and alſo her place the next day at noone, which is the 29. degree 0'. of the ſame ſigne. Now by taking 14. degrees 28'. out of 29. degrees, you ſhal finde the remainder to bee 14. degrees, 32'. with which difference you muſt reſorte to the table ſet downe in the 109. page. And by working as before is taught, you ſhal finde her place or longitude to be at fiue of the clocke in the 17. degree 23'. 6". 43'''. of Aries, which being counted vpon the Ecliptique line of the Globe, lay your ſemicircle of longitude and latitude to that point, and hauing learned her latitude by helpe of the Table of latitude before

fore mentioned, by which you shall find her latitude to be at that time 4. deg. 46'. to the Southward, then hauing counted that latitude vpon the semicircle of lōgitude & latitude, make there a mark vpon the Globe, for that is her very place at that instant that is to say, at 5. of the clocke in the afternoone the 26. day of May 1592

Now if you would know when she riseth and setteth, you haue no more to do but to bring her place before marked on the Globe vnto the Horizon on the East part, and the index of the houre whéele being rectifyed as is before said, will shew the houre of her rising aboue the Horizon, and by bringing her place to the west part of the Horizon, the Index will shew the houre of her setting, and by bringing her place to the brazen Meridian, you shal know at what houre she is full South. The like Tables you shall finde also in Stadius to find out the true place of euery other Planet at anie houre, as of Saturne, Iupiter, Mars, Venus, and Mercurie, which Tables doe beginne at the 128. page, and ende at the 143. page of his booke, the order of working by which Tables is like in euerie respect vnto that of the Moone last taught. By which Tables you may finde out the true place of anie other Planet in the Globe at anie houre of the day or night, and thereby to knowe the houre of his rising and setting, if you rightly obserue the rules before taught touching the vse of the foresaid Tables. And note that you may finde out the place of the Moone by the Globe without the helpe of any Ephemerides, by such wayes as here doe followe.

How to finde out the place of the Moone by the Globe, when she is aboue the *Horizon*, without the helpe of any E-phemerides or other *T*able whatsoeuer.

Proposition. 42.

Auing set the Globe at your latitude, and placed it so as it may rightly answere the foure quarters of the world, take the altitude of the Moone with your Astrolabe or quadrant, and marke therewith whether she be at that present East, or West, that is to say, on this side of the Meridian, or beyond the Meridian, that done,

take

The vſe of the Globe.

take the altitude of some fixed Starre, which you know, and is to be found in the Globe, and also at that time is aboue the Horizon, marking therewith in what part of the firmament the said Starre is, and with your crosse staffe take the distance betwixt the moone and that star. Then hauing these three things, that is, the altitude of the Moone, the altitude of the Starre, and also the distance betwixt the Moone and the knowne starre. Moue the globe together with the quarter of altitude to and fro vntill you haue made the Moone to haue the same altitude and like place in the Globe, that you founde it to haue in the firmament, and there make a marke vpon the Globe. Then bring the quarter of altitude towards the Starre, that the Star may haue the like altitude and like place in the Globe that it had in the firmament, and there hauing stayed the Globe that it may not moove, take with your Compasse vpon the Equinoctiall the distance betwixt the Moone and the starre before found by the crosse staffe, and keeping your compasses at that wydenesse, put the firme foote of your Compasses in the fixed Star, and cease not to moove the other foot together with the quarter of altitude towards the mark of the Moones altitude vntill you make them to meet, for there is the place of the moone.

Another way to find out the place of the Moone without taking the altitude of any Starre.

Propoſition. 48.

Seeke by your crosse staffe to knowe the distance betwixt the Moone and any two stars that you know & are to be found in the globe. That done, drawe vpon the globe on either starre an obscure circle according to the distance of the Moone from either of those two stars, which two circles will cut or crosse one another but in 2. points the one whereof is the place of the Moon, and which of those places it should be, your eie will easilie tell you. And by this meanes it were no hard thing (as Gemma Friſius ſaith) for our sea men in these daies to finde out all the starres that be in the nether Hemispheare and were vnknown

The vse of the Globe. 241

knowne to Ptolomie and to the other auncient Astronomers, and to cause them to be set downe in the Globe, yea and by this art the places of the rest of the Planets may be found out as well according to their Latitude as Longitude.

Moreouer Gemma Frisius sayth, that by knowing the true place of the Moone in the Zodiaque, you may also finde out therby the vnknowne Longitude of any Region.

How to finde out the Longitude of any Region.

The 49. Proposition.

Auing found out the place of the Moone in the Zodiaque, you must first know the very houre of her being in that place, and then learne by some Ephemerides or by the Tables of Alfonsus, at what houre the Moone doth enter into the selfe same degree of the Zodiaque in some other Region or Towne whose Longitude you alreadie knowe, and hauing reduced the houres to 24. take the lesser number of houres out of the greater, the remainder whereof must be reduced out of houres and minutes into degrees thus: Multiply the houres by 15. and the minutes of houres by 4. so shall you haue the degrees of the Equator contained betwixt the two Meridians. And such distance so intercepted is called the difference of Longitude, which difference you must adde to the knowne Longitude if the houres in that place were more in number, but if the houres were lesse in number, then you must subtract the foresaide difference from the knowne Longitude, so shall you collect the vnknowne Longitude of that place or Region which you seeke, and how far it is distant from the fortunate Iles.

Another way to find out the vnknowne Longitude of any place by the Globe.

The 50. Proposition.

Auing set the Globe at your Latitude and rectified the houre-wheele by the seuenth Proposition, bring the place whose Longitude you know to the brazen Meridian, and direct the Index of the houre-wheele to that houre

The vſe of the Globe.

houre in which the Moone doth occupie the former defined place in that Region or Towne: that done, leaue not to turne the Globe vntill the Index of the houre-wheele come to that houre in which you ſought the vnknowne place of the Moone, and the degrees of the Equator which the brazen Meridian cutteth, wil ſhew the vnknowne Longitude of the place which you ſeeke.

To the ende of this Treatiſe I haue thought good to adde a briefe deſcription of the two great Globes lately ſet foorth firſt by M. Sanderſon, and then by M. Molineux, and therewith to ſet down a briefe deſcription of Sir Frances Drake his firſt voyage into the Weſt and Eaſt Indies, and alſo the voyage of M. Tho. Candiſh both whoſe voyages and what courſe they helde are to be ſeene by helpe of two lines drawne on the terreſtriall Globe, of which lines the one is redde ſhewing the voyage of Sir Frances both outward and homeward. And the other line is blew ſhewing in like manner the voyage of M. Candiſh.

A briefe deſcription of the two great Globes lately ſet forth firſt by M. Sanderſon, and then by M. Molineux.

Theſe Globes doe differ in a maner nothing at all from the Globes of Mercator touching the circles before deſcribed, but only in the Horizon, for the Horizon of theſe great Globes is deuided into 13. ſpaces as followeth.

Whereof the firſt narrow & innermoſt ſpace next vnto the bodie of the Globe, containeth the degrees of the Zodiaque.

The ſecond containeth the numbers of the ſaid degrees proceeding from 10. to 30. in euery ſigne.

In the third ſpace are ſet downe the names and the characters of the 12. ſignes, and alſo the characters of the Planets that gouerne the ſaid ſignes.

In the fourth ſpace are ſet downe the letters of the dayes of the weeke.

In the 5. and 6. are ſet downe the numbers of the daies and names of the monethes according to the auncient Calender.

The vſe of the Globe.   242

In the ſeuenth the feſtiuall daies.

In the 8. and 9. ſpace the number of the daies and names of the monethes according to the new Romain Calender.

In the 10. & 11. are ſet down the numbers of the daies & names of the monethes according to the true Kalender lately calculated by a moſt excellent Mathematician and mine old acquaintance M. Dee of Mortlake, as I coniecture by the letters I. D. ſet downe vpon the ſaid Horizon.

In the 12. are ſet downe the Engliſh names of the 32. Rombes or windes of the Mariners compaſſe.

In the 13. and outermoſt ſpace are ſet downe the Latine names of the ſaid 32. windes.

But the Mappe which couereth M. Molineux his Terreſtriall Globe, differeth greatly from Mercator his terreſtrial Globe by reaſon that there are found out diuers new places aſwel towards the North pole, as in ẏ Eaſt & Weſt Indies which were vnknown to Mercator. They differ alſo greatly in the names, Longitudes, latitudes & diſtáces of ſuch places as haue bin heretofore ſet down not only in Mercators Globe which was made many yeres ſince, but alſo in diuers Maps more lately made, but who goeth nigheſt the truth I dare not iudge, becauſe I was neuer in thoſe places. But as touching the Map of ſtars which couereth the celeſtiall Globe of M. Molineux, I do not finde it greatly to differ from that of Mercator, ſauing that M. Molineux hath added to his celeſtiall Globe certaine Southerne images, as the Croſſe, the ſoutherne Triangle, and certaine other ſtars, whereof ſome do ſignifie Noes Doue, & others do ſignifie the image called Polophilax, the images whereof are not here ſet downe, but you ſhal find them deſcribed in Plancius his Map, made in the yeare 1592. whoſe longitudes, latitudes, and declinations, how truely they are ſet downe in the ſaid Globe or Mappe I am not able to iudge. He ſetteth downe alſo two cloudes nigh vnto the South pole, but not the vſe thereof. Moreouer to this briefe deſcription of M. Molineux his two Globes, I thought good to adde the firſt voyage of Sir Frances Drake, and of M. Thomas Candiſh, ſet forth by two lines, the one redde, and the other blew deſcribed in the Terreſtriall Globe of the ſaid M. Molineux, and alſo how farre Sir Martin Furboſher ſayled Northward as followeth.

Ii 2   The

The vſe of the Globe.

The firſt voyage of Sir Frances Drake by ſea vnto the Weſt and Eaſt Indies both outward and homeward.

IN the great terreſtriall Globe lately put foorth by M. Sanderſon and by M. Molineux, the voyage aſwel of Sir Fr. Drake, as of M.Th.Candiſh is ſet downe, & ſhewed by helpe of two lines, the one red, & the other blew, wherof the red line proceeding firſt frō Plymouth, doth ſhew what courſe Sir Frances obſerued in all his voyage, aſwell outward as homeward, and the blewe line proceeding alſo from Plymouth, ſheweth in like maner the voyage of Maſter Candiſh, and in that Globe is alſo ſet downe how farre Sir Martin Furboſher diſcouered towards the North parts. But firſt I wil deſcribe vnto you the voyage of Sir Fr. Drake that wrothie Knight & moſt Noble Neptune, accordinge as that red line directeth in ye ſaid Globe.

Firſt parting from Plymouth he ſayled with a North Northeaſt winde to an Ile called Mogodore, vpon the coaſt of Marocco, which place is not named in M. Molineux his Globe, and that place hauing in North Latitude 32. degrees, is diſtant from Plymouth according to that courſe, which the redde line ſheweth 780. leagues. In this Ile he built a little Pinnis or ſhallop, and from thence he ſayled to Cape Dalguere which is further Southward, and hauing in North Latitude 30. degrees, is diſtant from Mogodore about 40. leagues, and from thence hee ſailed to the Iles Canariæ, which are ſomewhat more Weſtward, and hauing 27. degrees in North Latitude, are diſtant from the Cape Dalguere about 100. leagues, and from thence he ſayled to Capo Blanco, which is more Weſternly, and hauing in North Latitude 21. degrees, is diſtant from the Canaries 120. leagues and ſomewhat more, and from thence he ſayled to the Iles of Capo Verde, which hauing in North Latitude about 14. degrees, are diſtant from Capo Blanco about 140. leagues. And frō thence to the great Cape of S. Auguſtine, which hauing in South Latitude about 8. deg. is diſtant from the Canaries 500. leagues, and from thence he ſayled more Weſternly vnto ye mouth of the Riuer called Rio de Platta, which hauing in South Latitude 36. degrees, is diſtant from Cape Saint Auguſtine 740. leagues, and from thence to the Port Saint Iuliano, hauing in South Latitude about 50. degrees, whereas one Doughtie was executed for conspiracie, and this Port is distant from Rio de Platta 380. leagues. And from thence he sayled to the Cape Virgine Maria, which hauing in South Latitude 52. degrées 3′0. is distant from Port S. Iulian 50. leagues, and from thence stræking in betwixt the Ile, whose Northeast Cape is called the Ile of the name of Jesus, and the Port Famine, he entred into the straight Magellane, which hauing in South Latitude 53. degrées 3′0. is distant from the Cape Virgine Maria 50. leagues, and from thence he passed through the Magellane straights to the Cape de Sancto spirito, which is a Cape of the South land, hauing in South Latitude 52. degrées 2′0. and is distant from Cape Virgine Maria about 150. leagues, from thence he sailed somewhat Westernly about 20. leagues, & there fetching a turne about certain Ilands called Las Anegadas, he tooke his course Northward alongst the West coast of America vnto y̆ Ile Lima, which hauing in South Latitude 12. degrées, is distant from the Ilands called Las anegadas 800. leagues, and from thence he sayled still Northward vnto Cape Guija, which hauing in South Latitude 1. degrée 3′0. is distant frō the Ile Lima 160. leagues, from thence still Northward he sailed to Cape S. Francesco, which hauing in North Latitude 1. degrée 30. minutes, is distant from Cape Guija 140. leagues, from thence he sayled stil Northernly to the Cape Mondecino, which is in the land called Quiuira, and this Cape hauing in North Latitude 40. degrées is distant by that course from S. Francesco 1740. leagues, from thence he sayled still Northward vnto a certaine Bay in the West part of Quiuira, which he named Noua Albion (that is to say) new Englande hauing in North Latitude 46. degrées. And this was the furthest part of his voyage outward, in which voyage hee sayled in all 6050. leagues, and from this Bay Sir Frances himselfe (as I haue heard) was of very good will to haue sailed still more Northward, hoping to find passage through the narrow sea Anian, which sea is not set downe by Maister Molineux in his Terrestriall Globe as a straight, but rather as a maine Sea, bearing in bredth 400. leagues, and so from thence to haue taken his course Northeast, and so to returne by the Iles Crocklande and Groynlande into England, but his Mariners finding the coast of Noua Albion to be very cold, had no good will to sayle any further Northward,

wherefore

The vse of the Globe.

wherefore Sir Frances was faine to come backe againe Southward to Mondecino, which (as hath béene said before) is distant from the foresaide Bay of Noua Albion 140. leagues. From thence he sayled in a manner right Southeast to the Iles Moluccas and touched at the Iles Terenate, Tidori, Machian, and Motill, which are nigh vnto the Ile Gilolo, which is right vnder the Equinoctiall, amongest which Iles he remayned a certaine time, of which Ilands the Ile called Terenate hauing about one degrée of North Latitude, is distant by that course from the Cape Mondecino 1180. leagues, and from thence hee sayled Southwest vntill he came to the West end of the Ile Iaua maior, which hauing in South Latitude nine degrées 30. minutes, is distant from the Ile Terenate by that course 530. leagues, and from thence he sayled still Southwest to the Cape di Buona Speranza, which hauing in South Latitude 35. degrées, is distant from Iaua maior 1630. leagues, then from Capo di Buona Speranza he making his course Northwest, sailed to an Iland called Serra Liona, which is vpon the coast of Afrique, and hauing in North Latitude 7 degr. 30. is distant from Capo di Buona Speranza 1090. leagues, then from thence he sailed towards the Iles of Capo verde vntill he came to the 12.degrée of North Latitude right vnder the first Meridian, which point is distant from the Island Serra Liona according to that course 300. leagues, & from thence he sayled Northward nigh to the Iles called Azores on the West side therof, which hauing of North Latitude 40. degr. are distant by that course from ye Iles of Capo verde 600. leagues from whence he directed his course Northeast to Plimouth, which hauing in North Latitude 51. degr. is distant by that course from the Azores 490. leagues, so as in his returne from Noua Albion to Plymouth he sayled in all 5960. leagues, which if you adde to the number of leagues of his outward voyage before set downe, which is 6050. leagues, you shall finde the totall summe of the leagues to be 12010. leagues, which is almost twice so much as the compasse of the whole world, which if you measure vpon the Globe by the Equinoctiall line containing 360. degrées, and doe allow for euery degrée thereof 60. Italian miles, you shall finde the number of such miles to amount to 21600. miles, which by allowing thrée miles to a league doe make no more but 7200. leagues,

.eagues. But if it might please Sir Frances to write a perfect Diarie of his whole voyage, shewing howe much he sayled in a day, and what watring places he found, and where hee touched, and how long he rested in any place, and what good Ports and Hauens he found, and what anchorage good or badde, and what maner of people, what trade of liuing, and what kinde of building and gouernement they vsed, in what aire they liued, and whether the ground were fertile or barren, dry or well watered with floods and fountaines, what mountaines, what mines, what woods or forrests, what beastes, fowles, or fishes, fruites, hearbes, plants, or other commodities he founde therein, and in what manner of seas he sayled, and what windes and currents were most rife in euery place: Also what rockes, sandes, sholdes, and all other places of daunger and perill, and by what markes such places are to be shunned. And finally what Moone doth make a full sea in euery Port where he arriued, and what windes doe alter any tide or Current, and all other necessary accidents most meete for sea men to knowe. In thus doing the saide Sir Frances I say should greatly profite his countrie men, and thereby deserue immortall fame, of all which things, I doubt not but that he hath alreadie written, and will publish the same when he shall thinke most meete.

The voyage of M. Candish vnto the West and East Indies, described on the Terrestriall Globe by the blew line.

This blew line as you see taketh his beginning from Plymouth like as the redde line doth, whereby you may plainely see, that M. Candish did not greatly differ in his course from that which Sir Frances helde, sauing that Master Candish hauing once passed the straight Magellane Westward, sayled not so farre Northward as Sir Frances did, for he sayled no further Northwarde but to the Port Saint Lucas, which is almost vnder the Tropique of Cancer, for that Port hath in North Latitude 22.degrees 3′0. keeping alwaies a nigher course vnto the maine, then Sir Frances did, and

The vſe of the Globe.

from the Port S. Lucas he ſayled in a manner Southeaſt about 320. leagues, and then he directed his courſe Eaſtward towards the Moluccas, & before he came to the Moluccas, he fetched about the Ile Catanaim, which is more Northward then Sir Frances went in thoſe partes, and from thence he ſayled full South to the Moluccas, from whence he directed his courſe Southweſt, and paſſed betwixt the Iles Iaua maior and Iaua minor, which two Iles in the ſaid great Globe are made to haue one ſelfe Latitude (that is to ſay about 10. degr. of South Latitude, counting from the Equinoctiall to the Parallel which paſſeth through both the Iles. But in all other Globes and Maps Iaua minor ſtandeth 10. degrees more to the South, and is placed behinde the South promontorie called the land of Beach. Then hauing paſſed betwixt the foreſaid two Iles, he held in a manner the like courſe that Sir Frances did in his returne to Plimouth, which the blew line doth ſhew ſo plainely as it needeth none other deſcription. Nowe as touching Sir Martin Furboſhers voyage, becauſe his owne deſcription thereof is in print, nothing is ſet downe in this Globe, but only the outermoſt end of his voiage named here Furbuſhers ſtraights, hauing in North Latitude about 63. degrees, in the mouth of which ſtraights is a little Iland called Hales Iland on the right hand, and on the left hand another little long Iland called Leceſters point. Truely this knight for his valourous venter, aſwell in this voyage as in diuers other his worthy ſeruices done vpon the ſea deſerueth great commendation, and I wiſh with all my heart that he and ſuch like might be much made of, and rewarded according to their deſert. And thus I leaue to ſpeake any further of M. Molineux his Globes, the vſe whereof is to be learned by thoſe Propoſitions which I haue heretofore ſet downe, ſhewing the vſe of Mercator his two Globes, for the practiſe and manner of working both by his Globes and by theſe Globes is all one.

A

A plaine and full description

of *Petrus Plancius* his vniuersall Map, seruing both for sea and land, and by him lately put foorth in the yeare of our Lord, 1 5 9 2.

In which Mappe are set downe many more places, aswell of both the Indies as of Afrique, together with their true Longitudes and Latitudes then are to be found either in *Mercator* his Mappe, or in any other moderne Map whatsoeuer, and this Map doth shewe what riches, power or commodities, as what kindes of beastes both wilde and tame, what plants, fruits, or mines any Region hath, and what kinds of marchandizes doe come from euery Region.

Also the diuers qualities and maners of the people, and to *whom they are subiect. Also who be the most Mightie and greatest* Princes of the world: A Mappe meete to adorne the house of any Gentleman or Marchant that delighteth in Geographie, and therewith this Booke is also meete to be bought, for that it plainly expoundeth euery thing contained in the said Mappe, Written in our mother tongue by M. *Blundeuill* *Anno Domini.* 1 5 9 4.

Imprinted at London by *Iohn Windet.*

IN DOCTISSIMI VERE-
QVE GENEROSISSIMI *THO.*
BLVNDEVILI, IN *PETRI PLANCII*
TABVLAM GEOGRAPHICAM ET
HYDROGRAPHICAM ELV-
cidationes, *Gualteri Hawghi*
Δεκατετράsιχον.

PLancius in tabula terras descripsit & vndas,
    Quæcunq; & toto condidit orbe Deus:
Multiplices horum partes, Regumq; per illas
    Sceptra superba, simul totius orbis opes.
Sed solis Gallis hæc Plancius, atq; Latinis
    Inter & hos, Doctis scripserat, haud alijs,
Tu Generose tuis patrio hæc Idiomate clarè
    Blundeuile refers, Cunctaq; nota facis.
Tu simul hæc auges, lucemq; hijs addis, & horum
    Vsum, multiplici non sine fruge doces.
Nec solum doctis, sed & omnibus hæc Idiotis
    Perspicua vt pateant, tu breuitate facis.
Plancius hæc alijs, tu nobis omnia tradis,
    Tuq; tuis Anglis Plancius alter eris.

A plaine and full description of Plancius his vniuersall Mappe, set foorth in the yeare of our Lord, 1592. written in our mother tongue by M. *Blundeuill*.

IN this Mappe, as you sée are drawne with redde inke two lines or Diameters crossing one an other with right Angles in the verie midst of the Mappe, whereof the perpendicular Diameter passing through the Iles Azores, and also through the Iles of Capo verde, signifieth the first Meridian and Arletræ of the world, at the vpper end whereof is set the North pole, and at the neather end the South pole. And the other ouerthwart Diameter signifieth the Equinoctiall, & also the line of East & West, that is to say, East on the right hande, and West on the left hand. And in this line are set downe the degrées of Longitude, which are to be counted from the foresaid first Meridian vpon the saide Equinoctiall towards your right hande from one degrée to 180. set downe in Arithmeticall figures thus, 10. 20. 30. and so foorth vntill you come to 180. which is the East Longitude of the world, and the West Longitude beginneth on your left hand whereas is set downe 190. and then 200. and so foorth vntill you come to 360. degrées, which is the whole circuit of the Equinoctiall, and of the whole earth. And in the first Meridian are set downe the degrées of Latitude, which doe procéde from the Equinoctiall to eyther of the Poles from one degrée to 90. written in Arithmeticall figures thus, 5. 10. 15. and so foorth to 90. which degrées doe procéde with equall distances from the Equinoctiall to eyther Pole, and the like degrées of Latitude are also set downe vpon the two outermost Meridians, aswell on the right hande as

The description and vse

on the left. And on ech side of the Equinoctiall are drawne with redde inke the two Tropiques, that is to saye, the Tropique of Cancer, and the Tropique of Capricorne, ech one being distant from the Equinoctiall 23. degrees 2′8. which is the greatest declination of the sunne, and towardes ech Pole are also drawne with redde inke the two Circles both Arctique & Antarctique, whereof the circle Arctique passeth through the Northernly part of Island, hauing in Latitude 66. degrees 3′0. and the Circle Antarctique passeth through the like degrees of Latitude towardes the South pole, and the distance of each of those Circles from eyther Pole is equall to the greatest declination of the sunne. And these foure Circles are Parallels to the Equinoctial, bounding the 5. Zones that is, the two cold, the two temperate, & the hotte Zone which lyeth in the midst of the world betwixt the two Tropiques. And on the left hand or West part of the Mappe are set downe from the Equinoctiall vpward towards the North pole, what number of miles and secondes of miles doe belong to euery degree of the North Latitude, proceeding from the Equinoctiall towards the North pole: for though that euery degree of the Equinoctiall being a great Circle containeth 60. miles, yet the further that you goe from the Equinoctiall towards either of the Poles, the lesser and lesser are your Parallel Circles in compasse, and therefore one degree of euery such Parallel must needes containe the fewer miles, but because the sea men doe commonly make their account on the sea by leagues, and not by miles: Plancius here right vnder the former Table of miles, setteth downe the number of leagues incident to euery degree or Parallel, in descending from the Equinoctiall to the South pole, appointing three miles to a league, and therefore he setteth downe vppon the very Equinoctiall 20. leagues, which is 60. miles, and next to that he setteth downe 19. leagues 5′9. and so proceedeth foorth diminishing still the quantitie of the leagues, euen till you come to the very South pole, and on the right hand of the Map hard by the outmost Meridian are set downe the nine Climes, and the longest day in euery degree of Latitude, proceeding aswell from the Equinoctiall to the North pole, as also from the Equinoctall to the South pole, from 12. houres to 187. dayes and 7. houres, which maketh the longest day to those that dwell right vnder the North Pole to be halfe a yeare

of Plancius his Mappe.

peare and the night asmuch. Besides the Circles and lines before mentioned, there are set downe in this Mappe certaine flyes of the Mariners Compasse, each one containing two and thirtie lines, which doe signifie the two and thirtie rombes or windes of the Mariners Compasse to knowe thereby howe one place beareth from another, and by what winde the Mariner hath to sayle to any place whereunto he would goe: In euery which fly the line of North and South may serue in stead of a Meridian, and the line of East and West may serue as a Parallel, by helpe whereof you may the more readily take with your Compasses the Longitude and Latitude of any place contained in the Mappe in such manner as is taught in my first Treatise of Uniuersall Mappes. Also in the very front of his Mappe he setteth downe the numbers of foure and twentie houres, euery houre containing fiftéene degrées of the Equinoctiall, which houres doe beginne on the right hande, and so procéede to the left, whereof the twelfth houre is placed at the ende of that Meridian which passeth through the Fortunate Iles. Nowe betwixt the 72. and 86. degrées of North Latitude hee setteth downe two long Ilandes extending from the West towardes the East somewhat beyonde the first Meridian, and from the saide Meridian more Eastwarde he setteth downe other two long Ilandes, affirming that the North Ocean sea breaking in betwixt these Ilandes with ninetéene gates or entrances, maketh foure straites and is continually carryed vnder the North Pole, and there is swallowed vp into the bowels of the earth, and he sayth further that right vnder the North pole there is a certaine blacke and most high rocke which hath in circuit thirtie and thrée leagues, which is ninetie and nine miles, and that the long Ilande next to the Pole on the West is the best and most healthfull of all the North partes. Next to the foresaid Ilandes more Southward he setteth downe the Ilandes of Crockland and Groynelande making them to haue a farre longer and more slender shape then all other mappes doe: At the East ende of the third long Ilande, is a straite hauing fiue gates or entrances, which by reason of their narrownesse and swift course of streame are neuer frozen, if this be true, I marueile howe any shippe durst enter through any of those straites to discouer the North sides

of

The description and vse

of any of those Ilandes, and how and where it came out againe. Moreouer at the East ende of the last Ilande somewhat to the Southwarde, he placeth the Pole of the Lodestone which is called in Latine Magnes, euen as Mercator doth in his Mappe who supposing the first Meridian to passe through Saint Marie or Saint Machaell, which are two of the outermost Ilands of the Azores Eastwarde, placeth the Pole of the stone in the seuentie fiue degræ of Latitude, but supposing the first Meridian to passe through the Ile Coruo, which is the furthest Ile of the Azores Westwarde, hee placeth the Pole of the Lodestone in the seuentie seuen degræ of Latitude, I thought good to seuer the Ilands last before mentioned from the bodie of the Mappe as partes belonging rather to the North pole, then to Europe, Asia, Africa, or America: for if Virgill did not let to say that Englande was Penitus exclusa ab orbe, mee thinkes that I may much more rightly say the like of these Ilandes, notwithstanding Plancius maketh the first two long Ilandes, and also Groynelande and Crockelande to be part of Mexicana, which mee thinkes is not meete, sithe they be deuided by the North Sea, likewise hee maketh the lande which is vnder the South pole, and is not as yet discouered, to bee part of Magellanica, which (in mine opinion) ought not to bee so, sith it is not onely deuided from Magellanica, by the straight Magellanicum, but also from Afrique, and from the East Indies by the great Southerne Ocean, and I beleeue that when the South lande shall bee all discouered, it will containe twice so much lande as Magellanica doth, and then I doubt not but that the Geographers will giue it some other name, and make many diuisions thereof: In the meane time I will followe Plancius his owne diuision of the worlde, which greatly differeth from that of Mercator, and of all other moderne Geographers, for they doe deuide the whole earth but into foure partes, that is to say, Europe, Afrique, Asia, and America, but Plancius by deuiding America into thrée partes that is, into Mexicana, Peruana and Magellanica, deuideth the whole earth into sixe partes, that is to say, Europe, Asia, Afrique, and into the thrée partes of America last mentioned, according to which diuision, he describeth the earth in the french
tongue

of Plancius his Mappe.   248

tongue in sixtéene pages, set downe at the foote of his Mappe, the foure last pages whereof doe onely containe the interpretation of the seuentie one little Tables or Inscriptions written in the Latine tongue, dispersed throughout the whole Mappe, expressing therein such thinges as hee thought most méete to bee noted in diuerse partes of the worlde, all which Tables or Inscriptions, I haue heere also set downe in our mother tongue: And although that to the foresaide Tables or Inscriptions Plancius hath attributed certaine numbers for the more easie finding out of the sayde inscriptions, yet not easie inough by reason that one selfe number is set downe in diuerse Tables, and therefore to the intent that you might the more readily finde out euery Table that is proper to the matter whereof it maketh mention, I haue here following ioyned to the number of euery such Table, his proper Longitude and Latitude, which with your Compasse you may quickly finde out, and more certainely, then by his numbers, if you remember the order thereof set downe in my Treatise of vniuersall Mappes. But now I will declare the contentes of the foresaide sixtéene pages in order as followeth.

The title of the first page is thus.

A briefe declaration of the diuision, forme or shape, and of the particularities of the world.

The Contents of the first page.

That the earth and the water doe make both together one rounde bodie, which the Cosmographers doe inuiron with fiue Circles, that is the Equinoctiall, the two Tropiques, and the two Polar Circles, and thereby doe deuide the worlde into fiue Zones, two colde, two temperate, and one extreme hotte: and though that the auncient Geographers doe affirme that thrée of those Zones were vnhabitable, the one for extremitie of heate, and the other two for extremitie of colde, yet within these hundred yeares last past

it is

The defcription and vſe

it is knowne by good experience that thoſe thre͡e Zones are well inhabited as the manifolde Countryes therein placed and greatly replenished with people of ſundry languages doe well teſtifie, of all which thinges I haue written at large in my Spheare. And therefore I make this page the ſhorter.

The Contents of the ſecond Page.

To knowe the true ſituation of the Prouinces and places contained in this Mappe, it is neceſſarie firſt to knowe their Longitudes and Latitudes. The degre͡es of Latitude or of the eleuation of the Pole, which is all one thing, are counted from the Equinoctiall to eyther Pole, which is 90. degre͡es, and the degre͡es of Longitude are counted vpon the Equinoctiall from the Iles of Capo Verde towardes the Eaſt, and ſo round about the earth vntill you come to the number of 360. degre͡es. The Prouinces and Townes that are ſituated vnder one degre͡e of Longitude, haue at one ſelfe time like houres of the day, but thoſe that are ſituate vnder diuerſe degre͡es of Longitude, doe differ in number of houres, for when it is in one Towne noonetide, it is in an other Towne that is diſtant from thence towardes the Eaſt 30. degre͡es two of the clocke in the afternoone, and ſo conſequently for euery 15. degre͡e of diſtance they differ one houre. Likewiſe they that dwell vnder one ſelfe degre͡e of Latitude haue equall quantitie of dayes and nightes, but yet ſo as they which dwell on the South ſide of the Equinoctiall haue the ſhorteſt day when wee haue the longeſt, and haue Winter when we haue Sommer. But thoſe that are ſituate vnder diuerſe degre͡es of Latitude, haue inequalitie of dayes and nights, for the nigher that any place is ſituate towardes any of the Poles the more houres the longeſt day of the yeare in that place containeth. But thoſe that dwell right vnder the Equinoctiall haue alwayes their dayes and nightes of like quantitie, and I vnderſtande here by the daye, the ſpace betwixt the ſunne riſe and the ſunne ſet, and you ſhall finde the quantitie of the longeſt day of the yeare in euery degre͡e of Latitude ſet downe in the Northeaſt part of this Carde. As for example,

of Plancius his Map.   249

ample to those that haue 30. degrées of latitude, the longest day is 13. houres 57'. and so the nigher that you go to the Pole the longer is the day, in so much as to those that dwell right vnder the Pole, the yeare is but a day and a night, that is to say, they haue 6. moneths day, & 6. months night. Moreouer the Geographers do diuide the earth into 9. climes for to distinguish therby the prouinces and regions by the quantity of the longest day, the middlemost parallel of euery clime increasing by halfe an houre, and you haue to consider that the degrées of latitude are in all places of like bignes euery degrée contayning 15. Almain leagues, or 60. Italian miles, but the degrées of longitude procéeding from the Equinoctiall towards any of the 2. Poles are vnequall, that is to say, euerie one containing fewer leagues or miles than other, but the degrées of the Equinoctiall it selfe are equall to the degrées of latitude, euerie one containing 15. Germaine leagues or 60. Italian miles, as you may plainly sée in the Table set downe in the Northwest part of this Map. And you haue to note that one Almaine league doth contain 4. Italian miles, and we haue described the degrées of longitude in the Southwest part of this Carde by the houres of the ships way, euery one decreasing lesse than other from the Equinoctiall to the Pole, whereby you may conceiue that two ships being right vnder the Equinoctiall 150. degrées distant one from another, and are to saile with like gate towards the North pole: when they shall come to the 60. degrée of latitude, their distance shall be no more but 75. leagues. And the further they goe towards the Pole the lesse distant they shall be one from another, in so much as when they be right vnder the Pole it self, they shall both méete, as you may sée in the 2. rounde figures contayning the description of the earth, and set downe in the 2. nether corners of the Map. This matter is to bee considered of the Mariners that they may thereby the better perceiue the imperfections of their sea Cardes. Moreouer, in the second page Plancius setteth downe the diuision of the earth as well according to the ancient as moderne Geographers, making first thrée generall continents or firme lands, whereof the first is so much as was knowne to Ptolomey and to the ancient Astronomers, as Europe, Afrique, and Asia, the second continent is called America, and the third continent is the South part

The description and vse

of the world, not yet fullie discouered, called of Plancius Magellanica, and he diuideth the second continent called America into three parts, that is, Mexicana, Peruana, and Magellanica, and by adding those three parts to Europe, Afrique, and Asia, he diuideth the earth into sixe partes, and first he setteth downe the description of Europe together with her boundes or limittes, and then the commodities thereof, as followeth.

Of Europe.

EVrope is farre lesse than all the rest, and yet exceedeth all others in noblenesse, in magnificencie, in multitude of people, in might, puissance, and renowne, the which in times past hath commanded both Asia and Afrique as Queene, by reason of the Monarchies of the Grækes and of the Romans, and at this day is of great force by the power of the Turkes and Muscouites. Moreouer it commaundeth many prouinces in Mexicana and Peruana by the power of the Spaniards and Portugals, and of other Christian Princes.

Europe is seuered from Asia and Afrique by the sea Mediterraneum, and by the sea called Marmagior, and by the Marish or sea called Palus Meotis, and by the Flood Tanais and Dwina. The chiefest prouinces of Europe are these, Almanie, Italy France, Spaine, Denmarke, Norway, Swethland, Muscouia, Polonia, Hungaria, Sclavonia, and Greece. The chiefest Ilandes of Europe are these, England with Scotland, Ireland, Sardinia, Corsica, Sicilia, Candia, Nigro Ponte sometime called Euboia, and Stalimene sometime called Lemnos. And in this second page hee setteth downe also the discription of Almanie thus.

Almanie is reputed to bee the greatest prouince in all Europe, and is situated in the middest thereof, which is bounded on the East with Polonia and Hungaria, on the South with Dalmacia, and Italia, and on the West with France, and on the North with the North Sea, and with the Sea called Mare Balticum. The inhabitantes of this Countrey warred in olde time with the Romanes for their libertie, and since manie hundred yeares past it hath holden the imperiall Scepter.

About

About the time of Christ his birth, it was a rude countrey, as Cornelius *Tacitus* saith, full of Wood, bushes, and marishes, but at this day it is so adorned with great magnificent Townes and well fortified, and is furnished with such a number of Castles, and Villages, and with such a number of people, and with such pollitique gouernement, as it is to bee compared to any prouince whatsoeuer in all the worlde. The soyle thereof is verie fruitfull both for corne and Wine, and hath manie nauigable Floods stored with plentie of Fish. It hath most excellent Fountaines, and hotte bathes, greate mines of Golde, of Siluer, copper, Tinne, Lead, and Iron. The inhabitants doe exercise as well nowe as they haue done in times past the Art military, and it hath manie learned men verie skilful in all sciences, and in Mechanicall artes, they were the inuenters of Artillerie, of Gunpowder, and of the noble Arte of printing, and of making artificiall dials and horologies.

The chiefe merchandizes that are transported out of Almanie into other countreys are these, Golde, Siluer, Copper, Tinne, Leade, vitrioll, Allum, Quickesiluer, Collours of diuers sortes, Slates to couer houses, Wheat, Wine, Fish, woollen cloath, Linnen cloath, Bombasine, Fustian, Suile, Armour, all sortes of workes made of Iron, or brasse, and other merceries.

The Contentes of the third page.

IN the thirde Page hee describeth the 17. Prouinces of the lowe Countries, also Italie, France, Spaine and Denmarke, as followeth.

The 17. Prouinces of the lowe Countries are counted a part of *Almanie*, by reason that the most part of the inhabitantes haue as well their originall as their language from the Almanes. But those of *Artoys*, of *H*enough Namures, and parte of the inhabitantes of Brabant, Flaunders, Lymburgh, and Liezenburgh doe speake the French tongue. These prouinces are situated partly within the ancient limits of *Almany*, that is to say, beyond the Orientall part of the flood Rhenus, & partly in Gallia Belgica, as also are the prouinces of the 4. princes electors

The description and vse

and of manie other prouinces of the Empire. In these prouinces are manie nauigable Floods and rich in fish, namelie the Rheane, Mosella, Mosa, and the Escaut, and there is great abounbance of all sorts of corne and cattell meete for mans vse, there is also a number of great Townes rich, mightie, and well peoplished, also of fortresses well fortified, faire Villages, but chieflie of braue and commodious portes and hauens, and an incredible number of shippes, and the continuall warres that they haue had and haue at this day do witnesse to all the world their great force, might, and riches. The inhabitantes in time of the Romaine Monarchie were and also are at this present greatlie renowmed for their skill in the arte militarie, and besides that, they are most excellent and industrious in all sciences and mechanicall artes, and they haue a great number of Mariners and Pilots well practized in the arte of Nauigation, and to these prouinces is attributed the inuention of the Mariners Compasse, according to the opinion of manie learned men, which trulie is one of the noblest inuentions that euer was found out since the world beganne. In the towne of Brughesse was inuented the Arte of Painting with Collours tempered with Oyle.

The Prouinces of Belgia doe sende vnto other Prouinces all sortes of Cloathes made of Woolle and Flaxe, linnen cloath of Cambria, Skarlettes interlaced with golde, Siluer, and Silke, Taffatas, Borattas, Grograines, single Buffin, Sayes of Leyden, and Howlscot, Worsteds, and halfe Worsteds, Fustianapes of Vellures, and of Wool, Bayes, silk, parchment lace, Sarcenet and inkle, all manner of twisted thred silke ready drest, purified Sugar, Buffe, Shamoyes, stripped Marokines, painted Pictures, Bookes, Armour, Cables, Ropes, and other Munitions belonging to Shippes, Kniues, Pinnes, and all sortes of Mercerie or Haberdash ware, and Fish dried and salted.

Italie being the mother of eloquence and of al Latine erudition doth extende it selfe like an arme towards the Southeast, lying betwixt the Tuscane sea and the gulfe of Venice, & is bounded on the west part with France, & on the north part with Almany

being

being seperated from the sayd two prouinces by the Flood Varo and the Alpes, and al the rest is inuironed with the sea, at the time of the natiuitie of Christ and since, shee most flourished, being adorned with the fourth Monarchie, and with the most mightie towne of Rome, which at that time was Quéene of many prouinces, of Europe, Asia, and Afrique, which citie in the time of the Emperour Vespasian, had in circuit 13. Italian miles and 200. paces, as Plinie writeth in his third booke and fift chapter. Flauius Vopiscus reciteth that this towne was inlarged by the Emperour Aurelius to 30. Italian miles, which is 10. houres of way or gate, allowing three mile for an houre. This Prouince hath brought foorth, as it doth at this present, inhabitants of great industrie and wit, and it contayneth many noble cities and of great renowme, as Rome, which hath bene sometime the head of the world, Venice the rich, Rauenna the ancient, Naples the gentle, Florence the beautifull, Genua the proud, and Millaine the great. In a town called Amalphe situate vpon the sea, betwixt Naples and Salerno, the Mariners Compasse was inuented, in the yeare 1300. according to the opinion of some, by one Iohn Goia, citizen of the same towne: notwithstanding Iohn Gorop Becanus doth attribute that inuention to Flaunders which séemeth the more likelie, for so much as al the Pilats & Mariners of France and Spaine and other places do name the 32. windes or rombes of the Compasse by the Belgicke names. The chiefe marchandizes that are sent out of Italie into other countries are these, Rice, Silke, Veluet, Satten, Taffatas, fine péeces of Linnen, Grograines, Rash, Stamin, Bombasins, Fustians, Feltes to make riding cloakes, plentie of rich armour, Wiar of golde and siluer, Allum, Galles, drinking glasses and looking glasses of Venice.

France hath bene alwayes estéemed to be the chiefest realme of all Europe, whose soyle is most fertile, and bringeth forth all kinde of Graine, and euerie other thing that is necessarie for mans sustenance, there is great store of wine, and great plentie thereof is distributed to other Prouinces nigh adioyning. The Prouince doeth abounde in oyle Oliue, and in Corrall, and in many other noble fruites. In France are many great

The description and vſe

great townes well walled, as Paris, Roan, Amiens, Orliens, Tours, Nantes, Poicters, Burghes, Tholous, Lyon, Narbona, & Marcelles. It hath 15. Archbishoprickes, and 108. Bishoprickes, and a great number of townes and villages, and 132000 parishes, it is greatlie peopled, and hath not such desertes or heathes as are in other prouinces of Europe, the French men haue bene, and are at this present renowmed in the Arte militare, and there be manie learned men in all faculties and sciences.

The chiefe Merchandizes that are caried out of France into other prouinces are these, Wheate, Rie, Beefs, Hogges Swine, and other cattell, Salt, wines, wilde Oliue, Chessnuts, Almonds, Prunes, Corrall, Diers wadde, Clothes, linnen, Canuas and Skinnes.

Spaine is inuironed round about with the Sea, sauing that on one side it is seperated from France by the mountaines Perenei. This countrey was sometime diuided into thrée Prouinces or kingdomes, that is, Taraconenſis, Lusitania and Betica, but now it is subdiuided into many Realmes, that is to say, Caſtilia, Aragon, Portugall, Gallicea, Lyon, Navarra, Toledo, Valentia, Murcia, Granado, Cordoa, and Algarbia, the which Realmes if they had bene reduced to one bodie of a Realme, as France is, and as they be at this houre subiect to one only king and Lord, it should bee without doubt one of the most mightie and puissant kingdomes of Europe. The inhabitantes of Spaine haue bene and are at this present much renowmed in the art militare, and in feates of warre, and it hath brought forth in times past many great Clearkes, as Seneca, Quintilian, Lucan and Martial, and in our time it had Iohannes Lodovicus Viues, Iohannes Oſorius, and Benedictus Arias Montanus. The prouinces of Spaine are become verie rich and mightie, by reason of their nauigatiō into America, Africa, Arabia, Persia, India, the Iles Moluccas and China, in which Prouinces (China excepted) the king of Spaine poſſeſſeth manie countries that be rich and of great power, and many townes and fortreſſes in a manner round about the earth.

The chiefe Merchandizes that grow in Spaine, and are caried into other Countries are these, wines, Oyles, Rice, all sorts

of Plancius his Map.   252

of fruites of Spaine, Liquoras, Silke, great quantitie of wolle, Lambe skinnes, Corke, Rosin, Steele, Iron and Armour.

Denmarke and *Norway* are verie great Regions, and are as large as the countries of Almany, bordering vpon Almany towardes the South, they extend towards the North to 71. degrée 30´. of North latitude, and towards the East they border vpon Swethland, and on the west and North side they are inuironed with the sea. These two Realmes are at this day vnder the gouernment of one only king, who also is lord of Island and of the Iles of Fero, *H*itland, and Gothland. Iuthland was sometime the habitation of the Cimbres, who in times past made cruell warres against the Romanes.

The Merchandizes sent from the two forsaid realmes into other prouinces are these, Oxen, Barly, Malt, stockfish, tallowe, nuttes and Filberds, hydes of Oxen, and Bucke skinnes, Mastes for shippes, planckes, and the toppes of Wainscot, Soliues, and fire wood to burne, pitch and tar, Sulphur and such other things.

The Contents of the 4. page, wherein he describeth *Swethland, Polonia, Hungaria, Sclauonia, Greece, England,* and *Scotland* as followeth.

Swethland is a great and mightie realme, bordering towards the East vpon Russia, and towardes the South vpon the East sea, called Mare Balticum, diuiding *Swethland* from *Almany* and Pomerania, and towards the west vpon Norway and Denmarke, and towards the North vpon Finmarke. Stockholme is the Metropolitane citie in this realme, wherein the king keepeth his court. From this realme is transported into other prouinces these Merchandizes, that is to say, Copper, Iron, lead costlie Furres, hydes or skinnes of Elkes, of Oxen, of Buckes, of goats, tallow, tar, barly, malt, nuts and Filbirds, and such like.

The description of Muscouia which should follow next, is set downe in the third table or inscription, which standeth in the verie front of the table, written in Latine, the interpretation whereof hereafter followeth in his place.

Kk 4   The

The description and vse

The kingdome of Polonia containeth Lituania, Podolia, the lesser Russia, Volhinia, Massouia, Samogitia, Prussia, and in a manner all Liuonia, which two last Prouinces did belong not long since to Almania. Polonia is bounded on the East with Muscouia and with the Tartaries, Perocopsiques, and on the South with Moldania, and Hungaria and towards the West with Almania, and towards the North with the sea Baltique and Muscouia.

The chiefe merchandizes that go out of this realme into other Prouinces are these, Wheate, Rie, and other graine, Spruse or Danske Beare, yellow Amber, Waxe honie, a certaine drinke made of Hony which we call Meade, hydes of Oxen dryed and salted, Flaxe, Hempe, Pitch, and Tarre, Ashes, Clauellees, wood Mazier, and of Cuvelier, and other such like merchandizes.

Hvngarie is a verie fruitfull Realme, rich and mightie, and it is bounded on the East with Moldauia, and Valachia, and on the South with Bosnia and Croacia, & on the West with Almania, and on the North with Polonia, it hath manie nauigable riuers, wherein are great store of fish, that is to say, Danubius, Dravus, Savus, and Tibistus. The chiefe townes are these, Buda, Gran, Weissenburgh, Rab, Prezburgh, Agria, Colocza, and Belgrada. The inhabitants of this countrey are warlike and hardie, and haue bene long time heretofore a most faithfull Rampyre and Bulwarke to all Christendome, but in the end by reason of their ciuill warres, the better part of them haue bene subdued in our time, and are made most miserable slaues to the Turke.

The Merchandizes which goe out of Hungarie into other Prouinces are these, Golde, Siluer, Copper, and diuers sorts of Collours, Salt, Wine, Wheat, Beefs, and fresh fish of the riuer salted.

Sclauonie is bounded on the East with Bulgaria, and Greece, and on the South side with the Gulfe of Venice, and on the West, with the North parte of Italie, and on the North side with Almanie and Hungarie. This Region containeth manie particular Prouinces as Liburnia, Croacia Bosnia, and Balmatia, the chiefe townes whereof are these, Raguza,

guza, Salona, Sabenica, and Zara. Sclauonie at this time is diuided into manie iurisdictions, for one great part thereof is subiect to the Turke, another part to the Emperour of Almany, and the rest situated vpon the sea coast is subiect to the Seniorie of Venice. At this day there is no tongue (the Arabie tongue excepted) that extendeth further than the Sclauonie tongue, for as it is the Uulgar tongue of Sclauonie, so is it familiar to them of Histria, Bohemia, Morauia, Sileucia, Polonia, and to the large Prouinces of the great Duke of Muscouia, Circassia, Perihoka, Georgiana, Mengrelia, Moldauia, Valachia, Bulgaria, Russia, Seruia, Albania, and to part of Hungarie, ~~that is~~ also familiar in the Court of the great Turke, and among his souldiers that serue in Asia and Afrique.

Greece sometime the mother of all science and erudition, is on the East, South and West side inuironed with the sea, but on the North side it is bounded with Seruia and Bulgaria, it hath in times past valiantly fought with and beaten the Monarchie of Persia for the libertie of their countrie, and finallie by Alexander the great hath triumphed ouer the same, and thereby erected the third Monarchie, by meanes whereof it came to passe that the Græk tongue was made common throughout Asia, Syria, and Ægypt, vntill such time as the Saracens and the great Turke did corrupt and change the same. The Emperours did rule in Greece from the time of Constantine the great vnto the yeare ~~1452~~ [1452], in which yeare Mahomet the great Turke forced the towne of Constantinople, and abolished the Empire of Greece in such sort, as euer since this magnificent and strong imperiall towne of the Christians, hath bene the seat of the Emperour of Turkie, and all the countrey made slaues to the Mahometanes.

The chiefe merchandizes that come from this countrie to other Prouinces are these, Gold, Siluer, Copper, Uitrioll, diuers sorts of collours, wines, Oyle, Ueluets, Damasks, Grograins, Turquesques and Wood.

England together with Scotland making both but one Iland is the greatest and mightiest of all Europe. And England

The description and vse

land is enuironed on all sides with the sea, sauing on the North side, which bordereth vpon Scotland. The aire according to the situation is indifferent temperate, for though it be more Northward than Flaunders, yet it is not subiect to such hard frostes and cold winters. The soyle is verie fruitfull, bringing foorth great plenty of wheate and of other corne, it hath great plenty of fruit trees, and there be many large and faire woods, sweete fountaines, floods and riuers full of fish, and a number of good hauens, also it hath many ritch Mines, as of Golde, Siluer, Lead, Iron, and chiefly of fine Tinne, wherefore it may be worthily counted amongst the most puissant and richest Ilandes of the world. This Iland nourisheth also a great nūber of cattel meet for mans vse, and chiefly of sheepe, which yeeldeth fine and good wooll, in which partly consisteth the profite and riches of the countrey, in such sort, as the golden Fleece ought to haue bene sought for in this Iland, and not at Cholcos. The inhabitants most commonly are tall of stature, beautifull and white of visage, couragious and meet for the warre, also they are ingenious and studious in the Arte of nauigation, in so much as in these dayes they haue traffique into verie farre countreyes, as into Greece, Natolia, Syria, Ægypt, Barbarie, Muscouia, and into manie other prouinces. London being situated vpon the Thames is the Metropolitane and chiefe towne of this Realme, and the Staple of the trade of Merchandizes, and the Courte royall, but Cambridge and Oxford are Uniuersities.

The Merchandizes sent from England into other prouinces are these, broad Cloaths, Carsies, Stamines, Bayes, Sayes, Saffron, Tin, Leade, Wheate, Barley, Malte, Beare, red Hearing, sea Cole and wood.

Scotland is the North part of this Iland, and is likwise inuironed round about with the sea, sauing on that side with which it bordereth vpon England. This Countrey is not so fruitfull as England, notwithstanding it is sufficiently prouided of all things that is needfull for mans nutriment, it is watred with diuers armes of the Sea, and is indued with many mountaines full of grasse, which serueth to feede their cattell. Edenburgh is the Metropolitane citie of this realme, wherein the

King

kéepeth his court. The Scottishmen are good Souldiers, which can endure scarsitie and the iniuries of the aire, and are very desirous to win honour. The inhabitants of the South part thereof doe speake the English tongue: but those of the North, and those of the Iles *Hebrides* doe vse the Irish tongue, and those of the Orcades doe vse the Norway tongue.

The Merchandizes which Scotland sendeth to other countries are these, course clothes, Karsies, Stamins, Freeses, Wool, Barlie, Malt, Fish, Hydes, leaden Owre, and Smithes cole.

The contents of the fift page.

N this page hée describeth Ireland, the Iles *A*-zores, Corsica, Sardinia, Sicilia, Candia, nigro Ponte, Stalimene, all which Ilandes doe belong to Europe, and in the latter end of this page hée beginneth to describe Asia.

IReland is nigh vnto England and Scotland, and is very rich in meddow ground, and hath great plentie of cattell as well tame as wilde, and fish as well of the sea, as of fresh riuers, and greate quantity of foule and birdes, but it hath scarsitie of corne by reason of the great moistnesse of the ayre. This Ile is frée from all venemous beastes, the inhabitantes are wilde people, great and strong, and swift in running, and by little and little they waxe euery day tamer than other, vnder the gouernement of the English men.

THe Iles of Azores are called of the Flemish Pilots & Marriners the Flemmish Iles, because those of Burghers were the first that discouered those Ilands, & albeit that at this present the inhabitants thereof are Portugales, there is yet a remnant of Flemmish families, as of the Bruines, of the Vltrickts & others These Iles are fruitfull, and bee 9. in number, that is to say, the Ile of S. Marie, S. Michael, Tercera, Graciosa, S. George, Pico, Fayal, Flores, and Coruo.

Tercera amongst all the rest is the strongest, & bringeth foorth diers Wad. The Ile of S. Michael bringeth foorth Sugar, and great aboundance of good Diers Wad.

Corsica

The description and vſe

Corſica is ſituate in the ſea Mediterraneum, and bringeth foorth moſt excellent wines, rough Horſes, and great hunting dogges: and this Ile is gouerned by the Geneueſes.

Sardinia is a verie fruitfull Ile, and chiefly of Wheat, which is tranſported from thence into Italie and into Spaine, likewiſe it hath very good Wine, both red and white, and verie good Salt, it hath alſo certaine mines of ſiluer, but not of ſo profitable yéld as in times paſt. The inhabitantes are ſtrong, and able to indure great labour and trauell. In great townes they ſpeake the Spaniſh tongue of Aragon, but in ſmall townes they ſpeake the vulgar tongue of the Ile.

Sicilia hath bene alwayes famous, and is called of Diodorus the Paragon of Iles, alſo the Grækes and the Latines haue greatly celebrated this Ile in their writings. This Iland hath great aboundance of Wheat and of al other grain, alſo of wine, Sugar, waxe, Honey, Saffron, Silk, and of all things els appertaining to the vſe of man. Wherefore this Ile, together with Ægypt was ſometime called the Grange of the Romanes. In this Iland is the hil Ætna, which alwayes burneth, and in the ſea of Sicill nigh vnto Drepano, as Plinie writeth in his 32. booke and ſecond chapter, there groweth verie faire redde Corall, in ſhape like to ſuch a trée or buſh as is here figured, which while it is vnder the water is grǽne and tender, but ſo ſoone as it commeth into the aire, it waxeth hard like a ſtone, and is red, there is found thereof alſo nigh vnto the ſea coaſt of Prouince, alſo in Italie nigh vnto Monte Alto, and to Naples, likewiſe in the red ſea, and in the Gulfe of Perſia, and there be thrée ſorts of Coral, that is, red, blacke, and white.

Candia ſometime called Creta, was in olde time enritched with the famous Labyrinth, and with a hundred cities, it had alſo a great number of good ſhips and expert Pilots, this Ile together with the others, as the Ile of Zante, Cephalonia, Corfue, and diuers others, be at this preſent gouerned by the Senate of Venice.

The Merchandizes tranſported out of Candia into other prouinces

uinces are these, noble wines as Malmsey, Muskadine, Corrants graine of Scarlet, Sugar, Chrystall of the mountaine, Cotton, and Buckeskins.

Nigro Ponte, sometime called Euboia, is a verie fruitfull Ile in Wheate, Oyle and Wine.

STalimene sometime called Lemnos, is an Ile which hath aboundance of wheate, and most excellent Wines. In this Ile they digge out in the moneth of August a certaine medicinable earth called of the Physitians *Terra sigillata*. There be manie other Iles besides these in Europe, as the Iles of Denmarke, the Iles of Zeland in Flanders, the Ile Frumentera, Iuica Maiorica and Minorica, and a number of Iles that are in Sclauonie and Greece.

ASia is seperated frō Europe by the floods *Tanais* & *Dwina* & from Afrique by the narrow part of land, which is nigh to Ægypt, betwixt the Mediterrane sea, & the red sea. Asia far excædeth in greatnes both Europe, Afrique and Pervana, and also in riches, as in pearles of great price, and precious stones and spyces it excædeth all the other countries of the world.

This region hath bene alwayes renowmed by the first and second Monarchie of the world, obtained by the *Syrians* & the *Persians*, as also it is at this day by the mightie Princes of China, and of Persia, and by the puissancie of the Tartarians.

In this part of the world man was created of God, placed in Paradice, seduced by Sathan, and redæmed by our Sauiour Iesus Christ, and in this region, were done in a maner all the histories and actes mentioned in the old Testament, and a great part of those of the newe Testament. The most celebrated prouinces of *Asia* are those that belong to the great Duke of Muscovia, also *Tartaria* and *China*, the rich prouince of *India*, as Guzarette, Corasan, Sigistan, Chirmania, Parthia, Persia, Media, Assyria, Armenia, Natolia, Syria, and Arabia. The principall Iles of Asia are these, Iapan, Luconia, Mindanao, Borneo, Sumatra, Ceilan, and Cypres, for as for the Iles of Gilolo, Moluccas Banda & Celebes, they belong to that part of the world which is

called

The description and vse

called Magellanica.

The most mightie Potentates of Asia are these, the king of China, the king of Persia, the great Turke, and the Emperour of Russia, otherwise called the great Duke of Muscouia, according to which Seniories all Asia is diuided into sixe partes, that is to say, the Asiaticall prouinces, belonging to the great Duke of Muscouia 1. Tartaria 2, China 3. the Indies 4. the Prouinces of the king of Persia 5. and those of the great Turk 6. And as touching the Asiatical Prouinces of the Emperour of Russia, and of the prouinces of Tartaria, we shall make mention thereof hereafter, when wee come to translate the Tables or Inscriptions written in Latine, marked with the numbers 3. and 4.

The Contents of the 6. page, wherein he describeth China, and the plant of pepper there growing, with the shape thereof.

China or Sina is the thirde part of Asia, somtime called of Ptolomy Sinarum regio, which on the East side is enuironed with the sea called of the ancient Geographers Oceanns Sicicus, or the east Ocean, and on the West it is bounded with the Indies and with Brumas, & on the North with Tartaria. This countrey is for many causes esteemed to bee the most ample, the richest & most mightie Realme of all the world, for it extendeth from the 18. degree to the 55. degree of North latitude, and it containeth in longitude 450. leagues of Almanie, and it is diuided into 15. great prouinces or Realmes, that is, Quincii, otherwise called Paquin, Xanton, Xiancii, Sancii, Suchuan, Honao, Nanquii, Chequiam, Foquiem, Cantam, Quancii, Suinam, or Huinam, Quiecheu, Fuquam, or Hucquam, and Quiancii. This Realme is adorned with manie nauigable Floods, and full of Fish, it is verie fruitfull, and bringeth forth great aboundance of all kinde of graine and amongst the rest, of Rice, euery yeare three or foure times in a yeare. It hath goodly woods & forrests, wherein do keepe a number of wilde Boares, Foxes, Hares, Conies, Sables, and Martins. The monntaines are full of grasse, seruing to feede infinite heardes or troopes of Cattell, both greate and small. There bee also manie mines of precious stones, of Golde, Sil-

uer

uer, Copper, Stéele, and Iron, and a great number of pearles, but not verie round, and great aboundance of silke. The townes there are very great, fortified, and well peoplished, which is easilie knowne by the greatnesse of Cantan, which is one of the least Metropolitane cities of the Realme, and yet it contayneth in circuite 12. Italian miles, and 350. Geometricall paces, which is more than foure houres iourney, not reckoning the suburbs, which are verie large and full of people. The principall Metropolitane towne where the King kéepeth his court, is named Paquine, or Suntie, that is to say in their tongue, the celestiall or heauenlie citie, touching the greatnesse whereof the Portugals and the Castilians doe write many incredible things, and according to the opinion of manie, that is the selfe same towne which *Marcus Paulus Venetus* calleth Quinzay, as that which hath diuers names in diuers languages. The like may be said of a towne in Flaunders, which the French men call Lile, the Fleminges Rufsill, and in Latine it is called Infulæ. In these prouinces bée manie good Portes and Hauens vppon the Sea, and a greate number of shippes: by reason whereof the Inhabitantes are mooued to say, that amongst them, there are as manie that dwell in Shippes vpon the Sea, as bee of them that dwell in houses vppon the land, and that their King might easilie make a bridge to passe from China to the towne of Malaccha, which is distant from them 300. *Almaine* leagues. But aboue all there is one thing worthie of great admiration, and that is a wall which hath in length 400. Spanish leagues, which the King of China caused to be built, to defende the countrey against the inuasion of the Tartarians, of which thing if the ancient men had had any knowledge, they woulde haue counted this worke amongst the seuen woonders of the world. The inhabitantes are men of Spirite, and geuen to labour. There was also inuented by them such a kinde of writing, that euery man of what nation so euer hée were, béeing some what exercised therein might pronounce in his mother tongue, euen as it were ciphered: They inuented also certaine Charriots, wherein they might sayle by the Winde vppon plaine grounde, as they doe in shippes vppon the sea. There are also men amongst them that are well learned in all Sciences

and

The description and vse

and especially in Architecture, wherein they excell all others, they are great louers of learning, and those that doe excell others therein are promoted to the most honourable estates: they haue good municipall lawes, and will suffer no Stewes, and they forbid that any man shall marrie any woman with whome he hath liued before in adulterie, and they greeuouslie punish all offences, & do forbid idlenes as the mother of manie euils, yea they constrain blind men to get their liuing, by turning with their hands, milles made to grinde corne, or any other things, and in their warres against the Tartarians, they get the victorie more by fine pollicie and stratagems, and by multitude of people, than by prowesse or feates of Armes. The Portugals doe report that the King bringeth to the field 300/000. footmen, and 200/000. horsmen. Now as touching their religion they be Paynims and superstitious Idolaters, sauing that ther are in many places some Christians, as Marcus Paulus Venetus testifieth.

The chiefest Merchandizes transported out of China into other prouinces are these, Golde, precious stones, Pearles, Muske, Rubarbe, the medicinable root, China, Purslane, aboundance of silke, Sugar, Rice, and all sorts of graine. The plant of Pepper is sowne at the roots of other trees, but specially at the roote of that Indian tree, which is called Faufell, and at the roote of the Date tree, to the tops whereof it climeth vp, much like as Iuie doth vpon a tree, or like to that which is called in Latine Clematis in English Perwinkle, which will winde about euery hearbe that groweth nigh it, the root is but small, and his leaues thin like vnto Citron leaues, but somewhat lesse, and sharpe pointed, greene and biting in the taste, the graines doe grow nigh one to another, like the long grape, and are alwayes greene vntil they be through ripe and drie, there be two kinds thereof, that is, white and blacke, but the plants of both are much like to our white and red Vines. Pepper groweth in places neere the sea side of Malacha, and in the Iles of Sunda and Cuda, situated nigh vnto the Ile Iaua maior, but the best kind of Pepper most plentifully groweth in the prouince of Malabar, betwixt the Cape Comori and the cape Canonar, but long Pepper is found in the realme of Bengala, and is another kind of plant, altogether vnlike to this.

The

of Plancius his Map.   257

The Contents of the seuenth page: wherein he first describeth the red figge tree of India, and setteth downe the shape thereof, then he describeth the East Indies, and last of all hee sheweth the nature of the Elephant, whose shape he setteth downe in the same page.

The Indian figge trée groweth round about Goa, the bodie thereof is high and great, and extending his braunches in a round forme, which like yellow or golden fillets do stoope down towards the earth, and so soone as it toucheth the earth, it bringeth foorth a newe generation of trées, which differeth nothing at all from the mother but onely in thicknesse, des Souches, and the branches of those doe bring foorth new trées in like manner, in so much as the mother with her offspring will in short time spread as much grounde as containeth an Italian mile in circuit, the fruites are small figges, and red as blood, as well without as within.

India tooke his name from the Flood Indus, which bordereth towards the East vpon the Realme of China, and towards the South vpon the great Ocean of India, and towardes the West vpon the sea of Arabia, and also vpon the Flood Indus, and towards the North vpon the sea Mare Euxinum, or Mar maior, and vpon Bramas.

This countrey is iudged at this day, as it hath bene long since to be the noblest and richest countrey in all the whole world, and it is diuided by the Flood Ganges into 2. partes, whereof the West part is called Indostan, or India intra Gangem, and the East part is called India extra Gangem, and it containeth many prouinces and Realms, as Cambaiar, Delli, Decan, Bisnagar, Malabar, Narsingar, Orixa, Bengala, Sanga, Mogores, Tipura, Gouros, Aua, Pegua, Aurea, Chersonesus, Sina, Camboia, and Campaa. These prouinces are watred with a number of goodly riuers, amongst the which Indus and Ganges are the most renowmed riuers of the world. Moreouer, these prouinces do abound in al things that may grow either within the earth or vpon the earth, except it be copper and lead, as Plinie affirmeth, also al manner of plants that grow there are very great, braue and excellent good

Ll   India

The description and vse

India exceedeth all other countries in precious stones and in spices, furnishing therewith almost all the worlde. It hath many rich mines of Gold, and great store of faire pearle, also greate multitude of all manner of cattell, horses onely excepted, which are brought thither out of Persia and Arabia. It is not long since that Callicute was the chiefe towne of Merchandize in India, but at this present Goa is the chiefe: there is also great trafique vsed at Dio, at Cananor, at Cochin, at Bengala, at Pegu, at Malacha, and at Sian.

Of the Elephant.

The Elephant amongst all other foure footed beastes is the greatest saue the Dragon and the Crocodyle, he is very ingenious, in so much, as it is incredible that which the ancient men write of him, and also the moderne, which haue sought more diligently to know his nature and disposition, hee is of force incredible, and meete to drawe shippes and boats both out of the water and into the water, and to drawe artillery and ordinance, hee is also meete for the warre, his teeth that shoot out of his mouth are Juorie, there is great number of them found in the Indies, and in Afrique, but the greatest and fittest for war, are found in the Jle of Ceiland, nigh to Calicute.

The Contents of the 8. page, in which he describeth the beast called *Rhinoceros*, and setteth downe his shape, and he describeth the Sinamon tree, shewing the shape of the trunke and of the leaf therof, and also the Muske Cat, with her shape, and in the latter end of the page he describeth the Realme of *Persia*.

The beast called Rhinoceros, is as long and as large as the Elephant, but not so high, for his legges are shorter, hee is armed not like a Tortoise (as *Plancius* saith) for that is couered all ouer with one shell, whereas this beast is armed with manifold strong, hard, & thick skales, which are yellow and spotted with purple, he hath a strong horne or bone vpon

vpon his nose, whereof he taketh his name, and hee hath another little horne vpon his backe, and he is a great enemie by nature vnto the Elephant, he is found in the Realmes of Cambaia, and of Bengala.

The Sinamon tree.

The Sinamon trée, is as big as the Oliue trée, the branches and griftes whereof are verie right, his leaues in collour are like to those of the Laurell trée, but in shape like to those of the Citron, his Flowers are white, and the fruites thereof are blacke and round, like a hazell Nut, the Sinamon it selfe is no other but the barke of the sayd trée, which groweth in the Prouince of Malabar, and in the Iles of Iava and Mindanao, but the best is found in the Ile of Ceiland.

The Muske Catte.

The Muske Cat, is like in shape to a common Cat, but shée is greater than either Catte or Foxe, her muzzell is somewhat long and armed with sharpe téeth, and with harsh haire, which haires (béeing angry) she will set vp as a Swine doth his bristles, she is in collour like a Wolfe, but that she is spotted with blacke spottes, the nether part of her Muzzel and the haires of her beard are white, her féete are blacke, her flanckes are whitish, and doe waxe whiter and whiter towardes her bellie, and next to her genitories, shee hath a little bagge like to a bladder or purse, into the which doeth fall the precious greace or humour, which they call Ciuet and Zibeth, which Ciuet is gathered out from thence with a spoone, if shee bee in mans kéeping, but when she is abroad and at her owne libertie, her bagge being full, shee will voyd that Ciuet of her selfe, and it will yéelde such a swéete sauour, as all they that sayle by that coast may smell it a farre off, as I haue heard. These musk Cattes are brought from the Realmes of Pegu and Tarnassary.

The description of *Persia*.

As those in Persia haue enioyed in times past the second Monarchie of the world, so at this present they be stil very

Ll 2   migh-

The description and vſe

mightie, for the king of Perſia is one of the greateſt Potentates in the whole world, as hee which commandeth all the great prouinces that doe border towards the Eaſt vpon the Flood Indus, and towards the South vpon the sea called Mare Caſpium, and vpon the Flood Oxo, within which limittes, are comprehended all the greateſt Realmes and landes, which the ancient Geographers were woont to call by theſe names, Aſſyria, Media, Sufiana, Perſia, Parthia, Hircania, Maigiana, Bactriana, Paropaniſa, Aria, Drangiana, Arachoſia, Caramania, and part of Armenia maior, the which at this preſent are called by other names as you may ſee in the Map.

The Perſians are a hardie and warlike people, and thought to be the beſt ryders or horſmen in all the worlde, they haue verie hard warres with the Turks, they be of moſt free and gentle nature, louers of ciuility, they make great account of learning and Sciences, they honour Nobilitie, wherein they greatly differ from the Turkes. Now as touching their religion, they be Mahometiſts, and yet in ſuch ſort, as both they and the Turkes doe count each one the other as Heretiques in that religion.

From the prouinces of Perſia are tranſported into other parts of the world theſe Merchandizes, ſtones called Turqueſſes, very faire and excellent pearles, great quantity of ſilke, Veluet, Damaſke armour, and a great number of moſt excellent horſes.

The Contents of the 9. page.

IN this page he deſcribeth and ſetteth downe the ſhape of the precious ſtone named Bezar, and the Dominions of the great Turk in Aſia, and the citie Aden in Arabia, alſo hee deſcribeth the beaſt called Cameleopardalis, and ſetteth downe his ſhape, alſo he ſetteth down the ſhape of three of the greateſt Pyramides that are in Ægypt.

Of the ſtone *Bezar*.

THe ſtone Bezar or rather Pazar (for that is his right name) groweth in Perſia in manner de Boucz, named Pazan, which are of diuers collours, but moſt commonlie redde.

This

This stone Bezar groweth in a concauitie in manner of a girdle about two handfull long and thrée inches broad, it is medicinable, and of great efficacie against all manner of poysons and venoms, and many other maladies, there is to be found of them in the entrie of Malacha and also in Pegu, but the best of them are in Persia.

Of the dominions which the great Turke hath in *Asia*.

The great Turke doth possesse in Asia, Natolia, sometime called Asia minor, and almost all Armenia, Mesopotamia, called at this present Diarbech, or Diarbekir, Syria, and a great part of Arabia, the most notable Merchant townes of this countrey are these, Trapezunda, Alepp, and a porte upon the sea called Tripoli, also Aman, Damasco, with his port Barutti and Mecha.

The Merchandizes that are sent from these Prouinces into other countries are these, great quantitie of silke, Ueluet, Damask Turkie Carpets, Cotton, and grain of Skarlet.

The Citie *Aden*.

Aden is the chiefe Merchant towne of the upper part of *Arabia*, which is gouerned by diuers Kinges, and this towne sendeth into other Prouinces of the fayrest Pearles, the true Baulme, Frankenscence, Mirrhe, and Horses.

The Beast called *Camelopardalis*.

This beast is called of the *Arabians*, Gyraffa, but the name Camelopardalis is compounded of Camel and *Pardale*, which is a Leoparde, hée hath a verie long necke like unto the Cammell, and is spotted with many spots, as is the Pardale or Leopard, he is a faire beast, and of gentle nature, as the shéepe, his head is like unto the heade of a Hart or Stag, but greater, his hornes are small topped, and couered with haire, and are about a handfull and a halfe long, hée hath eares tongue and féete like to an Oxe, his forelegges are long and tall

The description and vſe

tall, and his hinder legges are ſhort, whereby he ſeemeth alwaies to ſtand right vp, his head is ſomewhat higher than the Cammell, and this beaſt is to be found in Arabia, Æthiopia, and India.

The Pyramides.

IN Ægypt are manie Pyramides, whereof the two greateſt are counted amongſt the ſeuen woonders of the worlde, the greateſt of them (as witneſſeth Peter Belon) who moſt diligentlie viewed the ſame, is at the foote foure ſquare, and euerie ſquare containeth in length 324. paces, and in height 250. degrees or ſteppes, and euerie ſteppe hath in breadth 45. inches, which is three foote and 9. inches, but hee ſetteth not down what depth euerie ſteppe hath, which muſt not bee ouer deepe: for then how can anie man eaſilie mount vp to the toppe thereof, for hee ſaith it is plaine in the toppe, and ſo large as 50. perſons may ſtand thereon. It is found by writing that 360/000 men wrought 20. yeares in building this Pyramides.

The ſecond great Pyramides is ſomewhat leſſe, and ſmooth on the outſides without any degrees or ſteppes, and the toppe thereof is ſharpe pointed.

The contents of the 10. page.

IN this page hee deſcribeth the Crocodyle, and ſetteth downe the ſhape thereof, ſecondlie, hee ſheweth whereof the Mummie is bred. Thirdly, he deſcribeth the Unicorne, and ſetteth down his ſhape.

The Crocodyle.

THe Crocodyle is found in Ægypt, in the flood Nilus, and in India in the flood Ganges, and in the two Prouinces Mexicana and Pervana in manie riuers. This is a foure footed beaſt, which hath a horrible head, ſharpe teeth, a verie ſmall tongue, and a thicke tayle, and his ſkinne is hard and armed with hard ſcales, the nether parte of his mouth is
im-

immooueable, and the vpper part moueable, contrarie to all beastes, hée doth deuoure both men and beastes, and doth képe more in the water than on the land, and that which is greatly to be woondred at, hée is ingendred of an Egge, as great as a Goose egge, and hee groweth by little and little vntill hee come to the length of 18. cubites, or as some say to 22. cubites, which maketh 33. foote.

Of Mummie.

Mummie is made of bodies embaulmed, which they bring from Ægypt, whereas manie such embaulmed bodies were buried, about foure houres iourney beyond Cayre, whereas was sometime the great citie of Memphis, for before the natiuitie of Iesus Christ the Egyptians béeing Paynims did spare no cost to képe the bodies of their Parents from putrifaction, and therefore they are great palpable lyes, whereby fooles are perswaded, that the Mummie procédeth of those bodies which do perish in the sands that be in the deserts of Arabia, as though it were possible that those bodies could bee preserued in those sands without stench or putrifaction.

Of the Vnicorne.

The Vnicorne, as Lewes Vartiman testifieth, who saw two of them in the towne of Mecha, is of the height of a yoong horse or colt of 30. moneths old, which is two yeares and a halfe olde, hée hath the head of a Hart, and in his forehead he hath a sharpe pointed horne thrée cubites long, hée hath a long necke, and a mane hanging downe on the one side of his necke, his legges are slender, as the legges of a Goat, and his féete are clouen much like to the Goate, his hinder féete are hairy, and his haire in collour is like to a bay horse. This beast in countenance is cruell and wilde, and yet notwithstanding mixt with a certaine swétnes or amiablenes. His horne is of a merueilous greate force and vertue against Venome and poyson. The Vnicorne is founde in Æthiopia, like as the Indian Asse is found in India, which hath likewise one onely horne in his forehead,

Ll 4   The

The description and vse

The Contentes of the 11. Page, wherein hee first describeth Afrique, and then certaine fruites and spices, as Nutmegs, Mace, & cloues, & setteth down the shape of them, then he sheweth which bee the mightiest Princes in *Afrique*, and thirdly he describeth *Mexicana*, which is the first north part of *America*.

Afrique being the third part of the World, is seperated from Europe and Asia by the sea Mediterraneum and the red Sea, and by the land straight which is betwixt Ægypt and Palestina. The chiefe Prouinces of Afrique are these, Ægypt, Barbarie, Biledulgarid, Sarra, Æthiopia, Nubia, the large Prouinces of the Abassines, falsly called the lande of Prester Iohn, and also Monomotopa. The most renowmed Iles belonging to Afrique are these, Socotora, Madagascar, S. Thomas, the Iles of Capo Verde, and the Iles of Canarie and Madera.

The Nutmegge tree.

The Nutmeg trée groweth in the Ile of Bada, and differeth not much from the Peach trée, sauing that the leaues of the Peach trée are shorter and rounder: The fruit is couered with a thicke bark or huske, which when it is ripe cleaueth in sunder, and sheweth the Nut together with the shell, which is couered with Mace, the which at the first viewe is as red as Skarlet, and pleasant to behold, but when the Nut waxeth drie, the Mace do seuer from the Nut, and loosing by little and little their Skarlet collour, do waxe nigh vnto the collour of an Orange.

Of the Cloue tree.

The Cloue trée groweth in the Iles of Moluccas, which in greatnes and shape is like vnto the Lawrell trée, sauing that the leafe thereof is somewhat narrower. It hath many branches, and a great number of flowers, first white, afterward gréene, and then red, but being dried, they become black. The cloues do grow vpon the outermost ends of the branches, one hard by another and

whilest

whilest the flowers are gréene they excel all other flowers in sweet odour.

The chiefest Princes of Africa

The most puissant Princes of Afrique are these, the Emperour of the Moores or Ethiopians, which of the Arabians and of the Mahometistes is called Aticlabassi, and of his owne subiects, he is called Acegue and Neguz of the Abassines, that is to say, Emperor and King of the Abassines and Mores. Then the king of Monomotapa, the king of Morocho, the king of Fez and Sus. The great Turke also possesseth many prouinces in Afrique.

The chiefest Merchandizes that come from Afrique into Europe are these, Gold, Iuory, wood of Ebony, Aloes, Baulme of Ægypt, Mummie, Mirrhe, Anil feathers, Sugar, Ginger, Dates and wines of Madera, and of the Isle of Canarie.

Mexicana.

Mexicana which is the fourth part of the world, is on all sides enuironed with the sea, sauing that nigh vnto Nombre de Dios it is ioyned by a land strait to Peruana. The chiefe prouinces of Mexicana are these, the prouince of Mexico, otherwise called noua Hispania, terra Florida, Norum Bega, noua Francia, Estotiland, Saguenay, Chilaga, Toconteae, Marata, California, Tolm, Quivira, Agama, and Anian. The chiefest Iles lying on the North & Northeast part of Mexicana are these, Groynland, Crockland, Island, Freezland, Bacalaos, and Cuba.

The cheife Merchandizes that come from Mexicana into Europe are these, Gold, Siluer, Pearles, Cochenilles, to die with, Baulme, Salsaparillia, the root Mechoicana, Brimston, hydes of Oxen and Molue.

The contents of the 12. page.

In this Page he first describeth the beast called in that tongue Aiotochli, in Spanish Armadillio Then he describeth the 2. Prouinces Peruana, and Magellanica, then hee sheweth which bee the most mightie Princes of the world, and finally the diuers qualities of the people inhabiting the world.

The description and vse

The beast Armadillio is found in the Realme of Mexico, and he is no bigger than a cat, hee is headed like a Swine, and hath the féet of a Herison, and a long tayle, he is armed with scales, whereof he taketh his name, he keepeth for the most part within the ground, and as some suppose, doth liue by the earth, by reason that he is neuer séene to eat abroad out of his den, the bones of his tayle are medicinable, and do remedie the paine and deaffnes of the eares.

Though Plancius saith that this beast is armed with scales, yet my countryman William Greenway, who is a proper seruiter both by sea and land, and hath bene in the West Indies, and hath eaten of this beast, affirmeth his flesh to be white & verie delicate, and that he hath no scales, but that his skin is white and smooth like to a pig new scalded, and that sometime hée will shrinke vp the skin vpon his backe into diuers plates, and specially towards his fore partes and hinder partes, in such sort, as hee will make them almost to méete, and the former plates do hang downe vpon his shoulders like vnto two Poldrons, and his haire is white and short, growing thin here and there one, and he is eared and tailed like a rat, euen as he is here portraide, sauing that he is through-out of one selfe collour, and without scales.

Peruana.

Peruana being the fift part of the world, is also enuironed on all sides with the sea, saue wheras the foresaid landstrait doth ioyne the same to Mexicana, and the chiefe prouinces which it containeth are these, Brasilia, Tisnada, Caribana, Carthagena, Peru, Charchas, Chili, Chica, & the land of the Patagones: The most renowmed Iles are these, Hispaniola, otherwise called S. Domingo, Boriquen, & Margarita, which is the Ile of pearls.

The Merchandizes which are transported out of Peruana into Europe are these, gold, siluer, Emeralds, Pearls, the medicinable stone called Bezoar, Baulme, Ginger, Sugar, wood of Brasill, wood of Guaicum, called Lignum vitæ, long Pepper, Pepper of Brasill, Cassia solutiua and hides of Oxen.

Magellanica.

This is the sixt part of the worlde, which as yet is but little knowne, in such sort as we cannot write any thing touching the

of Plancius his Map.

the prouinces of the same, notwithstanding it is thought the prouince of Beach is verie rich, and hath abounbance of Golde, the chiefe Iles of Magellanica are these, Iava maior, and Iava minor, Timor, Banda, the Molucques, Romeros, the Iles of Salomon.

From the Ile Timor doth come into Europe, the white and pale medicinable simple called Sandalum.

From the Iles Banda doth come Nutmegs and Maces.

And from the Iles Molucques Cloues.

Which be the great Princes of the World.

The most mightie Princes of the world are these fiue, that is, the King of China, otherwise called the great Cham 1. the king of Persia 2. the great Turke 3. the Emperour of Æthiopia 4. the Emperor of Russia, otherwise called the great Duke of Muscouia 5. amongst which the king of China is a Pagan or Heathen: and the great Turke and the king of Persia are Mahometists: but the Emperour of Æthiopia and the great Duke of Muscouia do make profession of the Christian religion. Now as touching the king of Spaine, his puissance should bee much greater than it is, if his prouinces were not so seperated, and so farre distant on from another.

The qualities of diuers people in the World.

As touching the qualities of peoples, though God almighty hath created al men of one selfe bloud, and that all do take their beginning from the Arke of Noah, and that all men be of one selfe qualitie and shape of body, yet they differ in greatnesse, in proportion of members, and in collour: for the Patagones doe excæde all other creatures in greatnes. Againe, the men of China haue most commonly broad faces, litile eies, flatte noses, and little beards, and those that haue smallest fæte, are counted amongst them to be most beautifull, those of Africa haue grosser and thicker lippes than other people, the inhabitants of Agysimba, and of Guinea, and speciallie of the lands that be nigh vnto Cape de bona esperanza, are blacke, from whome the Orientall Indians do not much differ.

The

The description and vse

The Abassines or Moores of Ægypt, be of a duskish collour lyke to the Oliffe, the Inhabitants of Barbarie, be called white Moores, and those that dwell betwixt them and the Nigrites or blacke Moores be of a yellowish collour, the Spaniardes haue not found eyther in Mexicana, or in Pervana any Nigrites or Blackmoores, but onely in certaine villages nigh vnto Carque, the other nations vnder the hotte Zone, bee of collour browne bay, lyke a chessenut, and the nigher that they dwell to eyther of the Poles Arctique or Antarctique, the whiter most commonly they be, and as touching the rest all are like in qualities, shape, and fashion of bodie, as hath bene said before, wherfore they are meere lies that are woont to be told of the Pigmeans, in that they should bee but a foote and a halfe high, and like wise that which hath bene spoken of people, that shuld haue their heads their noses, their mouthes, and their eies in their breastes, or of those that are headed lyke a dog, or of those that haue but one eie, and that in their forehead, or of those that haue but one foote and that so great, as that it couereth and shadoweth all their bodie, or of those that haue greate eares hanging downe to the ground. All these are meere lyes, inuented by vaine men to bring fooles into admiration, for monsters are as well borne in Europe, as in other partes of the world.

Nowe in the foure pages following, hee setteth downe an interpretation of the Latine inscriptions dispersed throughout the Mappe, euerie which inscription hath his number added, which I doe also heare set downe in the same order, ioyning to euery number his longitude and latitude, to the intent you may the more easilie find out the said inscriptions, of which inscriptions there be in all 71. diuided into fiue partes, whereof the first parte conteineth 21. inscriptions, belonging partlie to Europe, but most to Asia, the second part containeth 12. inscriptions belonging to Afrique, the thirde part contayneth 11. inscriptions, belonging to Mexicana, the fourth parte contayneth 6. inscriptions belonging to Peruana, and the fift part containeth 21 inscriptions, belonging to Magellanica.

The

of Plancius his Map. 263

The 21. inscriptions belonging to Europe, but most to Asia.

The number of inscrip.	Longitude	Latitude	North or South.
1	5.	85. 30'.	North.

This arme of the sea doth take his course through 3. places running continually towards the North pole, and is frozen 3. moneths of the yeare, and it containeth in breadth about 37. leagues.

2	38 0'.	86. 30'.	North.

It is said that this Countrey is inhabited of Dwarffes called in Latin Pigmei, being in height foure foot as those be of Groinland, which are called Serelinges.

3	81. 0'.	77. 0'.	North.

Muscouia is bounded on the North side with the sea Pitzorique, called of the ancient Geographers Mare Glaciale, that is, the frozen sea, and towards the East it bordereth vpon Tartaria, and towards the South vpon the sea called Mare Caspium and also vpon the Turkes, and Tartaries Perocorliques, and towards the West, it bordereth vpon Lituania, Livonia, and vpon the Realme of Swethland: as touching their religion they obserue the Faith and ceremonies of the Græke Church, all their Bishops called in their language Vladiques, and also their Metropolitane are vnder the obedience of the Patriarch of Constantinople. The people is wise and subtill, and yet louing seruitude more than libertie or frædome, and they all confesse themselues to be the seruants and slaues of the great Duke, hauing seldome or neuer peace, for either they haue warres with those of Lituania, or with those of Livonia, or with those of Swethland, or with the Tartarians, or if they haue no warres, then they lie in garrison nigh vnto the 2. floods Tanais and Dwina, to defend their bounds from the depredations and inuasions of the Tartaries, they sharply punish robbers and stealers, and yet priuie theft and murder is seldome punished with death: their siluer money is not rounde, but hath the forme of an egge, their countrey is euery where ful of woods, and they haue great aboundance of rich furres, they send into all countries of Europe very

good

The description and vse

The number of inscrip.	Longitude	Latitude	North or South.

good Flaxe and hempe to make Cables and ropes, and a great number of hides, as well of Oxen as of Elkes, great balles of Waxe, much salted Fish, and Whales grease. The great Duke, called of his seruantes the Emperour of Russia, is rightly accounted amongst the most mightie Monarches of the world.

4	98. 0′.	82. 0′.	North.

This arme of the sea hath 5. mouthes or entries, and by reason that it is so straight and hath so violent a course, it is neuer frozen.

5	123. 30′.	58. 0′.	North.

Vnder the name of *Tartarie* at this day are comprehended all the prouinces that border towards the East vpon the sea of China & towards the South, are limitted with the prouinces of China, of India, with the flood Oxio, with the sea Mare Caspium, & with the lake or marish called *Palus Meotis*, & towards the West they are bounded with the flood Boristenes, & with the limits of Muscouia: for the Tartarians haue conquered all the countries which they possesse at this present, so as *Tartary* comprehendeth all that countrey which the ancient men were woont to call Sarmasia, Asiatica, and also the Scythias, that is, intra Imaum and extra Imaum, they began first to bee renowmed in Europe, in the yeare of Christ 1212. The Tartarians are diuided into certaine commonalties, and Colonies, called of them Hordes, but for so much as they dwell in diuers prouinces, that do extend far, and be farre distant one from another, they differ also in their manners and trade of life, they be men of a square stature, hauing broade and grosse faces, their eies hollow sunke into their heads, and looking somewhat a squint, and thick beards, they be strong of bodie and hardy, they eat horses and all other beastes howe so euer they are slaine, sauing hogges, from which they abstaine, they are able to endure hunger, thirst, watch, and all discommodities, and when they are distressed in their voyages with hunger and thirst, they let their horses blood, and with that blood quench their hunger and thirst, which kind of meat they cal in their language Besermannen, they call their Emperour Cham, that is to say, Prince, and therefore

of Plancius his Map. 264

The number of inscrip.	Longitude	Latitude	North or South.
for Cambalu is interpreted to be the seat or towne of the prince.			
6	125. 40´.	50. 0´.	North.

This round lake in the prouince of Sancii, tooke his first originall and beginning in the yeare 1557. by reason of an inundation or flood, which carried away 7. townes, besides villages and other places nigh adioyning, & a great multitude of people, whereof none were saued but onely one infant, sitting vpon a tree.

7	124 0´.	37. 0´.	North.

The inhabitantes of China, are of good spirit and ingenious, insomuch as they haue inuented certaine kinde of carts, wherein they may sayle vpon euen ground, hauing wind, and sayle as they doe in shippes vpon the sea.

8	171. 30´.	80. 30´.	North.

Plancius in this inscription setteth downe the opinion of Mercator, touching the beginning of longitude, and touching the Adamant stone, otherwise called the Lodestone, in Latine Magnes.

    Frances of Diep, a most skilfull Pilot, doth witnesse that the needle of the Mariners compasse, doth turn directlie to the North Pole, being in the Ilands of Capo Verde, that is to say, the Ile of Sal, the Ile of Bonauista, and the Ile of Mayo, whereunto those doe agree very nigh, which do say that the needle doth the like in the Iles of Tercera, and of S. Marie, which are part of the Flemmish Iles, otherwise called the Azores: but some others do affirme, that the needle sheweth the North pole best, being in the Ile of Corvo, which is the furthest Ile westward of the said Azores, and because the longitude of places by most liuely reasons, ought to take his beginning from the common Meridian of the world, and from the rocke or Pole of the Adamant stone, we here following the opinion of those that are most skilfull in this matter, haue set downe the first Meridian betwixt the Iles of Capo Verde, and the Azores, and because the needle in al other places, declineth either more or lesse from the Pole of the world, ther must needs be another Pole in some one place wherunto the needle doth

incline

The description and vse

The number of inscrip.	Longitude	Latitude	North or South.

incline from all coastes of the worlde, and I haue found by the declination of the nædle obserued at Ratisbone, otherwise called Regensberg, that is the same place, which I haue set down in the Mappe, and I haue likewise marked in this Carde, the situation of the Pole of the stone, in respect of the Ile Corvo, to the intent that according to the outermost places limitted by the first Meridian, the outermost boundes, betwixt which, this Pole ought to bee found, might be knowne, vntill the diligent and curious consideration of the Pilots shall bring vs something of more certaintie. Thus farre *Mercator*.

9	162. 0.	71. 0.	North.

These are the plaine fieldes of Bargu, whereof the inhabitants are called Mecriti.

10	154. 30.	68. 0.	North.

The Mount *Askai*, in which are to bee seene the sepulchers of the Kings of Tartary.

11	171. 30.	70. 0.	North.

Upon this Mountaine are set by the Tartarians two trumpetters of brasse, for a perpetuall memorie of their freedome gotten.

12.	175. 0.	67. 0.	North.

The Prouince Vng, which of our men is called Gogg.

13	170. 30.	65. 20.	North.

The Realme of *T*endus, which in the time of *Marcus Paulus Venetus*, which was in the yeare 1290. was gouerned by those Christians, which descended from the king Vncham.

14	165. 30.	65. 0.	North.

The Castell of King Vncham was builded in this place, against the inuasion of the Tartarians.

of Plancius his Mappe.   265

The number of inscrip.	Longitude	Latitude	North or South.
15	157. 0'.	65. 3'0.	North.

The Prouince Mongul, which of our men is called Magogg.

| 16 | 159. 2'0. | 66. 0'. | North. |

The desert Belgian, which is very great, all sandie and barraine.

| 17 | 159. 0'. | 61. 3'0. | North. |

In this Countrie of Cergutha, is found the best sort of muske, which groweth like as an impostume or bagge, nigh vnto the nauell of a certaine beast.

| 18 | 165. 3'0. | 58. 0'. | North. |

Campion is the Metropolitane Towne of Tanguth, wheras the inhabitants are partly Christians, partly Idolaters, and partly Mahometists.

| 19 | 158 3'0. | 53 0'. | North. |

This wall hath in length 400. Spanish leagues, and was built betwixt the mountaines by the King of China, against the inuasions and excursions of the Tartarians.

| 20 | 176. 0'. | 40. 0'. | North. |

Iapan hath three Ilands much renowmed, separated one from another by a straite of the sea, whereof the first and greatest is deuided into 53. Prouinces or Realmes, of which Meaco is the Metropolitane citie, the second is called Xima, which hath nine Prouinces or Realmes, the third is called Xicoca which containeth but foure Prouinces or Realmes.

| 21 | 71 0'. | 29. 0'. | North. |

Medina Talnabi, is the towne wherein is to be seene the Sepulchre of Mahomet.

The 12. Inscriptions belonging to Afrique.

| 1 | 67. 0'. | 26. 0'. | North. |

Coptos is a trim Marchant Towne, whereto are brought the Marchandises of India, and of Arabia, in which towne are dwelling many Christians.

The description and vse

The number of inscrip.	Longitude	Latitude	North or South.
2	58. 0'.	21. 0'.	North.

The flood Nilus, which by his inundations doth yearely water and fatte the Countrie of Egypt, and maketh it meruailous fruitefull.

| 3 | 55 0'. | 17 0'. | North. |

The flood Nubia, taketh his originall from the lake Nuba, as Ptolomie sayth.

| 4 | 50. 0'. | 15. 0'. | North. |

The flood Niger, here taketh his course, running vnder the ground 50. leagues.

| 5 | 62 0'. | 5. 0'. | North. |

In this place is the ample iurisdiction of the Emperour of Æthiopia, wrongfully called of those of Europe, the lande of Prester Iohn, the Arabs and the Moores doe call him Aticlabassi, but his owne subiects doe call him Acegue, and Neguz, that is as much to say, as Emperour and King.

| 6 | 58 3'0 | 2. 3'0. | North. |

The mount Amara, whereas are most carefully kept with continuall watch and ward of souldiers, the children and Grandchildren of the Emperours of Æthiopia.

| 7 | 53. 0'. | 7. '0. | South |

It is said that this countrie is inhabited of Amazones, which are women that make warre.

| 8 | 55. 4'0 | 15. 3'0. | South. |

In this place the King of Monomotopa, hath his great and ample iurisdictions.

| 9 | 42: 3'0. | 15. 4'0. | South. |

This South part of Afrique, vnknowne to the aunciet writers, is called by those of Persia and of Arabia, Zanzibar.

| 10 | 45. 3'0. | 1. 0'. | South. |

Heere is digged out of the Mines, great aboundance of golde.

Libia

of Plancius his Mappe.

The number of inscrip.	Longitude	Latitude	North or South.
11	30. 2′ 0.	20. 0′.	North.

Libia or inwarde Afrique, is at this day called Sarra, or the deserts.

| 12 | 19. 0′. | 15. 30′. | North. |

The people Azanagi, are of colour blacke graye, and they couer their mouthes as a member of shame, and doe neuer vncouer them but when they eate.

The 11.Inscriptions belonging to Mexicana.

| 1 | 301. 3′ 0. | 81. 0′. | North. |

This Ile is thought to be the best and most holesome of all the North parts.

| 2 | 301. 0′. | 74. 0′. | North. |

The Ile of Crockland, the inhabitants whereof say that they had their originall from Swethland.

| 3 | 311. 0′. | 61. 3′ 0. | North. |

In the yeare of our Lorde, 1500. one Gaspar Corteriale a Portingale, entred into these Regions, hoping to haue found some passage on the North towardes the Iles Molucques, but arriuing at the Riuer, which by meanes of the abundance of snowe there falling, is called Rio Neuado, which hath in North Latitude 62. degrées, he did leaue to sayle any further towardes the North, by reason of the great colde there, and turning to the South did fetch in all the sea coastes vntill hee came to Capo Razo, which hath in North Latitude about 48. degrées: and in the yeare 1504. the Britaines were the first that discouered all the sea costes of new France in America, nigh vnto the Golfe of Saint Laurence, which hath in North Latitude about 50. degrées and in the yeare 1524. Iohn Verazzan a Florentine, did part from the Port of Diep, the 17. of March, in the behalfe of Frances king of France, and sayled towardes the South sea coast of newe France, whereas he arriued at the 34. degrée of North Latitude, and from thence sayled towardes the East, viewing all the Sea costs vntill he came to Cape Britaine, which hath in North Latitude about 46. degrées.

The description and vse

The number of inscrip.	Longitude.	Latitude.	North or South.
3	311. 0'.	61. 30'.	North.

And in the yeare 1534. new France was againe visited by the Admirall Iaques Cartier, and in the yeare next following, it was conquered to the vse of the King of France, also in the yeare 1577. Martin Furbosher Englishman, arriued at ỹ north straite which is betwixt Groinland & Estotiland, that place hauing by this Map in the North latitude 65. deg. seeking passage by the North vnto Cathay, wheras he found certain Iles, & a mine of gold, wherwith hauing loden his ships, he returned into England with great hope of profite, but his successe was not answerable to his hope.

| 4 | 275. 2'0. | 70. 0'. | North. |

This daungerous beast is called Sucaratha, which being chased of Hunters, doth take her young ones vpon her backe, to saue both her selfe and them by flight.

| 5 | 271. 0'. | 64. 3'0 | North. |

This is a great lake or sea of fresh water, the limits whereof are vnknowne as they of Canada doe say, and as they haue heard by relation of those of Saguenay.

| 6. | 293. 3'0. | 52. 0'. | North. |

Alongst this Riuer a man may saile very commodiously, towardes the countrie of Saguenay.

| 7 | 287. 0'. | 47. 4'0. | North. |

All those that dwell betwixt Terra Florida, & Terra de Laborador be called by one common name Canadois, but there be many diuers Nations as those of Hochelada, Honqueda, & Corterealia, they are all very courteous to strangers, they liue commonly by fish, & they are clothed with skins of wild beasts, as they be also that dwell further towards the North, this countrie is also called new France, because the Britaines which are French men did first discouer it in the yeere of our Lorde 1504. the conquest wherof was atchieued by the Admirall Iaques Cartier in the yeare of our lord 1535. to the behoofe & vse of the French king. In the mountaines towardes the South, doe dwell many and diuers Nations, which be cruell people, liuing without any law, & do seeke by continuall warre, to vexe and oppresse one an other, as the people Auanares, Albardi,

of Plancius his Mappe.

The number of inscrip.	Longitude.	Latitude.	North or South.
7.	287. 0'.	47. 4'0.	North.

Albardi, Calicuares, Tagiles, Apalcheni, Mocole, Capatchi, Chilage, and many others, amongst which there be some of such agilitie and swiftnesse, as they may contend with horses, who can run fastest, they eate as those of Florida, certaine kind of Spiders, Ants or Pismires, Leazards, Adders & other venemous beasts. The land of Baccalaos, is so called of the fish Baccalaos, which is there taken. Terra Florida is so called of Easter day, because Iohn Ponce of Lion did discouer it on Easter day, in the yeare 1512. the which day is called of the Spaniards Pasqua du flores: this is a fertile countrie and rich in gold.

8	271. 0'	39. 0'.	North.

Marcus Nizza testifieth that this Prouince called the seuen Cities, is a very good countrey with whom Frances Vasker doth not agrée, for he saith they be places of no value, and like to Villages, and be vnder the iurisdiction of Ceuola, which at this present is called of the Spaniards Noua Granada.

9	211. 4'0.	82. 0.	North.

The North Ocean sea, entring with 19. mouthes betwixt the Iles of the North, doth make foure straites of the sea, and running floods, which continually take their course towards ye North, & there are swallowed into the bowels of the earth, like to springs of fountaines and floods.

In mine opinion this Inscription would haue béene one of the first of Mexicana, and not one of the last, sith it is no small leape, to turne so sodenly from the middest of America to the North pole.

10.	223. 0.	55.4 0.	North.

This Countrie is desert and plaine, in which are many wilde Horses, and Oxen with high backes like Camels, and wild shéepe like vnto those which Boetius writeth (in his description of Scotland) to be found in one of the Iles of the Hebrides.

11	240. 2'0.	40. 3'0.	North.

New Spaine was brought by force of armes vnder the obedience of the Spaniard, in the yeare 1518. by their generall, Ferdinando

The description and vse

The number of inscrip.	Longitude	Latitude	North or South.
11	240. 2′ 0.	40. 3 0.	North.

Ferdinando Cortes who conquered the same with great losse of his souldiours, but with greater ruine to the inhabitantes, who fought for the libertie of their Countrie, the soile is very fertill, and the Countrie is rich in golde and siluer, for in the floodes are found great sands of gold, and in the mountaines is drawne out of the mines great quantitie of siluer, and alongest the sea side they take to their great profite an infinite number of Oysters, wherein are found very faire pearles. In this Prouince there are many sault lakes, the water whereof by force of the sunne is turned into sault: there groweth also great abundance of Cassia fistula, and an other kinde of fruite, which the inhabitants call in their tongue Cacao, it is like to an Almonde, they haue it in great price, for of it they make a certaine drinke, which they loue meruailous well. The sea and floods which doe wash this Prouince, doe furnish them with great plentie of fishe, and in those floodes are many Crocodiles, whose flesh the inhabitants doe eate, and this beast will grow to be twenty foote long and aboue. This Countrie is full of great mountaines and high rocks, there is great diuersitie of languages, in so much as one vnderstandeth not an other without an interpretor. Mexico is the Metropolitane and Royall towne, or rather the Queene of all the principall Townes in the worlde, it is situated vpon the side of a lake or marish, yea the very foundation of it is a very marish, in such sort, as you can neither enter into it, nor come out of it, but by bridges, and it is aswell peoplished with inhabitants and Merchants, as any renowmed Merchant towne in Europe, the towne is very great, for it containeth in circuit 3. Spanish leagues.

The 6. Inscriptions belouging to Peruana.

1.	333. 0′.	3. 3′ 0.	South.

This great flood Maragnon, is called of some Oreigliana, and also the flood of the Amazones, it was discouered by Vincent Iohn Pinsonio, in the yeare 1499. and was sayled in a manner from

of Plancius his Mappe.

The number of inscrip.	Longitude.	Latitude	North or South.
1	333. 0'.	2. 3' 0'.	South.

from the head spring euen vnto the sea, by Franciico Oreigiano in the yeare 1542. the which voyage he perfourmed in eight monethes, hauing sayled 1660. leagues. This riuer kéepeth his water still fresh after it is entred certaine leagues into the sea, by reason that his course is so swift and violent.

| 2 | 330. 3'0 | 22. 0'. | South. |

Peruana is the South part of America, and is deuided by the Spaniardes into fiue goodly Prouinces, that is to saye, Castiliador, that is to say, the golden Castilia, Pompaiana, Peru, Chila, and Bresille: The Prouince Peru (before the arriuall of the Spaniardes) did extende a great deale further, when as their Countrie was yet vnder the gouernement of their naturall King, which then was called Ingas, as Giraua and others doe write. At this present it is limited on the North with the Towne Quito, and towardes the South with the Towne of Plata, that is to saye, the siluer Towne, and it was called Peruana as some write, of the flood and Port of the sea called Peru. It is at this present deuided (according to the situation of the Countrie) into thrée partes, that is to say, Sierras, Andes, and the flat or plaine countrie.

The playne Countrie is that which lyeth alongest néere to the Sea-coast.

Sierras is that part which is full of mountaines.

And Andes is that part of the Mountaines which tendeth towardes the East. Of all the Prouinces in the worlde this is the richest in golde, and in Emeraldes, the Metropolitane Towne of this Prouince, is the Citie Lyma, otherwise called the Towne of Kinges. Castillelledor tooke his name of the great abundance of golde that is there. The Prouince Popaiana tooke his name of the renowmed Towne Popaian, Chila is a colde countrie, by reason that it is so nigh vnto the South pole. The Prouince Brasilia tooke his name of the woode called Brasill, which groweth there in great abundance. To these Prouinces were méete to be ioyned, these other Prouinces, that is to say, Caribana, Charcas, Chica, and the lande of the Patagones.

Mm 4   This

The description and vse

The number of inscrip.	Longitude	Latitude	North or South.
3	329. 0´.	28. 0´.	South.

This beast is called of some Haute, but of a certaine people of Brasill, it is called Haye, which beast was neuer séene to eate or drinke, as some write, and therefore some thinke that she liueth without meate or drinke, onely by the aire.

4	328. 0´.	35. 2´0.	South.

This kind of beast is found in the Prouince of Parias, which in his fore parts is like to a Foxe, and behinde he is like to an Ape, sauing that he is fœted like a man, and hee hath the eares of an Owle, and vnder her ordinary belly, she hath also another belly, which openeth and shutteth, wherein she lodgeth her young ones, vntill they are able to get their owne liuing, and she neuer suffreth them to goe out but onely to sucke, Gesner calleth this beast an Ape-foxe, or a Foxe-ape.

5	329: 3´0.	36. 3´0.	South.

This is the land of the Patagones, the inhabitants whereof are Giants nine or ten foote high, which doe paint their visages with diuerse colours, made of herbes.

6	335. 0´.	51. 3´0.	South·

In the yeare 1582 the king of Spaine commaunded here to be built certaine fortresses, at the entry or mouth of the straight called Mare Magellanicum.

The 21. inscriptions belonging to Magellanica.

1	188. 3´0.	17. 20.	South.

This land is new Guinea, so called of the Nauigants and Pilots, because the sea coast and situation thereof, is like to that of Guinea in Afrique, and it was called by Andrew Corsali the Florentine, the land of Piccinacoli, and perhaps it is that which Ptolomy calleth the Ile Labadia, if it may be called an Ile, for it is not yet knowne whether it be any part of the South firme lande or no.

The

of Plancius his Mappe.   267

The number of inscrip.	Longitude	Latitude	North or South.
2	209. 0'.	31.3'0.	South.

It is not vnknowne to all those that are exerciled in Geography, that the degrées of Longitude do diminish and decrease from the Equinoctiall to either of the Poles, eyther North or South, whereby it falleth out, that the Prouinces which are next vnto the two Poles of the world, do differ greatly from that naturall shape which they haue by the roundnes of the earth, and for that cause we haue briefely drawne a description of the whole world in two round figures or Circles at the end of this Mappe, to the intent that euery man might sée their naturall situation so farre as may be shewed in Plano, that is to say, in flat forme.

3	231.	72.	North.

These Prouinces at this present are little knowne, yet it is sayd, that they are full of many kindes of wilde beastes. You shall finde this inscription in the round figure on the left hande nigh to the North pole, which indéede belongeth to Mexicana, and not to Magellanica.

4	247.2'0.	14. 0'.	South.

These two infortunate Ilandes were so called by Magellane himselfe, because he could find in them neither men, nor any thing else that was méete for mans sustenance.

5	283. 0'	61.2'0.	South.

These Prouinces were discouered by a Spaniarde, who being separated by tempest of sea, from the fléete or armie, ranne wandring here and there, through this great Southerne sea.

6	350.4'0.	85. 2'0.	South.

In the yeare 1493. when as the desire to saile into farre countryes increased more and more amongest the Castilians and Portugales, and that with great contention who should discouer most, Pope Alexander ordeined, that the Meridian which is 100. leagues distant towards the West, from euery one of the Ilands aswell of Capo Verde, as of those which they call Azores, should be the boundes and limits to eyther partie, for their nauigation, determining their rightes in such sort, as the Castilians should

haue

The description and vse

The number of inscrip.	Longitude	Latitude	North or South.
6	350.40′.	58. 20′.	South.

haue the West part of the worlde, to finde out vnknowne countries, and the Portingales the East parte, but there was such strife and contention betwixt them afterwarde touching the boundes of Nauigation, as this ordinance of Pope Alexander, pleased neither partie, and therefore in the yeare 1524. it was fully determined, that the Meridian which is distant 370. leagues towards the West from the Ile Saint Anthonio, being the most Westerne Ile of all the Iles of Capo Verde, should by the common bound of Nauigation to both parties.

7	19. 0′.	54. 30′.	South.

Here vnder the Latitude 42. degrees, & distant 450. leagues, from the Cape de bona speranza, and also 600. leagues distant from the Cape Saint Augustine, was found the Promontorie of the South lande, as Martin Ferdinando Denciso hath noted in the Epitome of some of his Geographie.

8	22. 0′.	65. 0′.	South.

In this our Geography, we haue in such manner described the circuit of the whole earth, as all the Countries are situated vnder their proper Meridians, which could not haue beéne done, without extending them more or lesse from the West towardes the East, notwithstanding beéing desirous to satisfie those that are practised in the arte of Nauigation, wee haue described the North Prouinces of Europe in this our Geographie, in such sort as their situation wholly agréeth with the perticular Mariners Cardes.

9	57. 20′.	54. 0.	South.

The 9. Inscription is intituled to the louers of Geographie.

We haue in this Inscription of the whole earth, employed all diligence to describe all the Seas and Prouinces, in such sort as euery place may haue his true Longitude and Latitude, in which matter, we haue spared neither labour nor cost, for we haue diligently conferred together the Sea Cardes, as well of the Castilians as Portingales, which they vse in their Nauigations to America and to India, and amongest others wee obtayned

rom

of Plancius his Mappe.

The number of inscrip.	Longitude	Latitude	North or South.
9	57. 2'0.	54. 0'.	South.

from Portingale a Mariners Carde, describing the whole earth very correctly, and besides that 14. other particular Cardes, in the which likewise all the Seas and Prouinces of the whole world, with their situation were comprised, all which Cardes being compared together, wee doe here nowe set foorth this newe description of all the Prouinces and Seas in the whole worlde, and that as correctly as may be, according to the consideration and obseruation which hath béen vsed by the most expert Geographers and Pilots, euen vnto this present houre: for in this Carde we doe describe all the sea coastes, Promontories, windinges in and out, Iles, Portes deapthes, sandes, showlds and rockes, also we haue added thereunto in place conuenient, the Mariners compasse, and the lines of the windes, which wee haue set downe as correctly as was possible for vs to doe, for the commoditie of Nauigation: and for so much as the true Longitude of the places can not be well obserued, without extending or enlarging too farre those Prouinces that are nigh vnto the North or South pole of the world, we haue therefore briefely comprehended the same in this our description of the world, in two rondles or circles, vnto the which we haue added another little Geographicall Carde, comprehending the north Prouinces of Europe, to the intent that in them euery man may sée with his eyes, the naturall situation of those Prouinces as well as it may be done in plano, as is more amply declared in other inscriptions of this Mappe. And this little Septentrionall Map last mentioned, is placed at the foote of this Map, in the very midst thereof comprehending all the North parts that lie betwixt the 52. degrée, and 7'2. of Latitude.

10	79. 3'0.	56. 0'.	South.

Here is a very strong current of the sea, which runneth East & West, betwixt Madagascar and the Ile Romeros, in such sort as Nauigation there is very troublesome and laborious, as Marcus Paulus Venetus testifieth in his third booke & 40. chap. whereof it must néedes follow of necessitie, that the sea coasts of this countrie are not far distant from Madagascar, in such sort as the great Orientall sea doth ebbe and flowe through this straite with great violence

The description and vſe

The number of inſcrip.	Longitude	Latitude	North or South.
10	79. 30.	56. 0.	South.

violence into the Weſt Ocean ſea, whereunto agræth the letter miſſiue of a Candiot, who was Embaſſadour for the Venetians to the king of Portingale. In which letter hee writeth that the men of this Countrie goe all naked.

11	91. 0'.	35. 0'.	South.

To this place arriueth a Portingale ſhippe called S. Paul.

12	212.	36	South.

The South firme lande, is called of ſome Magellanica, of Ferdinando Magellanus, who firſt diſcouered the ſame.

This Inſcription, together with the foure next following, are to be found in the ronde on the right hand.

13	146.	46	South.

Marcus Paulus Venetus, and Lewes Vartiman doe teſtifie in their bookes of perigrination, that here be very great and ample deſerts.

14	75	52	South.

Betwixt the Ile of Saint Laurence and the Iles Romeros, doe fall a moſt violent fluxe and refluxe of the Sea, Eaſt and Weſt.

15	46	55	South.

This is the lande of Popeniayes, ſo called of the Portingales, becauſe thoſe birdes in that Countrie are of incredible bignes.

16	15	46	South.

This Promontory of the South land, is ſituate 450. Spaniſh leagues from the Cape de bona ſperanza, and 600. leagues from the Cape of Saint Auguſtine.

17	148. 0'.	34. 0'.	South.

As we haue in the firſt ronde on the left hande ſet downe the deſcription of that part of the worlde, which extendeth from the Equinoctiall to the North pole: ſo in this other ronde, wee haue ſet downe a deſcription of that part of the worlde which extendeth from the Equinoctiall to the South pole, in ſuch ſort as this ronde containeth all Magellanica, and almoſt all Peruana together

The number of inscrip.	Longitude	Latitude	North or South.
17	148. 0'.	34. 0'.	South.

gether with a great part of Afrique, and a great number of the most noble and renowmed Iles of the world, and herein you may plainely sée with your eye, the naturall situation of those Prouinces that are nigh vnto the South pole.

18	148. 3'0	20. 3'0.	South.

The Realme of Maletur, which aboundeth in all manner of spices.

19	148 0'.	15. 4'0.	South.

The Countrie of Beach is rich in golde, but little frequented by Marchants of other Countries, by reason of the crueltie of the people.

20	161. 0'.	20. 0'.	South.

Iaua minor, bringeth forth diuerse spices which haue not yet béene séene in Europe, as Marcus Paulus Venetus testifieth in his third booke 13. chapter.

21	170. 4'0.	3. 0'.	North.

The Iles Moluccas are much renoumed for the great aboundance of spices, which are sent from thence into all countries of the world: The chiefest of those Iles are these, Ternarie, Tidoris, Motir, Machian, and Bachian, vnto which some doe adde Gilolo, Celebes, Burro, Amboino, and Bandar.

Besides all these Inscriptions, Plancius at the foure corners of his Mappe, setteth downe foure rondles, two aboue and other two beneath: and in that aboue on the left hand, representing the Northerne halfe of the celestiall Globe, he describeth al the North stars that are already, & in the other rondle on ye right hand representing the southerne halfe of the celestial Globe, he doth not only set downe such Southerne stars as were knowne to the auncient Astronomers, but also such Southerne starres, as haue béen found out of latter dayes by those that haue trauelled into the East and West Indies, as the Crosse, the Southerne Triangle, Noes Doue or Pigeon, and an other in the shape of a man, called Polophilax, and certaine others, touching which starres hee setteth downe nigh vnto the foresaid rondle, a certaine inscription written

The description and vse

written in the Latine tongue, which I haue here interpreted word for word in our mother tongue as foloweth.

We haue here set downe the fixed starres in their true places answerable to the yeare 1592. and not 95. as the Printer hath made it.

Of the South pole and of the starres that are about the same.

Least the South part of this Hemisphere or halfe Globe, should remaine voide and emptie, I haue taken these Southerne starres out of the obseruations of Andreas Corsalius Florentine, and haue diligently compared the same with the writings of Americus Vesputius, and of Petrus Medina, and haue reduced the saide starres into this forme or shape. But for so much as I haue seene nothing as yet to my satisfaction or contentment touching the Longitude, Latitude, Magnitude or nature of the said stars, I hartely pray all those that haue any more certaine knowledge of this matter then we, that they will enforme vs thereof, to the common good of all men: As touching the pole Antarctique, Corsalius writeth that there be two cloudie stars of a meane bignesse, which with a circular motion doe goe about another starre, that is distant from the Pole almost 11. degrees, and are sometime aboue and sometime beneath the said starre. Hitherto Plancius. But now to the intent that you may the better vnderstande all the foresaide foure rondles, I thinke it not amisse to describe the same vnto you, and to shew the vse thereof as followeth. You haue to note then, that the two vpper rondles: that which is on the left hand, signifieth the Northerne halfe of the celestiall Globe and the other rondle on the right hand, signifieth the Southerne halfe of the said Globe as hath bæne said before, and each one of these rondles is traced with certaine circles and lines: The outtermost Circle whereof being deuided into 360. degrees, and containing the Characters of the 12. signes, signifieth the Zodiaque, or rather the very Ecliptique it selfe, the Centre of which Circle, is the pole of the Zodiaque, which by continuall turning about, describeth another lesser Circle hard by it, signifying in

the

of Plancius his Mappe.

the North rondle the circle Arctique, and in the South rondle the circle Antarctique, the Centre of which lesser circle in eyther rondle, is the pole of the world, both which poles are distant from the pole of the Zodiaque 23. degrees 28. which is the greatest declination of the sunne. Moreouer in either rondle are drawne vpon ech pole of the worlde, two other circles, the largest whereof signifieth in both rondles the Equinoctiall, and the lesser thereof in the North rondle, signifieth the Tropique of Cancer, and in the South rondle the Tropique of Capricorne: besides these circles ech rondle is traced with 12. right lines, signifying those sixe Meridians or lines which passing through both the poles of the Zodiaque, doe deuide the Zodiaque into 12. equall parts, euery part containing 30. degrees: for so many degrees doe belong to the Longitude of euery one of the 12. signes, whereby the Zodiaque hath in Longitude 360. degrees, which Longitude is to be counted from the first point of Aries and so forth, according to the succession of the signes, and by helpe of these lines, you may know vnder what signe any fixed starre is: I shal not nede here to shew you, how the said fixed starres are situated, in either of the rondles, nor how they are named, because their images or shapes together with their names, are apparant to your eye. But if you would know the true place, the Longitude, Latitude, Magnitude and the nature of any fixed starre herein contained, then do thus: First to knowe the place and Longitude of any starre, lay a ruler or extend a threde, so as it may passe through the Pole of the Zodiaque, and also through the bodie of the starre, whose place and Longitude you seke, euen to the very Zodiaque, and somewhat beyond, and thereby you shall know in what signe, and in what degre thereof that starre is, for that is his place, and you shall knowe his Longitude by counting from the first point of Aries, vnto that degre, for that is his Longitude.

Now to know the Latitude of any starre, you haue to note that in each rondle there is a certaine right blacke line extending from the Zodiaque to the Pole, deuided by vnequall spaces into 90. degrees, which line is called the skale of the fixed stars Latitude, the vse whereof is thus: Set the firme foote of your Compasses in the very pole, and extend the other foote into the middest of the bodie of that starre whose Latitude you seke, and turne that foote

standing

The description and vse

standing at that widenes to the skale, and the number of degrées written vpon the skale, if you count from the Zodiaque vpward towards the Pole, will shew you the Latitude of that starre. Againe, to knowe the magnitude of any starre, Plancius setteth down in the North rondle the selfe same meane which Mercator also vseth in his celestiall Globe, that is to say, by making certain shapes of starres representing the bignesse of euery starre, according to his greatnesse, that by marking & comparing those shapes together, you might finde out, or rather coniecture the greatnesse of the starre which you séeke. Lastly he sheweth the nature of any starre by setting downe nigh vnto the star, the characters of those Planets, of whose nature that starre doth participate, all which thinges you shall more plainely vnderstande by this example here following.

Suppose that you would know the place, longitude, latitude, magnitude and nature of the starre called Arcturus: here because this is a North starre, you must therfore resort to the North rondle, and there séeke out the image Bootes, betwixt whose legs is the star called Arcturus which you séeke. And by extending a thréed which may passe through the pole, and also through the bodie of that starre euen to the Zodiaque, and somewhat beyond, you shall find his place to be in the 19. degrée of the signe Libra, and his Longitude counting frō the first point of Aries, vnto that degrée to be 199. degrées, and by obseruing the rule before giuen, touching the knowing of the Latitude of any starre, you shall with your Compasse finde the Latitude of this starre to be almost 32. degrées Northward, and by his shape you shall knowe that hé is of the first magnitude, and the characters of the two Planets Mars and Iupiter, placed hard by him, do shew that he is of their nature, that is to say, by participating of Mars he is extremely hotte and dry, and by participating of Iupiter he is hotte & moist, and looke what order is to be obserued in the North rondle, touching the North stars, the same is likewise to be vsed in the South rondle containing the Southerne starres: Amongest which you may sée the Images called the Crosse, whereby most Pilots in these dayes doe chiefely direct their course, being once past the Equinoctiall towardes the South pole, which Crosse, though Plancius doth here make to consist of fiue starres, yet I am sure that

Martin

Martin Cortes and Peter Medina, and all other late writers do appoint thereunto but foure starres, the shape and vse whereof, I haue set downe in my Treatise of Nauigation, according to the direction of Peter Medina. And those that haue trauailed into the Indies, doe all affirme that to the Crosse there doe belong onely foure starres and no more, wherefore I maruaile much, that Plancius doth set downe fiue, whereunto perhaps he is induced by the relation of some Spaniarde that neuer sawe them. Thus hauing described vnto you the two vpper rondles, representing together the celestiall Globe, and also shewed the vse therof, I will now describe the two neather rondles, whereof that on the left hande representeth the North halfe of the Terrestriall Globe, & that on the right hand the other halfe of the same Globe, towardes the South.

You haue then to vnderstand that the Centre or middle point in each rondle, signifieth the Pole of the world, that is to say, the North pole in the North rondle, and the South pole in the South rondle, and vpon each pole are drawne certaine Circles, the outtermost whereof, and furthest distant from the Pole signifieth the Equinoctiall, which is deuided into 360. degrees, euery degree containing 60. miles, which is the whole Longitude of the earth, from which circle at the end of euery tenth degree, are drawn certaine right lines to the number of 18. which doe meete in the very Pole, and doe signifie halfe Meridians, whereof that which passeth through the Iles Azores, and also the Iles of Capo Verde, is the first Meridian, from whence the longitude of the earth taketh his beginning, and there also endeth: which Meridian in the rondle on the left hand, is deuided into 90. parts, proceding from the Equinoctial to the Pole, signifying the North Latitude of the worlde, the like diuision and number of degrees of Latitude, hath also the first Meridian, in the rondle on the right hande, sauing that the saide Meridian tendeth vpwarde. Moreouer you haue to vnderstande, that in each of these rondles, are drawne nine Circles, equally distant one from another, called Parallels, which together with the Equinoctiall, doe make nine spaces, euery space contayning 10. degr. & besides these circles, there are drawne in ech rondle two other circles, the one greater, & the other lesser, the greater in the rondle on the left hand being

The description and vse

distant from the Equinoctiall 23. degrees 30. which is the greatest declination of the sunne, is called the Tropique of Cancer, and the lesser Circle being of like distance from the pole, is called the Circle Arctique, but the greater circle being of like distance from the Equinoctiall, in the rondle on the right hand, is called the tropique of Capricorne, and the lesser circle in the said rondle inuironing the Pole, is called the Circle Antarctique vpon which circles in each rondle you shall finde their names written. The chiefest vses of these two rondles are these: first to finde out the Longitude of any place, secondly the Latitude, and thirdly the distance betwixt any two places.

To finde out the Longitude of any place, you must doe thus: Extende a threede, so as it may passe through the pole, and also through the place whose longitude you seeke, euen to the very Equinoctial, and somewhat beyond, and holding the threed straight, the numbers of the degrees written vpon the Equinoctiall will shewe the longitude of the place.

And if you wil know the latitude of that place, or of any other, doe thus, Set the one foote of your compasse in the very pole, extending the other to the place whose Latitude you seeke, and keeping your compasse at that widenesse, bring the moueable foote to the first Meridian, whereon the degrees of latitude are marked, and there staying it, the number of the degrees, counting from the Equinoctiall vpward towards the pole, will shewe the latitude of the place. As for example, suppose that you would know the longitude and latitude of Lisbone, which is a famous towne in Portingale, here hauing first found out that towne in Spaine, which is nigh vnto the West Ocean, extende your threede from the pole through the middest of that Towne to the Equinoctiall and somewhat beyond, and you shall finde that the threed will cut the Equinoctiall in the 13. degree, which is the Longitude of Lisbone. Nowe if you would knowe the Latitude of the same place, set the one foote of your compasse in the Pole, and extende the other foote to Lisbone, and keeping your compasse at that widenesse, bring the moueable foote to the skale of latitude, and so you shall find that Lisbone hath in North latitude 38. degrees and 3′0. Now to knowe the distance betwixt any two places doe thus, Set the one foote of your compasse in the one place, and the

other

other foote in the other place, and apply that widenesse to the Equinoctiall, and looke howe many degrees of the Equinoctiall that widenesse comprehendeth, and by allowing 60. Italian miles to euery degree, you shall haue the distance by a right line, betwixt those two places, for by doing thus, you shall finde the distance betwixt Lisbone & Compostella to be 120. miles. Thus I haue sufficiently (I hope) expounded euery thing contained in Plancius his Map, his general skale made for the same onely excepted, whereof I come now to speake. In this skale are set downe the miles of Russia, of Italy, of Englande, of Scotland, the French leagues, the houre leagues, the Spanish leagues, the Germaine and Garscoyne miles, which two are all one, the miles of Sweuia in Germanie, of Scandia and of Swethland, which last three are likewise all one, the vse of which skale is thus. Take with your cõpasse the distance betwixt any 2. places which you desire to know, and apply the same to the skale of such miles as you would know, & so many miles the two places shal be distant one from another as the number of the skale doth shew: but if the distance betwixt the two places be longer then the skale, then hauing first taken the whole length of the skale with your compasses, looke howe manie times that widenes of your compasse measuring by a right line is contained in the distance betwixt the two places, and if there be any odde space left, streighten your compasse to that odde space, & apply that to the beginning of the skale, and adde the number of miles which you there find to the first great number, so shall you haue the totall summe. And lo here for each an example: first suppose that you would know the distance betwixt Cape S. Marie, & Cape finis terræ, which are two Capes or headlands in the West side of Spaine, both hauing in a maner one selfe Longitude, & do differ only in latitude by 6. degrees, for the one hath 37. & the other 43. in latitude, which distance if you take with your cõpasse by setting the one foot in the one place, & the other foote in the other place, and applying that widenes to the skale of Italian miles, you shal find the distance to be 390. Italian miles, but if you measure the same distance according to the Geographicall maner, which is to allow for euery degree of latitude 60. miles, you shal find y̆ distãce to be no more but 360. Italian miles. Let your other example be thus: Suppose that you would know the distance betwixt Compostella

The description and vse

postella in Spaine and Constantinople, which haue all one Latitude and doe differ onely in Longitude, heere because the distance betwixt these two places is longer then the skale, you must take with your compasse the whole length of the skale, and then to looke how many times that widenes is comprehended (measuring by a right line) betwixt the two said places, & you shall find that widenes to be comprehended in the distance betwixt those two places three times, wherefore if you multiply 840. by 3. it will make in all 2520. Italian miles. But if (according to the Geographicall kind of measuring) you doe multiply the difference of their Longitudes, which is 43. degrees, by the number of miles, which is also 43. belonging to the Latitude of both places, which Latitude is also very nigh 43. degr. you shal find that the distance betwixt those two townes is no more but 1870. Italian miles, which number of miles is not so great as that of the skale by 650. Italian miles. And therefore I can not thinke but that there is some error in the skale committed either by the Printer or else by the author through some negligence, and not for lacke of skill or knowledge how to make a true skale, being so excellent a Geographer as the Author by this & other his Maps heretofore made sheweth himselfe to be, or else there is some greater mistery therein, then I perhaps doe vnderstand, for in seeking to know the distance betwixt two places differing onely in Latitude, I finde the skale most times to agree with the Geographicall kind of measuring, but if the two places do differ either in longitude onely, or else both in longitude and also in latitude, then I find the skale to differ very much from the Geographicall kinde of measuring, wherfore I thinke it good briefely here to set downe certaine readie waies of finding out the distance of two places, differing eyther in latitude onely, in longitude only, or in both, which I doe shew also at large in the second part of my Spheare, Chap. 14.

How to finde out the distance of two places differing onely in Latitude.

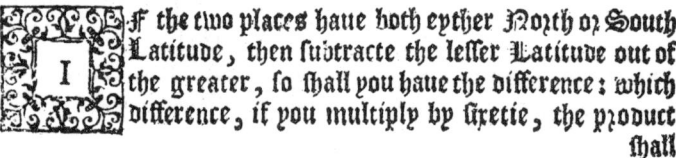F the two places haue both eyther North or South Latitude, then subtracte the lesser Latitude out of the greater, so shall you haue the difference: which difference, if you multiply by sixtie, the product shall

of Plancius his Mappe.

shall be the number of miles, and if to the whole degrées of difference there be annexed any minutes, then you must adde to the product for euery minute one mile. But if one of the two places haue North Latitude, and the other South Latitude, then you shall finde their difference by addition, and not by subtraction. As for example, suppose that you would knowe the distance betwixt a towne called Pasquali, which is the outermost towne in Moroa vpon the sea towards the South, hauing in North latitude 35. degrées 3′0. and a certaine towne in Afrique called Debsan standing nigh vnto the lake Zembre, which hath in South Latitude 12. degrées 3′0. here by adding these two latitudes together, you shall find the summe to be 48. degrées, and that is the difference of their Latitudes, which difference if you multiply by 60. the product will be 2880. and that is their distance.

How to find out the distance of two places differing onely in Longitude.

IF both places haue either East Longitude or West Longitude, then subtract the lesser out of the greater, so shall you haue the difference, which difference you must multiply by the number of miles belonging to their Latitude, which you shall finde on the Northwest side of the Mappe, or by the Table of miles answerable to one degrée of euery Latitude set downe hereafter in the end of this Treatise: and the product thereof shall be the number of miles whereby the one place is distant from the other. As for example, I finde Compostella in Spaine and Constantinople, hauing both almost 43. degrées of North Latitude, to differ only in East longitude, for Compostella hath in East longitude 13. degrées 3′0. & Constantinople hath in East longitude 56. degrées, the difference whereof by subtracting the lesser out of the greater you shall finde to be 42. degrées 3′0. here if you multiply 42. by 43. miles belonging to one degrée of the foresaid Latitude 43. you shall finde the product to be 1806. then to finde out the value of miles for the fraction 3′0, by the rule of proportion, you must say thus: if 60. require 43. miles what shall 30. require, and you shall finde in the quotient 21. miles, which you must adde to the

The description and vſe

former ſumme 1806. and it will make in all 1827. miles and $\frac{1}{2}$. which is the true diſtance betwixt the two foreſaid places. But if the one place haue eaſt Longitude, and the other Weſt Longitude, then you muſt finde the difference aſwell by Addition as by Subtraction. As for example, ſuppoſe that you would knowe the diſtance betwixt S. Domingo in the Ile called Hiſpaniola, and a certaine place in Afrique called Septem montes, nigh vnto the Ocean ſea, both places hauing 18. degrées of North latitude. And S. Domingo hath in Weſt longitude 310. degrées 3′ 0. and the place called Septem montes hath in Eaſt longitude ſeuen degrées. Here you muſt firſt ſubtract 310. degrées 3′ 0. out of 360. degrées, and there will remaine 49. degrées 3′ 0. whereunto you muſt adde the Eaſt longitude of Septem montes, which is ſeuen degrées, and it will make in all 56. degrées 3′ 0. which is the difference of their Longitude. Nowe if you firſt multiply 56. degrées by 57. miles belonging to 18. degrées of latitude, you ſhal find the product to be 3192. miles, and to finde out the value of miles for the fraction 3′ 0. you muſt ſay thus: If 6c. require 57. miles, what ſhall 3′ 0. require, and working by the common rule of thrée, you ſhall haue in the quotient 28. miles, and there remaineth $\frac{30}{60}$. which is one halfe mile. Now by adding 28. miles and $\frac{1}{2}$. to 3192. it will make in all 3220. miles $\frac{1}{2}$. and that is the true diſtance betwixt S. Domingo and Septem montes, but by Plancius ſkale you ſhall finde the diſtance to be 3410. miles, which differeth from the other almoſt 200. miles.

of Plancius his Mappe. 276

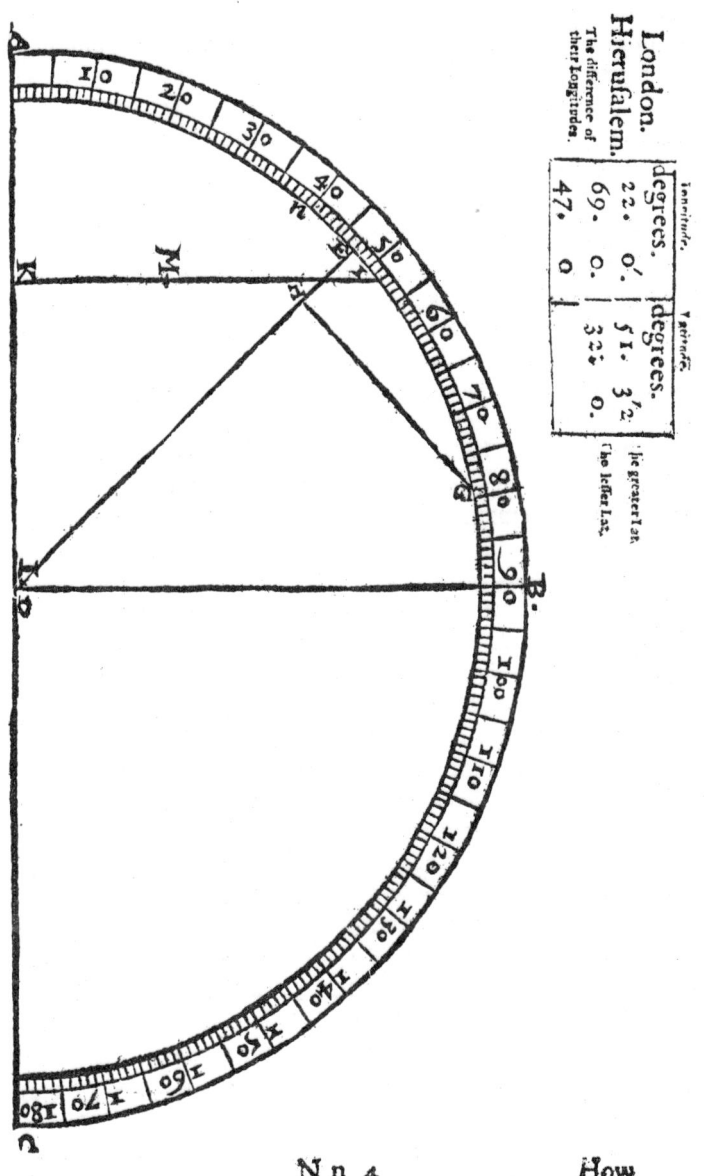

	Longitude.		Latitude.	
	degrees.		degrees.	
London.	22.	0.	51.	32. the greater Lat.
Hierusalem.	69.	0.	32.	0. the lesser Lat.
The difference of their Longitudes.	47.	0		

How

The description and vse

How to finde out the distance of two places differing both in Longitude and Latitude by helpe of a semi-circle deuided into a 180. degrees, which I had from my louing friend Maister Wright of Cayes Colledge in Cambridge, of whom I make mention aswell in my treatise of the Spheare as in that of Nauigation.

First drawe a Semicircle vppon a right Diameter, marked with the letters A.B.C.D. whereof let D. be the Centre like vnto this here described, and the greater that such semicircle is, the spaces of the degrees shall be the larger, and thereby the more easie to find out the minutes. Then hauing drawne your semicircle and deuided the same accordingly, suppose that by helpe therof you would find out the distance betwixt London and Hierusalem, which two townes doe differ both in Longitude and also in Latitude so much as is here set downe in the front of the figure according to such Longitude and Latitude as Plancius doth allow to eyther towne in his great Map. Now to find out the true distance of these two townes you must first take the lesser Longitude out of the greater, so shall you haue the difference of their Longitudes, which is 47. degrees, then count that difference vpon the Semicircle beginning at A. and so procæde to B. and at the end of that difference, make a pricke marked with the letter E. vnto which pricke draw a right line by your ruler from D. the Centre of the Demicircle: That done sæke out the lesser Latitude which is 32. degrees 0'. in the foresaid Demicircle beginning to account the same frō the pricke E. and so procæde towards the letter B. and at the end of the said lesser latitude set down another pricke marked with the letter G. from which pricke or point drawe a perpendicular line, which by helpe of your squire or compasses, may fall with right Angles vpon the former right line drawne from D. to E. and where it falleth, there set downe a prick marked with the letter H. That done sæke out the greater Latitude which is 51. degrees and 32'. in the foresaid Demicircle, beginning to account the same from A. towards B. and at the end of that Latitude set downe another pricke, marked with the letter I. from whence drawe another perpendicular line that may fall by helpe of your squire or compasses, with right

Angles

of Plancius his Mappe.

Angles vpon the Diameter A. C. and there make a pricke marked with the letter K. That done take with your compasse the distance that is betwixt K. and H. which distance you must set down vpon the said Diameter A.C. setting the one foote of your Compasse vpon K. and the other towards the Centre D. & there make a pricke marked with the letter L. then take with your Compasse the length of the shorter perpendicular line G. H. and apply that widenesse vppon the longer perpendicular line I. K. setting the one foote of your compasse at I. which is the end of the greater Latitude, and extend the other foote towardes K. and there make a pricke marked with the letter M. That done take the distance betwixt L. and M. with your compasse and apply the same to the Demicircle setting the one foote of your compasse in A. and the other towards B. and there make a prick marked with the letter N. And the number of degrées contained betwixt A. and N. will shew the true distance of the two places which you shall finde to be 39 degrées, which being multiplyed by 60. maketh in al 2340. miles, and whensoeuer you haue any minutes besides the whole degrées, remember to adde vnto the summe of degrées for euery minute one mile. By Plancius skale you shall find the distance betwixt London and Hierusalem to be 3040. miles which are 700. miles too many. But you haue to note by the way, that if the difference of the Longitudes doth excéede the number of 180. then you must subtract that excéeding difference out of 360. and the remainder shalbe the difference of the Longitudes, and then worke in all points as is before taught.

By this rule and the other two rules first declared, you shal easily trie the skale of any Mappe whether it be true or not, so as you first haue the true Longitude and Latitude of the two places whose distance you séeke to know. And thus I end with Plancius Map, hoping not to offend him with any thing that I haue added thereunto for the better instruction of those that haue not béene exercised in such matters.

And yet I had almost forgotten one thing, which is this, I haue here before as you haue read, made mention of drawing certaine perpendicular lines in the former figure of the Demicircle by helpe of your Compasses. Wherefore I thinke it necessary here to set downe the order thereof.

How

The description and vse

How to make with your Compasses, a perpendicular line to fal from any point giuen vpon another right line, making therwith right angles without the helpe of any squire.

Set the firme foot of your Compasse in the point giuen, and extend the other foote a little beyond the line right against the point giuen, & draw a secret Arch or portion of a Circle that may cut the said line in two points, and deuide that part of the Arch which lieth betwixt the two sections into two equall parts, setting a pricke in the very midst therof: then hauing laid your ruler to that pricke, and also to the point giuen, draw a right line, & that line wil fall vpon the other line with right Angles, as you may see by this figure.

the point giuen
the perpendicular
the arche

The Table of miles answerable to one degree of euery seuerall Latitude.

D	M	S	D	M	S	D	M	S	D	M	S	D	M	S
1	59	59	19	56	44	37	47	55	55	34	25	73	17	33
2	59	58	20	56	23	38	47	17	56	33	33	74	16	32
3	59	55	21	56	1	39	46	38	57	32	41	75	15	32
4	59	51	22	55	38	40	45	58	58	31	48	76	14	31
5	59	46	23	55	14	41	45	17	59	30	54	77	13	30
6	59	40	24	54	49	42	44	35	60	30	0	78	12	28
7	59	33	25	54	23	43	43	53	61	29	5	79	11	27
8	59	25	26	53	56	44	43	10	62	28	10	80	10	25
9	59	16	27	53	28	45	42	26	63	27	4	81	9	23
10	59	5	28	52	59	46	41	41	64	26	18	82	8	21
11	58	54	29	52	29	47	40	55	65	25	21	83	7	19
12	58	41	30	51	58	48	40	9	66	24	24	84	6	16
13	58	28	31	51	26	49	39	22	67	23	27	85	5	14
14	58	13	32	50	53	50	38	34	68	22	29	86	4	11
15	57	57	33	50	19	51	37	46	69	21	30	87	3	8
16	57	41	34	49	45	52	36	56	70	20	31	88	2	5
17	57	23	35	49	9	53	36	7	71	19	32	89	1	3
18	57	4	36	48	32	54	35	16	72	18	32	90	0	0

of Plancius his Mappe.

Though it be the common order of working to know by helpe of the former table, the distance of two places differing onely in Longitude, yet I thinke it a more sure way to find it out per Tabulas *Sinuum*, the rule whereof is thus.

First take the difference of the two Longitudes, by subtracting the lesser out of the greater, and the halfe of that shall be the Arch which you haue to seeke in the front of the Tables, then multiply the sine of that Arch by the sine of the complement of the common Latitude, and deuide the product thereof by the totall sine, the quotient whereof you must seeke out in the Tables amongest the sines, and the Arch of that sine is the one halfe of the distance, which being doubled shall be the whole distance contayning degrees of the great Circle, and euery such degree contayneth of Italian miles 60. and of German miles 15. and by working thus you shall finde the distance betwixt Compostella and Constantinople to be 1346. Italian miles, supposing the common latitude to be 43. degrees, and the difference of their longitudes to be 42. degrees 30. And by working by the common table you shall find the distance of those two places to be 1827. Italian miles as before, because the common Table hath no minutes of miles but onely seconds, which are not to be accounted of, & in working by Appian his Table hauing minutes of miles, you shal find the said distance to be 2184. Italian miles, and by Mercator his Map to be 1980. Italian miles, in whose Map the common Latitude of the said 2. places is 43. & the difference of their longitudes is 44.0'. And by the skale set down in Plancius his Map, you shal find the distance to be 2520. Italian miles, in which Map the common Latitude of the two foresaid places is 42. deg. 3' 0. and the difference of their Longitudes is also 42. degrees 3' 0. Truely I must needes confesse that it is not so easie to make a skale or trunke for a Mappe or a Carde drawne in plano, as for that which is drawne vpon a round bodie or Globe: and therefore it is no maruaile though the skales of Mappes drawne in plano, and likewise the trunkes set downe in the Mariners Cardes doe not alwayes shewe the true distance of places, which I beleeue is to bee done as truely and a great deale more readily by my friend Maister Wright his Semicircle before described, then by the rules of Gasparus Peucerus in his booke de dimensione ter-

rae,

The description and vse

ræ, which rules doe depend vpon the knowledge of the quantitie of the Angles and sides of Sphericall Triangles, which kinde of working is indéede more troublesome and tedious then readie or pleasant. But if Maister Wright would make his Demicircle an vniuersall instrument to find out thereby all the thrée kinds of distances as he promised me to doe, there were no way in mine opinion worthie to be compared vnto it, neither for the truenesse, easinesse, nor readinesse of working thereby.

A very brief and most

plaine description of Maister *Blagraue* his Astrolabe, which he calleth the *Mathematicall Iewell.*

Together with diuerse vses thereof, *and most necessarie for sea men,* written by Maister *Blundeuill.*

Imprinted at London by *Iohn Windet.*

To the Reader.

OF Astrolabes I haue neuer seene but three sorts, First that of Stofflerus, which for these hundred yeares past or there abouts, hath beene had in most price and estimation, as an instrument containing all the vses, or at the least the most part of all other Mathematicall Instruments, which because it requireth almost for euerie seuerall Latitude a seuerall Table: Gemma Frisius inuented since another kind of Astrolabe, hauing but one *Table* to serue for all Latitudes, and therefore he called it the *Catholicon*, that is to say, an vniuersall Astrolabe, which hath also been most esteemed & vsed many yeres. Since whose time, & of very late yeares, one of our owne countriemen a Gentleman of Reading besides London, called M. *Blagraue*, hath greatly augmented the saide *Catholicon*, and hath thereby as it were newly inuented a third kind of Astrolabe, which he calleth the Mathematicall Iewell, whereby are to be wrought more conclusions then by any other one instrument whatsoeuer, for which his most excellent inuention vsed therein, he deserueth great commendation. And to the intent that others which haue not beene exercised in such thinges, might the more easily attaine to the better vnderstanding of the said Iewell: I haue made a plaine description of all the partes belonging to the said Iewell, without committing any offence (I hope) to the Author therof, to the intent that euery Gentleman might haue the perfect knowledge and vse of so worthie an instrument: But broade Astrolabes though they be thereby the truer, yet for that they are subiect to the force of the winde,

and

The description and vse

and thereby euer moouing and vnstable, are nothing meete to take the altitude of any thing, and specially vppon the Sea, which thing to auoide, the Spaniards doe commonly make their Astrolabes or Rings narrow and waightie, which for the most part are not much aboue fiue inches broad, and yet doe waigh at the least foure pound, and to that end the lower part is made a great deale thicker then the vpper part towards the ring or handle: Notwithstanding most of our English Pilots that be skilfull, doe make their sea Astrolabes or ringes sixe or seuen inches broad, and therewith very massiue and heauie, not easie to be mooued with euery winde, in which the spaces of the degrees be the larger, and thereby the truer, of which kinde of Astrolabes or Rings, I shall speake hereafter in my Treatise of Nauigation, and also of the Mariners Crosse staffe. But whensoeuer you haue to take the altitude of the Sunne or of any other starre be it wandering or fixed, I would wish you to vse the Mariners heauie and massiue Astrolabe, which in mine opinion for that purpose, is the fittest and most assured instrument of all others and to finde out all other conclusions by helpe of Maister *Blagraue* his Iewell, or rather by helpe of the Celestiall Globe, which for Astronomicall matters is the perfectest instrument of all, in which are contained all the starres both southerne and Northerne that haue beene heretofore knowne, and those that are lately knowne or shall be knowne hereafter may be easily placed therein: which thing can not be so well perfourmed in any Astrolabe were it neuer so great. But because the Globe is comberfome and not portable, and therewith costly, and especially if it bee of any greatnesse, (for the greater the better) and also if it chance to be broken in any part of the bodie thereof, it can neuer bee made againe perfectly whole, it is therefore no meete instrument for euery Mariner to haue, but onely for such as be of good habilitie, in steede whereof to finde out manye necessarie conclusions, Maister *Blagraue* his Mathematicall Iewell may serue verie well, and specially if it had on the backe parte the like mater and *Rete* or Nette, to serue the South Latitude of the worlde as it hath in the fore parte to serue the

North

North latitude, for then I beleeue verilie, that the chiefeſt Sothern ſtarres, as the *Croſſe*, the Southerne Triangle, Noahs Doue or Pigeon, and another called *Polophilax*, latelie found out by ſuch as haue travelled by ſea on the South ſide of the Equinoctiall, might eaſilie be placed by Maiſter *Blagraue* in that *Rete*, hauing once learned of the skilfull Seamen the true longitude, latitude, and declination of the ſaide Starres, ſo ſhoulde his Iewell in mine opinion bee much more ſerviceable to the ſea men, then nowe it is, by reaſon that thoſe fewe Southerne Starres that be contained in his firſt Net, are nothing ſo nigh vnto the South pole, as the *N*ortherne Starres deſcribed in his firſt Net, are vnto the North pole, amongſt which Northerne Starres, it were verie neceſſarie for the Sea men, that the ſeuen principall Starres as well of the great Beare as of the little Beare, were all dulie placed in his firſt Net. Trulie if Maſter *Blagraue* his affaires would ſuffer him to take paine herein, I beleeue that his Iewell ſhould not be much inferiour to the celeſtiall Globe, which indeede it repreſenteth, and in ſo doing he ſhould greatly profite the ſea men, and deſerue thereby great good will and commendations at their handes.

What this word Aſtrolabe ſignifieth.

Efore I beginne to deſcribe vnto you the ſaid Inſtrument and all other parts thereof it ſhall not bee amiſſe to ſhew you what this worde Aſtrolabe ſignifieth.

This worde Aſtrolabe is as much to ſay as the handle or inſtrument of the Starres, by helpe whereof the manifolde motions and apparences of the heauens and of the Starres therein contained are known, and it is called of ſome a planiſpheare, becauſe it is both flat and rounde, repreſenting the Globe or Spheare, hauing both his *P*oles flatte both together, the ſhape or figure of which

The description and vſe

Inſtrument I doe not here ſet downe becauſe the Inſtrument it ſelfe is to be had for a ſmall price in diuers places of *London*, which if you lay before you when you mind to reade this my deſcription thereof here following, I doubt not but that you ſhall finde euerie part thereof ſo plainlie explaned as you ſhal need no other teacher to inſtruct you therin, or to help you to vnderſtand any of theſe Concluſions that are to be wrought by that Inſtrument, and ſpeciallie thoſe which I haue here ſet downe, ſuch as the *T*able following ſheweth, and as I thought moſt meet for ſea men to know, and being throughlie exerciſed in them, you ſhall the more eaſilie vnderſtand the manifold and neceſſarie concluſions ſet down by Maſter *Blagraue* himſelfe in his owne booke, which booke I would wiſh you to buy, and earneſtlie to ſtudie the ſame. But now I will firſt deſcribe the ſaide Inſtrument, and then ſhew the vſe thereof by ſo many concluſions as are contained in the table hereafter following.

The deſcription of Maſter *Blagraue* his Aſtrolabe, otherwiſe called the Mathematicall Iewell.

This Aſtrolabe is diuided into two parts, whereof the one is called the forepart, and the other the backepart. The foreparte containeth two principall partes, that is, the Mater, which is vnmoueable, and the Rete, which is moueable. Againe, the Mater is enuironed with a great circle, called the Meridian, paſſing through the 2. Poles of the world marked with the letters A.B. and through the two Solſticiall points of Cancer, and Capricorne marked with the characters belonging to thoſe two Signes, and therfore

map

of the Mathematicall Iewell. 282

may veriewell bée called the Colure of the 2. Solstices. And this circle is diuided by two crosse Diameters into 4. quarters euery quarter containing 90. degrées, so as the whole circumference of the circle is 360. degrées, which degrées doe procéede from 90. to 90.

Nowe as touching the 2. crosse Diameters, the one passing through the Center, and also through the two Poles, is the Axletrée of the world, signifying sometime the first Meridian, and sometime the right Horizon, and sometime the line of sixe houres as well for the Morning as Euening, and then it is the line of East and West, and sometime the line of North and South. *The 2. crosse Diameters in the Mater.*

And the other crosse Diameter drawne with red incke, signifieth most commonly the Equinoctiall, and sometime the line of East and West, and sometime the line of North and South, and especially when you haue to finde out the 12. houses of heauen, to which end are set down in the outermost Meridian these foure Latine words, first at the North pole marked with the letter A. Oriens, which is as much to say, as the East part: and at the south Pole marked with the letter B. Occidens, which signifieth the West part: and at the ende of the ouerthwart Diameter, drawne with red inke, and most commonly signifying the Equinoctiall, marked with the letter C. on the right hande is set downe Culmen Cœli, the highest part of heauen, which is the South, at which ende the Ringle or handle is fastened. And at the other ende of the saide Diameter, marked on the left hande with the Letter D. is set downe Imum Cœli, the lowest parte of heauen, which is the North. And by these two crosse Diameters the whole Mater is diuided into foure quarters, that is, Northeast, and Southeast, Northwest and Southwest. The Northeast quarter lyeth betwixt Imum Cœli, and the vpper ende of the Axletrée, whereon is written Oriens. And the Southeast quarter lieth betwixt the same point and Culmen Cœli, and the Southwest quarter lieth betwixt Culmen Cœli, and the lower endłof the Axletrée, whereon is written Occidens, and the Northwest quarter lyeth betwixt the point Occidens and Imum Cœli. *The Equinoctial.*

Besides these 2 crosse Diameters the Mater is traced with 180

The description and vse

The Meridians and houre lines in the Mater.

Meridians which doe passe through both the Poles, whereof the first passing through the center from Pole to Pole is the Colure of the Equinoxes, because it passeth through the first point of Aries, and the first point of Libra, and is otherwise called the Axletræ, and signifieth sometime the right Horizon, and somtime the first Meridian from which vpon the Equinoctiall are counted the degrees of longitude both East and West. And al the foresaid Meridians are most commonly vsed as houre lines, for euery 15th. Meridian, counting from the Limbe, doeth signifie one houre, for 15. degrees do make an houre, and 4. minutes do make a degree. And these hours at the end of euery 15th. Meridian are marked in the bodie of the Mater somewhat aboue the Tropique of Cancer with Arithmeticall Figures, proceeding from 1. to 12. forwarde towardes the right hand, and againe are marked somewhat beneath the Tropique of Capricorne with like Figures, proceeding backward towardes the left hand. Whereof the vpper numbers doe signifie the forenoone houres, and the lower numbers the afternoone houres, and are placed so as one selfe houre line doth passe through both numbers: for the line of the eleuenth houre in the forenoone serueth also to the first houre in the afternoone, and the forenoone houre of 10. serueth also to the houre of two in the afternoone, and the forenoone of 9. serueth to the houre of 3. in the afternoone: and the forenoone houre of 8. serueth to the houre of 4. in the after noone, and the forenoone houre of 7. serueth to the houre of 5. in the afternoone, and the forenoone houre of 6. serueth also to the 6. houre in the afternoone, and from thence forth towardes the left hande, 5. serueth to 7. 4. to 8. 3. to 9. 2. to 10. and 1. to 11. as before.

But whensoeuer you haue to finde out anie houre amongst the houre lines in the Mater, remember that the Axletræ which passeth through the Center and also through both the Poles is alwayes the sixt houre both of the morning and euening, from which you shall the more easilie finde out any houre that you seeke by allowing fifteene Meridians to euery houre, as is before saide. And you haue to note that all these Meridians doe sometime signifie euerie one a seuerall oblique Horizon, sauing that the first Meridian or Axletræ signifieth the right Horizon,

of the Mathematicall Iewell. 283

horizon as hath bene said before. And to find out in the Water the horizon of euery latitude, reckon from the axletree on both handes so many Meridians as may make vp the number of degrees of your latitude, and that Meridian shalbe your Horizon. In counting of which Meridians you shall finde that euery fift Meridian is drawne with a greater blacke line then the rest. Nowe ouerthwart these Meridians are drawne 180. Parallels, which proceede from 1. to. 90. whose nombers are set downe in the limbe of the mater thus, 10.20.30.40. and so forth both vpwarde and downeward, vntill you come to 90. and these Parallels are otherwise called the circles of declination as well of the Sunne as of any star. Amongst which there be 2. Parallels painted red, where of that towardes the North pole signifieth the tropique of Cancer, and the other towards the South pole the tropique of Capricorne. Then there is another oblique ouerthwart redde lyne, which passing through the center from Tropique to Tropique, signifieth the Ecliptique line, nigh vnto which on each side are set downe the Characters of the twelue Signes, whereof Aries and Libra are placed at the center, and Cancer at the one end of the Ecliptique on your right hand, whereas it toucheth the said Tropique, and Capricorne at the other ende of the saide Ecliptique, on your left hand, whereas it toucheth the said Tropique of Capricorne, and the rest of the signes are orderly placed betwixt the said Tropiques, euery two Signes one against another that haue like declination, as Gemini and Leo, Taurus and Virgo, which are placed betwixt the center and Cancer, then Pisces and Scorpio, and Aquarius and Sagittarius are placed betwixt the Center and Capricorne. And euery Signe contained in the Ecliptique line is diuided with slaunting black stræks into 15. spaces, euery space containing two degrées, which maketh 180. degrées, and bæing doubled, because the spaces are to bé counted both forward and backward, do make in all 360. degrées, which is the whole circuit or longitude of the Ecliptique, and the degrées of the said Ecliptique are larger towards the Limb than towards the Center, whereby the spaces betwixt Signe and signe are not equall, but some larger than other some.

And note that vppon the outermost Limbe of the Mater are drawne thrée circles making two reasonable spaces, both which spa-

Marginal notes: How to finde out the Horizon of euery seuerall latitude. — The Parallels of the Mater. — The two Tropiques. — The Ecliptick line of the Mater — The Limbe of the Mater.

The description and vse

spaces are diuided into 360. degr. both of them beginning at the North pole. And the degrees in the outermost space of the Limbe, beginning at the North pole do procéede from 15. to 15. downward towards your left hand, and so round about the instrument vntill you come to 360. which degrées do shew the houres of the Equinoctiall. And at the North pole, whereas the diuision beginneth, is set downe the figure of 6. signifying the sixt houre of the forenoone, then counting fiftéene degrées downwarde towards your left hand is set down in the foresaid space the figure of 5. then 4.3.2. and 1. And at the end of the Equinoctiall on the left hande is set downe 12. from which procéeding still downewards are set these numbers of houres, 11. 10. 9. 8. 7. and 6. which 6. standeth at the South pole, then from thence procéeding vpwardes are set downe these houres 5. 4. 3. 2. 1. and then 12. which standeth at the end of the Equinoctiall on the right hand, from whence procéeding still vpwards towards the North pole, are set downe these houres, 11. 10. 9. 8. & 7. and at the Pole it selfe is set downe the figure of 6. as before hath bene said, from whence you first began to count. And in the inner space of the same Limbe, the 360. degrées are to bee counted by 90. in such sort as the first 90. is placed at the North pole, the second 90. at the one ende of the Equinoctiall, marked with C. The third 90. at the South pole, marked with B. And the fourth 90 at the other end of the Equinoctiall, marked with D. Euery which quarter is to be counted both vpward and downwarde, according as you sée the numbers written with Arithmeticall figures, and are placed one aboue another in the saide space, whereof the vpper figures are somewhat lesser than the nether figures, both of them procéeding from 1. to 90. the greater figures from the Equinoctiall vpward to the verie Pole, and the lesser Figures from the Pole downward to the Equinoctiall, and are thus set downe. 10. 20. 30. 40.

of the Mathematicall Iewell.

A Description of the Nette, called in Latine *Rete*.

His parte is first inuironed rounde about with a great circle signifying most commonlie the Meridian or 90. Azimuth or Uerticall circle, and sometime it signifieth the Equinoctiall, and especiallie when the Center is taken from the Pole, which Circle is diuided by two crosse Diameters into foure quarters, euerie quarter contayning 90. degrées, which degrées are to be numbred in the verie Limbe of the Rete both vpwarde and downwarde, and of those two crosse Diameters, the one ende marked with the letter A. signifieth the Zenith, and the other end marked with the letter B. signifieth the Nadir or point opposite, and the other crosse Diameter marked with the letters C. D. signifieth the Horizon, which for distinctions cause is otherwise called the Finitor, because the Meridians before described in the Mater are to be called and vsed sometimes as Horizons, and this Finitor is a pretie broade Ruler, the verie edge whereof is diuided with small diuisions into 180. degrées, which being doubled by reckoning the same both forwarde and backwarde, (beginning at the Center) doe make vp thrée hundred and thrée score degrées.

This Finitor signifieth the Horizon of the Globe, the verie edge whereof being diuided by little short strækes into small portions or degrées, is alwayes to be applyed to any seuerall latitude when néede is. And the broader part thereof serueth onely to containe the numbers that are set therein both beneath and aboue to knowe thereby the number of euery Azimuth hereafter described, which numbers doe procéede from the Center to the right hand thus, 10. 20. 30. 40. 50. and so foorth to 90. Then backwarde towardes the Center are set downe 100. 110. 120. 130. 140. and so forth vntill you come to 180. which is placed at the verie Center. Then from thence towards the left hande are set downe these numbers 190. 200. 210. and

The Finitor.

The description and vſe

so foorth vntill you come to 270. and from thence turning againe towards the center are set downe 280. 290. and so foorth vntill you come to 360. placed at the Center.

The Almicanteraths.
And in the said Net are certaine circles, which are Parallels to the foresaide Finitor, procéeding towards the Zenith, and are in euerie respect like vnto the Parallels described in the Mater, and these Parallels are called Almicanteraths, that is to say, circles of Altitude, which beginning at the Finitor, doe procéede to the Zenith, marked with the letter A. from 1 to 90. And though there be cut out in the Rete but 30. Almicanteraths, yet for so much as euery space contained betwixt euery 2. Almicanteraths do containe 3. degrées, they make in all 90. for 3. times 30. maketh 90. which degrées you may sée set downe in the Limbe of the Rete on both hands thus, 10. 20. 30. and so foorth, till you come to 90. which standeth at the very Zenith.

The Azimuthes.
And these Almicanteraths are crossed with other circles called Azimuthes, that is to say, Verticall circles, which passing from the Finitor, do méet all in the Zenith, whereof though in this instrument there be set down but 12. being 15. degrées distant one from another, yet you must imagine that there be 180. which the degrees set downe in the Finitor, in such manner as is before described do shew. And you haue to note that if the Sunne bée in the very beginning of the Azimuthes, which is at the center, then is hée full East, and if he be in the 90. Azimuth, then he is full South, and when he is in the 180. then he is full West, and when hée is in the 270. Azimuth, then hée is full North.

How to count the Azimuthes according to the Mariners Compaſſe.
And if you would haue this account of the Azimuthes to aunſwere the Mariners compaſſe, then diuide the number of the Azimuthes wherein the Sunne is by 11. degrées and $\frac{1}{4}$ which is 15. minutes, and the quotient will shew the rombe or winde of the Mariners Compaſſe, so as in your account you procéed from the first Azimuth towardes your right hand, that is, from East to South, and from South to West, and from West to the North, and so from thence againe to the East point whereas you first began. Also you haue to note by the way, that these Azimuthes dee sometime signifie the circles of position, the vse whereof you shall find set downe hereafter in the 31. proposition of this Treatise.

In the meane time I wil procéed in describing the Zodiaque,

of the Mathematicall Iewell.   285

of the Rete, which in shape is like to an Egge, called in Latine *Figura ovalis*, the one halfe whereof extendeth towardes the Zenith, and the other towards the Nadir or point opposite to the said Zenith, in which circle are placed the 12. Signes, whereof the 6. Northern signes, that is, Aries, Taurus, Gemini, Cancer, Leo, and Virgo are placed in the nether halfe towards the Nadir, and the other sixe Southerne Signes, that is, Libra, Scorpio, Sagittarius, Capricornus, Aquarius, and Pisces, are placed in the vpper halfe of the Zodiaque towards the Zenith, and euerie one of these Signes are diuided into 30. degrees, which are set downe with Arithmeticall Figures in the said Zodiaque thus, 10. 20. 30. *The description of the Zodiaque of the Rete*

Moreouer, in the said Net are placed certaine fixed Starres to the number of 71. whose names here do follow. Caput Ophiuci, that is, the head of Serpentarius, Aquila, the Eagle. Caput Engon, the head of Hercules. Cuspis Sagittarii, the shaft of Sagittarius. Palma Ophiuci, the hande of Serpentarius. Cauda Delphini, the tayle of the Dolphin. Romboides Delphini, which is a starre in the Dolphins backe. Cor Scorpionis, the hart of the Scorpion. Frons Borealis, Media, and Australis Scorpionis, the Northern, Southern, and middle front of Scorpio. Lucida Lyræ, the bright Star of Lyra, or Vultur cadens. Lanx Chælæ Borealis & Australis, the North and South starre of the Balance. Corona Gnosiæ, the crowne of Ariadna. Præcedens & sequens caudæ Capricorni, that is, the former and follower in the tayle of Capricorne. Caput Draconis, the head of the Dragon Hastile Bootis, the Borespeare of Arcturus. Cauda Cigni, the taile of the Swanne. Fomahand, a star in the mouth of the Southerne Fish. Arcturus, a starre betwixt the legges of Bubulcus or Bootes. Humerus Bootis, the shoulder of Bubulcus. Dexter Humerus Cephei, the right shoulder of Cepheus. Crus Pegasi, the legge of the winged Horse. Spica Virginis, the Wheat eare in the hand of Virgo. Tres stellæ in cauda Vrsæ maioris, three starres in the tayle of the great Beare. Previndemiatrix, that is Virgo. Cauda Ceti, the taile of the Whale. Andromedæ Scapulum, the shoulder blade of Andromeda. Cingulum Andromedæ, the girdle of Andromeda. Humerus Vrsæ maioris, the shoulder of the great Beare. Corvi rostrum, the beake of the *Of Starres 71 contayned in the Rete.*

Crow

The description and vse

Crowe. Corviala dextra, the right wing of the Crowe. Cauda Leonis, the tayle of the Lyon. Ceruix Leonis, the necke of the Lyon. Cor Leonis, the hart of the Lyon, otherwise called Regulus. Lucida Hydræ, the bright Starre of Hydra. Capita Geminorum, the heades of the two Twins named Apollo and Hercules. Cancer, the Crabbe. Canis minor, the little Dogge. Canis maior, the great Dog. Canopus, a faire starre in the left oare of the Shippe Argus. Humerus dexter Aurigæ, the right shoulder of Auriga. Hircus, the Goate hanging at the backe of Auriga. Hedi, the two little Goats sucking behind at her paps. Humerus dexter Orionis, the right shoulder of Orion. Tres stellæ in cingulo Orionis, three stars in the girdle of Orion. Humerus sinister Orionis, the left shoulder of Orion. Pes sinister Orionis, the left foot of Orion. Oculus Tauri, the Bulles eie. Pleiades, the seuen little starres in the Buls necke. Extremum Eridani, the last end of the flood Eridanus. Caput Medusæ, the head of Medusa. Dextrum latus Persei, the right shoulder of Perseus. Cornu Arietis, the former Star of the Rams horne. Venter Ceti the bellie of the Whale. Iuba Ceti, the mane of the Whale.

How to know which stars be North or South.
And note that all these Starres in the Nette, whose longest tippes or pointes do point from the center outward, towardes the Limbe, are Northern Starres hauing North declination, and those whose longest tippes do point inwarde from the limbe towards the Center, are Southerne Starres, hauing Southern declination. All which Starres are set downe in a Table in the beginning of his third booke, which doth not onely shew their names, but also their longitudes, latitudes, and declinations, their natures, their right ascentions and magnitudes or greatnesse. And whereas both in the Mater, and also in the Rete one selfe circle is made to haue diuers significations, the cause thereof shall plainely appeare vnto you by the vse of the Astrolabe, in seeking to finde out thereby the propositions of the foresaid booke.

A description of the Labell.
And besides the partes before described, there is yet another part belonging to the forepart of the Instrument, called the Labell, the one end whereof is fastened to the Center of the Astrolabe, so as it may turne round about, and this Labell is diuided

of the Mathematicall Iewell.   286

uided into 90. degrées twise set downe therein with Arithmeticall Figures to bée rekoned aswell from the Center to the poynte of the Labell, as from the poynt thereof to the Center seruinge to dyuerse vses, yea the 90 degrées of the Label are sometimes to be repeated 4 times to make vp the number of 360. degrées, as shall hereafter plainlie appeare by the 12. proposition, shewing how to find out the right ascention of anie degrée or portion of the Ecliptique line.

And remember that the right line drawne from the Center of the Labell alongst the inward edge thereof is called the Fiduciall line, and is diuided into 90. small partes called degrées, which line or inward edge is alwayes to be vsed in any proposition, and not the outward edge or backe part of the Labell.

The fiducial line of ỹ label.

Thus much touching the fore part of M. Blagraue his Iewell with euerie particular part, with which forepart I wish you to be throughlie acquainted before you deale with the propositions here following, or with any other proposition contained in M. Blagraue his booke.

A briefe description of the backe parte of the saide Iewell.

IN the Limbe of the backe part is described the Theorique of the Sunne, to knowe therby in what Signe and degrée the Sunne is euerie day throughout the yeare, by laying the Diopter thereto, M. Blagraue calleth it a Ruler, which Diopter is made with two Pinules or square Tablets, each one pearced with two holes one greater than another, the lesser to take the height of the Sunne by his beame passing through the saide lesser holes, and the greater holes doe serue to take the altitude of the Sunne, béing something darkened in the day time, so as hée casteth no beame, or els the altitude of any Starre in the night season by looking with your one eie, the other béing shutte, through the two greater holes of the Pinules or Tablettes. And the middest of

The description and vſe

of this Diopter is faſtened with a pin to the center of the Aſtrolabe, ſo as the ſaide Diopter may turne round about, and the middle line of the ſaid Diopter is called the fiduciall line, becauſe it rightlie directeth the ſight of the eye to the foreſaid holes, the degrées of the altitude of the Sunne or of anie Starre are ſet downe in the outermoſt ſpace of the Limbe, diuided by 2. croſſe Diameters into foure quarters, euerie quarter contayning 90. degrées, the number of which degrées are ſet downe in the outermoſt ſpace of the Limbe of the ſaid backe part with arithmeticall figures right ouer the heades of the ſaide degrées of altitude. And you haue to note, that the perpendicular Diameter ſignifieth the Meridian, that is, the line of South and North, that is to ſay, the South point at the ring or handle, and the North at the point oppoſite, and the other ouerthwart Diameter ſignifieth the right Horizon, that is to ſay, the line of Eaſt and Weſt, the Eaſt being placed on the left hande, and the Weſt on the right hand. And vnder the Theorique of the Sunne you may drawe as manie circles of ſuch reaſonable diſtance one from another, as in the ſpaces thereof may be ſet downe, the yeares of our Lord, the Dominicall letter for the leape yeare, and the Dominicall letter for the common yeares, the Prime or golden number, the Epact, and on what day of March or of Aprill Eaſter day euery yeare falleth, in ſuch order as M. Blagraue hath himſelfe deſcribed in a Table made of purpoſe in the ſeconde booke of his Jewell the 11. chapter.

Thus hauing deſcribed euery particular thing as well in the forepart as in the backe part of the ſaide inſtrument, I will now ſhew you how to vſe the ſame, and howe to finde out thereby all the concluſions contayned in the Table here next following.

The

of the Mathematicall Iewell.   287

The Table contayning 32. neceſſarie concluſions to be wrought by this Aſtrolabe.

First, howe to finde out the place of the Sunne (that is to ſay) in what ſigne and degree thereof the Sunne is euerie day throughout the yeare, being not leape yeare, and alſo the oppoſite point of that degree. Propoſition. 1.

How to know the place of the Sunne in the leap yeare, and how to finde the leape yeare. Prop. 2.

How to take the altitude of the Sun or of any ſtar. Prop. 3.

How to take the Meridian altitude, that is to ſay, the higheſt or greateſt altitude of the Sunne or of any ſtarre. Prop. 4.

How to know the altitude of the Sunne at any houre without ſeeing the Sunne. Prop. 5.

How to know the Meridian altitude of the Sunne or of any ſtarre in the Net without ſeeing them. Prop. 6.

How to know the declination of the Sun, or of any ſtarre contayned in the Net. Prop. 7.

How to find out the latitude of anie Region diuers waies. Prop. 8.

How to know the houre of the day by the Sunne, and alſo in what part or coaſt of heauen he is at that inſtant. Prop. 9.

Howe to finde the riſing and ſetting of the Sunne euerie day in euerie latitude, and thereby the length of the day, and alſo in what coaſt or part of the Horizon he riſeth and ſetteth. Prop. 10.

How to knowe euery day at what houre the Moone riſeth and ſetteth, and how long ſhe continueth aboue the Horizon and alſo when ſhe is full South. Prop. 11.

How to find out the right aſcenſion of the Sunne, or of any degree or portion of the Ecliptique. Prop. 12.

Another more readie way to finde out the right aſcention of any degree or portion of the Ecliptique by the Rete. Prop. 13.

How to finde out the aſcentionall difference of the Sunne or of any degree or point of the Ecliptique. Prop. 14.

How

The description and vse

How to find out the oblique ascention of the Sunne, or of any point of the Ecliptique. Prop. 15.

How to find out the right ascention of any Arke or portion of the Ecliptique, and therwith to know what time it spendeth in rising in a right Spheare. Prop. 16.

Howe to finde out the oblique ascention of any Arke of the Ecliptique in any latitude, and what time it spendeth in his rising. Prop. 17.

How to finde out the oblique descention of any point of the Ecliptique in any latitude. Prop. 18.

How to find out the oblique descention of any Arke giuen of the Ecliptique, and therewith to know the time which it spendeth in his setting. Prop. 19.

How to knowe the height of any starre at any houre without seeing the starre, and thereby to finde out in the firmament all the starres that be described in the Net, and are to be seene with the eie. Prop. 20.

How to finde out the ascentionall difference of any starre. Prop. 21.

How to know the oblique ascention of any star. Prop. 22.

How to knowe what starres do neuer rise nor set in any latitude. Prop. 23.

How to know at what houre of the day or night any starre riseth or setteth. Prop. 24.

How to know howe long any starre continueth aboue the horizon in euery latitude. Prop. 25.

How to find out the starres houre, and therby to know the houre of the night. Prop. 26.

How to finde out the distance betwixt any two starres contayned in the Net. prop. 27.

Another way to know the distance of any two stars, their longitudes and latitudes being first knowne, and also by that meanes to finde out the distance betwixt any two places vpon the earth. prop. 28.

How to find out the degree of *Medium Cœli* at any houre of the day, that is to say, the degree of the Zodiaque that is in the Meridian at any houre that you seeke, and also the degree called *Imum Cœli*. prop. 29.

How

of the Mathematicall Iewell. 288

*H*ow to find out the horoscope or ascendent at any time of the day or night, and thereby to haue the foure principall angles of heauen. Prop. 30.

*H*ow to find the circles of position, and to know how much the *Pole* is elevated aboue euerie such circle in any latitude, without the knowledge whereof you cannot find out the 12. houses by this Astrolabe. prop. 31.

How to find out all the 12. houses of heauen, and therby to erect a figure at any houre of the day or night. prop. 32.

The vses of the *Astrolabe*, and first howe to finde out the place of the Sunne (that is to say) in what signe and degree thereof the Sunne is euerie day throughout the yeare, being not leape yeare, and also the opposite point of that degree.

The first. Proposition.

Ay the Diopter which is on the backside of the Astrolabe vpon the day of the moneth and that end of the Diopter with his fiduciall line will shew you in what signe and degrée thereof the Sunne is that day, and the other ende of the Diopter will shewe you the opposite point to that degrée, as for example: I would know the place of the Sunne the 17. of July, here by laying the Diopter vpon that day, I find the Sunne to be in the fourth degrée of Leo, and the opposite point thereof to bee the fourth degrée of Aquarius.

How

The description and vſe

*H*ow to know the place of the Sunne in the Leape yeare and how to finde the leape yeare.

Propoſition. 2.

WHen it is leape yeare, you muſt alwayes adde one degree more euerie day during that yeare vnto the place of the Sunne found by the firſt propoſition, which leape yeare you ſhall know by diuiding the yeare of the Lord by 4. for if there be no remainder left, then that yeare is leape yeare, ſo ſhall you finde the yeare of our Lord 1596. to be leape yeare.

*H*ow to take the altitude of the Sunne, or of any ſtarre.

Propoſition. 3.

TO take the altitude in any time of the day when the Sunne ſhineth, you muſt turne your face and alſo the left Tablet or Pinule of the Diopter towards the Sunne, holding the Aſtrolabe by the ring with your right forfinger or middle finger being ſomewhat bowed, in ſuch ſort as the Aſtrolabe may hang plumbe, and then with your left hand lift the Diopter vp and downe vntill the Sun with his beame doe iuſtlie ſtreeke through both the holes of each Pinule of the Diopter, ſo as you may ſee the ſhadow of the two holes of the vpper Pinule to play vpon the two holes of the nether Pinule, then marke vpon what degree of altitude the thinneſt edge or fiduciall line of the Diopter falleth in the outermoſt ſkirt or border of the backe of the Aſtrolabe, for that is the Sunnes altitude for that preſent. But if the Sun be couered with a cloud, ſo as it ſhineth not cleare inough to caſt any ſhadow, and yet ſo as it may be ſeene with the eie, then hang the Aſtrolabe by the ring vpon your right thomb, and turning your face towards the Sun, lift vp your hand with the Aſtrolabe ſo high, as by moouing the Diopter with your left hande vp and downe, you may with your right eie (the other beeing ſhut) ſee the Sunne through the greater holes of both the Pinules of the Diopter, and marke vppon what degree

of the Mathematicall Iewell. 289

of altitude the vpper ende of the Diopter falleth, and that is the altitude of the sunne at the time, and in this maner you must also take in the night season, the altitude of any starre.

How to take the Meridian altitude, that is to say, the highest or greatest altitude of the Sunne or of any starre.

The 4. Proposition.

Goe into some open place whereas the sunne shineth somewhat before noonetide, and there hanging the Astrolabe vpon your right fore finger or middle finger, take the altitude of the sunne in such manner as is before taught, at diuerse times with some pawse betwixt euery time to know therby whether such altitude increaseth or decreaseth, for if it increaseth, then the sunne is not yet at the Meridian, but if it decreaseth, then it is past the Meridian, and therefore you must watch diligently to take him when hee is at the highest. And you must do the like to take in the night season, the Meridian altitude of any knowne starre, sauing that then you must hange your Astrolabe vpon your thumbe before your right eye, and to doe as is taught in the last Proposition.

How to know the altitude of the sun, or of any starre at any houre of the day without seeing the sunne or starre.

The 5. Proposition.

Set the Finitor at your Latitude, and seeke out amongst the houre lines or Meridians of ye Mater in what point the Parallel or Circle of declination of the sun crosseth the houre line which you seeke, and the Almicanterath and Azimuth passing through that point do shew both the altitude and also the coast or part of heauen, wherein the sunne or starre is at that instant. This is a very necessary Proposition, for by knowing the height of the sunne at euery houre of the day in whatsoeuer signe the sunne is, you may make Tables for particular dials, as Cylinders, houre Quadrants, and such like to serue any latitude, yea rather this Instrument as M. Blagraue rightly sayeth is a Table of it selfe readie made to serue such purposes. Let

P p   the

The description and vse

the example of this Proposition be thus, Suppose that at 8. of the clocke in the morning the 21. of Aprill 1592. the sunne being then in the tenth degrée of *Taurus*, and his declination 15. degrées Northward, you would knowe the altitude of the sunne at that houre, which by working as the rule teacheth, you shall finde to be almost 30. degrées, and that hee is about 13. degrées distant from the East towards the South.

How to know the Meridian altitude of the sunne, or of any starre euery day throughout the yeare, without seeing eyther sunne or starre.

The 6. Proposition.

Irst séeke to know the declination of the sunne or star eyther by the seuenth Proposition next following, or else by some Table, & whether it be North or South, and knowing his declination, bring the Finitor to your Latitude, and staying it there séeke out in the limbe of the mater on your right hande the said declination, and there marke what Almicanterath toucheth that point, for that Almicanterath being counted vppon the limbe of the *Rete* from the Finitor, doth by and by shew the Meridian altitude of the sunne or starre for that day. As for example, I would know the Meridian altitude of the sunne the first of July 1592. at which time his Parallel or declination is 22. degrées and certaine minutes Northward. Here hauing layde the Finitor to the Latitude 52. I finde that the 60. Almicanterath toucheth the Parallel, and that is the Meridian altitude of the sunne that day. Againe suppose that I would know the Meridian altitude of the starre Oculus *Tauri*, in the Latitude 52. Here hauing brought the Finitor to the said latitude, I finde that the 53. Almicanterath toucheth his Parallel or declination, which is 15. degrées 4′9. Northward, so as I find his Meridian altitude in that latitude to be 53. degrées. But you haue to note that if the sun or star haue south declination, then you must count his parallel from the Equinoctial downward towards the south Pole, so shall you find the Meridian altitude of the great dogge called Canis maior, whose Southerne declination is 16. degrées to be 22. degr. & 3′ 0. And by knowing the Meridian altitude

tude of any ſtar, you may alſo knowe how farre he is diſtant from the Meridian or South line, if you ſubtract from his Meridian altitude his altitude taken at any other time of the ſame night, for the remainder will ſhew his diſtance from the Meridian, & if the ſtarre at the time of taking his altitude, be in the Eaſt part of the firmament, then he is ſo much ſhort of the Meridian, and if he be in the Weſt, then he is ſo much paſt the Meridian. As for example, knowing the Meridian altitude of Oculus Tauri to be 53 degrées, and his other altitude newly taken to be 30. deg. I find his diſtance from the Meridian to be 23. deg. aſwel by ſubtraction, as by counting vpon the limbe of the Rete the degrées contained betwixt the two foreſaid Almicanteraths. And becauſe the ſaid ſtarre was in the Eaſt part of the firmament when I tooke his altitude, I conclude that he wanted 25. degrées of arriuing to the Meridian, which maketh one houre and a halfe, and a little more.

How to find the declination of the ſunne, or of any ſtarre deſcribed in the net.

The 7. Propoſition.

This may be done two maner of waies, firſt knowing the place of the ſunne, ſéeke his place in the Ecliptique line of the mater, and looke what Parallel of the mater paſſeth through that degrée, and the number of that Parallel will ſhew the declination (that is to ſay) how farre the ſunne is diſtant from the Equinoctiall, counting from the Equinoctiall vpon the outermoſt Meridian or inner limbe of the mater, and according as the ſigne wherein the ſunne is be it Northerne or Southerne, ſo is the declination of the ſunne, and muſt bee counted eyther vpwarde or downewarde accordingly. The ſeconde way is thus, bring the Fiduciall line of the labell to the degrée of the ſunne in the Zodiaque of the net, and the number of degrées counted vpon the labell betwixt the limbe of the net, and the place of the ſunne which the labell toucheth vpon the outward edge of the ſayde Zodiaque will ſhewe his declination. By either of theſe two wayes you ſhall finde the declination of the Sunne, being in

The description and vse

the tenth degrée of *Taurus* to be 15. degrées Northward. And by this last way you may know the declination of any starre contained in the net thus. Hauing found in the net the starre whose declination you séeke, lay the Fiduciall line of the labell to the longest tippe of that starre, then count vpon the labell how many degrées are contained betwixt the limbe of the Rete, and that point whereas the labell toucheth the longest tippe of the saide starre, and that shall be his declination either Northerne or Southern according as the said tippe point either outward or inward: for if outward from the Centre, then it is Northerne, if inwarde towardes the Centre, then it is Southerne, as hath béene sayde before in the description of the Net, and by doing thus you shall finde that the starre called Canis maior, that is the greater dogge, hath in South declination 15. degrées 5′5. Againe you shall finde the first starre of the Rams horne called Cornu Arietis to haue in North declination 18. degrées.

How to find out the Latitude of any Region.
The 8. Proposition.

First you must know the place of the sunne, and also his declination, and hauing taken his Meridian altitude, recken the same amongst the Almicanterathes from the Finitor vpwardes, and turne about the Rete from the Pole arctique towards the Equinoctiall on your right hande, vntill it toucheth the Parallel of the sunne, for then looke on your left hand and you shall finde the Finitor to stand at that Latitude which you séeke. As for example, the 12. of Aprill 1591. the sunne being in the first degrée 20. minutes of *Taurus*, and his declination being then 12. degrées Northward, I finde his Meridian altitude to be 50. degrées, which I count vpon the limbe of the Rete procéeding from the Finitor vpwardes towards the Zenith, and then I turne the Rete vntill I haue brought that Almicanterath to the Parallel of the Sunne, which is 12. degrées, counting the same from the Equinoctiall on the right hande of the mater towardes the North Pole, and there staying the Rete, I finde that on the left hande the Finitor lyeth vpon the 52. degrée of Latitude counting from the North pole downe towardes the

Equinoc-

of the Mathematicall Iewell.

Equinoctiall. The common way of finding the Latitude is thus, if it be in the day time, then take the Meridian altitude of the Sunne, and if the Sunne be in any of the sixe Northerne signes, then subtract the declination of the sunne out of his Meridian altitude, and the remainder shall be the altitude of the Equinoctiall aboue your Horizon, which being taken out of 90. the remainder will shew the altitude of the Pole, but if the sunne be in any of the sixe Southerne signes, then you must adde his declination to his Meridian altitude, and the summe thereof shall be the altitude of the Equinoctiall, which being taken out of 90. the Remainder will shew the Latitude or eleuation of the Pole. But to know the latitude of any place in the night season, you must take the Meridian altitude of some knowne starre which both riseth and setteth, then after that you haue taken his Meridian altitude with your Astrolabe, you must learne to knowe his declination, and whether it be Northerne or Southerne, for if the starre haue North declination, then you must subtract his declination from his Meridian altitude, and the remainder shall be the Altitude of the Equinoctiall, which being taken out of 90. shall be the latitude or eleuation of the Pole, but if the declination of the star be Southernly, then you must adde his declination to his Meridian altitude, and that summe shal be the altitude of the Equinoctiall, which being taken out of 90. the remainder shall be the eleuation of the Pole. And there be diuers other wayes of finding out the latitude of any place, which I haue partly set downe in my Treatise of the two Globes about the latter ende thereof, and partly in my Treatise of Nauigation, whereas I speake of the North starre and of his Guardes.

How to know the houre of the day by the Sunne, and also in what part of heauen he is at that instant.

The 9. Proposition

Ake the altitude of the Sunne, and knowing the latitude of the place where you are, bring the Finitor of the Rete to that Latitude, and hauing stayed it there, looke in what point the Almicanterath or altitude of the Sunne crosseth the Suns Parallell

The description and vse

or Circle of declination in the mater, and the houre line passing through that point will shew the true houre, and at that instant you may also know in what Azimuth (that is to say) in what part of heauen, or as the Mariners terme it in what rombe or winde, as East, West, North, or South &c. the Sunne is at that instant, for that Azimuth which passeth through the foresaide point, is the Azimuth of the Sunne. As for example, the 21. of Aprill 1592. the sunne being in the tenth degrée of *Taurus*, and his declination being then 15. degrées Northward, I find by the Astrolabe, Quadrant, or Crosse-staffe, the altitude of the Sunne in the forenoone to be 30. degrées. Wherefore I hauing laide the Finitor to my Latitude which is 52. and stayed it there, I marke in what point the Almicanterath cutteth the sunnes foresaid Parallell in the mater, and I find that the houre line of eight in the forenoone cutteth that point, which sheweth that it was then eight of the clocke in the morning, and that the sunne was about 14. degrées distant from the East towards the South.

How to find the rising and setting of the sunne euery day in euery Latitude, and thereby the length of the day, and in what coast or part of the Horizon he riseth and setteth.

The 10. Proposition.

Ring the Finitor to your Latitude, and staying it there, looke in the mater where the Parallel of the Sunne doth cut the Finitor, and the houre line which crosseth that point, will shewe the houre of his rising and setting, and the number of houres betwixt his rising and setting is the length of the day, and the number of the Azimuthes betwixt that point of the Finitor, and the Centre or first Azimuth will shew you in what part or coast of the Horizon he both riseth and setteth. As for example, séeking by this rule to knowe at what houre the sunne riseth the 19. of June 1592. he being then in the sixth degrée 39. minutes of Cancer, and his declination then 23. degrées 1′ 9. I find that he riseth 12. minutes before foure, and goeth downe 12. minutes after 8. and thereby I find the length of the day counting from sunne rise to sunne set to

be

of the Mathematicall Iewell. 292

be 16. houres and 24. minutes, and that he riseth from the East towardes North 40. degrées, which according to the Mariners reckoning is Northeast and by East two quarters and somewhat more towards the North.

How to knowe euery day at what houre the Moone riseth and setteth, and how long she continueth aboue the Horizon, and also when she is full South.

The 11. Proposition.

First you must learne by some Almanacke or Ephemerides in what signe and degrée the Moone is, and whether it be a Northerne signe or a Southern signe, for if she be in a Northerne signe, then bring her place to the Horizon of your latitude in the Northeast part of the Astrolabe, but if she be in a Southerne signe, bring her place to the said Horizon in the Southeast part of the Astrolabe, & there hauing stayed the Rete bring the labell to the place of the Sun for that day, & the labell will point to the houre of the Moones rising in the limbe of the mater: but because the Almanack or Ephemerides do not set downe ye true place of the Moone but only at noone, you must therfore consider whether it be in the forenoone or in the afternoone that you séeke, for if it be in the afternoone you had néed to know how many houres are run from noone, and then for euery houre to adde halfe a degrée to the place of the Moone which you found at noonetide, but if it be in the forenoone, then you must subtract from her place at noone, for euery houre halfe a degrée, so shal you go very nigh to finde her true place in the Zodiaque for that houre, though you know not her latitude which is but 5. deg. at the most, & therfore can cause no great error in this matter. Now to knowe when she setteth you must do thus, if the Moone be in any Northern signe, then you must bring her place to the foresaid Horizon in the Northwest part of the Astrolabe, & by laying the labell to the place of the sunne, it will point to the houre of her setting, but if she be in any Southerne signe, you must bring her place to the Horizon in the Southwest part of the Astrolabe, and the labell being laide to the place of the Sunne will point to the houre of her setting. Now if you would knowe how long time she is aboue

P p 4   the

The description and vse

the Horizon, and also at what houre she is full South, then count the houres betwixt her rising & setting, and that shall be the time of her continuance aboue the Horizon, and the very middest of that is the true houre that she is full South. As for example, the fourth day of September 1592. in the latitude 52. at nine of the clocke at night the Sunne being in the 21. degrees 47. of Virgo or there abouts, & the Moone being in the 2. degree 30. of Capricorne, I am desirous to know when the Moone did rise that day, and by working according to the rule before set downe, I founde that she did rise about three of the clocke in the afternoone, and that she went downe at tenne of the clocke at night and halfe an houre past, and that she was full South or at the Meridian a little before seuen of the clocke in the afternoone. And as by this rule you may finde out the time of the rising and setting of the Moone, so may you finde the time of the rising and setting of the other fiue Planets, that is Saturne. Iupiter, Mars, Venus, and Mercury, any day throughout the yeare, so as you know their places in the Zodiaque, which the Ephemerides of Stadius doth shewe, not onely at noonetide, but also at any other houre of the day by helpe of certaine Tables made of purpose, the vse of which Tables I haue set downe in the latter end of my Treatise of the two globes.

How to find out the right ascention of the sunne, or of any degree or portion of the Ecliptique.

The 12. Proposition.

Ake the Rete and the label cleane from the Astrolabe, and seeke out in the Ecliptique line of the mater the signe and degree whose right ascention you would know, and marke what Meridian cutteth that point: that done, place the labell vpon the pinne which standeth in the very Centre of the Iewell, and make the fiduciall line thereof to lie right vppon, and alongest the Equinoctiall line eyther towardes your right hand or towards your left, according as the signe & degree whose ascention you seeke is placed in the mater. Then marke where the foresaid Meridian cutteth through the labell, and also through the

Equinoc-

Equinoctiall line, and the number of degrées contained in the labell, betwixt the Centre and that point of the Equinoctial is the right ascention of that degrée of the Ecliptique which you séeke, which number of degrées you must count vpon the labell in this maner. For if that signe and degrée be contained betwixt the first point of Aries and the first point of Cancer, then you must begin to count vpon the labell at the Centre, and so procéede forwarde towards Cancer, the right ascention of whose first point is 90. degrées, but if the signe and degrée which you séeke, be betwixt the beginning of Cancer and the beginning of Libra, which is at the very Centre right against Aries, then you must count vpon the labell backward from 90. to 180. by adding to euery tenth space of the labell 10. degrées, so as the first number procéeding towards your left hand shall be 100. and next to that 110. and so foorth vntill you come to 180. which is the right ascention of the first point of Libra, and from Libra you must count towards Capricorne 190. then 200. and so foorth till you come to 270. which is the right ascention of the first point of Capricorne: and from thence you must count 280. 290. then 300. and so foorth towards the Centre vntill you come to 360. which is the right ascention of the last point of Pisces, so as though there be set downe in the labell but 90 degrées both forward and backwarde, yet 90. being foure times repeated, doe make in all 360. which is the whole Longitude of the Equinoctiall. As for example, you would know perhaps the right ascention of the tenth degrée of Sagittarius, here by séeking in the mater you shall find that the Meridian passing through that degrée will cut both the labell being layde towards your left hande, and also the Equinoctiall in the 248. degrées 21. minutes, which is the right ascention of the tenth degrée of *Sagittarius*, in counting whereof remember to beginne from 180. that is from the Centre, and this ascention agréeth with the Table of right ascentions set downe in Stadius his Ephemerides the 44. page of his booke. Againe in séeking to know the right ascention of the 10. degrée of Taurus, if you beginne to count vpon the labell from the Centre, which is the first point of Aries, you shall finde that the Meridian which passeth through the 10. degrée of Taurus will cut the labell, and the Equinoctiall line in the 37. degrée 3′5.

Another

The description and vse

Another more readie way to finde out the right ascention of any degree or portion of the Ecliptique by the Rete.

The 13. Proposition.

LAy the Finitor euen with the Axletræ, signifying here the right Horizon, so as the first point of Aries may mœte with the East point of the saide Horizon, and lay the labell right vpon and alongest the Equinoctiall line, eyther towardes Cancer or towardes Capricorne, according as the signe and degrée which you sæke is placed in the Ecliptique of the Mater, and marke therewith what Meridian passeth through that degrée, and follow the same vntill it cutteth the Equinoctiall, and also the labell lying thereon, and by counting vpon the labell the number of degrées to that point in such order as is before taught, you shall haue your desire. As for example, if you sæke to knowe the right ascention of the tenth degrée of Leo, here hauing placed the Finitor as before is taught, sæke out in the Mater what Meridian passing through that degrée cutteth the said Equinoctiall, and in what point, and lay the labell to the said Equinoctial according as the signe requireth, be it towards Cancer or Capricorne, and by counting the degrées vpon the labell as is before taught, that is from 90. procæding from the limbe backeward towardes the Centre you shall finde the right ascention of the 10. degrée of Leo to be 132. degrées, which differeth from the foresaid Table but 2′7. a thing of small moment considering the narrowe spaces of the Meridians in the Mater.

How to finde out the ascentionall difference of the sunne, or of any degree or point of the Ecliptique.

The 14. Proposition.

BRing the Finitor to your Latitude, and there staying it looke amongst the Meridians in the Mater in what point the Parallel of the Sunne cutteth the Finitor, and the number of the Meridians contayned betwixt the

of the Mathematicall Iewell. 294

the Axletree, and that point shall be the ascentionall difference. As for example, in the Latitude 52. I would knowe the ascentionall difference of the Sunne being in the fourth degrèe of Cancer, at which time his Parallel of declination is 23. degrees Northward, here by working according to the rule I finde the ascentionall difference to be 33. degrèes 3′0.

How to find out the oblique ascention of the sunne, or of any point of the Ecliptique.

The 15. Proposition.

Auing found the right ascention and also the ascentionall difference by the former Propositions, consider whether the declination of the sunne or of any other point of the Ecliptique be North or South, for if it be North, then subtract the ascentionall difference out of the right ascention, and the remainder shall be the oblique ascention. But if the declination be South, then adde the ascentionall difference to the right ascention, and the summe thereof shall be the oblique ascention. As in the former example knowing the declination of the sunne to be 23. degrees Northwarde, I subtract the ascentionall difference which was 33. degrees 30. minutes, out of the right ascention of the sunne, which was 94. degrèes, and there remaineth 60. degrees 3′0. which is the oblique ascention of the sunne being in the fourth degree of Cancer, and his declination being 23. degrees Northward as is before supposed.

How to find out the right ascention of any arke or portion of the Ecliptique, and therewith to knowe what time it spendeth in rising in a right Spheare.

The 16. Proposition.

Ring the ende of the giuen arke founde out in the Zodiaque of the Rete, vnto the left ende of the Equinoctiall marked with D. and there staying it, lay the Fiduciall line of the labell vppon the beginning of

The description and vse

of the saide Arke, and the number of degrees in the innermost limbe of the mater contained betwixt the left end of the Equinoctiall, and the Fiduciall line of the labell shall be the right ascention of that Arke, and the houre set downe in the outermost limbe wherunto the labell pointeth, shall be the time which the said arke spendeth in his rising, so shall you finde the arke of the whole signe Taurus to be 30. degrees, and to spend in his rising two houres.

How to finde out the oblique ascention of any arke of the Ecliptique in any Latitude, and what time it spendeth in his rising.

The 17. Proposition.

Ind the oblique ascention both of the beginning and ending of the arke by the 15. Proposition, then subtract the said ascention of the beginning of the arke out of the ascention of the ending of the said arke, alwayes remembring that if the oblique ascention of the beginning be greater then the other, then to adde to the lesser 360. and out of that summe make your subtraction, for the remainder shall be the oblique ascention of the whole arke, then count the number of that ascention in the innermost limbe of the mater beginning at the left end of the Equinoctiall marked with D. and so procede vpwarde towardes your right hand, and where the said ascention endeth, there laye the labell, & the houre whereunto the labell pointeth sheweth the time which the saide arke spendeth in his rising in that Latitude. As for example, I would know the oblique ascention of the whole arke of the signe of Taurus, and what time that arke spendeth in his rising: first here hauing found the oblique ascention of the first point of Taurus to be about 13. degrees in the Latitude 52. and the oblique ascention of the last point of that signe to be about 30. here by taking 13. out of 30. there remaineth 17. degrees, which is the oblique ascention of the whole arke of the signe Taurus, which 17. degrees, I count in the limbe of the mater from the end of the Equinoctiall marked with D. vpward, and there laying the labell, I finde that it pointeth to one houre and two degrees, which maketh 8'. so as I conclude thereby that the whole arke

of the figne Taurus, spendeth in his oblique ascention one houre and eight minutes.

How to finde out the oblique descention of any point of the Ecliptique in any Latitude.

The 18. Proposition.

Auing found out the oblique ascention of the poynt opposite to the point giuen, adde thereunto 180. degrées, and the summe thereof shall be the oblique descention of the point giuen, alwayes remembring if the summe of addition doe excéde 360, to subtract out of that summe 360. and the remainder shall be the descention of the point giuen. As for example, you would knowe the oblique descention of the first point of Taurus, whose point opposite is the first point of Scorpio, and his oblique ascention according to Reynholdus his Tables is 222. degrées 36. minutes, whereunto if you adde 180. you shall make the totall summe to be 402. degrées 36. minutes, from which summe if you subtract 360. degrées according to the rule before giuen, there will remaine 42. degrées 36. minutes, which is the oblique descention of the first point of Taurus.

How to finde out the oblique descention of any arke giuen of the Ecliptique, and therewith to knowe the time which it spendeth in his setting.

The 19. Proposition.

Auing found the oblique descention of the beginning, and also of the ending of the giuen arke by the last Proposition subtract the descention of the beginning out of the descention of the ending, and the remainder shall be the oblique descention of the giuen arke, alwayes remembring if the subtraction can not be made to adde thereunto 360. that done, deuide that by 15. and the quotient will shewe the number of houres which the giuen arke spendeth in his setting: and remember that if in making that diuision there be any remainder

The description and vse

mainder left, to multiply that remainder by foure and so you shall haue the minutes. By working thus you shall finde the oblique descention of the whole Arke of Gemini to be 36. degrees 49. minutes, and in the Latitude 52. to spende in his going downe two houres and 24. minutes.

How to know the height of any starre at any houre, without seeing the starre, and thereby to find out in the firmament all the starres that be described in the net, and are to bee seene with the eye.

The 20. Proposition.

Ay the labell to the houre supposed vpon the limbe of the mater, and then bring the place of the sunne for that day to the Fiduciall line of the labell, and there hauing stayed the Rete, bring the Fiduciall line of the labell to the tippe of the starre whose altitude you seeke, and the labell will shew you vpon the limbe of the Mater how many degrees that starre is distant from the South. Againe by counting vpon the label, the degrees contained betwixt the point of the labell, and the tippe of the said star you shall haue his declination, which is North or South according as the starre is Northerne or Southerne. Then bearing in mind, aswell the stars distance from the South, as also his declination, work thus, Bring the Finitor to your Latitude, and vpon the stars Parallell or Circle of declination in the Mater count from the limbe of the Mater, the distance of the starre before found, and marke what Almicanterath crosseth that point, for there is the altitude of the star at that instant, and the Azimuth which cutteth that point, sheweth in what part or coast of heauen the starre is at that present. And remember to seeke the Parallell of the starre on that side of the Mater either on your right or left hand, so as it may fall amongst the Almicanterathes, for otherwise you shall not finde that you seeke. As for example the 26. of October 1591. at nine of the clocke at night, you would know the height of the first star of the Rams horne called Cornu Arietis the Sunne being then in the 12. degrees and 12. minutes of Scorpio. Here after that you haue laid the labell to the houre supposed, and brought the degree of the

Sunne

of the Mathematicall Iewell.   296

Sunne for that day to the Fiduciall line of the labell, and stayed the Rete there, bring the labell to the tippe of the starre in the Rete, and the labell will shewe in the limbe of the Mater how many degrées the starre is distant from the South, which you shall finde to be 28. degrées and 30. minutes, and the declination of the sayde starre being counted vpon the saide labell, to be 18. degrées, then kéeping those two numbers in minde, bring the Finitor to your Latitude supposing the same to be 52. degrées, and there staying the Rete, séeke for the Parallell of the starre which is 18. degrées on your right hande, and vppon that Parallel count by the Meridians from the limbe inwarde, the stars distance from the South, which was 28. degrées 30. minutes, and you shall finde that the 49. Almicanterath cutteth that point, so as you may conclude that the altitude of the starre called Cornu Arietis, was at that houre 49. degrées, and that the 47. Azimuth doth also passe through that point which sheweth that the starre is 47. degrées from the East towardes the South, which according to the Mariners account is Southeast, and somewhat more to the Southward. Now to finde out the saide starre in the firmament, if it be a starre light, you haue no more to doe but to lay the Diopter of your Astrolabe at that altitude, and to turn your face towardes that coast (that is to say) Southeast and somewhat more to the Southwarde, and the next bright starre which answereth in that coast to that altitude is the starre which you seeke.

How to finde out the ascentionall difference of any starre.

The 21. Proposition.

Ring the Finitor to your Latitude, and staying it there, looke in what point the Parallell of the starres declination cutteth the Finitor, and the number of the Meridians contained betwixt that point and the Centre, is the ascentionall difference, so shall you finde the ascentional difference of the Buls eye called *Oculus Tauri*, whose declination is 15. degrees 4'8. to be 21. degrees.

How

The description and vse

How to know the oblique ascention of any starre.
The 22. Proposition.

YF the declination of the starre be Northern, subtract the ascentionall difference out of the right ascention, and the remainder shall be the oblique ascention of the starre, but if his declination be Southerne you must adde the ascentionall difference to the right ascention, and that shal be the oblique ascention of the starre. As for example, because Oculus Tauri is a North starre, subtract his ascentional difference which is 21. degrees, out of his right ascention which as the foresaide Table sheweth is 62. degrees 3′0. and there will remaine 41. degrees 3′0. which is the oblique ascention of the sayd starre. But if it were a South starre as Spica Virginis, whose declination is almost 9. degrees Southwarde, and her ascentionall difference is 11. degrees 3′0. then you must adde the ascentionall difference to her right ascention, which is 195. degrees 5′1. so shall you finde her oblique ascention to be 207. degrees 2′1.

How to knowe what starres doe neuer rise nor set in anie Latitude.
The 23. Proposition.

YF the starre be a North starre hauing a greater declination be it neuer so little, then is the altitude of the Equinoctiall answerable to your latitude: As for example, suppose your latitude to be 52. which if you take out of 90. then the remainder is 38. degrees, which is the altitude of the Equinoctiall answerable to that Latitude, otherwise called the complement, then that starre neuer setteth in that Latitude: Againe if it be a South starre hauing greater declination neuer so little then the complement of your Latitude, then that star neuer riseth aboue your Horizon. As for example, in the latitude 52. the starre called Hircus, that is the Goate, being a North starre neuer setteth in that latitude because his North declination is 45.

degrees

degrées, which is greater then the complement of your Latitude by 7. degr. Also the star called Lyra, whose declination is 38. degr. like vnto the complement of your Latitude doth neuer set, but onely toucheth your Horizon aswell at his rising as setting, so contrariwise the star called Canopus, being a South star hauing 51. degr. 3'8. of South declination, neuer riseth aboue your Horizon in the foresaid latitude 52. And by this rule you may iudge in like maner of all the rest of the fixed starres both North and South.

How to know at what houre of the day or night any starre riseth or setteth.

The 24. Proposition.

First marke by the outwarde or inward shooting of the longest tippe of the starre, whether it bee North or South: for if it be North, then count amongest the Meridians in the Water so many Meridians as your Latitude amounteth to, beginning at the Axletrée, and so procéeding towardes the Northeast, which is betwixt the North pole and the Equinoctiall on your left hande, but if the starre be South, then count your Latitude procéeding from the saide Axletrée towards the Southeast, which is betwixt the ringle and the North pole on the right hande, and bring the tippe of the starre to that Meridian which now signifieth your Horizon, and there staying the Rete, bring the labell to the place or degrée of the sunne in the Zodiaque of the Rete, in which the sunne is that day you séeke, and the houre in the limbe whereto the labell pointeth, is the houre at which the starre riseth that day or night. Now to knowe when the same starre setteth, you haue no more to doe but to worke with the starre in the Northwest part in such order as you obserued before in the Northeast part of the Astrolabe. As for example, I would knowe at what houre the Buls eye called Oculus Tauri, doeth rise the last day of Iune the Sunne being then in the 17. degrée and 40. minutes of Cancer. Here because this starre is a North starre, I bring the longest tippe thereof to the 52. Meridian which is our Latitude counting from the Axletrée towardes the Northeast part of the Astrolabe

N q which

The description and vse

which is on my left hand, for that Meridian is alwayes the Horizon seruing the Latitude 52. and there staying the Rete, I bring the Fiduciall line of the labell to the place of the sunne which at that day is the 17. degreé 4′o. of Cancer as I said before, and I find that the label pointeth to one of the clocke 3′o. after midnight: wherefore I conclude that Oculus Tauri riseth that day at that present houre. Now to knowe at what houre that starre goeth downe the same day, I bring his longest tippe to the said Horizon towards the Northwest, and staying the Rete there, I lay the labell to the 17. degreé 4′o. of Cancer as before, so that the labell pointeth to foure houres and 20. minutes in the afternoone, at which time he goeth downe, so as he continueth at that time aboue the Horizon in the Latitude 52. 14. houres 48. minutes. And to knowe the abode of any starre aboue the Horizon, the next Proposition doth also shew.

How to knowe how long any starre continueth aboue the Horizon in euery Latitude.

The 25. Proposition.

B Ring the Finitor to your Latitude, and looke in what point the Parallel or declination of the starre cutteth the Finitor, and the number of the Meridians in the Mater contained betwixt the limbe and that point do shew the halfe time of his abode aboue the Horizon, which being doubled is the whole time of his abode aboue the Horizon. And in numbring the said Meridians whereof 15. do make an houre, remember to beginne to count from the right side of the Mater, proceeding towardes your left hand, and remember also that the middle Meridian or Axletreé, signifieth alwayes the sixth houre, so shall you not erre in your account. As for example, hauing brought the Finitor to the Latitude 52. looke in what point the Parallel of the foresaid starre Oculus Tauri cutteth the Finitor, & by numbring the Meridians proceeding from the limbe on your right hand towardes the left, you shall finde the Parallell of Oculus Tauri, being 15. degreés 4′9. to cut the eleuenth Meridian, which being doubled and then the summe thereof deuided by 15. maketh 14. houres and 4′8. as before.

How

of the Mathematicall Iewell.   298

How to finde out the starres houre, and thereby to knowe the houre of the night.

The 26. Proposition.

Irst hauing taken the starres altitude, set the Finitor to your Latitude, and looke at what houreline in the Mater that starres Almicanterath and the Parallell of his declination do meete, & hauing sought out the same houre in the limbe right against that point, bring the labell thereunto, for that is the starres houre, and there staying with your fingers end the very point of the labell, bring the longest tippe of the starre to the Fiduciall line of the labell, and staying there the Rete, turne the labell to the degree of the Zodiaque of the said Rete, wherein the sunne is that day, and the labell will point to the true houre of the night set downe in the outwarde limbe of the Mater. As for example, I suppose that the seuenth of October 1591. the sunne being then in the 23. degree 1′5. of Libra, I tooke the altitude of the starre Hircus, that is the Goate, which I found to be 20. degrees, here hauing set the Finitor to my Latitude 52. I looke in the Mater at what houre line that Almicanterath, and the Parallell of the same starre which is 45. degrees Northwarde doe meete or crosse one an other, and I find that they meete iust vpon the fourth houre line of the forenoone, wherefore I seeke right against that point the same houre in the outermost limbe of the Mater, for that is the starres houre, and hauing placed the labell at that houre, I stay it there with my finger vntill I haue brought the longest tippe of the starre Hircus vnto the Fiduciall line of the labell, and there staying the Rete, I turne the labell to the degree of the sunne which is 23. degrees 15. minutes of Libra, and I see that the labell pointeth to the seuenth houre of the night set downe in the limbe of the Mater and halfe an houre past. Maister Blagraue sayth that the sooner you take the altitude of the starre whereby you seeke to knowe the houre of the night, you shall haue the houre more truely.

The description and vse

How to finde out the distance betwixt any two starres contained in the Net.

The 27. Proposition.

Irst consider whether the declination of both stars be eyther South or North, or that the one bee South and the other North, for if both their declinations be South or North, then you must not leaue to turne the Rete to and fro vntill you haue brought the longest tippes of both starres to one selfe Meridian in the Mater, that done, count vpon the said Meridian howe many degrees are contained betwixt the tippes of the said starres, for that is the distance betwixt them. But if the one starre haue North declination and the other South, then turne the Rete to and fro vntill both starres doe lie vpon two such seuerall Meridians as ech of them is equally distant on ech hand from the Axletræ, then count how many degrees or Meridians are contayned betwixt the Axletræ and either of those stars (which you will) and that is their distance, so shall you finde the distance betwixt Oculus Tauri and Canis minor being both North stars to be 46. degrees 2'0. and the distance betwixt Oculus Tauri and Canis maior, whereof the first is a North starre, and the other a South starre to be iust 46. degrees.

Another way to knowe the distance of any two stars, their Longitudes and Latitudes being first knowne, and also by that meanes to finde out the distance betwixt any two places vpon the earth.

The 28. Proposition.

Irst seeke to knowe the difference of their Longitudes by subtracting the lesser Longitude out of the greater, then count that difference from the outermost Meridian of the Mater towardes the Center, and marke well that Meridian at which your account endeth, then number vpon the limbe of the Mater from the Equinoctiall the greater Latitude either

Northward

Northward or Southward according as the Latitude is, and to that point bring the Zenith of the Rete, then vpon the selfe same Meridian before marked count from the Equinoctiall the lesser Latitude, and looke what Azimuth passeth through that point, for the degrees which are contayned betwixt that point and the Zenith shalbe the distance: and thus doing you shal find the distance betwixt Oculus Tauri & Canis maior to be 46. degrees and 1′5. and the distance betwixt London and Venice to be 12. degrees and 2′0. which 12. degrees being multiplyed by 60. maketh 720. miles, whereto if you adde for the 2′0.20. miles, it will make in all 740. miles.

How to find out the degree of *Medium cæli* at any houre, of the day (that is to say) the degree of the Zodiaque that is in the Meridian at any houre that you seeke, and also the degree called *Imum cæli*.

The 29. Proposition.

FIrst seeke out the place of the sunne for that day in the Zodiaque of the Rete, and hauing laid the labell to the houre supposed vpon the limb of the Mater, bring the place of the sunne to the Fiduciall line of the label, and there staying the Rete, looke what degree of the said Zodiaque cutteth the noone line or Meridian at noonetide, which you shall easily find by laying the labell to the houre of 12. at noone, for the Fiduciall line of the label crossing the Zodiaque, will shew the degree of mid heauen at that houre. As for example, the 26. of June 1592. I would know the degree of Medium cœli, at eight of the clocke in the morning, the sunne being that day in the 14. of Cancer. Here by laying the labell to the said houre in the limbe of the Mater, I bring the place of the sunne to the Fiducial line of the labell, and there hauing stayde the Rete, I bring the labell to the twelfth houre at noone, and I find that the labell cutteth the Zodiaque of the Rete in the 18. of Taurus, which at that houre is the degree of Medium cœli, whose point opposite is the 18. degree of Scorpio, and that is at that houre Imum cœli which the labell being laid to the 12. houre of midnight will shew.

The description and vſe

How to finde out the Horoſcope or aſcendent at any time of the day or night, and thereby to haue the foure principall angles of heauen.

The 30. Propoſition.

Auing laid the labell to the houre giuen vppon the limbe of the Mater, ſtay it there with your finger vntill you haue brought the place of the ſunne for that day vnto the Fiduciall line of the labell, and there ſtaying the Rete, looke what degrée of the Zodiaque therof toucheth or croſſeth the Horizon anſwerable to your Latitude, for that is the aſcendent at that preſent houre. As for example, I would knowe the aſcendent at eight of the clocke at night the twelfth of October 1580. the ſunne being then in the 28. degrée 1′4. of Libra. Here hauing layde the labell vpon the ſayd houre, I bring the 28. degrée 1′0. of Libra to the Fiduciall line of the labell, and there ſtaying the Rete, I find by helpe of the labell that my Horizon which is the 51. 4′0. Meridian, counting on both hands from the Arletrée, that the firſt degrée of Cancer, doth croſſe my Horizon in the Northeaſt quarter, wherefore I affirme that to be the aſcendent or firſt houſe, whoſe point oppoſite being the firſt of Capricorne, is the deſcendent or ſeuenth houſe, then by bringing the Fiduciall line of the labell to the South end of the Equinoctiall at which the ringle hangeth, I finde by helpe of the labell that the 24. degrée 4′0. of Aquarius cutteth the Equinoctiall, which is the tenth houſe or Culmen cœli, whoſe point oppoſite being the 24. degrée 4′0. of Leo, is the fourth houſe, otherwiſe called Imum cœli, and thus you haue all the foure principall houſes of heauen for that houre, as you may ſée in this figure here following.

A

of the Mathematicall Iewell. 300

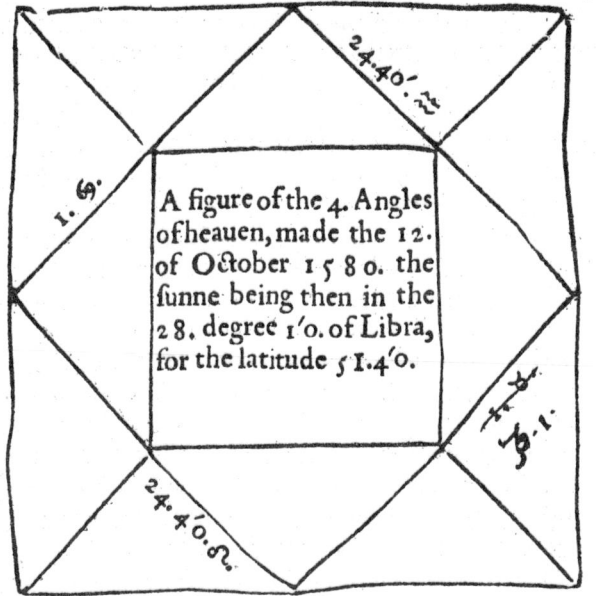

A figure of the 4. Angles of heauen, made the 12. of October 1 5 8 0. the sunne being then in the 28. degree 1′ 0. of Libra, for the latitude 51.4′ 0.

But the difficultie of finding out the true ascendent consisteth in knowing whether it is to be sought in the Northeast quarter, or in the Southeast quarter of the Iewell. The Northeast quarter is that which lyeth betwixt Imum coeli, and the North pole or East end of the Arletræ, because in this case the Arletræ signifieth the line of East and West, and the Equinoctiall signifieth the line of South and North, at the South end whereof is fastened the ringle or handle. But in the former example you may plainely se that the North part of the Zodiaque of the Rete doth cut the Horizon, as well in the Northeast quarter as in the Southeast quarter with two seueral degrees and signes, for in the Northeast quarter the Zodiaque cutteth the Horizon with the first of Cancer which is the ascendent, and in the Southeast quarter it cutteth the Horizon with the eight of *T*aurus, which is not the ascendent: for you haue to vnderstand that euery degre of the Zodiaque doth both rise and set either towards the North or towards the South, the first point of Aries and of Libra onely excepted, both which do rise right East, and goe downe right West, euen as the Equi-

Q q 4   noctiall

The description and vse

noctiall doth, whereof M. Blagraue doth gather a rule how to find out by his Iewell when the ascendent is to be sought either in the northeast quarter of the Water, or in the southeast quarter, which is thus: The ascention of any of the sixe Northerne signes is to be sought for in the Northeast part of the Iewell, & the ascention of any of the sixe Southerne signes is to be sought for in the Southeast part of the Iewell. For although that the North part of the Zodiaque of the Rete containing the 6. Northerne signes cutteth the Horizon answerable to your Latitude, aswel in the Northeast as in the Southeast part of the Iewell, yet you must séeke the ascendent in the Northeast part, and not in the Southeast part of the Iewell, because that euery degrée of any of the 6. Northerne signes riseth Northernly: so contrariwise if the South part of the Zodiaque containing the sixe Southerne signes doe cut the Horizon, aswell in the Northeast part as in the Southeast part of the Iewel, yet you must séeke the ascendent in the Southeast part, and not in the Northeast part of the Iewell. As for example, the second day of August 1592. the sun being in the 20. of Leo, I would know the ascendent at foure of the clocke in the afternoone: here hauing laid the labell to that houre, & brought the place of the sun to the Fiduciall line thereof, I find that the 13. degrée 3′0. of Aquarius doth cut the Horizon seruing to your latitude 52. in the Northeast part of the Iewell, and that the 19. of Sagittarius cutteth the said Horizon in the Southeast part of the Iewell, which must be the ascendent, because that euery degrée of any of the southerne signes riseth Southernly, and not Northernly.

How to find the Circles of position, and to know how much the pole is eleuated aboue euery such Circle in any Latitude, without the knowledge whereof you can not finde out the twelue houses by this Astrolabe.

The 31. Proposition.

Irst bring the Zenith of the Rete to your Latitude, so shall the Azimuthes become circles of position, then vpon the Parallel of declination of the point giuen, according as the declination thereof is north or south: count from the limbe on your right hand, the number of y houres giuen

of the Mathematicall Iewell.

giuen amongst the Meridians in the Water, & the Azimuth which the Meridian of that houre cutteth shalbe the Circle of position, which had, you shall find the eleuation of the pole aboue the Circle of position thus: Count your Latitude amongst the Almicanteraths from the Zenith vpon the Circle of position found, & to that point whereas your Latitude endeth, bring the Fiducial line of the label, & reckon vpō the label how many degrees are cōtained betwixt the limbe of the Rete & that point, for the number of those degrees is the eleuation of the Pole aboue the Circle of position. As for example vsed by M. Blagraue himselfe, the 12. of October 1580. the sunne being then in the 29. degrée 1′0. of Libra, and his South declination 11. deg. 1′0. I would know at 8. of the clock at night in what Circle of position the sunne at that time was, here I bring the Zenith to the Latitude of Reading, which is 51. deg. 4′0. then for 8. of the clocke I count 8. houres amongst the Meridians in the Water, from the South part of the Iewell vppon the Parallell of the sunne being then 11. deg. 3′0. and by attributing to euery houre 15. Meridians, I find that the Meridian whereas the 8. houre endeth doth crosse the 28. & ½. Azimuth, counting the Azimuths from the Zenith line, and that is the Circle of position wherin the sun was then vnder the earth in the Northeast quarter. Now to knowe how much the Pole is eleuated aboue that circle of position. I do count from the Zenith the foresaid Latitude 51. degrées, 4′0. vpon the said Circle of position, and to the point where that Latitude endeth I bring the labell, & counting thereon from the limbe of the Rete the degrées contayned betwixt the said limbe, and the foresaid point, I find the number of them to be 43. degrées 3′0. wherfore I conclude that the Pole is eleuated aboue the said Circle of position 43. degrées 3′0.

How to find out all the 12. houses of heauen, and thereby to erect a figure at any houre of the day or night.

The 32. Proposition.

You shall vnderstande the order of this better by this one example giuen by M. Blagraue himselfe, then by manifold rules which be also so plainely expressed and obserued in this example, as you néede none other instruction

The description and vſe

ſtruction to erect the like figure in any Latitude, and at any houre of the day or night. Suppoſe then that you would erect a figure for one that was borne the twelfth of October 1580. at eight of the clocke at night in the Latitude 51.4′0. the ſunne being then in the 28. degrée 1′0. of Libra, and his declination being at that time 11. degrées 1′0. Southward: Here firſt ſet downe in your figure alreadie drawne the foure Angles before found by the 30. Propoſition anſwerable to the ſaid day & houre, then to find out the reſt of the houſes do thus, bring the Zenith of the Rete to the latitude preſuppoſed, which is 51. degrées 4′0. & ſtaying it there, marke what number of Azimuthes or Circles of poſition doth cut euery 30. degrée of the Equinoctiall, counting the degrées of the Equinoctiall by helpe of the Meridians from the limbe towardes the Centre, but you muſt count the number of the Azimuthes or Circles of poſition from the Zenith line towardes the limbe, of which Circles of poſition there be no more but ſixe at the moſt (that is to ſay) thrée betwixt the limbe and the Centre on the one ſide, and as many on the other ſide of the Centre, the eleuation or rather depreſſion of euery one whereof you muſt ſéeke to knowe by the laſt Propoſition. But M. Blagraue ſayth that you néede to knowe the eleuation or rather depreſſion, but only of two Circles of poſition, that is of that which ſheweth the eleuenth and third houſe, and of that which ſheweth the twelfth & ſecond houſe, and ſo ſayth Stadius alſo in the beginning of his Ephemerides, treating of the 12. houſes, for hauing them you haue all the reſt. The order of working according to M. Blagraue his rule is thus: hauing brought the Zenith to the foreſaid Latitude, looke what Circle of poſition cutteth the Equinoctiall in the firſt 30. degrées next to the limbe, and you ſhall finde that the 47. Azimuth or Circle of poſition (counting from ẏ Zenith line) cutteth that point of the Equinoctiall, and therewith ſerueth to the eleuenth houſe, and alſo to the third houſe, and the eleuation of this Circle is 32. deg. then from thence tell vppon the Equinoctiall 30. degrées more, which do make in all 60. degrées, and through that point you ſhal finde the 19. Azimuth or Circle of poſition to paſſe, whoſe eleuation is 47. degrées and this Circle ſerueth to the twelfth houſe & to the ſecond houſe. Now kéeping well in mind thoſe two laſt eleuations, that is, 47. and 32. worke thus: Lay your labell to the

eigth

of the Mathematicall Iewell. 302

eigth houre of the morning, which is 30. degrées distant from the right Horizon or Axletrée, and bring to the Fiduciall line thereof the Culmen cœli or 10. house first set downe in your figure which is the 24. of Aquarius, and staying it there, looke what degrée of the Ecliptique cutteth the 32. Horizon which serueth to the 11. house counting from the right Horizon towards your left hande. And by following that Horizon vp towards the North pole, you shall finde that the 25. degrée 3'0. of Pisces cutteth the same Horizon very nigh vnto the Pole, but by my instrument I finde it to be the 27. degrée of Pisces, which perhaps is not truely made, and therefore set downe in the eleuenth house of your figure 25. degrées 3'0. of Pisces. Then bring the foresaid Culmen cœli together with the labell to the tenth houre of the morning, which is 30 degrées further towards the South, and staying it there, looke what degrée of the Ecliptique cutteth the 47. Horizon, seruing to the twelfth house, and you shall find that the 20. degrée of Taurus cutteth that Horizon, wherefore set downe the 20. of Taurus in the twelfth house of your figure, then bring the Culmen cœli together with the labell to the 12. houre at noone, and you shall sée the ascendent which is the first of Cancer to cut the oblique Horizon, which is 51. degrées 4'0. and from thence bring the label and culmen cœli to two of the clocke in the after noone, and staying it there you shall sée the 21. degrée of Cancer to cut the 47. Horizon which sheweth the second house and also the twelfth house as before, wherefore set downe 21. of Cancer in the second house of your figure, that done, remooue the labell together with Culmen cœli to foure of the clocke in the afternoone, and there staying it, you shall sée the seuenth of Leo to cut the 32. Horizon which sheweth both the third house and the eleuenth house as before, wherefore set the seuenth of Leo in the third house of your figure, so haue you eight of the houses, that is to say, the first, the second, the third, the fourth, the seuenth, the tenth called Culmen cœli, the eleuenth and the twelfth, so as there want onely foure, that is, the fifth, the sixth, the eight and the ninth house. The fifth is opposite to the eleuenth, the sixth to the twelfth, the eigth to the second, and the ninth to the third, whose opposite signes ech one hauing like number of degrées are to be placed in the foure houses of your figure that be wanting. As in the fifth house the 25. degrée 3'0. of Virgo,

in

The description and vſ-

in the ſixth houſe the 20. degrée of Scorpio in the eigth houſe the 21. degrées 3′0. of Capricorne, and in the ninth houſe the 7. degrée 1′0. of Aquarius, ſo haue you all the twelue houſes. And by this meanes M. Blagraue ſayth that you may make the places of the twelue houſes to ſerue for euer in any Latitude, ſo as you doe diſtinguiſh frō the reſt with ſome colour thoſe two Horizons whereof the one doth ſhew the eleuenth and the third houſe, and the other Horizon ſheweth the twelfth and ſecond houſe, yea and by this meanes you may (as he ſayth) make as true Tables to finde out the twelue houſes in euery Latitude as thoſe that be calculated of purpoſe.

The figure of the foreſaid twelue houſes.

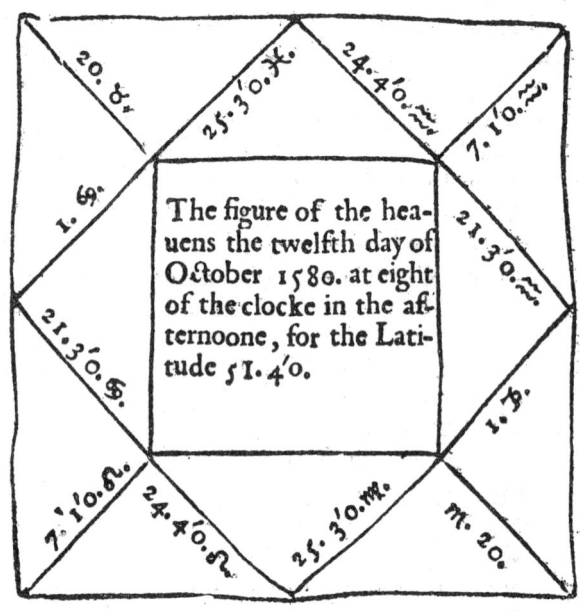

A nevv and neceſſarie

Treatiſe of Nauigation containing *all the chiefeſt principles*
of that Arte.

Lately collected out of the beſt Mo-
derne writers thereof by M. Blundiuile, and by him
reduced into ſuch a plaine and orderly forme of
teaching as euery man of a meane capacitie
may eaſily vnderſtand the ſame.

They that goe downe to the Sea in ſhips, and occupie their buſines in
great waters: Theſe men ſee the workes of the Lord and his wonders in the deepe. Pſalme. 107.

Of Nauigation, what it is, and with what order the principles thereof are here taught by Master Blundeuill, according to the rules of the best moderne writers of that Arte.

NAVIGATION is an Arte which teacheth by true and infallible rules, howe to gouerne and direct a ship from one Port to another safely, rightly, and in shortest time: I say here safely, so farre as it lieth in mans power to performe. And in saying rightly, I meane not by a right line, but by the shortest and most commodious waye that may be found: And by saying in shortest time, I meane thereby according as the ship is good of sayle, and according as both winde and tide shall serue.

The meanes to attaine to this Arte as to all other Artes are two parts, whereof euery Arte consisteth, that is Methode and Practise, which is as much to say heere as instruction and experience: For of the chéefe points of this Arte, some are to be learned by instruction, and some onely by experience: for what instruction I pray you wil serue to make a good coaster, that is to say to know any Cape or Cliffe when he séeth it, vnlesse he hath first séene it & hath before taken good markes thereof. Also to know the currents in all places, the deapth and qualities of waters, sandes, flats, or shoulds, and such like, the perfect knowledge whereof consisteth chiefely in experience. Of which experience I minde not here to treate, because I knowe that a Dutchman called Wagoner or Aurigarius hath lately written of these things notably wel, & especially for these our seas and the Northeast seas, whose booke I doubt not but that some of our learned sea men will set foorth in our mother tongue, with some augmentation of their owne epperience for the East and West Indies, for the profite of their countrie and to their owne praise & commendation. In the meane time

I will

The Arte of Nauigation.

I will procéde with instruction, the chiefe pointes whereof are these here following: first to know the vses of such instrumentes as euery skilfull sea man ought to haue with him that mindeth to sayle any long voyage, which are these here following: First a perfect Kalender or Ephemerides, the Mariners ring or Astrolabe, the Crosse-staffe called of the Spaniardes Balla stella, the two Globes both Celestiall and Terrestriall: Of all which things I haue in a manner alreadie written seuerall Treatises. Item an vniuersall Horologe, to knowe thereby the houre of the day in euery Latitude, and a Nocturnlabe to know thereby the houre of the night. And it were necessarie for him that sapleth long voiages to carry with him a Topographicall instrument to describe therby those straunge Coasts & Countries wherby he sapleth, the vse of which instrument is plainely taught by William Borne in his booke called the Treasure of Trauailers, and by Maister Digges in his booke called the Pantometria, wherefore I shall not néede to speake any further thereof: Also the Mariners compasse, whereof I do not only make a plaine description and shew the vses thereof, but also how to finde out the variation of the same, according to the precepts of the best moderne writers thereof. And lastly the Mariners Carde, whereof whilest I speake, I doe not onely shewe the making thereof, and how to drawe the Parallels in better sort then they are vsed to be drawne in the common Cardes, but I set downe therewith the chiefest vses of the Card, amongst which is taught how to know by help of the Card and certaine Tables made of purpose, what way your shippe hath made in sayling by any rombe or wind, yea in sailing right East & West, which heretofore (as Cogniet sayth) hath béene thought a thing vnpossible to be knowne, but only by coniecture: And being out of your way how to know in what place you are, & to marke the same in your Card that you may the more readily direct your ship againe to the place whereunto you would go. But all these instruments serue to little purpose, vnlesse you know also the North star with his guards, & diuerse other stars situated aswel towards the North as towardes the South pole, together with their Longitudes, Latitudes, declinations, and greatnesse, to knowe thereby the Latitude of any place, & the houre of the night: also the course of the sunne and his declination, by helpe whereof you may know

the

The Arte of Navigation.   305

the Latitude of any place, the times and seasons of the yeare, the houre of the day, and the length of the day and night in euerie latitude. And finallie you must know the course of the Moone, whereon dependeth the knowledge of the tydes in all places: Of all which thinges I mind here to treat in such order as the Chapters hereafter following doe shew.

A generall Kalender or Almanacke for euer, containing these 9. Chapters next following.

HOw to find out the golden number euery yeare. Chap. 1.
    *H*ow to finde out the Epact in euery yeare. Chap. 2.
   *H*ow to know the Epact by the Mariners rule vpon your thombe. Chap. 3.
   *H*ow to knowe the age of the Moone in everie moneth throughout the yeare. Chap. 4.
   *H*ow to know the change, ful, and 4 quarters of the Moon, euerie Moneth throughout the yeare. Chap. 5.
   *H*ow to know in what signe and degree the moone is euery day throughout the yeare. Chap. 6.
   *H*ow to find out the moouable feastes everie yeare, onelie by knowing the day of coniunction in the moneth of Februarie. Chap. 7.
   *H*ow to finde out the circle of the Sunne, called in Latine *Cyclus Solaris*, and thereby the Dominicall Letter in euery yeare. Chap. 8.
   *H*ow to find out the number of indiction. Chap. 9.

A briefe description together with the vse of the diurnall Table or Almanacke of *Iohannes Stadius*. Chap. 10.
   *H*ow to find out the place of any Planet by the Ephemerides

The Arte of Navigation.

des. Chap. 11.

Of the Mariners Ring or Astrolabe, and of his crosse staffe Chap. 12.

A briefe description of M. *Hood* his crosse staffe, and of all the partes thereof. Chap. 13.

How to set the partes of M. *Hoods* staffe together to serue such Astronomicall vses as do chiefly belong to the Mariner. Chap. 14.

The shape or figure of the foresaide staffe, hauing all his partes set together to serue for Astronomicall vses. Chap. 15.

How to take the altitude of the Sunne at any houre that he is to be seene with the eie by M. *Hoods* staffe. Chap. 16.

Howe to take the altitude of any Starre with M. *Hoods* staffe. Chap. 17.

How to take the distance betwixt two stars with M. *Hoods* staffe. Chap. 18.

Of the Mariners Astrolabe. Chap. 19.

A briefe description of the Mariners Astrolabe, and the vse thereof. Chap. 20.

A brief description of the Mariners crosse staffe. Chap. 21

The vses of the Mariners crosse staffe. Chap. 22.

Of the Wind, what it is, and of the divers kindes and names thereof. Chap. 23.

A briefe description of the Mariners Compasse, and vse thereof. Chap. 24.

Of the Lodestone, and of the variation of the Compasse in Northeasting and Northwesting. Chap. 25.

How to finde out the variation of the Compasse in everie latitude. Chap. 26.

Of the Mariners Card, and of the making therof. Chap. 27

The shape and figure of the first liniaments of the Mariners Carde drawne after the olde maner, and how to set downe the places of the land or sea therein. Chap. 28

A Table to drawe thereby the Parallels in the Mariners carde, together with the vse thereof in truer sort than they haue bene drawne heretofore. Chap. 29.

The draught of the Meridians and Parallels of the Mariners Carde or nauticall Planispheare according to the former

The Arte of Navigation.

table. Chap. 30.

The foure chiefest vses of the *Mariners* card. Chap. 31

How to know the way of your ship, and how many leagues are to bee accounted for one degree of latitude in euerie rombe whereby you saile. Chap. 32.

How to account the leagues in sayling directlie East or west without changing latitude or altitude of the Pole. chap. 33.

A Table to helpe you to know what way your shippe hath made in sayling right East or West without chaunging your latitude together with a brief description & vse therof, chap. 34.

An example of counting the way of your shippe in sayling right West. chap. 35.

An other example of counting the way of your ship in sayling right East. chap. 36.

To know how much you goe out of your way in sayling by one wrong rombe or by more. chap. 37.

Of the North Starre, otherwise called the Lodestarre, and of his guards, and how to know the same. Chap. 38.

The vses of the North starre and of his guards. chap. 39.

To knowe by helpe of a little Table made according to the Mariners rule touching the 8. principall rombes, shewing how much and when the Lodestarre is either aboue or beneath the Pole, that you may know thereby the true altitude of the Pole in taking the height of the Lodestarre with your Astrolabe or crosse staffe. chap. 40.

How to make an Instrument which wil shew at any houre of the night how much the Lodestar is either aboue or beneath the Pole in euerie other rombe as well as in the 8. principall rombes, and also the true houre of the night. chap. 41.

How to know by the foresaid twofold instrument as wel the mounting and descending of the North star as the true houre of the night both at one instant and also the elevation of the Pole, Chap. 42.

What stars are to be obserued by those that sayle beyond the Equinoctiall vnder the South pole. chap. 43

Of the Sunne, and of his motion, and of the chiefest apparances belonging to him. chap. 44

A Table shewing the declination of the Sunne euerie day through-

The Arte of Navigation.

throughout the yeare, and the vse thereof. Chap. 45.

Of the foure seasons of the yeare, that is, Spring time, Summer, fall of the leafe, called otherwise Autumne, and winter. Chap. 46.

How to knowe when the Sunne riseth and setteth in everie latitude, and thereby the length of the day and night, and also in what rombe or wind he riseth and setteth, and how much he declineth everie day from the Equinoctiall either Northward or Southward. Also howe to know the elevation of the Pole, otherwise called the latitude of any place, by knowing the Meridian altitude of the Sun, & his declination. Chap. 47.

Of the shaddow of the Sunne, and how to know therby the houre of the day in any latitude by help of an vniversall diall. Chap. 48.

Of the Moone and of all her diuers motions. Chap. 49.

How to know in what signe the point *Auge* of the Moone is in any yeare. Chap. 50.

When the Moone is saide to be in Coniunction with the Sunne, or to be at the full, and what her greatest latitude is aswell frō the Ecliptique line, as from the Equinoctial. Chap. 51.

How to know in what part of the Zodiaque the head of the Dragon is euerie yeare. Chap. 52.

How to know the tydes in any place by the Moone. Chap. 53.

How to know by helpe of an instrument the tydes at any place. Chap. 54.

How a generall Rutter shewing the tydes in al places shuld be made. Chap. 55

A

A general Kalender or Almanacke for euer.

What this word Kalends signifieth, & from whence it is deriued is before set downe in the firste booke of my Treatise of the Sphere the 45. chapter. But because there be generall rules to knowe the coniunction of the Moone with the Sunne, her full, and all her foure quarters, and also the moueable feasts and Dominicall letter, and such like things easilie to be learned without the help of any particular Kalender, I thought good first to set downe this general Kalender, contayning 9. propositions, as followeth.

How to find out the Golden number euery yeare.
Chapter. 1.

The Golden number is the number of 19. proceeding from 1. to 19. and so to begin again at 1. And it is so called because it was sent in golden letters frō Alexandria in Ægypt to Rome: For in 19. yeares the Moon doth make all her sundry motions & changes, and returneth again to the place where she first began. And to finde the foresaid number the way is thus. Adde 1. to the yeare of the Lord wherof you inquire, and diuide the same by 19. and the remainder shall be the Golden number for that yeare, as for example, being desirous this present yeare 1590. to knowe the Golden number, I adde 1 to the said yeare, and so make it 1591. which being diuided by 19. there remaineth 14. which is the Golden number of this present yeare.

But when there is no remainder, then 19 is the Golden number, and remember that the Golden number beginneth alwaies at the first of January, and the Epact the first of March.

The Arte of Navigation.

How to finde out the Epact in everie yeare.

Chapter. 2.

He Epact is a number not exceeding 30. because the Moone betwixt change and change, neuer passeth 30. dayes, and thereby the common Lunar yeares consisting of twelue Moones is lesser than the Solar yeare by 11. dayes, for to euerie Moon are attributed no more but 29. dayes and a halfe, which make in all 354. dayes, so as the common Solar yeare consisting of 365. dayes excedeth the Lunar yeare 11. dayes, from whence the Epact taketh his originall, which Epact is found thus. Multiply the golden number of the yeare by 11. the product whereof if it be vnder 30. then it is the Epact. But if the product be aboue 30. then diuide the product by 30. and the remainder shall be the Epact. As for example, to know the Epact in the yeare 1590. the golden number being 14. as before: here hauing multiplyed the same by 11. and diuided the product thereof by 30. I find the remainder to be 4. which is the Epact of the said yeare. Also by knowing one former Epact you shal haue it euer after, by adding thereunto 11. and if the number doe excede 30. then you must diuide the same by 30. and the remainder shall be the Epact, as by adding 11. to 4. I know the Epact shall be the next yeare, which is 1591. the number of 15. and by adding 11. to 15. the Epact in the yeare of our Lord 1592. shall be 26. and so forth.

How to knowe the Epact by the Mariners rule vpon your thumbe.

Chapter. 3.

Irst, you must suppose the inside of your left thumbe to be diuided into three spaces, and the nethermost space to containe 10. the middle space 20. and the highest space towards your thumbes end, to containe 30. and knowing first the golden number, begin to tell the same at the nether space saying there 1. at the middle space 2. at the thirde space 3. then

be-

The Arte of Navigation. 308

beginne againe at the lowest space, and there say 4, and so continue your account still after that manner, vntill you haue the full summe of the Golden number, and marke vpon what space the full summe falleth, for the Golden number being added to the number of that space, doeth shewe the Epact, so as the totall summe doth not excæde 30. for then you must subtract 30. and the remainder shall bee the Epact: as for example, in the yeare 1591. the Golden number is 15. which bæing counted vppon your thombe in such order as is before taught, it will fall vpon the highest space, which is 30. to which if you adde the Golden number, which is 15. it will make in all 45. from which summe if you subtract 30. there will remaine 15. which shall bee the Epact for that yeare, so as the Epact and Golden number in that yeare are like numbers. For euery thræ yeares they are alwayes like, as when the Golden number is either 3.6.9.12.15. or 18. the Epact hath also like number.

*H*owe to knowe the age of the Moone in everie moneth throughout the yeare.

Chapter. 4.

Adde to the Epact the number of monethes from the beginning of March, together with the moneth wherein you sæke, and also the number of the dayes past of that moneth, wherein you sæke, and the summe of such addition will shewe you the age of the Moone, as for example, I would know the age of the Moone the sixt day of December in the yeare 1590. Here knowing the Epact of that yeare to bee 4. I adde thereunto the number of the moneths from the beginning of March, which are 10. moneths, and also the number of daies of December, which are sixe, and the summe thereof is the age of the Moone, if the summe bee lesse than 30. But if the summe of such addition do excæde 30. then you must subtract 30. and the remainder shall bæ the age of the Moone, so as the moneth wherein you sæke haue 31. dayes, for if it hath lesse than 31. dayes,

Rr 4 then

The Arte of Navigation.

then you must subtract but 29. and the remainder shall bee the age of the Moone, as for example, suppose you seeke the age of the Moone the 22. of Nouember in the yeare 1589. in which yeare the Epact was 23. here by adding to the Epact first 9. Monethes, and then twentie two dayes, it maketh in all 54. out of which by subtracting 29. because Nouember hath but 30. dayes, the remainder is 25. which was the age of the Moone in that moneth, day, and yeare.

How to know the change, the full, and foure quarters of the Moone euerie moneth throughout the yeare.

Chapter. 5.

Martin Cortes in his booke called the Art of Nauigation teacheth a rule to find the day of the change, in euerie moneth by knowing the age of the moone that day you seeke, and then by reckoning from that day backward the number of the dayes that went next before, or els by taking the age of the Moone out of the dayes of the moneth that went next before, as for example, the last of October 1592. I finde by the fourth proposition the age of the Moone to be 5. which being taken out of 31. (for so many dayes October hath) there remaineth 26. which day by this meanes should be the day of the change, but the very day indeed was the 25. of the said moneth, for the former rule to know the age of the Moone is not so true it selfe, but that sometime it will fall a day either ouer or under. But Gemma Frisius teacheth to find out the day of the change in euery moneth thus. Adde to the Epact the number of the monethes from the beginning of March whereof that Moneth wherein you seeke the change, must be counted as one, and then subtract the product or sum of that addition from 30. and the remainder will shew the day of the change, which rule I finde to be true so long as the summe of the addition doth not exceed 30. but when the sum of the addition is more than 30. which will commonlie chance when the Epact is a great number, as 26. or 29. he teacheth no rule for that, wherefore I thinke it best to take such summe out of 59. and the remainder shall be the day of change, or at the most but one day ouer

The Arte of Navigation.   309

ouer or vnder, and hauing the day of the change, you shall haue the day of the full by adding 15. dayes more to the day of change: And hauing the change and the full, you shall easily haue all her foure quarters by adding or subtracting 7. dayes.

Howe to know in what signe and degree the Moone is euerie day throughout the yeare.

Chapter. 6.

Ome doe set downe rules to know in what signe and degree the Moone is euerie day, but such as are not true. And to say the truth the Moone hath so many diuers motions as it cannot be done but by speciall Tables calculated of purpose. And by what rule soeuer you work, you must first know the place of the Sunne, which the Ephemerides most trulie sheweth, and in looking for that, you shall also find hard by it the place of the Moone, that is to say, in what signe and degree shee is euery day, and so orderly the place of any other Planet, and therefore I leaue to speake any further thereof.

How to find out the moouable Feastes euerie yeare, onely by knowing the day of coniunction in the moneth of Februarie.

Chapter. 7.

He way to finde out the Coniunction in euerie moneth is before taught in the fift proposition. And hauing the day of the coniunction in Februarte, you may assure your selfe that the next tuesday following is alwayes Shrouetuesday, for though the coniunction it selfe doe fall vppon a Tuesday, yet the next Tuesday after that, shall bee Shrouetuesday, and the next Sunday after that, is called Quadragesima, which is the first Sunday in Lent, and sixe weekes next after that is Easter day, whereunto if you adde fiue weekes more, that is to say, 35. dayes, you shall haue Rogation Sunday, and 4.

daies

The Arte of Navigation.

dayes next after that is Ascention day, and tenne dayes next after Ascention day is Pentecoste or Whitsunday, and seuen dayes next after that is Trinitie Sunday, and foure dayes next after that is the Festiuall day called Corpus Christi. And you haue to note that the Sunday called Adventus Domini, which wee call the first Sunday in Aduent, is alwayes the fourth Sunday before Christmasse day, and the Sunday called Septuagesima is the third Sunday before Quadragesima, otherwise called the first Sunday in Lent, and betwixt Septuagesima and Quadragesima there are other two Sundayes, whereof that which is next before Quadragesima is called Quinquagesima, and the next Sunday before that is called Sexagesima.

How to finde out the circle of the Sunne, called in Latine by a Greekish name *Cyclus Solaris*, and thereby the Dominicall Letter in euery yeare.

Chapter. 8.

This Circle was inuented more to finde thereby the Dominicall Letter than to shew any greate chaunges of the Sunnes motions therein: And because there bee seuen dayes in the weeke, commonlie signified by seuen letters, A. B. C. D. E. F. G. This Circle therefore is made to containe twentie eight yeares, for foure times seuen doe make twentie eight, by helpe whereof is knowne the true order of the letters, whereof A. signifieth alwayes the first day of euerie yeare, and for euery leape yeare are appointed two Dominicall letters, whereof the one continueth from the beginning of that yeare vntill Saint Mathias euen, and the other from thence to that yeares ende, as you shall more plainlie perceiue by that which followeth. And first I will shew you how to finde out the iust number of the circle of the Sunne euery yeare, which is done thus.

Adde to the yeare of the Lord giuen or supposed 9. and diuide that product by 28. and the remainder shall be the number of

the

The Arte of Navigation.

the foresaide Circle: As for example, I would know the number of the saide Circle in the yeare 1590. whereunto by adding 9. I make the summe to bee 1599. which being diuided by 28. there remayneth 3. Nowe to finde out the Dominicall letter by the number of the sayd Circle, you must resorte to the figure following, consisting of three Circles, making two spaces, in the vpper space whereof are set downe the aforesaide seuen letters, and in the nether space the numbers of the Sunnes Circle, in such sorte as euerie number hath his proper Dominicall letter standing right ouer his heade, and for euerie leape yeare there are set down two Dominicall Letters. Al which letters are to be counted backward, & not forward as they are placed in the Almanacks. As for example, hauing found the number of the Sunnes circle for the yeare 1590. to bée 3. in the lower space, you shall finde the letter D. standing ouer his head, which is the Dominicall letter for that yeare. And note that when 28. is the number of the Sunnes circle A. is alwayes the Dominicall Letter, from which number you must beginne to count againe at 1. and so procéd backward to 28.

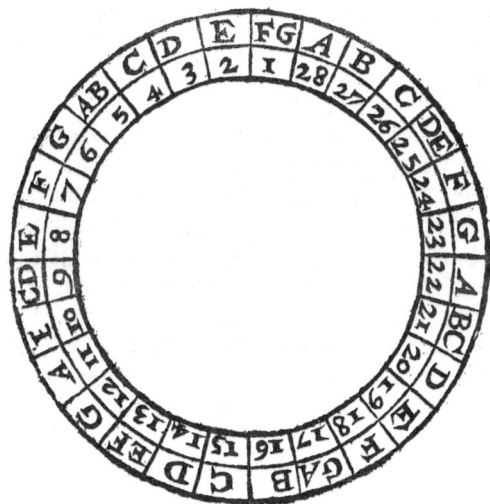

How

The Arte of Navigation.

How to find out the number of Indiction.

Chapter. 9.

This number consisteth of 15. yeares, and is commonly set down in al the Charters of the Bishops of Rome, and in the instruments and writings of their pronotaries, and therefore is called Indictio Romana, wherein they séeme to followe the auncient Romanes, which vsed the like indiction of yeares, but to other purposes, whereof I make mention in my Spheare: The way to find out this number euerie yeare is thus, Adde to the yeare of the Lorde giuen 3. and deuide the product thereof by 15. and the remainder shall be the number of the said indiction. But you haue to vnderstand that this Indiction is to be counted from September, and not from March as is the Epact.

But the surest, and therewith the most generall Kalender to serue in al places is the Ephemerides or dayly Almanack and especially that of Iohannes Stadius, which wil serue for 14. years yet to come: And before those yeares be all expired, I doubt not but that the like will be set forth by some one or other of our learned Astronomers, amongst whome for many years past the Germanes haue bene most famous. I chose the Ephemerides of Stadius because he is more portable, and of lesse price than that great Ephemerides of Leovitius. The chiefest vses whereof, and most méete for Mariners, though I haue already set downe in the latter end of my Treatise of the Globes, to the intent you might by the helpe thereof find out in the Globe the true place of the Moone and of euery one of the other 5. Planets that hath both longitude and latitude, yet I thinke it good once againe here in this place, most properlie requiring the same, briefly to shewe the vse of the saide Ephemerides, and speciallie of the diurnall Table which beginneth at the yeare of our Lord 1583. endeth with the yeare 1606.

A briefe

The Arte of Navigation.   311

A briefe description togetherwith the vse of the diurnall *Table* or *Almanacke* of *Iohannes Stadius*.

Chapter. 10.

THis Table beginneth at the 202. page of his booke called the Ephemerides or dayly Almanacke. And euerie page of the said Table that is on the left hand is diuided into 9. collums. In the first collum whereof on the left hand are set down the daies of the moneth. First, the Gregorian dayes according to the Romane account, and next to that the dayes of the moneth according to our English account, and then in the front of euerie other collum are set downe the characters, first of the Sunne, and then of the other sixe Planets, that is to say, of the Moone, Saturne, Iupiter, Mars, Venus, Mercurie, and last of all, the head of the Dragon figured thus ☊. And right vnder these seuen Planets, and also vnder the heade of the Dragon are set downe the signes and degrees wherein euerie of these is euery day of the moneth throughout the yeare at noonetyde. And in the foote of the saide Table is set downe the latitude of euery one of the fiue planets, proceeding by the dayes of the moneth diuided into three parts. And in the margent of euerie left page are set downe the chiefest feasts and Saints dayes that fall in euery moneth throughout the yeare.

And moreouer, in the Table on the right hande right against the left Table are set downe first the dayes of the moneth, and then what coniunction or any other aspect the Moone hath with anie of the other sixe Planets, that is, with the Sunne, with Saturne, with Iupiter, with Mars, with Venus, and with Mercurie, which Planets are set downe in the front of the said Table, and vnder them the Characters of such aspectes as the Moone hath that day with any of the other Planets. The characters of which aspects are these here following.

$$\;\;\;\;\; ☌ \;\; 8 \;\; \triangle \;\; \square \;\; *$$

Whereof the first signifieth a Coniunction, the second an opposition, the third a trine aspect, the fourth a quadrat aspect, and the fift a sextile aspect.

Two

The Arte of Navigation.

Two Planets are saide to bee in a coniunction when they are both in one selfe signe.

And to be in an opposition when they are in two seueral signes distant one from another, that is to say, 6 signes distant one from another.

And to be a trine aspect when they are distant one from another by foure signes:

And to be in a quadrat aspect when they are distant one from another by three signes.

And to be in a sextile aspect when they are distant but 2 signes one from another.

How to find out the place of any planet by the Ephemerides.

Chapter. 11.

Now to find out the place of any Planet, or of the head of the Dragon by this diurnall Table, you must first seeke out the day of the moneth in the first Collum of the left Table, and right against that on your right hand in the said left Table, you shall find in the common angle right vnder the Planet or Dragons head, which soeuer of them you seeke, the signe and degree wherein the saide Planet or Dragons head is the said day at noontyde.

And to find out the aspectes which the Moone hath with anie of the Planets the same day, you must resort to the other Table on the right hand, obseruing like order as before.

The Example.

As for example, the 21 of Aprill 1592 which is the first of May according to the Romane accompt, I find by the Table on the left hand, the Sunne to be in the 10. degree, 15. minutes of *T*aurus, the Moone to be in the third degree 47 minutes of *C*apricorne, Saturne to be in the 8. degree 10 minutes of *C*ancer, Iupiter to be in the 18. degree 28 minutes of *S*agittarius, Mars to be in the 12 deg. 6 minutes of *G*emini, Venus to be in the 2 degree 0 minutes of *A*ries, and *M*ercurie to be in the 6. degree 20 minutes of *T*aurus, and the head of the Dragon to be in the 29 degree 45 minutes of *G*emini. And right against this in the Table on the right hand, you shall find the Moone to bee in a trine aspect with the Sun, to be in an opposition with Saturne, to be in a trine aspect with Mercurie. Thus

The Arte of Navigation.

Thus much touching general Kalenders or Almanacks. Now as for particular Kalenders I take that which Robert Norman hath set downe in his new Attractiue to be the fittest for our countrie men, which containeth many necessarie things, as the contents thereof together with the Table next following the same doth shew, which Kalender I think it superfluous to be set down againe here, and the rather for that I wish euerie Mariner not to be without that booke, and especially contayning besides the Kalender many other good precepts touching nauigation, to which booke is also ioyned a verie learned and Mathematicall discourse touching the variation of the Mariners Compasse made by Maister William Borough controller of her Maiesties Nauie, who in mine opinion is one of the skifullest men in the Art of nauigation that is in this Realme.

Of the Mariners Ring or Astrolabe, and of his crosse staffe.

Chapter. 12

Now according to the order set down in the beginning of this Treatise, I must needs speak somewhat of the common Mariners Ring or Astrolabe, and of his crosse staffe which serue not to so many purposes as that of Stofflerus or of M. Blagraue before described, but onlie to take the altitude of the Sunne, or of any Starre or Planet, neither are so many conclusions to bee wrought by their common crosse staffe as by that of Gemma Frisius, or by that which Maister Hood hath latelie inuented. Sith our Mariners for the most part doe vse their common crosse staffe to none other ende but eyther to take the altitude of the Sun or of any fixed star or planet, or els to take the distance betwixt two stars, the making of which staffe is plainly set downe by Martin Cortes in his arte of Nauigation, and also the making of their Astrolabe.

But Cogniet and Wagoner do set downe a new kind of Crosse staffe, hauing 3. transoms or crosses euery one longer & shorter than another by the one halfe, affirming that so many conclusions may be wrought therby as by that of Gēma Frisius, but in mine opiniō they are not to be cōpared in al respects to that of Gemma Frisius which though by reason that the yard therof is of so great a length

as

The Arte of Navigation.

as it is not maniable in a shippe, yet vpon the land it is most seruiceable, nor yet to that of Maister Hoods inuention, which is most maniable, and therwith verie light of carriage, by whose staffe I thinke verilie that as many things may be wrought as by anie other kind of crosse staffe whatsoeuer, and with lesse trouble, yea, and in matters of Astronomie more trulie by reason that the yard and transome be of one selfe length, and thereby the degrees bee larger throughout the whole quadrant than they be in the common crosse staues, for in them the degrees from 50. vpwards to 90. are verie small and haue narrower spaces than M. Hoods staffe hath. Againe, whereas in vsing the Mariners crosse staffe in such latitude, as the Sunnes beames be of great force, they are faine to haue glasses made of purpose to saue their sight, and in some places all too little. But in vsing M. Hoods staffe they shall not need to behold the Sunne it selfe at all, but onely to marke vpon what degree of the yard the shadow of the Uane streketh. Moreouer, when the Sunne or starre is 50. or 60. degrees high, they are fain to vse their Astrolabe and not their staffe, which Astrolabe in mine opinion, as I haue said before, is the best instrument of all others to take the altitude of the Sunne in the day, or of any starre in the night, and because I haue here commended vnto you M. Hoods staffe, I will first set downe a plaine description thereof together with those few Astronomicall vses which do chieflie belong to the Mariner, without committing anie offence I hope to the author thereof, and then I will describe vnto you the Mariners ring and his common staffe together with the vses of the same.

But as I was about to describe vnto you M. Hoods staffe, a friende of mine comming in the meane time desired mee that I woulde first set downe the making and vse of the crosse staffe with three transomes, which Wagoner and Michaell Cogniet doe so much commend, and as I heare, is vsed of manie sea men in these daies, whose request I could not well denie, and therfore loe here followeth both the making and vse thereof.

The Arte of Nauigation.   313

Of the croſſe ſtaffe hauing three Tranſames or Curſours, commonly vſed in theſe dayes, the vſe and making whereof doth hereafter followe according to the deſcription of *Wagoner* and *Michaell Cogniet*.

First prepare a right ſtaffe of firme woode that is ſquare euery way, bearing in thickneſs three quarters of an inch, and in length three or foure foote, for which you muſt alſo prepare three Tranſames or Curſours euery one ſhorter then another by the one halfe, for the longeſt would containe in length twelue inches, the ſeconde ſixe inches, and the ſhorteſt or leaſt three inches. And euery one of theſe Tranſames or Curſours muſt be cut with a ſquare hole in the very middeſt, ſo as they may be made to runne iuſt vpon the ſtaffe to and fro, theſe things being prepared, you muſt deuide three ſides of the ſtaffe into certaine degrees to ſerue the three ſeuerall curſours as followeth.

First you muſt vpon a ſmooth ſquare Table ſomewhat longer euery way then the ſtaffe, and for want of one ſuch Table you may ioyne two Tables that may ſtand eauen together, drawe by helpe of a true long ſquire a right Triangle marked with the letters A. B. C. and let A. be the right Angle, and the Centre as you ſee in the figure following, and let both the ſides of the ſaide Triangle be of like length to your ſtaffe, then putting the one foote of your compaſſe in the Centre A. and the other foote in B. or C. drawe a quarter of a circle from C. to B. and hauing deuided that quarter into two equall partes, make a pricke in the middeſt marked with the letter D. and laying your ruler to that pricke and to the Centre A. drawe a right line which ſhall be A. D. then deuide the halfe Quadrant D. B. into nine equall partes making 90. equall degrees proceeding from B. to D. ſo as the firſt degree may be at B. and the 90. at D. that done, take with your compaſſe the iuſt halfe of the longeſt Tranſame, and keeping your compaſſe at that wideneſſe, ſet the one foote in the Centre A. and the other foote in the line A. C. ſo farre as that wideneſſe will extende, and there make a pricke marked with the letter E. then from that point drawe a right line that may be a Parallell

The Arte of Nauigation.

rallell to the line A. B. which line shall be E. F. then take with your compasse the halfe of the middle Transame, and kéeping your compasse at that widenesse, set the one foote in A. and the other foote in the line A. C. so farre as that widenesse doth extend and there make a pricke marked with the letter G. and from that pricke drawe a right line that may be a Parallell to A. B which line shall be G. H. Thirdly take the one halfe of the least Transame with your compasse and transferre that widenesse to the line A. C. setting the one foote of your compasse in A. and make a pricke with the other foote in the line A. C. as before, marked with the letter I. and from that pricke drawe another right line that may be a Parallell to the line A. B. which shall be the line I. K. Now to graduate the first side of your staffe to serue the longest Transame, you must lay the ruler to the centre A. and drawe right lines from thence to euery degrée of the circumference contained betwixt 90. and 30. and those lines shall deuide the line E. F. into so many vnequall spaces as doe belong to the first side of the staffe, for you must lay the first side of the staffe to that line to be marked according to the diuision of that line, the neather section whereof towardes the lower ende of the staffe must be marked with 90. and the vpper section with 30. so is the first side of your staffe truely deuided to serue the longest Transame, now to serue the middle Transame you must deuide the line G. H. by drawing right lines from the Centre A. to euerie degrée of the circumference contained betwixt 30. and 10. which lines will deuide the line G. H. into vnequall spaces, of which spaces the lowest must be marked with 30. and the highest with tenne, according to which diuisions you must marke the seconde side of your staffe by laying the side close to the line G. H. so shall that side be marked to serue the middle Transame, then lay your ruler againe to the Centre A. and drawe right lines to euery degrée of the circumference contained betwixt 10. and the first or second degrée next to B. and those lines shall deuide the line I. K. into vnequall spaces, the lowest whereof is to be marked with 10. and the highest with 2. or 5. according as your instrument will beare, and according to those sections you must deuide the thirde side of your staffe by laying the same close to the saide line I. K. and remember to make your lines of diuision

The Arte of Nauigation. 314

tion so finely as is possible, so shall your staffe be the more truely graduated. And note that in graduating your staffe, you shall not néede to drawe right lines from the Centre to euery degrée of the circumference contayned betwixt the 90. degrée and the 60. degrée, for when the sunne or any starre is higher then 60. degrées, you must vse your Astrolabe and not your staffe which will not serue you to looke so right vp to take so great an altitude eyther of sunne or starre.

A figure of the foresaide Triangle.

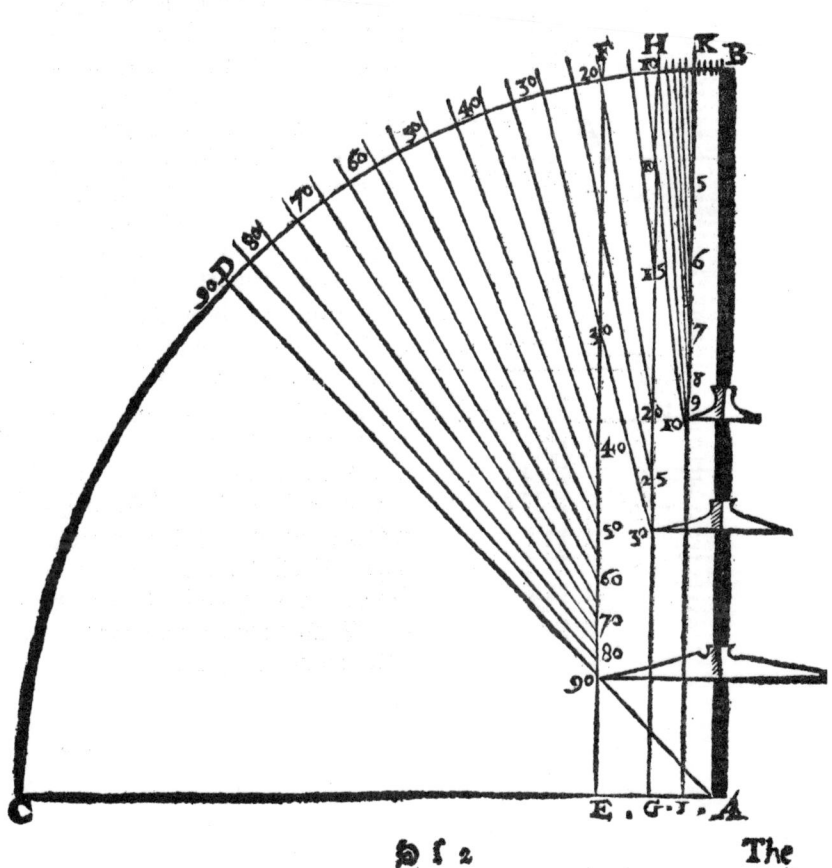

The Arte of Nauigation.

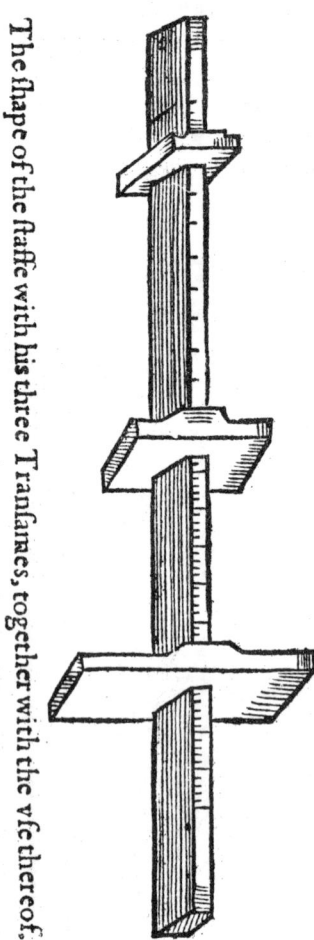

The shape of the staffe with his three Transames, together with the vse thereof.

Whensoeuer you would take the Altitude of the Sunne or of anie starre, you haue first to consider whether the sunne or starre be 30. degrées or more high: for then you must place the longest Transame vppon your Staffe. And set the lower end of your staffe marked with ninetie to your eye, which is alwayes to be done howe high or lowe so euer the sunne or starre be, and you must moue the Transame eyther forwarde or backewarde vntill you may sée by the vpper ende of the Transame the bodie or middest of the sunne or starre, and with the neather ende of the Transame the Horizon, and then looke in what degrée the Transame cutteth the staffe, for that is the Altitude of the Sunne and starre at that present, but if the sunne be not 30. degrées high, then you must put on the middle Transame, and if hée bée lesse then tenne degrées high, you must put on the shortest Transame, and then do as before. Thus much touching the Crosse staffe with 3. Transames, nowe I will describe vnto you Maister Hood his staffe, and shew you the vse thereof.

A briefe

The Arte of Nauigation.

A briefe defcription of Maifter *Hoode* his croffe ftaffe, and of all the partes thereof.

Chap. 13.

The figure or fhape of euery part of the faid ftaffe.

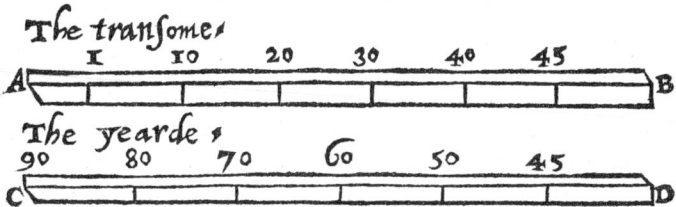

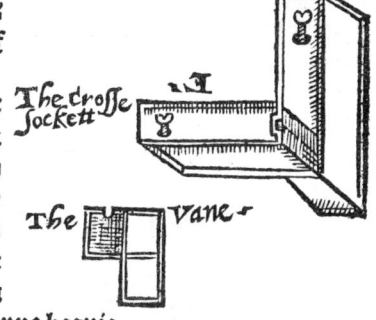

THis inftrument as you fee confifteth of foure parts.

Firft of 2. fquare rulers ech one bearing in thickneffe three quarters of an inch or there abouts euerye way fquare, and in length I would wifh ech ruler not to containe aboue one yarde, for then in fome vfes they would waxe toppe heauie.

Of which two rulers the one marked with the letters A. B. is called the Tranfame, which is deuided on the one fide into 45. degrees beginning at one and fo foorth to 45. And euery degree is deuided againe into fixe leffer parts making 60. minutes, for fixe times ten maketh 60. which ferueth for Aftronomicall vfes.

And on the oppofite fide to that, the faid Tranfame is deuided into 1000. equal parts beginning at 25. and fo increafeth by 25. vntill you come to 1000. Euerie which 25. partes is deuided into fiue leffer partes, and euery one of thofe againe into fiue partes, which maketh in all 25. for fiue times fiue is 25 and this ferueth for Geometricall vfes.

The Arte of Nauigation.

The other square ruler called the yarde & marked with the letters C. D. is deuided into degrees like to those of the Transame procæding from 45. to 90. which together with those degrees that be in the first side of the Transame doe make vppe the number of a iust Quadrant to serue Astronomicall vses. And the side opposite to that is deuided into 1000. degrees procæding from 25. to 25. like in all respectes vnto the opposite side of the Transame to serue Geometricall vses.

The third part belonging to this instrument is a double socket of brasse marked with this letter E. ioyned together with right Angles, and standing crosse one to another. And ech socket hath a skrewe to kæpe the same fast to his staffe at any degræ that you list to set the same: And also a long notch to the intent it may be layde close to any degræ of the ruler whereto it belongeth.

The fourth part belonging to this instrument is a vane of brasse marked with the letter F. the vpper edge whereof is pearced with a little round hole for the beame of the sunne to passe thorough the same. And this vane is made with a socket, and with a skrew to hold it fast to that ruler whereon it is set. And for certaine Geometricall vses it were necessarie (as some thinke) that there were two such vanes.

Thus hauing described all the partes of the foresaide instrument, I wil now shew you how to set those parts together: which to serue Astronomicall vses is thus done.

How to set the partes of Maister *Hoodes* staffe together to serue such Astronomicall vses as doe chiefely belong to the Mariner.

Chap. 14.

First put the neather ende of the Transame marked with the 45. degræ into the double socket, so as the Transame may stand right vppe, and that the notch of the socket may mæte euen with the sayde 45. degræ, and turne the skrewe that the socket may stand fast at that degræ, that done put the yarde into the crosse socket, so as the notch of the sayd crosse socket may lie iust vppon the 45. degræ of the yarde, and there make

The Arte of Nauigation. 316

make it fast by turning the skrewe: then put on the vane, which to serue Astronomicall vses must be set at the highest ende of the Transame, in such sort as the vpper edge of the vane pearced with the hole may stande eauen with the first point or strœke of the saide Transame, and when the Transame standeth on the left hande of the yarde, then the vane must be placed on the right side of the Transame: But if the Transame doe stande on the right hande of the yard, then the vane must be placed vpon the left side of the Transame: And the yard and Transame would be so set together as the degrees of the Transame procœding downeward from one to 45. may looke towards you, that is to say, may stand right before your face: And the degrees of the yarde procœding from 45. to 90. would lie vpwarde, so as the ende of the yarde marked with 90. may point to your breast or be set to your eye as occasion shall require, for such Astronomicall vses as heere doe followe.

The shape or figure of the foresaide staffe, hauing all his partes set together to serue for Astronomicall vses.

Chap. 15.

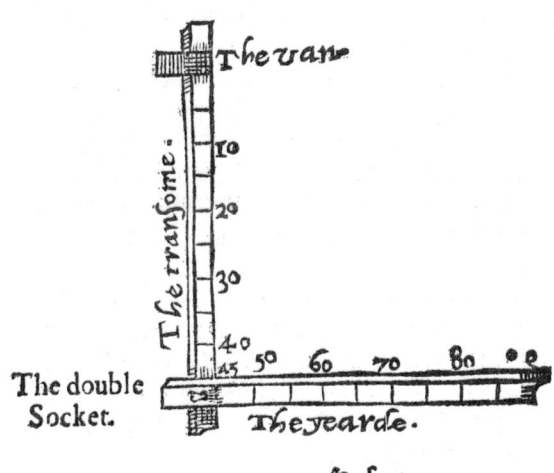

Ss 4

The Arte of Nauigation.

How to take the altitude of the sunne at any houre that he is to be seene with the eye by M. Hoods staffe.

Chap. 16.

The Staffe being set in such order as is before taught, goe into some open place whereas you may see the sunne, and turning the ende of the yarde marked with ninetie towardes your brest, holde the yarde so leuell as you can, that it may be a iust Parallell to the Horizon, and turne both your face, and also the vane of the transame towardes the sunne. Then you haue to consider whether the sunne at that present be eyther iust 45. degrees high, or more, or els lesse then 45. degrees, which you shall easily knowe thus: For if he be iust 45. degrees high, then the shadowe of the vpper edge of the vane will stréeke iust vpon the 90. degré of the yarde lying at your brest: But if he be more then 45. degrees high, then the shadowe of the vane will stréeke short of the 90. degré: And if it be lesse it will cast no shadowe at all vppon the yarde, but stréeke cleane beyond it ouer your shoulder. Nowe knowing by this meanes whether the altitude of the sunne be more or lesse then 45. degrees, you shall take his true altitude thus:

If his altitude be lesse then 45. degrees, then make the Transame to sinke downe through his socket vntill the vpper edge of the vane standing vppon the Transame doe caste his shadowe iust vppon the nintieth degré of the yarde. Then looke at what degré the notch of the Socket wherein the Transome standeth, doth cut the saide Transame, for that is the true altitude of the sunne at that instant.

But if the altitude of the sunne be more then 45. degrees, then drawe the double socket vpon the yarde nigher towardes your brest, vntill you sée the shadowe of the vane to fall iust vppon the nintieth degré of the yard, that done, looke vpon what degré of the yarde the notch of the double socket cutteth, and that is the true altitude of the sunne at that instant.

How

The Arte of Nauigation.

How to take the altitude of anie starre with Master *Hoodes* staffe.
Chap. 17.

THere is no difference betwixt the order of the taking the altitude of the sunne and of anie starre, but onely that in taking the altitude of any starre, you must set the ende of the yarde marked with the 90. degrée to your eye, staying the same vppon the vpper bone of your chéeke, and that in such sort as you may sée the starre by a right line passing from the ende of the yarde, lying at your eye to the vpper edge of the vane at one instant, being alwayes sure that the yarde lie leuell, making a iust Parallell to the Horizon, which Horizon is more easie to be séene vpon the sea, then vpon the lande, for on the sea there be neither hils nor trées to hinder the sight thereof.

How to take the distance betwixt two starres with Maister *Hoodes* staffe.
Chap. 18.

THe instrument being set in such order as was first taught: Set the ende of the yarde marked with 90. at your eye: and let the other end of the yarde directly point to one of those starres whose distance you séeke, and turne the edge of the vane to the other star be it on the right hande or on the left, according as you thinke good your selfe: Then you must consider whether the distance of the two starres be iust 45. degrées or more or else lesse, for if it be iust 45. degrées, then the outward edge of the vane, and the ende of the yarde marked with 90. will be answerable in your sight to that distance without any more adoo. But if the distance be lesse then 45. degrées, which you shall easily perceiue by your eye (for then the starre that is on your right or left hand, will appeare within the vane and fall short thereof) then you must thrust down the Transame in his socket vntill you may sée the two starres by a right line passing from the ende of the yarde marked with 90.

and

The Arte of Nauigation.

and lying at your eye, right foorth to the outwarde edge of the vane, that done, looke in what degrée the double socket cutteth the Transame, and that shall be the true distance of the two starres. But if the distance of the two starres be more then 45. degrées, which you may knowe also by your eye, because the starre which is on your left or right hande, will appeare cleane without the vane: then drawe the double socket together with the Transame towardes your eye, vntill you may sée the two starres by a right line as before, and the degrées which the notch of the double socket cutteth on the yarde, shall be the true distance of the two starres.

Thus much touching the Astronomicall vses of M. Hoodes staffe méete for the Mariner. Now as for setting of the said staffe to serue Geometricall vses, and to finde out the manifolde conclusions to be done thereby, I wholly referre you to M. Hoodes owne booke, thinking it good now to speake somewhat of the Mariners Astrolabe or Ringe, and of his vsuall Crosse-staffe,

Of the Mariners Astrolabe,

Chap. 19.

Michaell Cogniet setteth downe the making of a new kinde of Astrolabe, which he calleth an vniuersall Astrolabe, onely because the declination of the sunne euery day is added in manner of a Table to the one ende of the Diopter of the saide Astrolabe, which is of no great importance sith the sunnes declination doth alter in the space of 30. yeares and lesse, which being expired, the declination is to be new calculated: In the meane time that Table which Norman hath set downe in his Attractiue, and also that which is more lately calculated set downe by my selfe in the first part of my Sphære, and likewise in this Treatise whereas I treat of the motion of the sunne and of his declination, may serue these 20. yeares and more without any great error: Wherfore leauing to speake any further of Cogniet his vniuersall Astrolabe, I will describe here the common Astrolabe and Crosse-staffe vsed by most Mariners, the shape or figure of both which here foloweth.

The Arte of Nauigation.

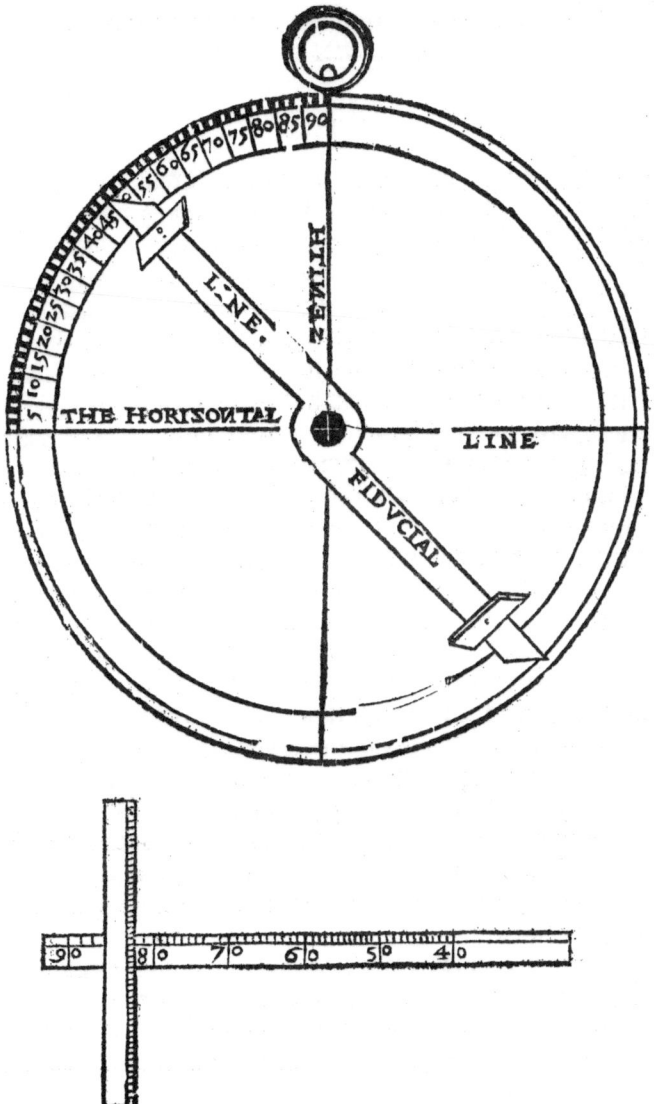

The Arte of Nauigation.

A briefe description of the Mariners Astrolabe, and the vses thereof.

Chap. 20.

He limbe of the Mariners Astrolabe is traced as you seē with three Circles, making two spaces to containe therein the degreēs and numbers of altitude: and these Circles are deuided by two crosse lines called Diameters, and cutting one another in the very Centre of the Astrolabe into 4. equall parts, of which quarters the vppermost on the left hand towards the Ringle is only marked with degreēs and numbers, as you may seē in the figure: In which figure the perpendicular Diameter signifieth the Zenith, or the line of South and North, at the vpper end whereof is fastened the Ringle or handle. And the other ouerthwart Diameter signifieth the Horizon the one end whereof on the left hande signifieth the East point, and the other end on the right hande the West point. And to this Astrolabe (as to all other) doth belong a ruler or Diopter, which as you seē hath at ech end a square tablet pearced with two holes, the one greater and the other lesser, the greater to looke through with your eye, to take the altitude of any starre or of the sunne being so darkned by some cloud as though it casteth no shadowe, yet it may be seēne with the eye: And the lesser hole is for the sunnes beame to passe thorough when he shineth cleare, and some Astrolabes are deuided in like maner of the Zenith on both sides, and haue two Diopters, whereof the one is pearced with a great hole, and the other with a smaller hole to serue to such purposes as is aboue saide.

The vse of this Astrolabe is onely to take the altitude of the sunne at any time of the day, or of any fixed starre or Planet in the night, in such sort as is before taught in my description of Maister Blagraue his Astrolabe, in the third and fourth Proposition thereof.

A briefe

The Arte of Nauigation.

A briefe description of the Mariners Crosse-staffe.
Chap. 21.

THis staffe consisteth of two foure square rulers of woode, whereof the one is called the yarde, and the other the Transame or Crosse: And the yard containing in length most commonly three quarters of a yarde is the longer péece, and is deuided into 90. vnequall degrées: for from 10. to 90. the degrées doe grow lesser and lesser, and the Transame containing for the most part but one third part of the yard, is cut with a square hole in the middest, so as it may runne vp and downe vpon the yard, & some Transames haue two running vanes to be set at what widenes you list, to take thereby the distance of any two starres.

The vses of the Mariners crosse-staffe.
Chap. 22.

THe vse of the Mariners staffe chiefly consisteth in two points, that is, to know therby the altitude of the sun in the day time, or of any starre in the night season, and the other is to know thereby the distance betwixt any two starres.

The first is thus done, hauing put the staffe or yarde through the square hole of the Transame, set the end of the yard which is marked with 90. to your eye, laying that end vpon the vpper point of your chéeke bone kéeping your legs close together, and hauing directed the other end of the yard towards the sunne or star whose altitude you séeke, moue the Transame to and fro or vp & downe, vntill you may sée with the one eye (winking with the other) the one end of the Transame to méete iust with the Centre or middest of the sunne or starre, and the other end to touch the Horizon both at one instant, and that degrée of the yard which the square edge of the Transame cutteth, wil shew how high the sunne or starre is at that present: And in taking the altitude of the sunne or of any star, remember alwaies to holde the yarde so leuell as it may be a iust Parallel to the Horizon, for so shall you take the altitude the more truely. Now as touching the second point, which is to knowe

the

The Arte of Nauigation.

the true distance betwixt any two starres you must doe thus.

Set the end of the yarde marked with ninetie to your eye as before, and moue the Transame to and fro vntill you may sée the one end of the Transame to answere the one starre, and the other end thereof to answere the other starre, for that degrée of the yard which the Transame cutteth will shewe the distance betwixt the two starres: And if the two starres be so nigh together, as they wil not conueniently answere the two ends of the Transame, then you may vse the two moueable vanes, which (as I saide before) you may set at what widenesse you list but yet so as they may both stand equally distant from the Centre of the Transame or from both ends thereof.

Thus hauing shewed the vse of the Mariners Ring and crosse staffe, I will now procéde to the other instruments, whereof the two Globes are part, which Globes I haue alreadie described and shewed the chiefest vses thereof in a Treatise by it selfe, and as for the vniuersall Horologe to know thereby the equall houre of the day in euery Latitude, and also the Nocturnlabe to knowe the houre of the night, I shall speake of the one when I come to speake of the sunne, & of the other whereas I treate of the North starre and of his guards, for there I will set downe the figure and shape of ech instrument, and shewe how to vse the same: But the two chiefest instruments belonging to a Mariner, are his compasse and his Carde, whereof I come now to speake.

But sith the Fly of the Compasse representeth the 32.windes, I thinke good first briefely to define what the wind is, and to shew howe many windes the auncient Mariners did vse, I say here briefely because I haue alreadie spoken sufficiently thereof in the latter end of my Spheare.

Of the winde, what it is, and of the diuers kindes and names thereof.

Chap. 23.

He winde according to Aristotle, is an exhalation hotte and dry, engendred in the bowels of the earth, and being gotten out is carried sidelong vpon the face of the earth. And of windes there be foure principall

The Arte of Nauigation.

cipall which doe take their names from the foure quarters of the earth from whence they blowe, that is North, South, East and West: And though the Grækes and Latines deuide euery quarter into three partes, and thereby make in all but twelue windes, whose names both Græke and Latine are set downe in the latter ende of my Spheare, yet our latter Sea men to be the most assured of their Rowtes and Courses, doe deuide euery quarter of the Horizon into eight windes, so as in all they make 32. windes, giuing them such names as are set downe in this figure heere following representing the Flye of the Mariners compasse.

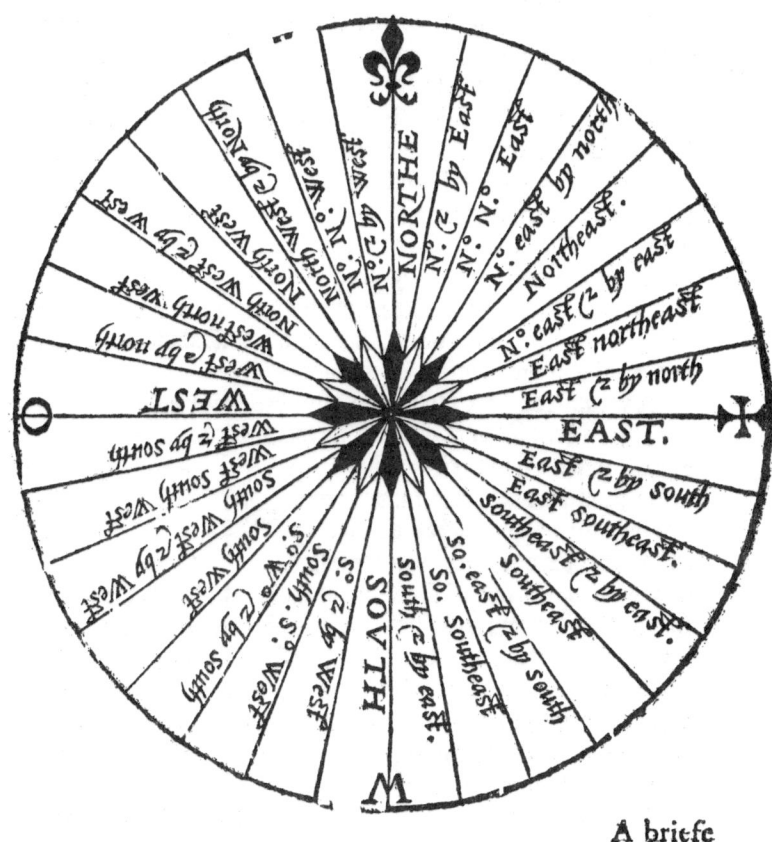

A briefe

The Arte of Nauigation.

A briefe description of the Mariners compasse, and the vse thereof.

Chap. 24.

The Mariners Compasse may be very well deuided into two essentiall partes, that is, the Fly, and the wyars touched with the Lodestarre called in Latine Magnes. And first you haue to vnderstand, that the flie is a round white Carde traced with 32. lines all passing thorough the Centre of the Circle, which lines doe signifie the 32. windes in such sort as the figure before set downe doth shew, of which lines that which is marked with the Flower-deluce signifieth the north, whose point opposite is the South, and that which is marked with a Crosse signifieth the East, whose point opposite is the West, and the outtermost Circle of the said Fly signifieth the Horizon, which circle is also deuided into 360. degrees like vnto those of the Equinoctiall, so as euery space betwixt point and point containeth 11. degrees and 15. minutes, which is the fourth part of a degree. Moreouer, this Circle is deuided in 24. houres by allowing to euery point three quarters of an houre, which is 45. minutes, for an houre containeth 60.ᵐ and halfe an houre containeth 30.ᵐ and the quarter 15.ᵐ. Moreouer, the common Mariners doe deuide euery point of the Compasse into foure quarters to make the more exact account of their Routes or Rombes. But now as touching the other essentiall part of the Mariners Compasse which is the wyars, you shall vnderstand that they are of yron or steele, and made in this forme, and being touched at the one end with the Lodestone, they are fastened to the backe side of the Fly, eyther right vnder the line marked with the Flowre-deluce, or else somewhat distant from the same, either towardes the East or else towardes the West, for such cause as shall be declared in the next Chapter, for those wyars being thus touched, doe make that part of the Fly that is marked with the Flowre-deluce, to stand alwayes towardes the North, and to that ende the Fly hauing in the Centre a latten socket, is put into a turned boxe, in the middest whereof is a

sharpe

The Arte of Navigation. 321

sharpe pointed Latten pinne, vpon which the Flie turneth about, and that turned boxe is couered with glasse, partlie to kéepe the Flie cleane, but chieflie that the winde should not mooue it to and fro. And this turned boxe is hanged by two rounde narrowe plates of Latten in another square boxe of thinne wainscot boord, so as it may alwayes hang leuell, howsoeuer the shippe swayeth or inclineth on either side, and though of the Mariners Compasse there be diuers vses, as you shall perceiue hereafter, yet the chiefest end and vse thereof is to shewe the North part of the world, whereby the Shipmaister knowing what course hée hath to holde and how the port or place to which hee goeth, beareth from the place from which he departeth, may by looking alwayes to his Compasse, know how to direct his ship accordinglie.

Of the Lodeſtone, and of the variation of the Compaſſe in Northeaſting and Northweſting.

Chapter. 25

The Lodestone or Adamant, called in Latine Magnes, & in Italian Calamita, hath two marueilous great and secret properteies or vertues, the one to drawe stéele or Iron vnto it, and the other to shew the North and South parte of the world, of which stones some bée of more force than others according to the place from whence they come. For those are counted best which are found in the East Indies vpon the coaste of China and Bengala, which is no shell but a whole stone of sanguine collour like to Iron, and is firme, massie, and heauie, and will drawe or lift vp the iust waight of it selfe in Iron or stéele. And as Norman saith in his newe Attractiue, such stones are commonlie solde for their waight in Siluer.

Next to this, Norman commendeth the redde stone of Arabia, hee commendeth also the stone of Almaine, which is in collour like to Iron, but spongeous, and thereby lighter than the other. And as for the blacke and white stone of Elba, which is an Ilande, not farre from Piombino, hée saith that the vertue thereof is but of small force, and of no long continuance.

T t But

The Arte of Navigation.

But the worste are those that come from Norway, whose collour is mixt with graie, I haue hearde that there bee of them here in England, but howe good I knowe not, it is maruell that Norman maketh no mention thereof, whose Treatise called the newe Attractiue, together with Maister Borough his addition I would with all that be studious in this Arte to reade most diligentlie, for truelie in mine opinion the secretes of this Stone and variation of the Compasse, was neuer better disciphered, nor by more experiences tried than it hath bene by these two men last named. The booke is most necessarie for nauigation and of an easie price, méete for euerie poore mans purse, though perhaps in some pointes not méete for euerie mans vnderstanding. Now whensoeuer the wyers are to be touched with this stone, they must be made very cleane and void of all rust, to the intent that the Iron may the more firmelie receiue the vertue of the stone. And it is well knowne by good experience, that by vertue of the Lodestone the North point of the Compasse declineth alwayes from the true North, either to the East, or West more or lesse according to the latitude of the place wherein you are vnlesse you be right vnder the Meridian of the Azores. And therefore, most men in these partes of the worlde doe vse to set the North point of the wiers not right vnder the Floure de Luce signifying the North part of the world, but rather somewhat inclining towarde the East halfe a point or thereabout to auoide the Northeasting and Northwesting of the Compasse, the cause of which declination by diuers learned men hath bene diuersly taught, but not rightlie as *M*ichaell Cogniet saith, vntill *M*ercator found out the true cause, who first learned by the experience of one Francis of Deepe, an excellent Pilot, that in sayling vnder the Meridian which passeth thorough the Iles called the Azores, the néedle doth decline neither East nor West, whereupon Mercator by calculating the variation of the Compasse at Ratisbone, found that the Pole of the Lodestone ought to be put in that Meridian which passeth through the foresaid Iles so as it may be distant from the Pole of the world sixtéene degrées and ½ or rather as Maister Borough saith, sixtéene degrées and twentie two minutes, and by that calculation Cogniet hath found the variation of the Compasse at the towne of Antwerp to be 9. degrées.

And

The Arte of Navigation.

And Maiſter Borough by the helpe of the newe inſtrument of variation firſt made by Robert Norman, and afterward perfected by himſelfe, hath found the variation for London to be 11. degrées and ſiftéene minutes, which is a whole point from North to Eaſt. But whenſoeuer you depart from the foreſaid Meridian of the Azores, bée it neuer ſo little, either towardes the Eaſt or Weſt, the néedle will varie and decline accordinglie. And his greateſt declination is when you come to a full quarter of that Parallell wherein you ſaile, for from thence it declineth leſſe and leſſe vntill you come againe vnder the foreſaid Meridian, which thing Cogniet doeth plainelie demonſtrate by this Figure here made of purpoſe.

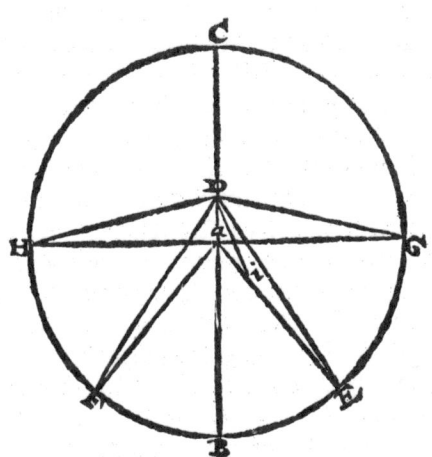

In which Figure the letter A. which is in the verie center of the Circle repreſenteth the north Pole, and the right line, marked with the letters C. B. ſignifieth the Meridian which paſſeth through the Iles Azores, in which line is a point or pricke marked with the letter D. ſignifying the Pole of the Lodeſtone, diſtant from the North Pole ſixtéene degrées and a halfe. Nowe, if your ſhippe bee in the point B. then your needle declineth on neither ſide, but pointeth right to the North

Tt 2   Pole,

The Arte of Navigation.

Pole, and also to the Pole of the Lodestone. But if you sayle Eastwarde and arriue to the point E. then the right line of the North is E. A. but your needle declineth on the right hande towardes his owne Pole D. which is to the East, so much as the angle A. E. D. doth shewe. Likewise if you saile from B. Westwarde to the point F. the right line of the North shall be F. A. but your needle will decline to his owne Pole D. towardes the West, so much as the angle A. F. D. doth shew. And to bee short, the needle doeth neuer shewe the right North, but onelie in C. or B. to eyther of which pointes the nigher that you approch the lesse your needle declineth, and the more that you goe from anie of these two pointes the more your needle declineth either East or West, but the greatest declination thereof is in H. or G. for then you are a iust quarter from the foresaide Meridian. Also by this Figure he plainelie sheweth that of two sundrie townes hauing one selfe Meridian, that which is nærest to the Pole of the worlde hath greatest declination. As for example, suppose the letter I. in the foresaide Figure to signifie that towne which is nigher to the Pole, and the letter E. to signifie that towne which is more distant from the Pole, both townes bæing vnder one selfe Meridian E. A. Here you sæ that the Needle being in E. doth shewe the Pole of the Lodestone by the line E. D. and bæing in I. it sheweth the Pole of the Lodestone by the line I. D. Nowe according to the doctrine of Euclyde, the angle A. I. D. is greater than the angle *A*. E. D. whereby it followeth that the needle declineth more in I. than in E. But whereas Mercator affirmeth that there should bæ a mine or great rocke of Adamant, whereunto all other lesser rockes or needles touched with the Lodestone doe incline as to their chiefe fountaine, that opinion sæmeth to mæ verie strange, for truelie I rather belæue with Robert Norman, that the properties of the stone, as well in drawing stæle, as in shewing the north Pole, are secrete vertues giuen of God to that stone for mans necessarie vse and behoofe, of which secrete vertues no man is able to shewe the true cause.

How

The Arte of Navigation.

How to find out the variation of the compasse in every latitude.

Chapter. 26.

WIlliam Boorne in his Regiment of the Sea teacheth diuers wayes how to finde out the variation of the Compasse, as well by the Sun in the day time, as by the North starre in the night season. First, thus mark at what point of the Compasse the Sunne riseth and setteth, for if hæ riseth at the East point of the Compasse, and goeth down at the West Northwest point, then the Compasse is varied one whole point, that is to say, the North point of the Compasse standeth North and by East. But because the aire is seldom cleare at the rising and setting of the Sunne, you may do thus, take with your Astrolabe or crosse staffe the altitude of the Sun in the forenoone, the sooner the better, at some iust point of the compasse, & take again his altitude in the afternoone, when he is in like degræ of altitude, and mark therwith at what point of the Compasse the Sunne hath such altitude, and by the difference thereof you shall know the variation of the Compasse. As for example, you finde by your Astrolabe or crosse staffe, that the Sunne is 20. degræs high at the Southeast point of the Compasse, and obseruing the same againe in the afternoone you finde his height to bée 20. degræs at the West Southwest point of the Compas, wherby you sée that the compasse is varied one whole point, that is to say, that the North point standeth North and by East, and the South point is South and by West. He teacheth it also another way, which is thus.

Take with your crosse staffe or Astrolabe the height of the Sun at noontyde, which is called the Meridian altitude of the Sun, and thereby you shall haue the true Meridian of the place where you are, with which Meridian if the South point of your Compasse doth agré, then your Compas hath no variation at al, but if the south point therof do swarue or incline on either side from the said Meridian, marke how much it differeth, and that difference will shewe you the variation. Nowe to knowe the variation of the Compasse by the North starre, doe thus, set your Compasse with the North starre, and if you finde them to agré, then there is no

Tt 3   varia-

The Arte of Navigation.

variation, but note that this is to be done when the two guardes or pointers of Charles waine are right ouer or right vnder the North starre, for if these two starres be West from the North star, then the North star is a third part of a point vnto the East-ward of the North Pole, and if the said two starres be right East from the North starre, then the North starre is a thirde part of a point vnto the Westward of the north Pole.

The compasse (as Bourne saith) doth varie most in sayling long voyages East and West, and though it varieth two or three points, yet you may know what course to holde without alteration of the Wiers any maner of way. As for example, suppose the Northeast point to stand right North, and your course is to goe right West, here in this case you may vse the Southwest point in stead of the West point, whereby you may perceiue that it maketh no great matter which point standeth due North, so as you take that point of the Compasse for the North which directly pointeth to the North. But Robert Norman and Maister Borough by the help of their new inuented Instrument of variation, doe shewe how to finde out the variation of the Compasse much more exactlie than euer it hath bene heretofore taught, which instrument together with the booke, I would wish all sea men to haue, and therby to learne the perfect vse of the instrument, which vse that booke teacheth both plainlie and learnedlie, in which booke is also proued by diuers demonstrations that there is no such attractiue point as some haue dreamed, but rather a respectiue point whereunto the needle of your Compasse wil alwayes turne in what part of the world soeuer you saile. But of the place where that respectiue point should be, diuers learned Pilottes haue had diuers opinions, for some haue imagined it to be in the heauens, and some aboue the heauens, if it were in the heauens, then the needle would dayly turne about, and alter according to the motion of that heauen, wherein the said point is, which is nothing so. And to bee aboue the heauens, it is contrarie to the old rule of Philosophie, which saith that Extra Cœlum non est locus, and therefore Norman and Maister Borough haue great reason to say that it is in the bodie of the earth beneath the Horizon, for they haue tried by diuers liuely experiments that the North parte of the needle of his owne accorde and nature would alwaies

decline

The Arte of Navigation.

decline downwarde if it be not otherwise counterpoised or letted, & by their demonstrations supposing the Meridian of the Azores to be the first or common Meridian, and also knowing the altitude of the Pole at London to be 51. degrées, 32. minutes, they find that according to that latitude the Pole of the Lodestone, béeing in the upper face of the earth, and right vnder the foresaid common Meridian is 25. degrées 44. minutes distant from the Pole of the worlde. And the point respectiue to bée distant in a right line from vnder the Horizon of London 71. degrées, 50 minutes and the variation of the Compasse to be as hath bene said before, 11. degrées and 15. minutes, but to find out the true place of the respectiue point in euerie latitude, they say that no certaine rules can as yet be set downe, by reason that the compasse doeth vary more sodenly in one place than in another, for in some place it will varie more in sailing 200. leagues than in another place in sayling 400. leagues. Againe, it will sometimes be retrograde, for M. Borough in sayling betwixt the North Cape and Vaigates, towards the Northeast, and looking by his computation that the néedle should haue increased his variation towards the East, hée found that it was sodainly turned backward towarde the West, notwithstanding both Norman and M. Borough doe affirme that though the Compasse hath in seuerall Horizons seuerall variations, yet in any one Horizon the néedle alwaies respecteth one onely point without alteration, and that in his declining it kéepeth the like order and that certainlie in euery place. And although the néedle of the Compasse by reason of the waightines of the Flie cannot decline downward according to his own propertie, but onlie sheweth the point respectiue alwayes vpon the Horizon, which indéed, as they say, is most necessarie for nauigation, yet by such meanes and conclusions as are set down in the foresaid book, a diligent Pilot hauing with him a perfect stone may by exact obseruations find out the increasing or decreasing of the declnation of the néedle, which declination you shall find, as they thinke to bée more or lesse according as the point respectiue is more or lesse distant from the place whereas the triall is made, which being diligently obserued in sundrie places, with the certaine variation of the néedle from the Meridian, the true place of the point respectiue may be found out as they thinke.

Tt 4   Of

The Arte of Navigation.

Of the Mariners Card, and of the making thereof.

Chapter. 27.

The Mariners Carde, which some cal a nautical Planisphear, is none other thing, but a description made in plano vpon paper or parchment of the places that be in the sea, or on the land next adioyning to the sea, as points, Capes, Bayes, Portes, Floods, Ilandes, Rockes, Sandes, and such like. And such Cardes are either vniuersall or particular, the vniuersall Cards are those wherin are described the most parts of the world, such as is the vniuersall Carde or Map of Mercator, or of Plancius. The particular are those wherin some speciall partes of the sea and land are described. And both these kindes of Cardes are traced with certaine lines, whereof some are called Meridians, some Parallels, and some the Lines of the Mariners compasse, shewing the 32. winds before described in the beginning of this Treatise. The order of making which cardes in times past was woont to be thus.

First drawe with a paire of compasses a secrete circle that may be put out, so great as you shall thinke meet for your carde, which circle shall signifie the Horizon, then diuide that circle into foure equall quarters, by drawing two Diameters crossing one another, in the center of the foresaide circle with right angles, whereof the perpendicular line is the line of North and South, and the other crossing the same is the line of East and West, at the foure ends of which crosse Diameters you must set downe the foure principall windes, that is, East, West, North and South, marking the North part with a flower deluce in the toppe, and the East part with a crosse, as you may see in the figure following. Then diuide euerie quarter of the saide circle with your compasses into two equall partes, setting downe prickes in the middest of euerie quarter, through which prickes, and also through the center of the circle, draw two other crosse lines, which must extende somewhat beyond the circumference of the Horizon, which two crosse lines together with the first two crosse lines shall diuide the circle into 8. partes, and thereby you shall haue

the

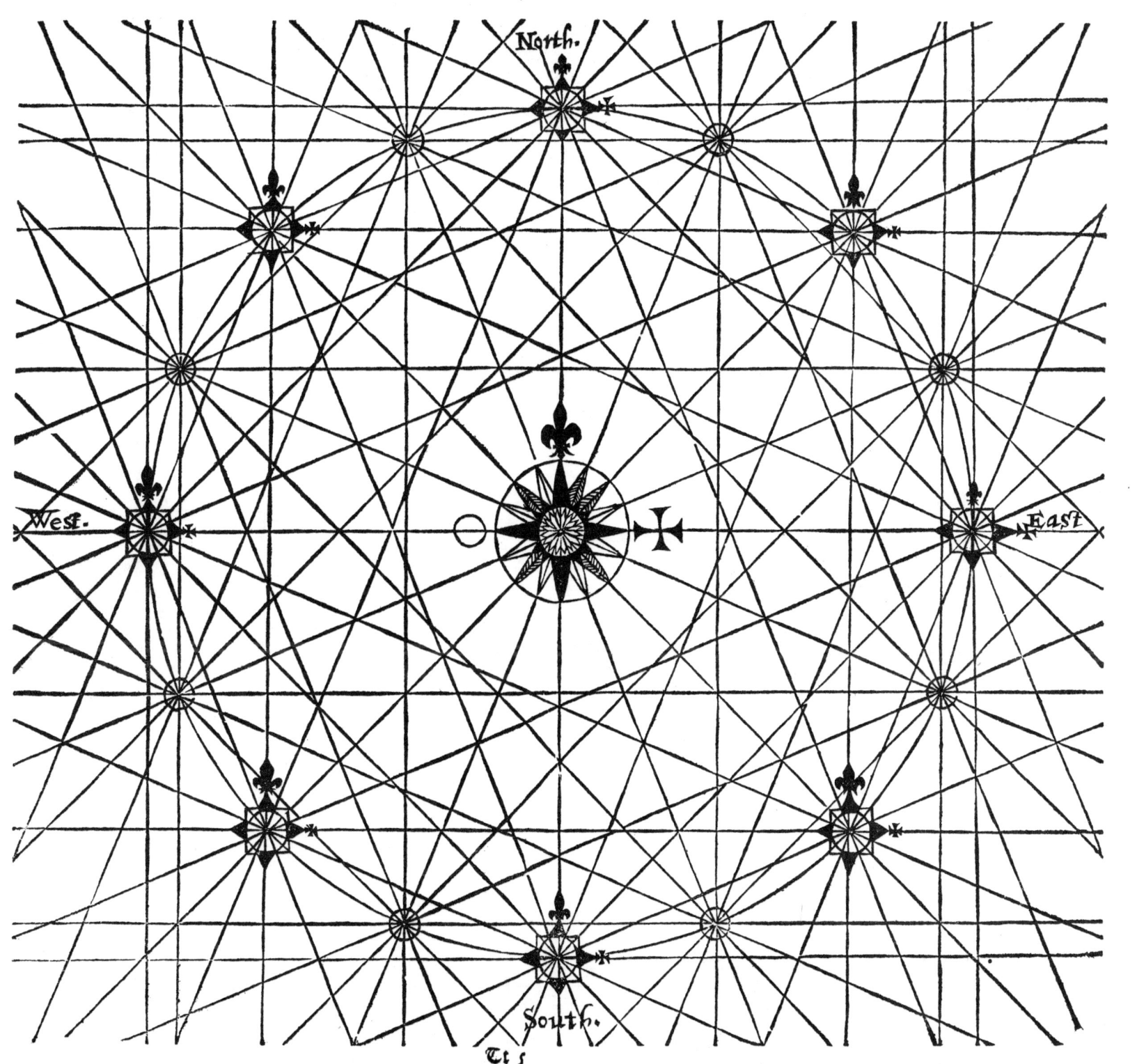

The Arte of Navigation.

the eight principall Windes. That done, diuide euerie eight parte of the saide Horizon into two equall partes by drawing other two crosse lines through the center, and extending somewhat beyond the circumference of the Horizon as before, wherby the whole circle shall be diuided into 16. partes, which shall suffice without making anie more diuisions, which woulde cause a confusion of lines, & at the end of euery one of these 16. lines you must drawe a little circle, whose center must stande vpon the circumference of the Horizon, euerie one whereof must bee also diuided into 16. parts by the helpe of 16. lines, diuersly drawne from the center of one little circle to another, in such order as the figure here placed more plainely sheweth to the eie, than I can expresse the same by mouth. And these little circles do signifie 16 litle Mariners compasses, the lines wherof signifying the winds, do shew how one place beareth from another, and by what winde the shippe hath to saile. But besides those little circles there is woont to be drawne also another circle somewhat greater than the rest vpon the verie center of the Horizon, which Circle by reason of the 16. lines that were first drawne passing through the same, is diuided into 16. parts, and the Mariners doe call this circle the mother compas.

Here place the Mariners Carde.

The shape and figure of the first liniaments of the Mariners Carde drawne after the olde maner, and how to set downe the places of the land or sea therein.

Chapter. 28.

Ow the true setting downe of the places in the Mariners carde, as Points, Capes, Bayes, Floods, Ilands, and such like is to be done by knowing what latitude and longitude euery place hath, which is to be learned either by modern tables made of purpose: for the ancient Tables make no mention of any longitudes or latitudes of such places as are in the new found land, which land to them was neuer knowne, or els by such Mariners cardes already made

made as doe shew the true longitudes and latitudes of those places which you woulde describe in your Carde, of which longitudes and latitudes, and especially the longitudes of the places in the West Indies, few or none are as yet truely set down,

Moreouer, to knowe the distance of places, (that is, how many leagues or miles one place is distant from another there is woont to be set downe in the Mariners Carde a scale, otherwise called by the Mariners a Tronke, the making wherof is plainly taught by Martin Cortes in his Arte of Nauigation, in the second chapter of his third booke, and also how to graduate the Cardes, to shew what latitude euerie place hath, and there also hée teacheth how to translate one Carde into another, and howe to reduce a greater Carde into a lesser, and contrariwise. To which booke I referre you, and the rather for that it is in English, translated many yeares since out of Spanish by M. Richard Eden.

But for so much as the sea and the earth doe make together one whole rounde body, the lines of the 32. rombes in the Carde being drawne right, and made to signifie great circles, can neuer shew the true course that the ship hath to hold, which Michaell Cogniet proueth by a Figure demonstratiue, and thereof gathereth thrée conclusions. First, that a man may sayle right north and South round about the worlde, if the Sea in all that course be nauigable, and so returne againe to the Porte from whence he first departed. Secondlie, that making the Equinoctiall his Parallell hé may sayle East and West round about the worlde, and so returne to the porte from whence hee departed, Also if hé sayle East and West in any other Parallell that is distant from the Equinoctiall, hee may returne by the same Parallell to the porte from whence hé first departed, and yet not about the whole worlde, for that cannot be done, but onelie when the Equinoctiall is his Parallel, Thirdlie, that whosoeuer sayleth by any other rombe than by one of the foure principall, hé by often changing his Meridian and Horizon must needs saile by a line Spirall, which is neither perfectlie right, nor perfectlie rounde, and thereby hé may well approch the Pole, and also goe rounde about it, but yet with vnequall distance, so as he shall bee nigher beyonde it than on this side, by meanes whereof hee cannot returne to the place fromwhence hé came,

as

The Arte of Navigation. 326

as you may plainelie perceiue by this Figure demonſtratiue here placed.

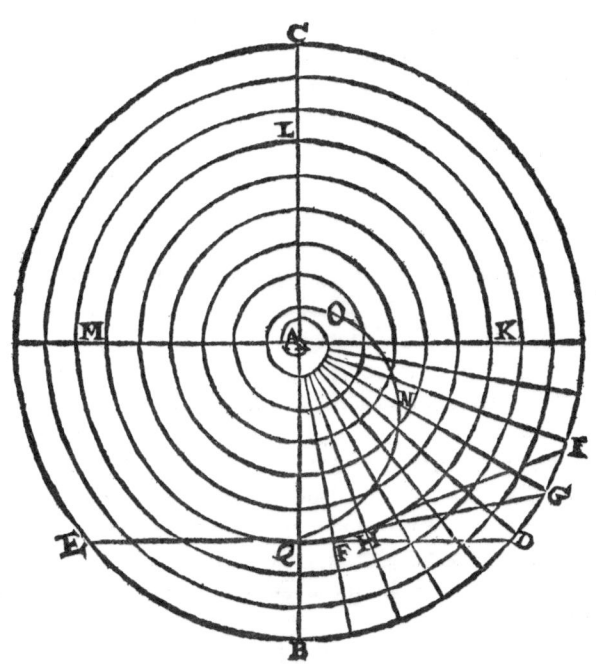

In which Figure the letter A. doeth ſignifie the North Pole, and the letters B. C. the Meridian paſſing through the Pole A. then ſuppoſe your ſhip to beé in Q. whereas the Pole is eleuated 30. degreés and Q. to be your Zenith, and the right line G. Q. D to bee your right line of Eaſt and Weſt cutting the foreſaide Meridian with right angles, and let D. be the Eaſt point, and E. the Weſt point.

Nowe you may ſayle from Q. towardes the North with a South winde, and from A. you may ſaile againe Southwarde with a North winde vntill you come to the South Pole, and from thence you may ſayle againe Northwarde with a South wind

The Arte of Navigation.

wind vntill you come againe to Q from whence you first departed, and so you shall haue gone rounde about the worlde vnder one selfe Meridian B. A, C. But in sayling East or West, you shall continuallie change your Meridian, and thereby change your East point, for in sayling from Q. towardes D. you come immediatelie to the Meridian A. F. whose right East point is G. and in sayling further Eastward, you come to the Meridian A, H. whose right East point is I. and so foorth from one Meridian to another, notwithstanding in keeping still in one selfe parallell marked with K. L. M. you may saile rounde about the Pole A. which is the center, and so come againe to the Port Q. from whence you first departed, but not about the whole worlde: for that you cannot doe vnlesse the Equinoctiall were your Parallell, as hath bene saide before, but if you sayle by any other rombe than by one of the foure principall, that is, East, West, North, or South, your course shall neither be by right line, nor yet by true circle, but by a spirall line, which is partly right, and partlie round, so as you cannot with like course returne to the place from which you departed, by reason that you change so often both Meridian and Horizon, as by the Spirall line Q. N. O. drawne in this Figure you may easilie perceiue, which approcheth nigher to the Pole beyond it than on this side. Moreouer the sea men by making the Meridians and parallels in their cardes all of equall distance, they make some countries far greater than they should be by the one halfe. Also by that meanes he that sayleth East and West round about the Pole in the Parallell whose latitude is 60. degrees, should make as long a voyage as hee that sayleth East and West alongst the Equinoctiall, the voyage whereof is twise as long as the other, for the redresse and remedie of which faultes, Cogniet hopeth to finde out some more perfect rule of making cardes when opportunitie of time shall serue: in the meane time to reforme the saide faults, Mercator hath in his vniuersal carde or Mappe made the spaces of the Parallels of latitude to bee wider euerie one than other from the Equinoctiall towards either of the Poles, by what rule I knowe not, vnlesse it be by such a Table, as my friende M. Wright of Caius colledge in Cambridge at my request sent me (I thanke him) not long since for that purpose, which Table with his consent

The Arte of Navigation.

sent, I haue here plainlie set downe together with the vse there-of as followeth.

The Table followeth on the other side of the leafe.

A Table

The Arte of Navigation.

A *T*able to drawe thereby the parallelles in the Mariners Carde together with the vſe thereof in trewer ſort than they haue bene drawne heretofore, and the vſe thereof.

Chapter. 29.

Degrees of the Meridian, beginning from the Equinoctial circle.	Equall partes of the Meridian in the Mariners Card, of which partes euery degree of the Equinoctial contayneth 60 miles.	Degrees of the Merridian beginning at the Equinoctiall circle.	Equall parts of the Meridian in the Mariners Card, of which partes euery degree of the æquinoctial containeth 60. miles.
1	60	27	1684
2	120	28	1752
3	180	29	1820
4	240	30	1889
5	300	31	1959
6	361	32	2029
7	421	33	2100
8	482	34	2173
9	542	35	2245
10	603	36	2319
11	664	37	2394
12	725	38	2470
13	787	39	3546
14	849	40	2624
15	911	41	2703
16	973	42	2783
17	1035	43	2865
18	1098	44	2948
19	1162	45	3032
20	1225	46	3118
21	1289	47	3205
22	1354	48	3294
23	1419	49	3385
24	1484	50	3477
25	1551	51	3572
26	1617	52	3668

Degrees

The Arte of Navigation.

Degrees of the Meridian. &c.	Equall partes of the Meridian. &c.	Degrees of the Meridian. &c.	Equall partes of the Meridian &c.
53	3767	67	5482
54	3868	68	5639
55	3972	69	5804
56	4078	70	5976
57	4187	71	6156
58	4299	72	6346
59	4414	73	6547
60	4532	74	6759
61	4655	75	6985
62	4781	76	6226
63	4911	77	7484
64	5046	78	7764
65	5189	79	8067
66	5331	80	8399

The vse of this Table for making the sea Carde is thus. Ouerthwart the middest of the plaine superficies wherein you would drawe the lineamentes of the Carde, describe a right line marked with the letters A. B. C. whereof B. is the verie middest or center, and this line representeth the Equinoctiall, the ende whereof on the right hand, marked with the letter C. signifieth the East, and the other end on the left hand marked with the letter A. signifieth the West, which Equinoctiall line must bee diuided into 36. degrees, then crosse the same squirewise with perpendicular lines, passing through euerie tenth or fift degree, as you see in the example following. Then take with your compasses the length of halfe the Equinoctiall, that is, 180. degrees, and set one foote of your Compasses in the mutuall intersection of the Equinoctiall, and of that perpendicular or Meridian which passeth through the East ende of the Equinoctiall line, marked with the letter C. and with the other foote make a pricke in the same perpendicular or Meridian, and marke that pricke with the letter D. that done, diuide the space contayned betwixt this pricke and the E-

qui-

The Arte of Navigation.

quinoctiall firste into thrée equall partes, and euerie one of those into other thrée equall partes, so haue you 9. partes. And againe, euerie one of those into thrée, so haue you twentie seuen partes, and diuide euerie one of those partes into foure partes, so shall you haue a hundred and eight partes, and if there bee space inough, diuide again euerie one of those into ten, so shal you haue a 1080. partes, and if it be possible, diuide againe euerie one of those partes into other tenne, so shall you haue in all 10/800. partes, but this can hardlie be done, vnlesse the Carde be verie large, wherein euerie degree of the Equinoctiall is neere an inch long, which happeneth verie seldome, and therefore 1080. partes shall suffice.

And for the easier numbring of these partes, set to them Arithmeticall figures with blacke leade, which may afterwarde be put out when your worke is done, beginning at the Equinoctiall and so procéde from thence both Northward and Southwarde, then looke what number standeth right against euerie degrée in the former Table, which degrées doe extende from 1. to fourescore degrées, and omitting alwaies the first figure on the right hande of that number which you finde, for so you must alwaies doe when your diuision containeth no more but 1080. partes, count that number which remayneth vpon the line of diuision, and there make one pricke, and make another pricke of the same distance from the Equinoctiall vpon the outermost Meridian on the left hand, through which two prickes drawe a right line, and that shall be your first Parallell of latitude, and so procéde with all the rest first Northwarde, and then Southward, if you mind to make an vniuersall Carde. But the example following containeth no more but the one halfe of an vniuersall Carde procéding from the Equinoctiall line Northwarde vnto the 80. degrées of latitude: for further Northwarde than 8½. degrées is no land as yet discouered or knowne, nor yet all that, so farre as euer I could learne.

The

The Arte of Nauigation.

IN this last figure or Table is first drawn (as you see) the Equinoctiall line marked with the letters A. B. C. and that line is deuided into 360. degrees, and therein also are drawne perpendicular lines, as well thorough the beginning and ending of the sayde Equinoctiall lyne, as also through euery tenth degree thereof, which be the Meridians, and are euery where equidistant each one from other, then take halfe the length of the Equinoctiall which is A. B. or B. C. with your Compasses and setting one foote in the end of the Equinoctiall marked with C. make with the other foote a pricke at D. in the Meridian or perpendicular line marked with the letters C. D. E. then deuide the space contained betwixt C. & D. into 1080. parts in such sort as before hath beene shewed, and set the figures vnto them as here you see, to the intent that you may the more readily number the parts. Then looke in the first Table what number answereth to euery 10. degree of the Equinoctiall, & casting away the first figure of that number on the right hand, find out the parts answerable to the number remaining in the line C. D. and at those parts set prickes in both the outermost Meridians through which prickes you shall drawe the Parallels. As for example in the first Table you see that the number right against 10. degrees is 60. (the first figure 3. towardes the right hande being reiected) therefore looke 60. in the line C. D. and by that part draw the first Parallel distant 10. degrees from the Equinoctiall. And after this maner all the rest of the Parallels are to be drawne.

 Many doe vse to paint the vacant places in their Cardes with ouer many flags, and the compasses thereof with diuerse and superfluous colours which William Borne misliketh, wishing that in stead thereof they would shewe by letters or other Characters what moone doth make a full sea, in such places as are necessary to be knowne, and also to drawe the true shape and fashion of euery Cape or headlande that is needefull about the coast, and at what point of the compasse the land riseth of this or that fashion, for being neare the land it wil seeme to be of one fashion, & being far off to be of an other fashion, and to mistake any place on the sea is very dangerous to the Mariner. But aboue al things let him that saileth by Card & Compasse be sure that the needle of his compasse haue the like declinatiõ that his needle had which made the Card.

V v   For

The Arte of Nauigation.

For Cogniet reporteth that certaine Mariners being in the west Indies, and seeing the North starre to be Northeast, maruailed thereat very much not knowing the cause of that error, which indeed was for that the needle of their compasse was made to decline Northeast, wheras it should rather haue declined Northwest, for the West Indies do stand Westernly from the Azores. It is necessarie also to set downe in the Card such places as are daungerous, as sands, flats, or shoulds, rockes, and such like things as are not alwayes to be seene with the eye, to the intent that the Mariner being aduertised thereof may shunne the same. All which things before mentioned, Wagoner hath in all places contayned in his booke very well obserued. Thus hauing spoken sufficiently of making the Mariners Carde, Now I thinke it good to shewe the most necessary and chiefest vses thereof.

The chiefest vses of the Mariners Carde.
Chap. 31.

The chiefest vses of the sea Card are these foure here following:

The first is to know thereby how that place whereunto you would sayle, beareth from the place or Port from whence you set off, or depart. And that is to be knowne by the lines of the Mariners Compasses painted in the Card in this manner following: Take a paire of Compasses and hauing opened them, set the one foote thereof in the very place from whence you depart, and the other foote in the next line of that Compasse which is nearest vnto your place of departing, I meane such line as doth most rightly direct you to the place to which you would go, and your Compasse being opened at that fit widenesse to serue that line, draw them from the place of your departing vnto the place whereto you would go, suffering that foote of the Compasse which standeth vpon the line of the winde whilest you draw it forward, not to swerue one iot from that line, and that line will eyther rightly direct you to the place assigned or fall short thereof, or else ouerreach the same, if it fall short, then take another line nearer to the place from which you departed, and if it ouerreach, take some line that is further off from the place of your departing

ting, and hauing found a line that pointeth directly to the place, consider what winde or rombe it is, for by that winde the place assigned beareth from you, and the rombe or winde opposite to that is the winde whereby you haue to sayle.

The second vse is to knowe by the Carde how farre the place whereto you go is distant from the place of your departing, which is done by helpe of the skale or trunke set downe in the Carde thus: Take the iust distance betwixt the two places with your compasses by setting the one foote in the one place, and the other foote in the other place, and apply that widenesse of the compasses to the skale or trunke, and the trunke wil shew how many leagues the one place is distant from the other, and if the distance betwixt the places be longer then the trunke, then take first the length of the trunke with your Compasses, and looke how many times that is contained in the space betwixt the two places, and if there doe remaine any odde measure, then hauing taken that odde measure with your compasses by setting thē at such widenes as is answerable to that odde measure, apply that widenesse to the first part of the trunke, so shall you know the iust measure of the whole. And this rule serueth to take the true distance of any other two places whatsoeuer set downe in the Card.

The third vse is to know by the Card what Latitude or altitude of the Pole any place set downe in the Carde hath, which is done by helpe of the line of degrées of Latitude, otherwise called the Graduation of the Carde in this manner following. Set the one foote of your compasses in the very place whereof you would know the Latitude, and the other foote in the line of East & West, which is next vnto that place, and kéeping that foote still vpon that line, draw your compasses forward vntill you come to the line of degrées, and marke what degrée of the said line the foot of the compasse which was first set in the place doth crosse or touch, for that is the degrée of Latitude for that place, numbring from the lowest degrée of Graduation vpwarde, so shall you finde the Latitude of Lisbone in Portingale by Mercator his vniuersall Carde, and by the Carde set downe in Martin Cortes booke, and also by Medina his Carde drawne in his booke of Nauigation to be 38. degrées 30. minutes and somewhat more. But by the Tables of Ptolomie you shall finde it to be 40. degrées 24. minutes, and

The Arte of Nauigation.

by Appian his Tables to be 39. degrées and 38. minutes.

The fourth vse chanceth when you are driuen out of your right course by stormes or tempest, which stormes how to foresée and to prognosticate, is plainely taught by Martin Cortes in the 19. chap. of his second booke. Also you may be driuen by force of contrary windes, by surging of the sea, or by ouerthwart tides, currants, & such like impediments, so as you can not lay your course right to the place assigned, for remedy whereof you must séeke in what place you are, & to note the same in your Card which as the Mariners terme it, is to make a pricke in the Carde, which to doe truely in time of néede many things are to be known & well obserued and kept in memorie. First to know what latitude the place from whence you first departed hath, then to kéepe in minde what way your shippe did make good at euery shift of winde, that is to say, how many leagues, and in how long time you sayled by euery seuerall winde: and then not knowing well where you are, nor how farre you are distant from the place whereto you would go, learne to knowe by helpe of your Astrolabe or crosse staffe, in such sort as is before taught, the altitude of the Pole in that place where you are, which if you finde to be all one with the Latitude of the place of your departure, then you may assure your selfe that you haue sayled by the line of East and West without altering your Latitude, but if you finde the Latitude of the place where you are, to be more or lesse then the Latitude of the place from whence you departed, then resort to your Card, and take two payre of Compasses opened at such widenes as the one foote of the one Compasse may stand in the place from which you departed, and the other foote of the same Compasse to stand in the line of the rombe whereby you sayled: and let the one foote of the other Compasse stand in that degrée of Latitude which you last found, and the other foote of the same Compasse in the next line of East & West, and holding the Compasses so ordered in ech hande one paire: draw them both so as they may méete together, taking good héede in drawing them, that the foote of that compasse which was placed in the line of the winde, may at no time swarue from that line, nor the one foote of the other Compasse to swarue from the line of East and West, wherein it was first placed, and whereas the two féete of those Compasses doe méete,

that

The Arte of Nauigation.

that is to say, that foote of the one compasse which was drawne from the place of your departing doe méete with that foote of the other compasse which came from the degrée of Latitude last found, where these two féete (I say) doe méete, there make a pricke or marke in your Carde, for that is the place where your shippe is at that instant: And from thence you must take your right course againe to the place whereunto you would go. But because it is necessarie aswell at this time as at all other times, to knowe what way your shippe hath made, and that the same is not in mine opinion, so plainely nor so commodiously taught by any one that I haue read, as by Michaell Cogniet, I minde therfore in the two chapters next following to set downe his way not only how to find out the way of your shippe when you saile South and North vnder one selfe Meridian, or in any other place where you are to change in your gate the latitude or altitude of the Pole, but also how to finde out the way of your shippe in sayling right East and West, without changing the altitude of the Pole, which way as he sayth, was neuer heretofore knowne to any Pylot but to him selfe first author and inuentor thereof.

How to know the way of your ship, and how many leagues are to be counted for one degree of Latitude in euerie Rombe whereby you sayle.

Chap. 32.

First you haue to vnderstande that in saylinge iust North and South, you do alwayes abide vnder one selfe great Circle called the Meridian, vnder the which, when you haue sayled so farre as the altitude of the Pole is changed one degrée, then haue you gone 17. Spanish leagues, and a halfe, and you haue to note that euery Spanish league containeth 2857. fathams, and that our English league containeth no more but 2500. fathams, so as the Spanish league is more then our English league by 357. fathams, & euery fatham containeth vi. foote. Againe in sayling East and West, you do alwayes remaine vnder one selfe Parallell, by meanes whereof the altitude of the Pole

The Arte of Nauigation.

Pole doth neuer alter, and therefore no true account can be made of the leagues, but by such meanes as Cogniet teacheth in the Chapter following. But if in sayling North and South, you decline one rombe either towards the East or West, and go so farre as the altitude of the Pole is changed by one degrée, then you haue made somewhat more then 17. Spanish leagues and a halfe, and to be short, the more rombes that you decline towardes the East or West, the more leagues in number doe belong to one degrée of altitude, as you may plainely sée by this figure demonstratiue here following in which the letter A. signifieth the place or point in the Card, from which you depart situated in the Parallel A. B.

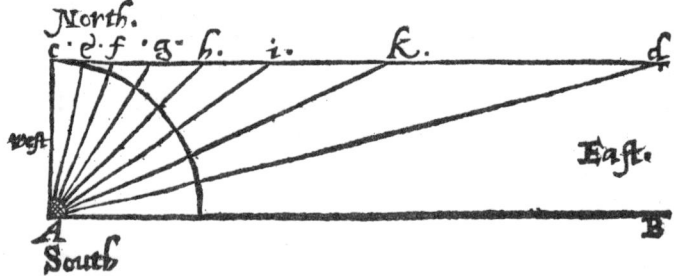

Then suppose C. D. to be another Parallell equally distant from A. B. by one degrée of altitude. Now if you saile right East or West, then you shall alwayes remaine in the Parallell A. B. equally distant from the Pole. But if you saile from A. right North so farre as the altitude of the Pole is augmented one degrée, then your ship shal be in C. and if you sayle by the first rombe towardes the East so farre as the Pole doth altar in altitude one degrée, then your shippe shall be in E. and thereby your way must néedes be longer, and so consequently the more rombes you decline from North to East, the longer is your way, and the more leagues must be accounted to one degrée of altitude, as the lines drawne from A. to E. F. G. H. I. K. and D. do shew. And what so euer is saide here of the quarter from North to East, the same is to be vnderstode in all the other thrée quarters, that is, from North to West, from South to East, and from South to West. But if you will knowe how many leagues doe belong to

euery

The Arte of Nauigation.

euery degrée according to the rombe whereby you sayle, then consider well this Table here following.

The first Table.

For in sayling from North or South towardes East or West so farre as you come to change one degrée of altitude of the Pole, the saide degrée doth require for the first rombe, &c.

Rombes.	Leagues.
1	17 $\frac{5}{6}$
2	18 $\frac{14}{15}$
3	21 $\frac{1}{20}$
4	24 $\frac{1}{4}$
5	31 $\frac{1}{2}$
6	45 $\frac{3}{4}$
7	89 $\frac{2}{3}$

Now then to know how much way you haue made in sayling, you must first know aswell the Latitude of the place from which you departed as of that place whereunto you be arriued: then by the foresaid Table séeke to knowe how many leagues doe belong to a degrée of that rombe whereby you haue sayled, for in multiplying the number of the leagues by the degrées of the difference of the two Latitudes, the product thereof will shew you how many leagues you haue sayled, notwithstanding sith the way may be made longer or shorter by changing or shifting of the winde, it is néedefull that the Pylot haue consideration thereof, who by skilfull coniecture must sometime eyther adde to, or take fro according as néede shall require. Moreouer by the foresaid figure marked with letters, you may also easily vnderstande how much you change in Longitude, that is to say, how much you are distant from the Meridian of that place from whence you departed, be it eyther towardes the East or West, for he that sayleth from the point A. as is aforesaid right North or South, he remaineth alwayes vnder one selfe Meridian: but he that sayleth by the first rombe towards the East or West so farre as he changeth one degrée of Altitude, and arriueth to the point E. is now distant from his first Meridian so much as is the space betwixt C. and E. which we finde by computation to be thrée leagues and a halfe for one degrée of Latitude, which amounteth to twelue minutes of a degrée, and so of the rest of the rombes, as appeareth by this Table here following.

The Arte of Nauigation.
The second Table.

For in sayling from North or Southe towardes East or West so farre as you change one degrée of altitude of the Pole, you change also your Meridian, and thereby your Longitude, the quantitie whereof answerable to euery rombe, is set downe on the right side of this Table.

Rombes	leagues	Degrees and minutes of Longitude	
first	3 ½	0	12
second	7 ¼	0	25
third	11 ⅔	0	40
fourth	17 ½	1	0
fift	26 ⅕	1	30
sixt	42 ¼	2	25
seuenth	88 0	5	2

And to make this more plaine by example suppose that you sayle from Lisbone, which is a famous port in Portingale, by the winde Southwest and by west, which is the fift rombe from south to West, so farre as you finde the altitude of the Pole to be 18. degrées lesse then at Lisbone. Now if you would know how many leagues you haue sayled, and also how much the Meridian of that place is more westward then the Meridian of Lisbone, then doe thus. Looke in the first Table, and you shall finde that to one degrée of the fift Rombe doe belong 31. leagues and a halfe, which leagues being multiplyed by 18. doe make in all 582. leagues and ¼. which you haue sayled. Then looke in the second Table, and you shall finde for the fifth rombe one degrée and 30. minutes of Longitude, which being multiplyed by 13. and a halfe, do make 27. degrées and ¾. of a degrée, for by so much is Lisbone more Eastward, then the place where you are. And whereas the first Table is made according to the proportion of right lines such as are commonly drawne in Mariners Cardes, Cogniet maketh another Table according to the proportion of circular lines, which for that it differeth very little or nothing from the first Table, I omit here to set it downe. But now because the first Table doth chiefely serue those that sayle either East or West in any Parallell betwixt the Equinocttall and the 60. degrée of Latitude or Altitude of the Pole: And that from thence forth by reason that they sayle by more oblique and spirall Circles doe make the longer voyage, Cogniet thought good to adde a thirde Table shewing how many leagues be answerable to one degrée of altitude to those that sayle eyther East or West in any Parallell

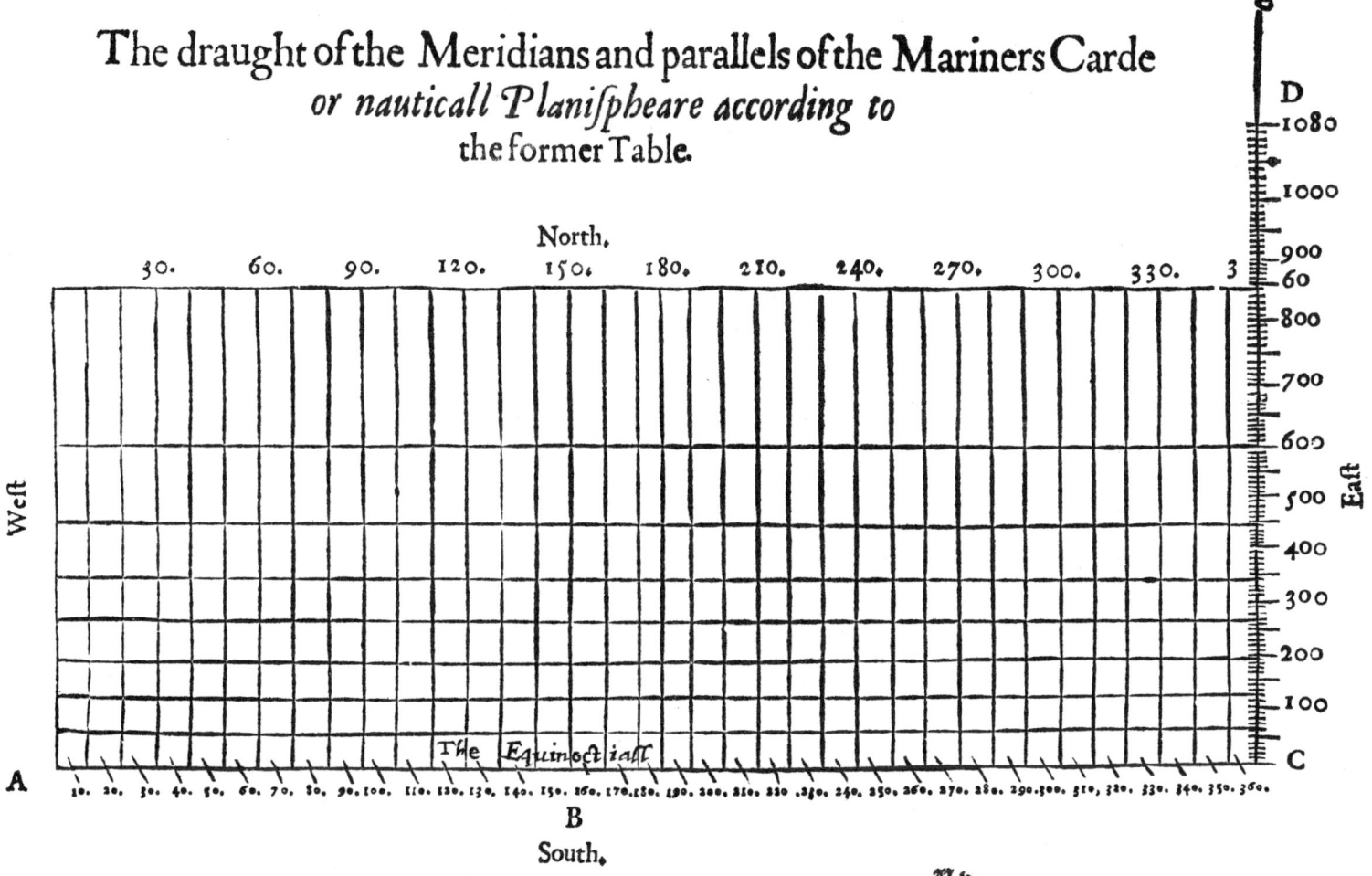

rallel that is betwixt the Pole and the 60. degrée of altitude, which Table differeth not much from the others in the foure first rombs, but in the thrée last, that is in the 5. 6. and seuenth rombe it differeth greatly, and most in the seuenth as you may easily perceiue by comparing this and the first Table together.

The third Table.

	Deg	M		Leagues	
The first rombe hath	1	1	Which according to the proportion of 17. leagues & a halfe, for one degree do make for euerye rombe so many leagues as this Table sheweth.	17	49/14
the second rombe	1	5		18	23/14
the third rombe	1	12		21	0
the fourth rombe	1	26		25	1/12
the fift rombe	1	50		32	1/12
the sixt rombe	2	42		47	1/4
the seuenth rombe	5	44		100	1/3

Howe to account the leagues in sayling directly East or West without changing the latitude or altitude of the pole.

Chap. 33.

IN sayling right East or West, you continue still in one selfe Parallel without making any charge of Latitude of the Pole. Most men therefore thinke it impossible for you to make any true account of the leagues but onely by coniecture, for remedie whereof Cogniet hath inuented a rule most certaine, as he saith, the foundation whereof is thus: First you must suppose the Meridian of that place from whence you depart to be a firme and fixed point, and to be the very beginning of that Parallell wherein you saple, then you must knowe what houre it is aswell at the place from whence you first departed, as also at the place whereunto you are arriued; and hauing the difference of the houres, you must knowe howe many leagues euery houre ualdeth, according to the Parallell of that altitude of the

Pole

The Arte of Nauigation.

Pole vnder which you sayle, so shall you easily knowe how manie leagues you haue sayled. This foundation being first laide, Cogniet setteth downe his generall rule in this manner following: When soeuer you haue to sayle (sayth he) right East or West, you must first prouide your selfe of these two thinges, the one is of an Astronomicall ring that may iustly shew the houre at anie time: In steede whereof I thinke it better to haue an vniuersall Dyall, such a one as is described by William Borne in the one and twentieth Chapter of his Regiment: the other thing is to haue a true houre glasse that will run continually foure and twentie houres, and therefore had neede to be three times more long and large then common houre glasses be, whereof the glasse makers can quickly prouide you, and because the ship leaneth sometime on the one side, and sometime on the other, it shall be needefull to hange the saide houre-glasse with rings of brasse or latten, to the intent that it may alwayes hange leuell like as the Mariners Compasses are wont to be hanged in their boxes. Now being thus prouided of these two instruments, you must at your departing prepare your houre glasse, that is to say, you must set it a running iust at noone when the sunne is right South, not forgetting to turne it in your voyage once a day in the very instant that it is readie to runne out. Then hauing sayled certaine daies, and being arriued at the place where you would knowe what way you haue made, looke to your houre-glasse, and tarry vntill it be full runne out, at which instant seeke to know what houre it is by your Astronomicall ringe, or rather by the vniuersall Dyall, for if you sayle East, you shall find it to be past noone, but if you sayle West, then it will want of noone, and keepe those houres in minde that you may know how much it is more or lesse then noone, for those houres by shewing the difference of the houres that are betwixt the Meridian of the place from whence you departed, and the Meridian of the place whereto you are arriued, will certifie you with the helpe of this Table here following, what quantitie of way your shippe hath made.

A table

A Table to help you to know what way your ship hath made in sayling right East or West without changing your Latitude, together with a briefe description and vse thereof.

De. of Latit.	leagues.	De. of Latit.	leagues.	De. of Latit.	leagues.	De. of Latit.	leagues.
0	262 1/2	23	241 1/2	46	182 1/4	69	94 1/3
1	262 5/12	24	239 3/4	47	179	70	89 5/6
2	262 1/4	25	238	48	175 1/2	71	85 1/2
3	262 1/8	26	236	49	172	72	81
4	262	27	234	50	168 1/2	73	76 2/3
5	261 1/2	28	231 7/8	51	165	74	72 1/3
6	261	29	229 1/2	52	161 1/2	75	68
7	260 1/2	30	227 1/4	53	157	76	63 1/2
8	259 7/8	31	225	54	154 1/4	77	59
9	259	32	222 1/2	55	150 1/2	78	54 1/2
10	258 1/2	33	220 1/4	56	146 2/3	79	50
11	257 1/2	34	217 1/2	57	143	80	45 1/2
12	256 2/3	35	215	58	139	81	41
13	255 3/4	36	212 1/3	59	135	82	36 1/2
14	254 2/3	37	209 3/4	60	131	83	32
15	253 1/2	38	206 7/8	61	127	84	27 1/2
16	252 1/4	39	203 5/6	62	123	85	22 3/4
17	251 1/8	40	201	63	119	86	18 3/8
18	249 2/3	41	198	64	115	87	13 3/4
19	248 1/4	42	195	65	111	88	9 1/4
20	246 3/4	43	192	66	106 3/4	89	4 2/3
21	245	44	188 3/4	67	102 1/2	90	
22	243 1/4	45	185 1/2	68	98 1/4		

This Table as you sée consisteth of foure collums, in euery one whereof are set downe on the left hande the degrées of altitude of the Pole, and next to that on the right hande the leagues answerable to euery degrée, the vse whereof is thus: First séeke out in

The Arte of Nauigation.

in the said Table the degrée of altitude belonging to that Parallell vnder which you sayled, and next to that on the right hande you shall finde the nūber of leagues incident to that degrée: which number of leagues, if you multiply by the number of houres before found, the product thereof will shew you how many leagues you haue sailed. And if there be any minutes annexed to the hours, then multiply also the number of the foresaid leagues found in the Table, by those minutes, and deuide the product therof by 60, and the quotient shall be leagues, which you must adde to the former leagues, and the summe thereof will shew how many leagues you haue in all sayled.

An example of counting the way of your shippe in sayling right west.

Chap. 35.

Vppose that you haue to saile from the Cape Saint Vincent in Spaine right West, and therefore you prepare your houre glasse as it may begin to run iust at noone, and hauing sailed 8. or 9 dayes, (not forgetting once euery day to turne your houre-glasse) you doe arriue at one of the Iles of the Azores called S. Mary, and there hauing tarried vntill your houre-glasse be cleane run out, and séeking to know at that instant by your Astronomicall ring, or by some vniuersall Dyall what houre it is, you finde that it is a 11. of the clocke and 10. minutes past, which wanteth 50 minutes of iust noone, and that is the difference betwixt the two Meridians, that is to say, the Meridian of Cape S. Vincent, and the Meridian of the Ile S. Mary, both which places being in one selfe Parallell must néedes haue one selfe Latitude or Altitude, which is 37. degrées, which altitude being found in the Table in the second collum on the left hande, you finde there also hard by it on the right hand, 209. leagues and $\frac{1}{4}$. of a league, for euery houre of that Parallell, which number of leagues, if you multiply by the foresaid 50. minutes, the product shalbe 10/487. $\frac{1}{2}$. which if you deuide by 60. you shall finde in the quotient 174. leagues and $\frac{12}{24}$. parts of a league, which is the iust quantitie of your whole voyage.

An

The Arte of Nauigation.

Another example of counting the way of your shippe in sayling right East.

Chap. 36.

Suppose then that you haue to sayle from the newe found lande right East in the Parallel of 50. degrees: and hauing caused your houre glasse to begin to run iust at noone, you set foorth and sayle the space of 15. dayes, not forgetting euery day once to turne your houre glasse. Now if at the 15. dayes end, you would know how many leagues you haue sayled, to finde in your Card in what place you are, you must first tarrie vntill your houre-glasse be cleane runne out, and at that instant seeke by your Astronomicall ring, or by some vniuersall Dyall, to know what houre it is, which because you haue sayled right East, you finde to be two houres and twelue minutes afternoone, then resorting to the Table aforesaide, you finde in the third colum the 50. degrees of altitude, together with the number 168. leagues and a halfe, annexed to the saide degrée for one houre of that Parallell, which summe being multiplyed by two houres, maketh 337. leagues, then multiply once againe 168. by the odde 1/2. and deuide the product thereof by 60. so shal you find in the quotient 33. leagues and somewhat more, which being added to the former summe 337. leagues will make in all 370. leagues and a little more, and that is the true quantitie of your voyage. And Cogniet sayth that by practizing this way of counting, you may know euery day, yea euery houre, what way your shippe maketh in sayling right East or West.

To knowe how much you goe out of your way in sayling by one wrong rombe or by more.

Chap. 37.

If in sayling to any place you direct your course eyther too high or too lowe by one rombe you loose in euery 100. leagues 19. leagues and 3/5. of a league, so as you go out of your way almost the fift part of your voyage. And if by two rombes, then in euery hundred leagues you loose 39. leagues, that is to say you go so far out

The Arte of Nauigation.

out of the right way: And if you fal thrée rombes out of your right course, then in euery hundred leagues you loose 58. leagues: and if you fall foure rombes, then you loose in euery hundred leagues 76. leagues, and the more rombes that you mistake in your direction, the more you wander out of your right course. Thus much touching the Mariners compasse and his Carde: nowe wee will speake of the North starre, and of his guardes, and then of the sunne and of the Moone, and so ende.

Of the North starre, otherwise called the loadstarre, and of his guardes, and how to know the same.
Chap. 38.

Though euery Mariner knoweth the North star so soone as he séeth it in the firmament, because it is his chiefe guide to direct thereby his shippe in the night season in all the North partes of the world: yet euery man that is no sayler knoweth him not, and therefore minding here to treate thereof, I thinke it not amisse first to teach him how to knowe it, and in what part of the heauen it is placed, and how farre it is distant from the North Pole.

The Poets do faigne that of the 48. Images of the fixed stars that be in the heauen, which are otherwise called Constellations, there be two Beares hauing tayles, whereof the one is called the great Beare, and the other the little Beare, in euery of which are seuen principall stars: And that bright starre which is in the very tippe of the little Beares tayle, is commonly called of our English Mariners the Loadstar, of the Latines Stella polaris, of the Grækes Cinosura, and of the Arabians Alrucuba. But to know this star, you must first find out the seuen stars of the great Beare by looking towards the North part of the firmamēt, of which stars foure being placed in his bodie, do make (as it were) a 4. square, and the other thrée stars being placed in his tayle, which is somewhat rising in the midst, doe represent the portion of a Circle, the little Beare hath also the like shape sauing that her féete & belly do turne vpwards, so as the tippe of her taile is answerable to the two hindermost starres of the great Beare, as you may sée in these figures following.

Some

The Arte of Nauigation.

Some do call the great beare great Charles Waine or Wagon, and the little Beare little Charles wain, because in each figure the foure square stars do signifie the four whéeles, & the other thrée stars the 3. horses: Nowe to finde out the Loadstarre, imagine

a right line to passe thorough the two hindermost starres otherwise called the guardes or pointers of the great Beare, as you sée here in this figure, and that line will rightly direct your eye to the Loadstarre which standeth in the tippe of the little Beares taile, for there is no bright star that can be séene betwixt the two gards of the great Beare and the Loadstarre. Notwithstanding some do affirme that there is another starre nigher vnto the Pole then the foresaide Loadstarre, and that star is as they say no more but 50. minutes distant from the Pole: I haue heard that there is a Priest an Astrologer with the King of Denmarke, who is woont to finde out this starre by the helpe of an instrument whereby hee getteth first the sight of sixe other stars at one instant, which way I feare mee is more troublesome then profitable, and if there were any such starre indéede, and therewith so néedefull and méete to bee vsed as is the Loadstarre, I doubt not but that some of our learned Pilots would haue found it out, and also haue had some vse thereof long ere this time: But leauing this matter, I will returne to the description of our Mariners common Loadstarre, and shew of what bignesse it is: also what Longitude, Latitude, declination, and motion it hath, and finally the chiefest and most necessarie vses thereof.

This

The Arte of Nauigation.

This starre according to the prutenicall Tables is of the third bignesse, and hath in Longitude 53. degrées 30. minutes, counting from the first starre of the Rams horne, whose place in these our dayes is in the 27. degrée 30. minutes of Aries, and by this meanes the place or Longitude of the Loadstarre is in the 21. degrée of Gemini, & his Latitude, counting from the Ecliptique line towardes the North Pole of the Zodiaque is 66. degrées 0. minutes, and his declination counting from the Equinoctiall towardes the North pole of the worlde is 86. so as he is in these dayes distant from the North pole 4. degrées, some say 4. degrées and 9. minutes, but Cogniet sayth, that he is distant from the pole but 3. degrées and a halfe, who findeth by the Astronomicall Tables that this starre at the birth of Christ was 12. degrées 36. minutes distant from the Pole, and euer since hath gone decreasing, and shall decrease euery day more then other, vntill it come to be but 26. minutes distant from the Pole, which is as nigh as it can approch to the Pole, for then his distance shall beginne againe to encrease. This starre maketh his daily reuolution from East to West in 24. houres, as all other starres doe by vertue of the first mooueable, but his circuit is so small, and his gate so slow, as in the space of 24. houres he goeth not much more then 24. degrées of the Equinoctiall, and in his turning round about he maketh as it were a little Circle, in the middest whereof is the North pole it selfe, which is inuisible & can not be séene.

And though the image or constellation of the little Beare containeth many stars, yet the seuen before mentioned and set foorth in figure are most obserued, of which seuen starres in the little beare, two stars called the gards of the North star are to the Mariners most familiar, for diuers respects hereafter declared. And of those gards the one is Northerne & the other Southern, and both are said to be of the second bignesse, & yet the Southerne gard séemeth to the eye both lesser and darker then the other. Of which two stars I thinke it not amisse to set downe here the Longitude, Latitude, and declination as I haue done before of the Loadstarre.

The Longitude of the South garde is 100. degrées 30. minutes, whereby his place is in the ninth degrée of Leo. And his latitude is 72. degrées 40. minutes, and his declination is 73. degrées, so as he is distant from the North pole 17. degrées.

The

The Arte of Navigation.

The longitude of the North guard is 109. degrées, 30. minutes, so as his place is in the 20. degrée of Leo, and his latitude is 74. degrées, 50, minutes, and his declination is 76. degrées, so as his distance from the North Pole is 14. degrees. Thus hauing shewed the greatnesse, the longitude, the true place, the latitude, and the declination aswell of the Lodestarre, as of his two guardes, I thinke good now to set downe the chiefest vses that sea men haue of them.

The vses of the North starre, and of his guardes.

Chapter. 39.

The first vse is to knowe thereby the variation of the Mariners Compasse.

The second is to know by his guardes when the North starre is aboue or beneath the Pole.

The third is to know by the North starre and his guardes with the help of an instrument called a Nocturlabe the houre of the night.

The fourth is to know the eleuation of the Pole. And you haue to note that the North starre and his guardes are alwayes aboue our Horizon, and doe neuer goe downe in any place whereas the North Pole hath any eleuation, be it neuer so little. But besides these starres there be also in this our latitude diuers other images of starres that are alwayes aboue our Horizon, and doe neuer go downe, as the Dragon, the great Beare, the image of Cepheus, of Cassiopeia, the image of Auriga, hauing the Goat at his back which is a faire bright starre of the first bignesse, and many others, which I leaue to name, because I haue heretofore described them all at large in my treatise of the Globes, by which starres by reason that with vs they neuer goe downe, the latitude of any place and also the houre of the night may bee knowne so well as by the Lodestarre.

But now as touching the vses of the foresaide starres before mentioned, the first whereof shewing how to find out by the Lodestarre the variation of the Compasse, is taught before in the 25. chapter, according to William Boorne his rule set downe in his
Regi-

The Arte of Navigation.

Regiment, and as for the other three vses they are plainly taught in the two chapters next following.

To knowe by helpe of a little *Table* made according to the Mariners rule touching the 8. principall rombes, shewing how and when the Lodestarre is either aboue or beneath the Pole, that you may knowe thereby the true altitude of the Pole in taking the height of the Lodestarre with your Astrolabe or crosse staffe. chap. 40.

Chapter 40.

TO know this you must alwayes haue due consideration of the two guards of the little Beare, for when those guards are iust Southwest from the Lodestarre, then is the Lodestarre at his highest in the verie Meridian, and therefore it is right aboue the Pole, and when those guards be iust Northeast from the Lodestar then the Lodestarre is againe in the Meridian at his lowest, and thereby right vnder the Pole, and in both places his distance from the Pole is but three degrees and a halfe, as Cogniet saith, according to which account he setteth downe this Table following.

This Table is diuided into two collums, whereof that on the left hand contayneth the eight principall rombes or windes, and that on the right hand contayneth the degrees and minutes of distance of the Lodestarre from the Pole, being either aboue or beneath the Pole.

If the guards be	The rombes or windes.		The degrees and minutes of the declination of the Lodestarre from the Pole.		
	West. Southwest South Southeast.	Then the Lodestar is	1 3 3 1	½ ½ 0 0	Aboue the Pole.
	East Northeast North Northwest		1 3 3 1	½ ½ 0 0	Beneath the Pole.

This

The Arte of Navigation.   338

This table differeth in one point from the Mariners common rule set downe as well by Medina, and by Martin Cortes, as also by William Boorne touching the guards, for they all appoint but halfe a degrœ of declination of the North starre from the Pole, the guardes being either in the rombe Southeast, or Northwest, to both which rombes Cogniet appointeth one whole degrœ of declination. The vse of which table in seeking to know, the eleuation of the Pole is thus.   First, hauing taken with your Astrolabe the altitude of the Lodestarre aboue your Horizon, obserue immediatelie in which of the 8. former rombes his guardes be. For if they be in any of the 4. vpper rombes of the Table, then you must subtract from the height of the Lodestarre taken with your Astrolabe, so much as is set downe for that rombe in the collum on the right hand of the said table, and the remainder shall be the true altitude of the Pole, but if the guardes bee in any of the foure nether rombes, then you must adde so much to the height of the Lodestar, and the summe thereof shall be the altitude of the Pole. But because this table serueth only for the 8. principall winds and for no more. Cogniet therefore setteth downe the making of a twofolde instrument, whereby you shall not onely know (as he saith) how much the Lodestarre is aboue or beneath the Pole in euery other rombe as wel as in the 8. principall rombes, but also you shall know therby the true houre of the night more exactlie than the Nocturlabe of Munster or Appian doth shew.

How to make an Instrument which wil shew at any houre of the night how much the Lodestar is either aboue or beneath the *Pole* in euerie other rombe as well as in the 8. principall rombes, which are only contained in the former *T*able, and also the true houre of the night, the shape whereof followeth.

Chapter 41.

The Arte of Navigation.

The shape or figure of the Rectifier of the North star.

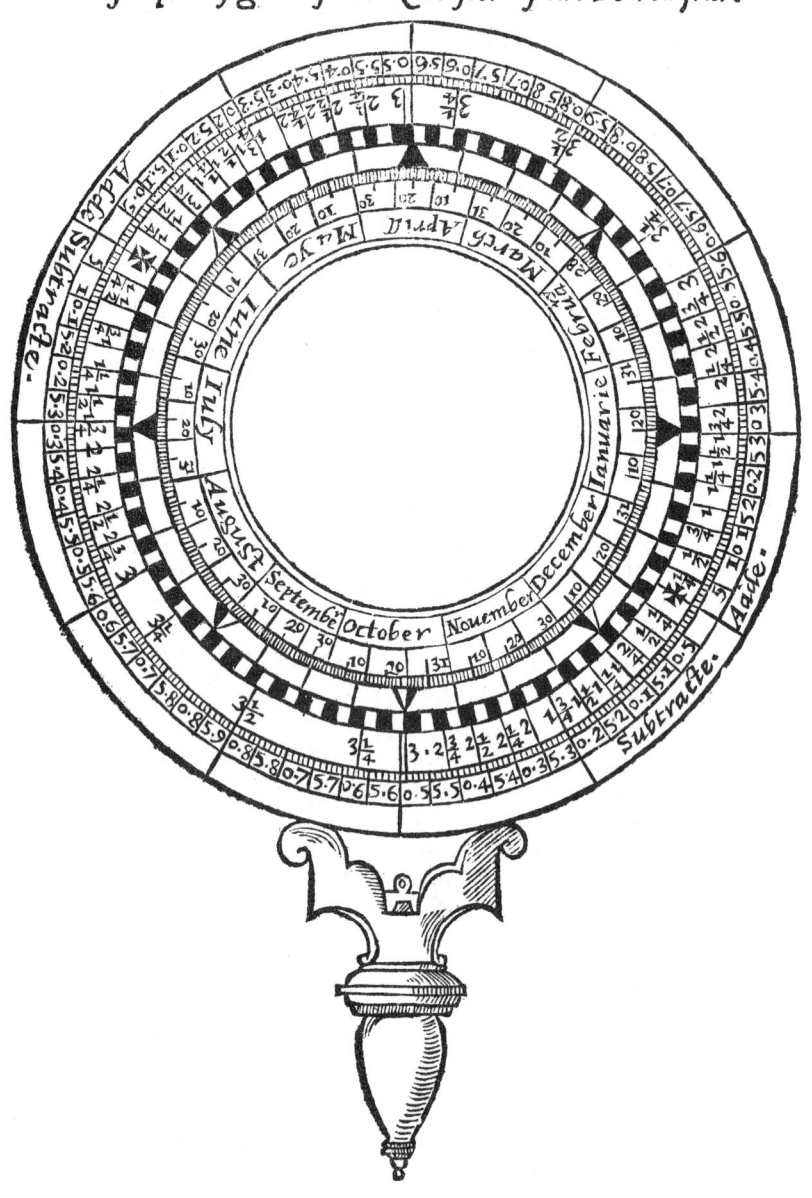

The Arte of Navigation.

Ogniet calleth this instrument Rectificatorium Stellæ Polaris, that is to say, the Rectifier of the North starre, ioyning thereunto a Nocturlabe differing nothing at all from the Nocturlabe of Munster and of others, but onelie in placing the 21. of October in stead of the 28. of the same moneth vppon the line of North and South towards the handle of the instrument for such cause as is hereafter declared.

The making of the rectifier of the Lodestarre, is thus: Vpon a smooth péece of boord of firme wood, or vpon a péece of polished plate of brasse or Latten being sixe or seuen inches broade, hauing a handle, as you sée in the former figure, draw a circle diuided into 4. quarters by help of two crosse Diameters cutting one another in the center with right angles, the perpendicular whereof shall signifie the Meridian line, that is to say, the line of North and South, at whose vpper end set North, and at the nether end South, and the other ouerthwart Diameter shall bee the line of East and West, hauing East marked on the right hand and West on the left hand, and in euerie quarter of the foresaide circle you may if you will place the like rombes that are in the Flie of the Mariners Compasse, as you may sée done in the foresaid figure. Moreouer, you must diuide the vpper quarter on the left hand into thrée equall partes, and hauing taken two of those partes with your Compasses, measuring from the North point downe toward the West, and there make a pricke, marking the same with a little blacke crosse, and from that crosse drawe a right line that may passe through the Center vnto the circumference of the Circle, and there make another little blacke crosse, and this line shal diuide the circle into two equall partes, which line if you crosse with another right line passing through the center, and making thereby right angles, you shall diuide the circle into other foure quarters differing from the first foure quarters, though not in quantitie, yet in place, of which foure last quarters you must diuide that which is on the left hand into 90. equall degrées, beginning your account at the little blacke crosse on the left hand, and so procéede downward towards the South point of the circle.

Now to know how much the North star in euery rombe mounteth or descendeth and how to place the same vpon the instrument

The Arte of Navigation.

so as you may knowe how much to ad to the altitude of the Lodestarre, or to subtract from the same, you must vse this Table following, which Table consisteth of two collums, whereof that on the left hand containeth the degrees and minutes of the declination of the Lodestarre, proceeding from one quarter of a degree to another, vntill you come to 3. deg. and a half, which is (as Cogniet saith) the greatest declination of the Lodestar, from the Pole: and the other collum on the right hande containeth the degrees and minutes of a quarter or quadrant, diuided by diuers and sundrie proportions of numbers into 90. degrees which proportional degrees and minutes are to bee reckoned in the first quadrant diuided into 90 equal degrees, & next descending from the first crosse on the left hande thus.

The degrees and minutes of the declination of the lodestarre from the pole.	The degrees and minutes of the Quadrant, diuided by divers portions of numbers into 90 degrees.	
	Deg.	Min.
¼	4	6
½	8	13
¾	12	22
1 0	16	36
1 ¼	20	55
1 ½	25	23
1 ¾	30	0
2 0	34	51
2 ¼	40	0
2 ½	45	35
2 ¾	51	48
3 0	59	0
3 ¼	68	13
3 ½	90	0

First, beginning at the said crosse, and descending downwarde, tell out 4. degrees, and 6. minutes, & right against that set downe ¼. then tell in the same quadrant from the saide crosse 8. degrees, and 13. minutes, and right against that set downe ½ and so proceede according as the Table directeth you, vntill you come to 90. against which you must set. down 3. deg. & ½ And as this quarter is diuided, so you may deale with the other 3. quarters, and in each of the 2 quarters which are beneath the 2 crosses forget not to write this word Subtract, and in each of the 2. quarters aboue the 2. crosses write this word *Ad*, as you see in the former figure. Then ther shal want nothing but only a ruler or index, which must be fastened to the center of the instrument, so as it may be turned round about a pinne hauing a round hole in it, through which you may see the Lodestar, and also at the same instant by lifting vp and

downe

downe the index you may sée on the outside of the instrument the foreguard of the said star appearing euen with the right edge of the index called Linea fiduciæ, or the fiducial line, drawn from the center of the instrument alongst the inward edge of the index, and so is the instrument for that purpose fully perfected. But if you wold ad therunto a nocturlabe, then you must draw vpon the center therof diuers circles next to the space cõtaining the winds: and the spaces betwixt those circles must be some wider, & some narrower For the first vpper space must be narrow, containing the daies of euery moneth, and the next space somewhat wider, containing the number of those daies set down in Arithmeticall figures, and the third space wider than that, containing the names of the moneths then next to the lowest circle of the nethermost space place a little rundle fastened to the instrument with the foresaid hollow pin, so as it may turne round about the same. And this rundle must be diuided into 24. hours, that is to say, 12 for the day, & 12. for ỹ night, which rundle would be made with téeth wherof one must be longer than al his fellowes signifying alwaies ỹ 12. houre of the night, which is alwaies to be laide vpon the day of the moneth wherein you séeke to know the houre of the night. And remember in distributing the dayes of the moneth, that you alwayes set the 21 day of October beneath towards the handle in the verie line of North and South passing right through the middest of the handle, so shal the instrument shew the houre of the night more trulie than when the 28. day of October standeth beneath vpon the line of South and North, as it doth in the common Nocturlabes, the makers whereof had respect to the Pole it selfe which is inuisible, and not to the North starre which is apparant to the eie, by meanes whereof the guards of the Lodestarre sometime doe shew sooner or latter by 7. degrées and 18. minutes, if you count from the Pole, which is almost halfe an houre difference, and for that cause also Cogniet maketh his account in his foresaid rectifier to begin at the little blacke crosse, seuen degrées more forwarde than it ought to doe, if he should count from the Pole, and not from the North starre.

Xx 4   How

The Arte of Navigation.

How to know by the foresaid twofold instrument aswel the mounting and descending of the North star as the true houre of the night both at one instant and also the elevation of the Pole,

Chapter 42.

Irst then hauing layed and stayed the great tooth of the moouable rundle marked with 12. vppon the day of the moneth wherein you séeke, and holding the instrument by the handle with your one hand right before your face, leaue not to put that hand forward from you, or to bring it backwards towards you, vntill you may sée with the one eie, winking with the other, the North starre through the hole of the pin, which is in the center of the Instrument: and so soone as you sée the North star, lift with your other hand the index vp and downe vntill you sée also at that instant the North guard of the Lodestarre on the outside of the instrument appearing euen with the fiduciall line or inwarde edge of the said index. Then staying the index there, looke vpon what houre it falleth, for that shall be the houre of the night. And look also at that instant vpon what degrée of distance it falleth in the outermost border of the instrument wherein those degrées are set downe together with these wordes adde and subtract. For from the point marked with a little blacke crosse nigh vnto the Northwest descending through the nether moyetie or halfe deale of the instrument vntill you come to the other blacke crosse, placed nigh vnto the Southeast, the North starre is alwaies aboue the Pole so much as the index sheweth, which you must alwaies subtract from the height of the North starre to knowe the eleuation of the Pole, hauing first taken with your Astrolabe the height of the North starre. But if the index doe fall vppon any degrée or part of a degrée in the vpper moyetie or halfe deale of the instrument then to know the eleuation of the Pole, you must adde so much as the index sheweth vnto the altitude of the Lodestarre, as the words Adde and Subtract written in the outermost border of the instrument doe plainely shew.

But now if you would know at any houre of the day or night,

The Arte of Navigation.

in what rombe the foresaid guardes be without seeing them and also how much the Lodestarre is declined from the Pole, you néede doe no more but to lay the great tooth of the moueable rundle vppon the day of the moneth, and then to bring the index vnto the houre which you require, and the said index will shewe in the border of the instrument in what rombe the guardes be, and how much the North starre is aboue or beneath the Pole.

I haue found by often triall that this instrument wil shew the true houre of the night, and also in what rombe the guardes bée, and thirdlie, how much the North starre is at any time either aboue or beneath the Pole, and by adding the degrée of distance found in the limbe of the instrument, or by subtracting the same according to the rule before giuen from that altitude of the north star, which I haue before taken with my Astrolabe, I haue found at all times the true eleuation of the Pole wheresoeuer I haue made triall thereof. But sith other Starres may perhaps appeare when the North Starre with his guardes shal be hidden, I woulde wish all carefull Mariners to acquaint themselues with as manie bright Starres as they can, and especiallie with those which do both rise and set, and also to learne by some Table the declination of euerie such Starre and whether it be Southernlie or Northernlie, for by taking the Meridian altitude of a nie such starre and by adding to, or by taking from the altitude thereof, his declination according as the same declination is either Northernlie or Southernlie (for if it be Northernlie, then you must subtract his declination, and if it be Southernlie, you must adde the same to the altitude of the Starre) you shall finde thereby the altitude of the Equinoctiall, which being taken out of 90. the remainder will shew the eleuation of the Pole. The meetest starres for this purpose in these our North partes of the worlde are these, Hircus the Goate, Canis minor, the little Dogge, Canis maior, the great Dogge, Dexter humerus Orionis, the right shoulder of Orion. Cingulum Orionis, the girdle of Orion, Cor Leonis, the Lyons hart. Bubulcus, the Beareward, Spica Virginis, the wheat eare in the hand of Virgo. Aquila volans, the flying Eagle. Caput Andromedæ, the head of Andromeda. Ras Algol, the head of *Medusa*. Oculus Tauri, the Buls eye, and diuers others.

The Arte of Navigation.

And in anie case consider whether the declination of the starre be greater or lesser than his Meridian altitude, for if his declination be greater then his meridian altitude, as of those stars which are nigh the Pole, then you must take his Meridian altitude with your Astrolabe at two seuerall times, that is to say, when he is at his highest in the meridian, and also when he is at the lowest point of the same Meridian called the depression, and hauing added those Meridian altitudes together, take halfe thereof, and that half shall be the eleuation of the Pole, which way of finding out the eleuation of the Pole is nothing meet for a Mariner that is vnder saile. But now to procéede according to Cogniet his direction in this matter, hee saith, that if in sayling you approch so nigh vnto the Equinoctiall as the eleuation of the Pole is not aboue 17. degrées, then the guardes are not so easilie séene, wherefore it shall be néedfull to take som other starre whereby you may both know the houre of the night, and also how much the Lodestarre is eyther aboue or beneath the Pole, and Cogniet thinketh none of the starres more meet for that purpose than one of these two, that is, either the starre called Caput Medusæ, that is the head of Medusa called of the Arabians, Ras Algol, or els the starre called Hircus, that is, the Goat, both which are faire bright starres and of the first bignesse: and the cause why he appointeth the head of Medusa, is for that this starre is directlie opposite to the former guard, in such sort as this starre is alwayes aboue the Horizon, when both the guardes are vnder the Horizon. Wherefore hauing prepared your instrument, that is to say, hauing laid the longest tooth of the mooueable rundle vpon the day of the moneth wherein you séeke, and hauing found out thereby in such order as is before taught, the north star and also the star called the head of Medusa, both at one instant, marke where the index falleth, and immediately turne the index from that point to the point opposite, abating twelue houres, and so the Index shall shew you thrée things at once. First, the rombe wherein the guardes are at that present. Secondlie, the houre of the night. And thirdly how much the North starre is aboue or beneath the Pole. But because this starre is not knowne perhaps to all Mariners, Cogniet would haue you to take the other bright Starre called Hircus, which (as hée saith) goeth 9. houres and $\frac{1}{2}$. before the guardes, in such

sort

sort, as when the former guarde is East, you shall finde this Starre, counting from the North starre, to be Northwest, and almost 45. degrées distant from the Pole, with which starre if you worke, as you did before with the heade of *Medusa*, sauing that you shall not néede to turne the index vnto the opposite point, but onelie to rebate from the point on which it falleth 9. houres and ½ you shall knowe all the thrée thinges last mentioned. And this rule (as he saith) is so generall as you may haue your desire by working in like manner with any other starre that is to you certainlie knowne, & is at that time aboue the Horizon.

What starres are obserued by those that sayle beyond the Equinoctiall vnder the South Pole.

Chapter. 43.

He ancient Astronomers, as *Ptolomey*, *Timochares*, *Hypparchus* and others did neuer describe any star to be more nigh vnto the South pole, than that which is called *Canopus*, which is a faire bright starre of the first bignesse, and according to the Tables of *Copernicus*, is distant from the South Pole 38. degrées, and ¼. But those that haue sayled in the South seas of later dayes, haue founde out other starres vnknowne to the ancient Astronomers, which are much néerer vnto the said Pole. For *Albericus Vesputius* writeth of thrée Starres, making together a Triangle Orthogonall, that is to say, hauing one right angle, now called the southern Triangle, the middle star wherof is distant from the south Pole 9. degrées, ⅖.

There be also latelie found out diuers images of other starres nigh vnto the South pole, as that which is called *Noah* his Doue, or Pigeon, and another called *Polophilax*, made in the shape of a man, whose longitude and latitude hath not as yet bene rightlie set downe by any that I haue read.

But the sea men of these present dayes doe most commonlie obserue foure great stars, which according to the shape and forme thereof they call the Crosse, imagining the greatest starre of the foure to be the foot, & that which standeth right ouer him to be the head of the crosse, & the other two to be the 2 armes: and when they

sée

The Arte of Navigation.

sée that the head doeth directlie answere the foote, then they say that the foote of the saide crosse is right aboue the Pole, and distant from the same 30. degrées. Therefore, hauing taken the altitude or height of that star aboue ye Horizon with their Astrolabe they subtract 30. degrees from that altitude, and the remainder is the eleuation of the Pole. Medina in his fift booke and eleuenth chapter setteth downe the shape of the foresaide crosse in this maner.

And he saith that these starres are neither anie of those starres that are appointed to the twelue Signes in the Zodiaque, nor yet any of the 36. Images or constellations that bee in heauen. Moreouer he saith, that in taking the altitude of the great Starre called the foote béeing in his right place, that is to say, when hée is directlie opposite to the head, and that you finde his altitude to be 30. degrées, then you may assure your selfe that you are right vpon the Equinoctiall. And if you finde his altitude to bee more than 30. degrées, then you are past the Equinoctiall towardes the South Pole. But if you find it to be lesse than 30. degrées, then you are still on the North side of the Equinoctiall.

Besides the starres aboue mentioned, our Mariners in these North parts of the world are woont to obserue diuers other stars, to the number of 32. whose longitude and declination together with their bignesse and also when they rise and set, and when they are mounted to the Meridian, that is to say, are iust South, is plainlie set forth by Tables collected of purpose out of the Astronomicall Tables by William Boorne, which Tables you shall finde in the 20. chapter of his book called the regiment of the sea, And Robert Norman doeth also set downe the like tables in his booke called the Attractiue, and therefore I thinke it superfluous to repeat the same againe here, and speciallie sith I haue described vnto you all the starres that bee in the firmament that were knowne to the ancient Astronomers, and haue shewed you how to find out by the Globe their longitudes, their latitudes, their declinations,

The Arte of Navigation.

nations their greatnes and all other accidentes belonging to the Starres in my treatise of the Celestiall Globe, which I wrote of purpose to further yong sea men.

The knowledge of the Stars serueth sea men chiefly to knowe thereby the latitude of any place, and also to knowe the houre of the night: And thirdlie, to coniecture by their manner of rising and setting, and other their aspectes what weather is like to followe, either foule or faire: the rules whereof to teach trulie bebelongeth to Astronomers, yet many sea men by diligent obseruation doe attaine to right good iudgment therein. Wherefore leauing to speake any further of the Starres, I will now briefly speake of the Sunne and of his motions, of his rising and setting in euerie latitude, and of his declination from the Equinoctiall, and of other his like apparances. I say here briefly, because I haue alreadie spoken of him at large in the 1. part of my spheare wheras I treat of the Zodiaque.

Of the Sunne, and of his motion, and of the chiefest apparances belonging to him.

Chapter 44.

He Sunne according to the moouing of the first mooueable which is from East to West maketh his dayly reuolution in 24. houres, as all other Starres doe, but according to his owne motion, which is from West to East in going through the twelue Signes of the Zodiaque he spendeth a whole yeare, for his daylie mouing vpon his own center, called the excentrick, because it is out of the center of the world, is little more then 59. minutes, 8. secondes, making thereby the whole yeare to consist of 365. dayes, fiue houres 49. minutes, 8. seconds, 19. thirdes, 37. fourths, 24. fiftes, and this is called of Copernicus, the equall tropicall yeare, which taketh his beginning from the first point of Aries, otherwise called the Vernall Equinoxe, into which point the Sunne entreth not euery yeare at one selfe day of the moneth, for sometime hee entreth into that point the tenth day, and sometime the eleuenth day of March,

which

The Arte of Navigation.

which day of his entring is alwayes trulie set downe in euerie Ephemerides, and because the beginning of the tropicall yeare is so vncertaine, the Astronomers doe make their yeare called the syderal yeare, to begin at the former star of the Rams horne, and thereby doe make the yeare to consist of 365. dayes. 6. houres 9. minutes and 39. secondes, according to which yeare they alwaies rectifie or bring to equalitie, aswell the equall as the vnequal tropicall yeare, of both which I haue spoken in the first part of my Spheare, the 38. and 39. chapters.

Moreouer, the Sunne hath three motions, that is, slowe, swift and meane. His slowe motion is when he is in the point called Auge or Apogeon, which is a point imagined to bee nigh vnto the outermost edge of the circle which carrieth the bodie of the Sunne called Deferens Solis, and is furthest distant from the center of the world, which point in these our daies is in the 9. degree of Cancer, or there abouts. And beeing in that point, he goeth little more than 57. minutes in 24. houres. Againe, his swift motion is when he is in the opposite point to the Auge, called Perigeon, which point in these dayes is in the 9. degree of Capricorne, and being in this point, he goeth one whole degree and almost two minutes in 24. houres, which is almost fiue minutes more than he maketh in his slow motion. His meane motion is when he is in the midst betwixt the two foresaid points, wheras in 24. houres he goeth one whole degree & somewhat more than two minutes. And these three sundrie motions doe cause the Equinoctiall pointes not to be of equall distance. For the Sunne spendeth seuen dayes and somewhat more in going from the Equinoxe of March, to the Equinoxe of September, than he doth in going from the Equinoxe of September to the Equinoxe of March. For if in this present yeare 1592. which is leape yeare, you count the dayes from the Uernall Equinoxe, which is the eleuenth of March, vnto the Autumne Equinoxe, which is the 13. of September, you shall find the number of the dayes to be 186. and the other number from the 13. of September to the eleuenth of March to be but 180. which is lesse than the first number by 6 dayes, and if it were not leape yeare the difference would be seuen dayes, because that Februarie in the leape yeare hath 29. dayes.

It is necessarie that sea men haue some vnderstanding of the
three

The Arte of Navigation.

thꝛée foꝛesaide motions to the intent that they may the better know the true place of the Sunne and thereby his true declination. And note that no calculation of his declination can continue without errour aboue twentie foure yeares. For as often as the leape yeare commeth about, which is euerie foure yeares, the Sunne is vppon the Equinoctiall sooner by halfe an houre. But as foꝛ the true place of the Sunne, and especially euery day at nooutide, the Ephemerides doth most trulie shewe, and hauing his place, you shall easilie finde his declination by this Table following, which will serue foꝛ these twentie yeares and moꝛe, the like whereof, together with the vse of the same is set downe in the first part of my Spheare the 13. chapter.

A Table

The Arte of Navigation.

A Table shewing the declination of the Sunne euerie day throughout the yeare, and the vse thereof.

Chapter 45.

Degrees of the Signes.	♈ ♎			♉ ♏			♊ ♐			Degrees of the Signes
	D	M	S	D	M	S	D	M	S	
1	0	23	53	11	50	6	20	22	57	29
2	0	47	46	12	10	56	20	35	7	28
3	1	11	39	12	31	34	20	40	55	27
4	1	35	30	12	51	59	20	58	20	26
5	1	59	20	13	12	12	21	9	21	25
6	2	23	8	13	32	12	21	19	59	24
7	2	46	54	13	51	58	21	30	13	23
8	3	10	37	14	11	30	21	40	3	22
9	3	34	18	14	30	48	21	49	29	21
10	3	57	54	14	49	51	21	58	29	20
11	4	21	28	15	8	40	22	7	6	19
12	4	44	57	15	27	13	22	15	17	18
13	5	8	22	15	45	30	22	23	3	17
14	5	31	42	16	3	32	22	30	24	16
15	5	54	57	16	21	17	22	37	19	15
16	6	18	6	16	38	44	22	43	48	14
17	6	41	9	16	55	55	22	49	50	13
18	7	4	6	17	12	48	22	55	27	12
19	7	26	57	17	29	23	23	0	38	11
20	7	49	40	17	45	40	23	5	22	10
21	8	12	16	18	1	39	23	9	39	9
22	8	34	45	18	17	18	23	13	29	8
23	8	57	5	18	32	37	23	16	53	7
24	9	16	16	18	47	38	23	19	50	6
25	9	41	19	19	2	18	33	22	19	5
26	10	3	12	19	16	37	23	24	22	4
27	10	24	56	19	30	36	23	25	57	3
28	10	46	30	19	44	14	23	27	5	2
29	11	7	53	19	57	30	23	27	46	1
30	11	29	5	20	10	25	23	28	9	0
	♍ ♓			♑ ♒			♋ ♌			

The Arte of Nauigation.

The vse of this Table is thus: If the sunne be in any of the signes set down in the front of the Table, then séeke his degrée (first found by the Ephemerides) in the left collum, the degrées whereof do descend from one to 30. and the square Angle answerable to that signe and degrée will shew his declination. But if the sunne be in any of the signes that are in the foote of the Table, then séeke his degrée in the right collum, the degrées whereof do ascend, and the square Angle answerable to that signe and degrée will shew his declination.

There be other things also méete for Sea-men to knowe, touching the sunne as these: First to knowe the foure seasons of the yeare, secondly to know by his declination, the length of both the day and the night in euery latitude, and how it doth encrease and decrease: Item to know in what rombe or winde, and at what houre he riseth and setteth, and also his Meridian altitude, that is, when he is right South euery day, to finde out by the helpe of that, and by knowing his declination, the true Laticude of any place, & by his shadow to know the houre of the day, which are two chiefe pointes that the Mariner hath most néede to knowe. Of all which thinges I minde here to treate both plainely and briefely.

Of the foure seasons of the yeare, that is, Spring time, Summer, fall of the leafe, called otherwise Autumne, and winter.

Chap. 46.

IN this our Clime the spring is said to begin when the Sun entreth into the first point of *Aries*, which is about the xi. of March, and continueth vnto the last point of *Gemini*, which time is saide to be hot and moyst, and therefore is likened to childhoode: And sommer beginneth when the Sun entereth into the first point of *Cancer*, which is about the 12. or 13. of June, and endeth when he is in the last degrée of *Virgo*: and this time is said to be hotte and dry, and therefore is likened to Adolescencie. Then Autumne or fall of the leafe beginneth when the Sunne entreth into the first point of *Libra*, which is about the 13. or 14. of September, and endeth when the Sun is in the last degrée of *Sagittarius*:

P p   and

The Arte of Nauigation.

and this time is said to be colde and dry, and therefore is likened to manhood: Finally winter beginneth when the sunne entreth into the first point of Capricorne, and endeth when he is in the last degree of Pisces: and this time is said to be colde and moyst, and therefore is likened to olde age: notwithstanding Galen in his first booke de Elementis, sayth that the spring is temperately hot and moyst, and therefore a most wholsome time: And sommer is more hotte then colde, and more dry then moyst, and therefore is sayd to be hotte and dry: And Autumne is also saide to be dry because it is more dry then moyst, and yet neither hotte nor colde, but vnequally mixt, and thereby infectiue and causing sickenesse. And winter is sayd to be colde and moyst, not because it is colder or moyster then any other season, but because that in winter moysture excæedeth drinesse, and coldnesse excæedeth heate. But you haue to vnderstãd that these foure seasons haue not like qualities in all the 5. Zones: For in the burnt Zone, and specially to those that dwell right vnder the Equinoctiall the sunne being in Aries or Libra causeth greatest heat, & thereby two sommers because he is then right ouer their heads, & being in either of the Solstices, that is in the beginning of Cancer or Capricorne he causeth two winters because he is then furthest from them as I haue declared vnto you in the second part of my treatise of the Spheare the 20. chapter, whereas I treate of the seasons and shadowes incident to diuerse Climes and Parallels whereunto I referre you, and so I ende with this matter.

How to knowe when the Sunne riseth and setteth in everie latitude, and thereby the length of the day and night, and also in what rombe or wind he riseth and setteth, and how much he declineth everie day from the Equinoctiall either Northward or *Southward*. Also howe to know the elevation of the *Pole*, otherwise called the latitude of any place, by knowing the Meridian altitude of the Sunne, and his declination.

Chap. 47.

The most part of all these thinges haue bæne taught before in my Spheare, in my Treatise of the Globes, and also in my treatise of the Astrolabes. And in the first two Treatises I shew also how to finde out the Longitude
of

of any place, and therefore néedeth not here to be rehearsed: But if you would know how to handle the declination of the sunne being vpon the sea, then reade 7. 8. 9. and 10. chapters of William Borne his booke called the Regiment of the Sea, and you shal be fully instructed therein: The whole effect of all which Chapters Robert Norman setteth downe in few words in his new Attractiue in this manner as followeth.

First learne whether the sun haue South declination or North declination, which you shall know by his being in any of the Northerne or Southern signes: Then marke what shadow he casteth, and whether it striketh towardes the Pole whereunto he is néerest, or to the contrary. For if the sunne casteth his shadowe the same way that he is from the Equinoctiall, he shalbe betwixt you and the Equinoctiall, & then hauing taken his Meridian altitude subtract the same from 90. & adde vnto the remainder the suns declination for that day, and the summe thereof shalbe the eleuation of the Pole or the distance of your Zenith from the Equinoctiall otherwise called the Latitude which is alwaies equal to the eleuation of the Pole: But if the sun casteth his shadow to the contrary side of the Equinoctiall, that is to say, being in his North declination casteth his shadow Southward, or being in his south declination casteth his shadow Northward, then either the Equinoctiall shalbe betwixt you & the sunne, or you in the Equinoctiall, or else you shalbe betwixt the Equinoctial and the sunne, which you shall know thus: Adde the declination of the sunne for that day wherein you séeke vnto his Meridian altitude, and if the summe of the addition be lesser then 90. degrées, then so much as it wanteth of 90. degrées shall you be distant from the Equinoctiall on that side on which the shadow stréeketh: but if it amounteth iust to 90. degrées, then you shalbe right vnder the Equinoctial. Againe if it be more then 90. degrées, then so much as is the ouerplus, so much shall you be from the Equinoctiall towards the sunne, at which time you shall be also betwixt the Equinoctiall and the sunne.

And if you finde the Meridian altitude of the sunne to be euen with your Zenith, then looke what declination the sun hath at that instant, and so much shall you be from the Equinoctiall on that side wherein the sunne is: But if the sunne haue no declination, then shall you be right vnder the Equinoctiall line.

Of

The Arte of Nauigation.

Of the shaddow of the Sunne, and how to know therby the houre of the day in any latitude by help of an vniuersall diall.

Chap. 48.

MY former order now requireth that I should speake somewhat of the shadow of the Sun, and of the diuersitie therof, according to the Clime or Parallel vnder which you sayle. But for so much as I haue shewed you in the second part of my Sphere from the 20.to the 27. chapter of the same, what diuerse shadowes the Sunne yeeldeth in diuerse Climes and Parallels, and also haue shewed what Vmbra recta and Vmbra versa is in the 40. Chapter of my Treatise of the Astrolabes: I minde not here therefore to make a new recitall therof, but only to shew how you shall find by his shadow the true houre of the day in euery Latitude by a generall Dyall, made to serue in all Latitudes, of which Dyals, though I haue seene diuerse and of diuerse shapes, yet none liketh me better then that which William Borne setteth downe in his Regiment the 21. Chapter, and calleth it the Equinoctiall Diall, which serueth not onely to know the houre of the day by the shadow of the sunne, but also the houre of the night by the shadowe of the Moone when she shineth cleare, which Diall being of small charge, I would wish all Mariners to haue: The making and vse whereof is so plainely set downe by himselfe in the foresaid Chapter as I thinke it superfluous to set it downe againe here. And thus ending with the the sunne, I will now turne my pen to the Moone.

Of the Moone and of all her diuers motions.

Chap. 49.

THe Moone is a round thicke and darke bodie, hauing no light of her selfe, but onely such as she receiueth from the sunne, and shee maketh her dayly motion from East to West as all other starres doe in 24. houres, according to the mouing of the Primum mobile. But according to her owne motion, which is from West to East, she goeth but 13. degrees 12. minutes in the space of 24. houres,

and

The Arte of Nauigation.

and that is according to her meane motion. And she passeth thorough the twelue signes of the Zodiaque in 27. dayes and eight houres: during which time the sunne by his natural motion which is also from West to East, is remooued from the place of coniunction almost 27. degrées, so as the Moone not finding the Sun there spendeth two dayes foure houres, and 44. minutes more in ouertaking him, which being added to 27. dayes and 8. houres doe make in all 29. dayes 12. houres and 44. minutes. Notwithstanding by reason that the Moone hath aswel as the Sunne thrée motions, that is, swift, meane, and flowe, she may change sometime sooner and sometimes later then in 29. dayes 12. houres 44. minutes, and yet one change counted with an other shall make vpp the selfe same summe. And note that this her thréefold motion dependeth vppon two pointes, the one called the Auge, and the other the point opposite to the Auge.

The point Auge of the Moone is when she is furthest distant from the earth, and the point opposite to that is when shée is nighest to the earth: For when she is in the point Auge, she goeth little more then twelue degrées in 24. houres: but when she is in the opposite point she goeth almost 15. degrées in 24. houres. And in her meane motion which is in the middest betwixt the two foresaide points, shée goeth 13. degrées and 12. minutes in 24. houres. Now because the Mariners doe account the moouing of the Moone by the pointes of their Compasses, they may thereby vnderstand that she goeth not alwaies in 24. houres one point and thrée minutes as they reckon, but sometime more and sometime lesse: For when she is in her slow motion, she goeth little more then 12. degrées in 24. houres, in which time the Sunne goeth one degrée, so as the Moone is distant from the Sunne but 11. degrées, which is but 44. minutes of an houre, which wanteth 4. minutes of a whole point, whereto is attributed 4'8. as hath béene said before: and in her swift motion she goeth 15. degrees, in which time the sunne goeth one degree, so as she is distant from the sunne 14. degrees, which is more then a point and 3. minutes of the compasse. And you haue to note that the point Auge of the Moone is mooueable and passeth through the Zodiaque in the space of 19. yeares, and thereby sometime causeth the full of the Moone to happen sooner or later.

The Arte of Nauigation.

How to knowe in what signe the point Auge of the Moone is in any yeare.

Chap. 50.

IF you would know in what signe the Auge of the Moone is in any yeare, then you must consider the prime or golden number of that yeare, for when the prime or golden number is one, then her *Auge* is in Aries. And if the Moone be also then in Aries she is in her slowe motion: and being in the point opposite which is ♎, she is in her swift motion. And sith this Auge of the Moone goeth through the 12. signes in 19. yeares as hath bene said, it must nedes fall out that in 9.yeares and a halfe her Auge commeth to be in Libra: And then the Moone being there she is in her slowe motion: And being in the point opposite which is Aries, she is in her swift motion: Again when the prime is 5. then her point Auge is in Cancer: and the Moone being there she is in her slowe motion, and when she commeth to be in the point opposite which is Capricorne, she is in her swift motion: And when the prime is 14. or 15. then her point Auge is in Capricorne, where if the Moone be also, then shee is in her slowe motion: and being in the point opposite which is Cancer, she is in her swift motion. And note that when the Moone is in her swift motion she maketh her change, or full, or any other aspect the sooner: And contrariwise when she is in her slow motion she maketh her change, or ful, or any other aspect the later. Moreouer you shal see her at the time of her change eyther sooner or later, according to the time of the yeare, for from January to June you shal see her within 24. houres after her change, because she hath during those monethes North declination from the sunne, & maketh a greater arch then the sunne doth. But from July to December you shall not see the Moone scant three dayes after the change. But you may see her within 24. houres before her change, because that during those monethes she hath South declination from the sunne: And note that when the Moone is three dayes and 18.houres, which is the halfe quarter of the Moone, the sea men doe call that time the prime day, because the Moone is then 4.points to the Eastward of the Sunne, which is three houres, for to euery point is attributed

three

The Arte of Nauigation.   348

thrée quarters of an houre as hath béene saide before. Moreouer it is necessary for sea men to knowe when the Moone riseth & setteth, and in what part of the Horizon in euery Latitude: and how long she shineth and when she is full South, and also what Latitude she hath, and whether it be South or North euery day and houre thoroughout the yeare. All which thinges are most easie to be found by the Globe, and by helpe of the Ephemerides in such sort as is before taught in my Treatise of the Globes, and also by helpe of M. Blagraue his Astrolabe, her true place in the Zodiaque being first knowne by the Ephemerides. It is méete also to know when she is in coniunction with the sunne, or at the full, and the rest of her quarters, which is easily knowne by the Ephemerides.

When the Moone is saide to be in Coniunction with the Sunne, or to be at the full, and what her greatest latitude is aswell from the Ecliptique line, as from the Equinoctiall.

Chap. 51.

She is said to be in coniunction with the sunne when the sunne and she be both in one selfe signe and like degrée: But when she is at the full, then she is opposite to the sunne, and distant from him 6. signes, which is the one halfe of the Zodiaque, containing 180. degrées: And in euery quarter she is distant from the sunne thrée signes which is 90. degrées: Moreouer the Moone is said to haue Latitude both Northerne and Southerne from the Ecliptique line, which line ye deferent of the Moone crosseth in 2. points, and thereby maketh two intersections, whereof the one tending towardes the North is called the head of the Dragon, and the other intersection towardes the South is called the tayle of the Dragon, so as when the Moone is passed 90. degrées from the Dragons head towardes the North, then her Latitude is 5. degrées Northward: And when she is distant 90. degrées from the tayle of the Dragon towardes the South, then her Latitude is also 5. degrées Southward, which is the greatest Latitude that she hath on eyther side of the Ecliptique line, whereof I haue written more at large in the first part of my Spheare the 15. chap.

in

The Arte of Nauigation.

in which you shall finde a figure representing the saide Dragon both head and tayle: But her Latitude is to be considered in two respects, that is not onely from the Ecliptique line, but also from the Equinoctiall, for from the Ecliptique line her greatest Latitude is but 5. degrees on eyther side of the Ecliptique, as hath bin said before. But from the Equinoctiall her greatest Latitude is 28. degrees and a halfe on either side of the Equinoctiall, which in mine opinion might be more rightly called her greatest declination, which exceedeth by 5. degrees the declination of the sunne, for that is but 23. degrees and a halfe, or rather 2/3. on either side of the Equinoctiall. But this great Latitude of the Moone is onely to be vnderstood when the prime is one, and that her Auge is in Aries: For when the prime is betwixt 9. or 10. yeares or more, the Moone declineth not from the Equinoctial on either side aboue 18. degrees and a halfe at the most.

How to know in what part of the Zodiaque the head of the Dragon is euerie yeare.

Chap. 52.

IF you would know in what part of the Zodiaque the head of the Dragon is, then you must consider the prime, for when the prime is one then the Dragons head is in the first point of Aries, euen as the point Auge is. And in 19. yeares it passeth thorough the twelue signes aswell as the point Auge of the Moone, but with contrary course for the point Auge of the Moone mooueth according to the succession of the signes, that is, from Aries to Taurus, Gemini and so foorth: But the head of the Dragon hath a contrary motion, that is, from Aries to Pisces, and so into Aquarius and so foorth, so as in 9. yeares and a halfe it meeteth iust with the point Auge of the Moone in the signe Libra. Thus you see that by knowing the prime, you learne also to knowe in what signe the Auge of the Moone, and also the Dragons head is, and what Latitude the Moone hath, aswell from the Ecliptique line as from the Equinoctiall. But one of the chiefest points to be had by the Moone, is to know thereby the tides, that is when the sea floweth and ebbeth in any place: whereof we come now to speake.

How

How to know the tydes in any place by the Moone.

Chap. 53.

BEfore that you enter into any Hauen or Riuer, it is necessary to knowe the true tydes of that place, which tydes are subiect to the motion of the Moone, for she causeth at one place or other alwaies in one certaine rombe full Sea. As for example it is alwayes full sea at Antwerp when the Moone is either right East or West: and it is ebbe or lowe water there when she is North or South, and because the Moone doth passe through all the rombes of the Mariners Compasse in 24. houres, they allow to euery rombe ¾. of an houre, which is 45. minutes, and that being multiplyed by 32. do make iust 24. houres, wherefore if the two rombes, North and South do yéeld ech of them 12. houres, then the first rombe must néedes yéelde ¼. of an houre, the second one houre and ½. and the third two houres and ¼. and so foorth of the rest: For by adding to euery rombe ¾. you shall find that the East and West doe yéelde alwayes 6. houres: but then you must note that according to the age of the Moone the tydes do fall euery day later and later, wherfore to know the true time of the tyde in any place, you must first learne by some Rutter or by the relation of others that can tell in what rombe the Moone causeth full sea in that place, & then at what houre it is full sea, the Moone being eyther in the change or at the full, which you shall know by allowing to euery rombe ¾. of an houre in such manner as is before set downe. But if you would knowe at what houre it is full sea in that place euery day, then you must first vnderstand that the Moone in 30. daies slacketh 24. houres, which amounteth to ⅘. of an houre which is 48. minutes for euery day, for so much she declineth euery day from the sunne, then looke how many dayes the Moone is old, and hauing multiplyed the same by ⅘. that is to say by 48. minutes, adde the product thereof to those houres at which it is full sea, and you shall haue the true time of full sea euery day. As for example, suppose that it was full sea at the last new Moone at some place here

The Arte of Nauigation.

in England at three of the clocke in the afternoone, and nowe I would know at what a clocke it shall be full sea 5. dayes after the new Moone. Now if you multeply 5. being the age of the Moone by 48. minutes, and adde to the product therof three houres, which was the time of the last change, the full sea shall be at that place at seuen of the clocke. But if the summe of such addition be aboue 12. houres, then you must cast away the 12. and the remainder shall shew you the true houre of full sea.

How to know by helpe of an instrument the tydes at any place,

Chap. 54.

There is also an easier way to knowe the tydes euery day in any place by the helpe of an instrument set downe by Cogniet, wherof both the making and vse here followeth. First vpon some bord well playned and made smooth, draw a Circle, and deuide the same into 30. parts signifying the dayes or age of the Moone, setting the number of 30. aboue in the toppe of the instrument, and place all the other numbers as 1.2.3. and so forth towards your right hande, that done, make a moueable ronde which may turne about within the verdge of the first Circle, and deuide that into 24. houres, and also into 32. rombes, setting the North point marked with the Flower deluce at the twelfth houre aboue, & the South point at the twelfth houre beneath, and the East point at the sixth houre on your right hand, and the West point at the sixth houre on your left hande, and so shall your instrument be perfect, the vse whereof is thus:

First you must knowe what rombe of the Moone in that place which you seeke maketh a full sea, and also the age of the Moone by some Almanacke or some other rule before taught, then hauing these two thinges turne the inward ronde of the houres & rombes vntill the foresaid knowne rombe doth iustly answere to the 30. day of the great ronde, and there staying it firme with your finger, seeke in the outermost border of the greater ronde the age of the Moone, and that will shew you in the ronde of houres the very houre of full sea that day in that place.

But

The Arte of Nauigation.   350

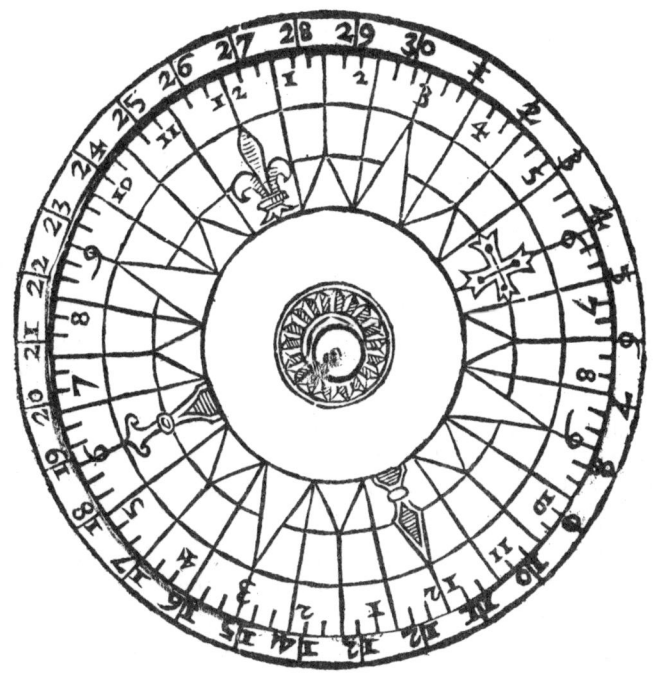

But one chiefe thing touching the knowing of the tydes is to be noted as Borne sayth, which is that the sea will flow more by one point of the Compasse in the spring tydes, then in any of the quarters of the Moone called Nep tydes in euery riuer that hath any indraft, and is of some reasonable distance from the sea: As for example, it floweth at Graues end at the change or full of the Moone when she is South Southwest: but in any her quarters it skant floweth when she is South and by West, and this rule as he sayth is generall for euer.

How

The Arte of Nauigation.

How a generall Rutter shewing the tydes in al places should be made.

Chap. 55.

WIlliam Borne doth set downe what Moone doth make a full sea, aswell here in the most parts of our English coasts as in some other parts of France & Spain, and so doe many others whose Tables touching the tydes are called Rutters, whereof some are truely set downe and some are false: But sith such Rutters do serue but for a few particular places, I would wish that some learned Pilot that hath sayled many and sundry long voyages, to make a generall Rutter that might serue for al places if it were possible, or at ye least for so many places as are knowne in these daies: And I would wish such generall Rutter to be made in manner of an Alphabet: and that euery place might haue his true Longitude and Latitude added thereunto, to the intent that euery place might be the more easily found out in any Map or Card that is graduated with degrees of Longitude and Latitude: And then to shewe what Moone doth make a full sea in euery such place, which thing who soeuer would performe, he should in mine opinion deserue great commendation. And thus I end this Treatise, praying all the learned Seamen not to be offended or greeued with mee for that I doe make young Gentlemen our owne Countrimen partakers of their most worthy knowledge, whereof the ignorant are not able to iudge, nor to yeelde them that prayse which they deserue: yea rather I hope that they will help to perfect what so euer I haue herein left vnperfected, to make the young Gentlemen the more skilfull, and thereby the more seruiceable to their Countrie, and in so doing they shall procure to them selues great good will and infinite thankes.

FINIS.

THE PARTS OF THE VOLVELLES TO BE
CUT OUT AND FIXED TO THE CENTRE
OF THE WOODCUT BY MEANS OF A PIECE
OF THIN TWINE.

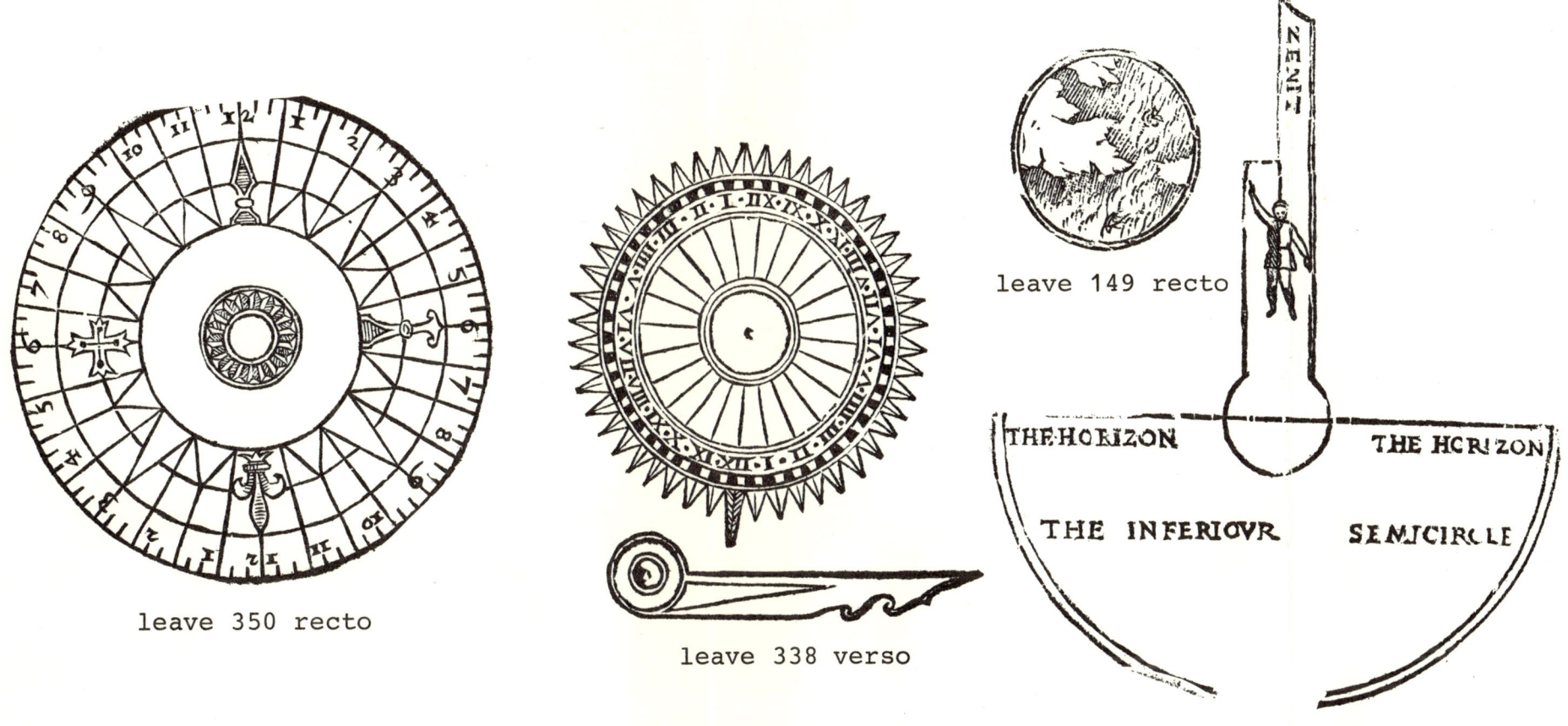

leave 350 recto

leave 338 verso

leave 149 recto

leave 149 recto

QB
41
B59
1594a

SEP 29 1972